CROP BREEDING

A CONTEMPORARY BASIS

Other Pergamon publications of related interest

BOOKS

BUCKETT
An Introduction to Farm Organisation and Management, 2nd Edition

DILLON
The Analysis of Response in Crop and Livestock Production, 2nd Edition

KENT
Technology of Cereals, 3rd Edition

LOCKHART & WISEMAN
Introduction to Crop Husbandry, 5th Edition

NASH
Crop Conservation and Storage, 2nd Edition

VOSE
Introduction to Nuclear Techniques in Agronomy and Plant Biology

JOURNALS

CURRENT ADVANCES IN PLANT SCIENCE
JOURNAL OF STORED PRODUCTS RESEARCH
OUTLOOK ON AGRICULTURE

CROP BREEDING

A CONTEMPORARY BASIS

Editors

P. B. VOSE

*International Atomic Energy Agency, Vienna, at
Centro de Energia Nuclear na Agricultura
Brazil*

S. G. BLIXT

*Weibullsholm Plant Breeding Institute
Sweden*

PERGAMON PRESS

OXFORD · NEW YORK · TORONTO · SYDNEY · PARIS · FRANKFURT

U.K.	Pergamon Press Ltd., Headington Hill Hall, Oxford OX3 0BW, England
U.S.A.	Pergamon Press Inc., Maxwell House, Fairview Park, Elmsford, New York 10523, U.S.A.
CANADA	Pergamon Press Canada Ltd., Suite 104, 150 Consumers Road, Willowdale, Ontario M2J 1P9, Canada
AUSTRALIA	Pergamon Press (Aust.) Pty. Ltd., P.O. Box 544, Potts Point, N.S.W. 2011, Australia
FRANCE	Pergamon Press SARL, 24 rue des Ecoles, 75240 Paris, Cedex 05, France
FEDERAL REPUBLIC OF GERMANY	Pergamon Press GmbH, Hammerweg 6, D-6242 Kronberg-Taunus, Federal Republic of Germany

Copyright © 1984 Pergamon Press Ltd.

First edition 1984
Reprinted (with corrections) 1984

Library of Congress Cataloging in Publication Data

Main entry under title:
Crop breeding, a contemporary basis.
Includes index.
1. Plant-breeding. I. Vose, Peter B. II. Blixt,
S. G. (Stig G.)
SB123.C799   1983   631.5'3   83-13284

British Library Cataloguing in Publication Data

Crop breeding.
1. Plant breeding
I. Vose, Peter B. II. Blixt, S. G.
631.5'3   SB123
ISBN 0-08-025505-1

Printed in Great Britain by A. Wheaton & Co. Ltd., Exeter

PREFACE

This book is not a book on crop breeding *per se*, and there are no chapters on genetics, cytogenetics, experimental design or statistical treatment of data, which have done so much during the last 50 years to put plant breeding on a scientific basis. What the book attempts to do is to bring together in an integrated manner various physiological and technical aspects of plant breeding that have developed in recent years.

This is surely a very exciting time for those interested in crop improvement, as at no time has the plant breeder had greater possibilities open to him. Much more information has become available to us on carbon assimilation and partitioning, and response to stress factors etc., while ancillary techniques such as mutation induction, plant tissue culture, and the application of computers are providing new approaches. The challenge of possibilities also brings problems: how best can we incorporate the new physiological knowledge in a practical plant breeding programme? Do the newer techniques offer a better possibility of achieving the breeding objectives? Faster? More cheaply?

We had problems with some potential topics. For instance, we felt that any book concerned with plant breeding could not ignore disease resistance, because in so many crops disease susceptibility is a major limiting factor. Thus, the *Phaseolus* dry bean crop in Latin America could be doubled immediately if disease were reduced, without any further breeding for yield factors. However, breeding for disease resistance is a very broad topic and could occupy the whole book, we therefore compromised by inviting our contributor to discuss the strategy of breeding for disease resistance, in terms of general philosophy and approach. The history of breeding plants for insect resistance is much shorter than that for disease resistance. A subsequent book of this type may be able to include a section on the strategy of breeding for insect resistance. We considered it for this book, but concluded that despite the progress that is being made, for example in breeding for insect resistance in rice, it seems too soon to try and develop a general statement.

We would have liked to have included a chapter on quality factors, which are extremely important in many crops. We reluctantly came to the conclusion that the only common factor in quality attributes is that they are all different! Therefore, a short general treatment seemed of little value. The exception is the quest for higher protein, which both in objective and techniques is common to a number of crops of World importance.

The last chapter is mostly a self-indulgence, written for our own enlightenment, under the guise of providing an epilogue! However we hope that the general theme will stimulate readers to develop their own ideas on the topic, whether or not they choose to agree with us.

We should like to extend our sincere thanks to our contributors, who have all put substantial effort into their chapters, often at considerable cost to their existing workload. Without their enthusiasm the book would not have been possible, and we are very grateful to them. We would particularly like to thank Mrs. Diva Athié for her invaluable help with the preparation and manuscript.

P. B. VOSE

STIG BLIXT

CONTENTS

CHAPTER 1

NEW PARAMETERS AND SELECTION CRITERIA
IN PLANT BREEDING

S.K. Sinha
Indian Agricultural Research Institute, New Delhi, India

M.S. Swaminathan
International Rice Research Institute,
Los Banos, Philippines

THE CHANGING SITUATION

The human race has entered the space age. Twenty years of space exploration has led to the realization that first, *Homo sapiens* will have to depend only on the Earth for producing food and secondly, the finite resources of 'spaceship Earth' cannot be exploited in an exponential manner. Hence, as we approach the twenty-first century, plant breeders will have to examine how to meet the new challenges ahead. To conduct such an exercise scientifically, it will be necessary to anticipate the kinds of changes which are likely to take place in the major farming systems of the world in the years to come. For example we need to consider the ecological effects of increased use of fossil fuels, the need for renewable energy sources, the necessity to improve diet, and the changing outlook for employment, with dwindling land resources. All this against a steadily rising World population.

It is in the context of these challenges that the currently used parameters and selection criteria in plant breeding will have to be reviewed. Before discussing the 'course corrections' that may be necessary in the coming years, it would be useful to review briefly the experience of the past.

1

Historical developments

Efforts to select more desirable plants among mixed populations and to hybridize plants to combine parental characteristics in offspring in a well-planned manner first started in Europe and later in North America. It is not surprising therefore, that the crops which attracted the attention of plant breeders most in the late nineteenth and early twentieth centuries were potato, wheat, sugarbeet, oats, rye and flax. These crops together met the requirements of food, sugar, fiber and forage. In addition, vegetable crops also received attention. The emphasis in plant breeding concentrated on the improvement of an individual crop. Since the climatic conditions in Europe did not favour the cultivation of more than one crop per year in the same plot of land, the concept of breeding for multiple or mixed cropping was not even discussed. (See Hayes *et al*[46], Elliot[35], Brewbaker[18]).

Subsequently, when plant breeding started in the United States and Canada, it was the efforts of individual plant breeders that fixed parameters for crop improvement. Since the land-man ratio was very favourable and crop yields could be improved through fertilization, consumer quality and organoleptic properties attracted much attention. In addition to yield, resistance to pests and diseases were the other major objectives.

The following inferences can be drawn about the parameters and objectives until the mid-nineteen fifties:

1. The objectives in plant breeding were determined by ecological parameters and consumer preferences.
2. Although improvement in yield was one of the objectives, there was no reference point in terms of upper limits to measure the extent of success or achievement. This is reflected in the following statement of Allard[5], 'There is of course no way of estimating how close modern production methods have brought us to the maximum yield possible. It does appear certain, however, that great advances remain to be made'. He further emphasized, 'there is every reason to believe that the exploitation of the variability of most domesticated species that has occurred up to the present has not taken us even near to the maximum productivity that is theoretically possible. If this is true, plant breeding has greater contributions to make in the future than it has made in the past'.
3. Breeding for resistance to diseases and pests received considerable stress. The following serious epidemics underlined the importance of resistance breeding:
    -   The Irish famine of the 1840's due to the potato late blight epidemic (*Photophthora infestans*).
    -   The wheatless days of 1917 in the USA, due to stem rust epidemics (*Puccinia graminis*).
    -   The shortage of food in Bengal Province of India in 1943 associated with the brown spot disease of rice (*Cochiobolus miyabeanus*).
    -   The complete elimination in the mid-1940's of all oats derived from the variety Victoria in the USA due to Victoria blight disease (*Cochiobolus victoriae*).
    -   The Southern corn leaf blight epidemic (*Cochiobolus heterostrophus*) of 1970–71 on all US maize hybrids carrying T-type cytoplasmic male sterility.
    -   The rapid shift from the rice brown planthopper (*Nilaparvata lugens*) biotype 1 to biotype 2 during 1974–76, when large areas in the Philippines and in Indonesia were planted to a few semi-dwarf types.

By the middle of the current century emphasis was placed on yield per acre, nutrient uptake, quality and resistance to adverse or extreme conditions (Hayes et al.[46]). The objectives and parameters in plant breeding continued to be those which were important in temperate countries. The special problems of the tropics and sub-tropics were, however, highlighted by Swaminathan[90] who stated 'Plant breeding all over the world and more particularly in the tropics and sub-tropics has entered a new phase. The pressure of population and the consequent increase in demand for food on the one hand and depletion of the area under cultivation due to growing urbanization on the other have stimulated research on the scientific destruction of ceilings to yield. Altogether new parameters of selection such as measuring productivity per day, stability of performance, utilization of solar energy, photo-insensitivity, suitability for mechanical harvesting and processing characteristics, protein quality, profile of amino acids and freedom from toxins have been introduced in breeding research'. Frankel[41] defined plant breeding as genetic adjustment to the physical, biological, technological, economic and social components of the environment.

NEW PARAMETERS

Thus far, plant breeders have mainly been concerned with bringing about a continuous improvement in the productivity of that part of the plant which is of interest to man, security of the crop through in-built resistance to pests and diseases and nutritive and organoleptic or other desired quality characters. Hereafter, additional considerations will have to be added to breeding parameters:

(i) Ecology

There is evidence that the increase in atmospheric CO_2 now taking place due to the burning of fossil fuels and denudation of forests could lead to a global warming of about $2°C$ at the beginning of the twenty-first century (Manabe & Wetherald[59]). If this happens, this would lead to a reduction in maturity period. Per-day productivity will then become an important selection criterion in the temperate part of the world also. On the other hand, a higher CO_2 in the atmosphere could help in enhancing the productivity of several crops, particularly of C_3 crops. Another aspect of the ecological changes now in evidence relates to facing the problems of salinity, alkalinity, acidity and various forms of soil erosion and desertification (i.e. all man made changes which either destroy or diminish the biological potential of soil).

(ii) Need for renewable energy sources

Energy is becoming a major limiting factor in economic development. The advance of a country is linked with the extent of consumption of fossil energy. In countries such as the United States, Japan, West Germany, etc. the consumption of non-renewable energy is more than 95% (Calvin[20]) but in developing countries where industrialization has picked up only recently, the consumption is less than 50%, as in India. Nonetheless, the problem has already become acute and biological conversion of solar energy appears to be a major alternative in the present crisis. In years to come the requirements of the developing and developed countries are likely to be different. In the former, generation of energy at the village level involving insignificant cost in energy input would be one of the requirements, whereas in countries like Australia, USA, etc. liquid fuel for transport would continue to be a major necessity.

Advances in productivity improvement have so far been accompanied by an increase in the consumption of non-renewable forms of energy. Obviously, this pathway will lead to a blind alley as the fossil fuel reserves get depleted. Hence, it will be essential that the energy costs of improving the yield potential of crops are taken into consideration in breeding programmes. Genetic pathways of improved cultural energy conversion by plants will have to be studied and followed. At the same time, the new leads now opened up through (i) protoplast fusion and organelle transfer, (ii) DNA and phage induced transformation and (iii) recombinant DNA cloning, for incorporating characters like biological nitrogen fixation, improved nutrient uptake and utilisation and greater photosynthetic ability and resistance to moisture stress will have to be followed up vigorously.

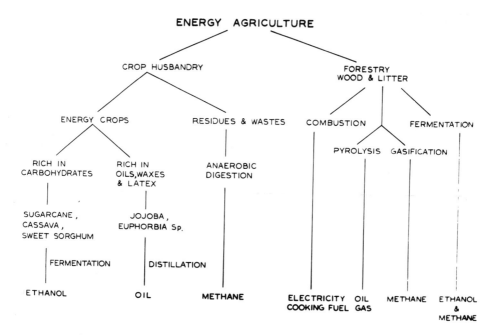

Fig. 1.1. Energy agriculture: the processes through which fuel can be obtained.

Photosynthetic products will increasingly become a major source of energy supply in the future. There are several possibilities (Fig. 1.1) such as:

1. Use of agricultural waste and plant residues
2. Production of crops capable of producing liquid fuel
3. Energy plantations and their subsequent use.

Plant breeders and plant scientists will have to play a significant role in all the above options which individually or in combination may be found more suitable at a particular place. Some of the important objectives would be as follows:

1. Efficient bacterial strains for gasification of agricultural waste.
2. **Sugarcane, cassava** and **sweet sorghum** have been identified as most promising crop plants in the tropics, which could be utilized for liquid fuel production.

What is most important is the identification of the agro-ecological regions where the production of a particular plant would be most profitable.

3. **Eucalyptus, Casuriana** and **Salix** are among the species which have been identified for energy plantations (NAS, US[71]). There are several species amongst these genera which •may have different adaptive qualities. Identification of suitable types according to an agro-climate would be necessary. Dry matter accumulation would be the most important single criterion for selection.

(iii) Nutrition, quality and diet

Zwartz and Hautvast and others[106] have reviewed the nutritional objectives for plant breeders. It was recognised that, while in developing countries the problem is of deficiencies in energy, protein, energy-protein, vitamins and sometimes of excess of toxins causing various diseases, in developed countries the problem is of excess nutrition. The excess nutrition consists of high energy diet comprising fats and proteins. On the basis of this understanding several areas of quality improvement in individual crops were identified. Apart from what has been repeatedly emphasized, two important points need consideration:

1. When efforts to change the composition of grains are made, is it necessary that the improvement occurs without influencing total yield or quality in some other constituents?
2. Diets in all parts of the world do not consist only of one or two grains or cereals. In fact there are various diets which can be classified as follows:

    (a) Meat based diet
    (b) Rice based diet
    (c) Wheat based diet
    (d) Cassava and yam based diet
    (e) Millet based diet, etc.

In almost all the above diets with adequate purchasing power, there is a combination of fats and vegetables. Therefore, an analysis of the diets in different parts of the world is necessary to put emphasis on a particular crop or vegetable.

Swaminathan[91] pointed out that efforts were being made to alter the amino acid composition in favour of essential amino acids, without a clear knowledge of their biosynthetic pathways. Indeed, it could be that a balance among amino acids derived from aspartate is maintained and increase in one is at the cost of another. Therefore, what is important is the greater flow of carbon through aspartate to increase the amounts of lysine, threonine, isolencine and methionine. Therefore, it is suggested that now genotypes should be evaluated on the basis of total carbon present among a group of biosynthetically related amino acids.

Bressani[17] made an extensive survey of diets in some developing countries. Most of the diets were based on cereals such as rice, maize and *sorghum* or cassava. Addition of legumes enhanced the diet efficiency, both in terms of the protein efficiency ratio (PER), as well as increase in body weight. An important fact was that processing of pulses improved their utilization efficiency. In the past, the PER or diet efficiency of many pulses was determined on the basis of raw beans. Nowhere in the developing world are pulses used raw. In China, India and many South East Asian countries, soaked and germinated seeds of mung bean and chickpeas are commonly used. Recent studies in India have shown that this simple process enhances the value of PER as well as diet. Furthermore, the use of grain legumes or pulses is important, not only from the point of view of protein source but they provide vitamins and also quickly available sugars.

In conclusion, improvement in nutritional quality of individual crops should now be linked with the improvement of total diet as preferred by different sections of society at different places.

(iv) Employment

The farming systems of the world are becoming polarised into major groups — 'super farms' and 'small farms'. The super farm system of farming involves the cultivation of farms exceeding 1000 ha in size through a high degree of mechanisation. The super farm technology may involve aerial seeding and fertilization, land levelling through laser beams and combine harvesting. Thus machines will replace human labour and cropping schedules will have to be designed by computers. The small farm operation, characteristic of the farming systems of many developing countries, involves the cultivation of holdings less than 10 ha in size. Such a small farm may have to support a family of five to six members. Hence, the farm should provide opportunities for labour intensive operations. Examples of plant breeding operations designed to achieve this objective are the production of hybrid cotton seeds in India through hand pollination and emasculation (Swaminathan[94]), and the raising of potato crops in China from seedlings derived from sexual seeds. Also, the small farm operation will invariably involve crop-livestock integrated production systems or agriculture-aquaculture systems or enhanced cropping intensity through multiple, relay and inter-cropping. In gardenland crops, small farm technology will involve planning combined crop canopies which can help the farmer to derive maximum economic benefit from cubic volumes of air and soil.

In addition to the above, the fact that we will have to depend upon Earth alone to meet the food needs of the future implies that land will become a diminishing resource for agriculture. More and more land will go out of agriculture for industrial, communication and domestic needs. In other words, productivity improvement and increased intensity of agriculture through raising several crops a year on the same plot of land will be the only mechanisms available for increasing production. The plant breeder will, therefore, have to bring about a continuous improvement in production per units of land, time, water, energy and air space. Also, the undulations caused in production by weather aberrations and pest epidemics will have to be minimised to the maximum extent possible.

(v) New parameters: summary

The new parameters for plant breeding which are of immediate importance, as well as having long-term objectives both in the developing and developed countries, can be stated as follows:

1. Increasing availability of food containing desirable nutritional value and organoleptic properties.
2. Improvement in the efficiency of food production for each unit of cultural and solar energy invested.
3. Stability of crop production through resilience to weather behaviour and resistance to the triple alliance of weeds, pests and pathogens.
4. Identification and improvement of plants suitable as sources of biomass and renewable energy.

5. Identification of more efficient plant types for crop-livestock and agriculture-aquaculture production systems.
6. Identification of plant types suitable for multiple, relay and inter-cropping and for 3-dimensional crop canopies, where annuals and perennials are grown in a co-operative farming system.
7. Identification of genotypes for optimum response to high, low and zero input conditions.
8. Breeding relatively photo and thermo-insensitive varieties characterised by a high per-day phytomass production so as to develop contingency cropping patterns to suit different weather probabilities.

THE BASIS OF YIELD

(i) Theoretical, potential and actual yield

Considerable improvement in the yield of cereals such as wheat, rice, maize, sorghum and other crops has occurred in the last few decades. The maximum yields reported in temperate and tropical regions are given in Table 1.1. These yields are still far below those projected as theoretical yields by de Vries et al.[25] and Loomis and Williams[57]. Therefore, the question remains as to what would be the upper yield limit a plant breeder could attempt to achieve. Buringh et al.[19] made an analysis of the absolute production potential in terms of a grain equivalent based on soil, climate, precipitation and radiation characteristics in different parts of the world. They concluded that the absolute production potential was 49,830 million tonnes/year. Sinha and Swaminathan[86] using a similar methodology, but also considering the

TABLE 1.1. Average, best and world record yields

Food crop	Average 1979 (US)	Best farmers (US)	World record (US)	Ratio record/average
		(Metric tons/ha)		
Maize	6.9	14.7	22.2	3.2
Wheat	2.3	6.9	14.5	6.3
Soybeans	2.2	3.6	5.6	2.6
Sorghum	4.0	17.8	21.5	5.4
Oats	2.0	5.6	10.6	5.3
Barley	2.7	9.0	11.4	4.3
Potatoes	31.0	68.0	95.0	3.1
Rice (crop/112 days)	5.1	9.1	14.4*	2.8
Sugarcane	84.0	140.0	250.0	3.0
Sugar beets	44.0	80.0	120.0	2.7
Milk/cow x 1000 lbs	12.0	35.0	50.0	4.2
Eggs/hen/year	235.0	275.0	365.0	1.5

* 28 Tons/ha in the Philippines.

Wittwer[104]

duration for which total interception of light is possible, calculated the absolute production potential in terms of a gram equivalent of India as 4,572 million tonnes per year. De Wit et al.[26] have defined production potential of a crop as the growth rate of closed green crop surface, optimally supplied with water and nutrients, in a disease and weed free environment, under the prevailing weather conditions. Accordingly, a crop capable of retaining a closed crop canopy for 100 days could produce 35,000 kg dry matter ha^{-1}. A 50% harvest index from such a crop would give 17.5 tonnes ha^{-1} of dry matter or 20.0 tonnes ha^{-1} of grains with 12% moisture. It is thus clear that there is considerable scope for improvement in situations where water and nutrients are not limiting.

Recognising the fact that the potential yield of a crop is determined, not only by the genotype but by the environment in which it grows and the management it receives, FAO has started an exercise of deliniating areas in Africa and Asia which are most suitable for different crops (FAO[38]). Therefore, it would be appropriate that the breeding objectives are fixed in relation to the expected potential of different crops based on the exercise intiated by FAO.

(ii) Energy value of different crops

The yield of different crops is generally compared on the basis of weight (de Vries et al.[25], Wittwer[104]). The weight is also estimated on the basis of calculating harvest index (Donald[28], Hamblin and Donald[44], Jain & Kulshrestha[49]). While this is a simpler method of comparing yield and harvest index, it is not necessarily the best. Since the composition of various grains differs, the energy content differs for the same weight for different crops and sometimes even cultivars (Table 1.2). On the basis of energy content, it is obvious that 0.6 kg of an oil containing seed would be equal to 1.0 kg of a cereal grain. Therefore, in terms of weight, it would not be realistic to expect similar yields of an oilseed crop like groundnut and a cereal like wheat (Table 1.3). Furthermore, the harvest index of some of the oilseed crops is as good as those of high yielding cereals when harvest index is calculated on the basis of partitioning of energy rather than of grain weight (Table 1.1).

Consequently, if the change in composition of grains is one of the objectives in favour of proteins with a particular amino acid profile or in favour of lipid content, while retaining the same weight, this would require additional energy (Bhatia & Rabson[16], Mitra & Bhatia[63]). Sinclair and de Wit[81] calculated the requirements of photosynthates in case nitrogen assimilation is enhanced for protein content. Most of these estimates are based on the glucose utilization efficiency through different metabolic pathways (Penning de Vries[74]). In these estimates, it is expected that all the biochemical reactions would occur in synchrony, which often does not happen, and consequently the glucose utilization efficiency could be poorer than expected in natural systems. Nonetheless, these studies are useful to a plant breeder because he is made aware of the biochemical constraints imposed by the system. Therefore, he has either to seek a compromise between yield and quality, or he has to be prepared to improve photosynthate availability simultaneously. If this factor is not recognised, it will be found that improvement in calorie yield may invariably result in a protein or lipid penalty and vice versa.

(iii) Breeding for yield improvement

Breeding for yield is one of the principal objectives in different crops in all parts of the world. Development of a new concept or a new technique has usually been responsible for significant advances in yield. Among these the phenomenon of heterosis and the introduction of dwarfing genes could be considered important landmarks. However, after the quantum jump

TABLE 1.2. Total dry biomass, grain yield, their energy content and harvest index

Crop species	Total biomass g/m^2	Grain yield g/m^2	Harvest index	Total biomass energy K Cal m^2	Grain energy K Cal m^2	Harvest index
Wheat (Kalyansona)	1558	583	37.4	6104	2387	39.1
Wheat (Cv Moti)	1375	596	43.3	5410	2441	45.1
Triticale (DTS-141)	1520	484	31.8	6062	1997	33.1
Barley (Cv. Ratna)	1867	445	23.8	6986	1798	25.7
Chickpea (Cv. JG.62)	1027	327	30.5	3992	1412	35.3
Pigeonpea (Cv.Prabhat)	1008	310	30.7	3994	1379	34.5
Brassica Compestris (Yellow)	1160	377	29.0	5055	2249	44.4
B.Compestris (Brown)	1380	421	26.0	5828	2636	45.2
B.Juncea (Mustard)	1820	486	26.0	7608	2959	38.9

TABLE 1.3. Relative grain yields (weight) of different crops on the basis of their composition. Values are expressed as 1.0 kg wheat equivalent

	Composition (% of dry weight)				Seed g/g photosynthate	Wheat equivalent
	Carbohydrate	Protein	Oil/lipid	Ash		
Cereals						
Wheat	82	14	2	2	0.71	1.0
Rice	88	8	2	2	0.75	1.05
Sorghum	82	12	4	2	0.70	0.98
Pulses						
Chickpea	68	23	5	4	0.64	0.90
Pigeonpea	69	25	2	4	0.66	0.93
Mung bean	69	26	1	4	0.66	0.93
Oil Seeds						
Groundnut	25	27	45	3	0.43	0.60
Sesame	19	20	54	7	0.42	0.59
Rape and Mustard	25	23	48	4	0.43	0.60

in yield potential associated with the exploitation of hybrid vigour and the dwarf and fertilizer responsive plant type, a yield plateau has been reached in crops like wheat and maize. Consequently, current emphasis in these crops is on breeding for disease resistance so that the high yield potential can be combined with stability of performance. Grafius[42] presented evidence to show that genes for yield *per se* did not exist in barley. Therefore, the question remained that in a case where there were no specific genes for yield, how was this character expressed at the phenotypic level? A clear understanding of the genetic basis of yield is obviously essential for launching a successful breeding programme.

(iv) Physiological basis of yield

The yield of a crop is the result of a large number of processes involving germination, growth, differentiation, development and senescence. Therefore, identification of a particular process which limits yield in a given environment becomes the primary consideration. Nonetheless, an important fact about yield is that a crop must produce an adequate amount of dry matter and appropriate partitioning of the phytomass for it to give good yield. This brings in the following questions:

1. Has there been any improvement in photomass (dry matter) production in the course of improvement of various crops?
2. What is the relation between dry matter and yield?
3. Can dry matter production be improved if it limits the yield?

Recent studies using old and new varieties of wheat released in India and Britain, and of peanuts released in the United States suggest that there has been no significant change in dry matter production (Austin *et al.*[9], Sinha *et al.*[87], Duncan *et al.*[30]). A major change in partitioning of dry matter was mainly responsible for improved yield. In peanuts, this change was due to change in growth pattern. These studies lead to two important conclusions:

1. In those crops where no major improvement has occurred as in many leguminous and oileferous crops, improvements in yield in the short-term could be achieved through changing the pattern of growth and the partitioning of phytomass. Efforts to improve dry matter production could be a long-term objective.
2. The crops in which significant improvement has already occurred, such as wheat, rice, sorghum, maize, soybean, etc., improvement in dry matter production per unit area without disturbing partitioning as obtained now, could be an important objective.

The above conclusions lead us to the following questions:

1. How is dry matter obtained and how could it be improved?
2. What determines the partitioning of dry matter and when it is determined?
3. What is the effect of the preceding phase on partitioning?

The components of processes leading to dry matter production were described by Swaminathan[93]. These include leaf area development, leaf area index, the rate of photosynthesis, the rates of respiration and photorespiration, nutrient uptake, nitrate assimilation and water use. Genetic variability in many of these characters both at the species as well as varietal level has been established. However, a brief discussion of some of these characters may be of value:

(v) Selection for the rate of photosynthesis

In most crops it has been shown that a leaf area index of four to six is the ceiling for dry matter production and yield. Therefore, increase in total dry matter production can come about only by improving the rate of photosynthesis per unit area while retaining the upper limit of the leaf area index. This has prompted many a worker to look for genetic variability in this character so that improvement could be obtained. At the species level in wheat it was shown by Khan and Tsunoda[53] and Evans and Dunstone[36] that the primitive species of wheat have higher rates of photosynthesis than the cultivated *Triticum aestivum*. These studies were based on a very limited number of genotypes and may not be indicative of the variation available within a species of *Triticum* itself. A more important fact which emerged from these studies was that a smaller leaf size was associated with higher photosynthesis rate (Dunstone and Evans[31]). Earlier Wilson & Cooper[101] had also shown that small cell size was associated with higher photosynthesis rate in *Lolium*. In fact they were able to characterize parents on the basis of mesophyll cell size and obtain desirable segregates which improved dry matter production in *Lolium*. In another study Takano and Tsunoda[97] obtained a curvilinear regression of the leaf photosynthesis rate on leaf nitrogen content in *Oryza* species and developed a model to predict the optimal level of leaf nitrogen to maximising photosynthesis rate of leaf canopies (Kishitani, Takano and Tsunoda[54]). Variability in photosynthesis rate at the varietal level has been observed in many crop plants (Dornhoff and Shibles[29], Sinha[83], Bhagsari and Brown[14], Wallace *et al.*[99]). Indeed it is difficult to distinguish the basis of differences in photosynthesis rate even when there are large differences as reported by McCashin and Canvin[60]. Since the rate of photosynthesis is dependent on many components (Fig. 1.2) it is also difficult to expect a

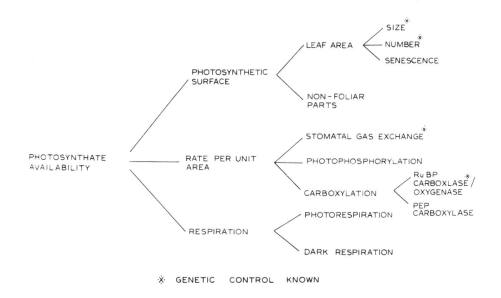

Fig. 1.2. Components of photosynthesis.

simple segregation for this character. When efforts were made to establish inheritance of C_4 and C_3 characteristics in *Atriplex* species, the results could not establish any clear pattern of segregation.

The question still remains whether it is possible to improve the rate of photosynthesis and what could be the short-term and long-term objectives in this respect? The question is discussed in some detail in Chapter 7, but from the plant breeder's point of view the following points need consideration:

Under field conditions, a crop plant experiences changes in the micro-environment with respect to temperature, humidity, water deficit and other factors. All these are known to adversely influence the rate of photosynthesis (Begg & Turner[12], Hsiao[47]). Adaptation in the rate of photosynthesis to these situations may be a desirable objective.

Extension studies on adaptation of photosynthesis to water deficit and temperature have proved that some phenotypic characteristics may provide advantage (Ehleringer & Bjorkman[33], Ehleringer[32], Mooney et al.[64]). Simulation models developed by Ehleringer[32] showed that C_4 plants had the advantage up to $40°$ latitude, but beyond this C_3 plants had a distinct advantage. However, even within this latitudinal limit the advantage was temperature dependent. Pubescence and leaf spectral characteristics were considered important in the adaptations of C_3 plants (Ehleringer & Bjorkman[33]). Similarly the reflectance properties of leaves were considered important. Therefore, there appears considerable scope for improvement of photosynthesis in relation to adverse environment.

It is at present necessary to develop a methodology for determining variation in photosynthesis rate under field conditions. Some parameters, including stomatal conductance, leaf nitrogen, mesophyll cell size, the number of veins per unit area, leaf size and leaf thickness may serve as useful criteria.

(vi) Nutrient availability and dry matter production

Nutrients, particularly nitrogen availability is one of the important factors determining agricultural production. Hageman et al.[43] concluded that the enzyme nitrate reductase (NR) can be a limiting factor in the utilization of nitrate and consequently in dry matter production.

Subsequently Croy and Hageman[22] showed that the activity of nitrate reductase is an index of reduced nitrogen. The reduced nitrogen in the form of ammonia was shown by Murata[67] to have a linear relationship in rice with (a) leaf expansion rate, (b) total leaf area development, (c) tiller number, (d) number of spikelets differentiated and (d) photosynthetic efficiency. Eilrich and Hageman[34] further showed a linear relationship between NR activity and reduced nitrogen in the vegetative phase. This relationship was extended to the yields of grains and grain protein in maize (Deckard et al.[27]). In recent years, several studies have supported this relationship partly or fully (Abrol and Nair[1], Johnson et al.[52], Dalling & Loyn[23], Mishra et al.[62]). On the basis of these studies it is advocated that NR activity could be used as a selection criterion. However, the genotypes that have high NR do not necessarily give more dry matter or reduced nitrogen and protein if a large number of genotypes are included in the study (see data of Dalling & Loyn[23]). This point was discussed at some length by Sinha and Nicholas[85] and it was suggested that NR activity in combination with the leaf area potential could provide a better index than NR activity alone.

Genetic variability in NR activity is known and it has been shown that this enzyme in maize is controlled by two genes (Warner et al.[100]). In a study of inbreds and hybrids it was observed that only the low $\times$ low NR types gave heterosis whereas high $\times$ low or high $\times$ medium resembled the better parent (Hageman et al.[43]). In a recent study using seven male parents and three female parents in sorghum, Mishra et al.[62] observed that the high NR parents showed no heterosis with either low or medium parents. This study shows that the high NR character has relatively high heritability.

In conclusion, NR activity has been shown to be associated with dry matter production and reduced nitrogen. It is time now to show the utility of this character as a selection criterion by using it alone or in combination with leaf area. This could be done by identifying the desirable types in a large germplasm collection and assessing their performance.

(vii) Nitrogen harvest index

Emphasis has often been given to harvest index in improving grain yield. Nitrogen harvest index, showing partition of nitrogen between grain and the remaining plant parts, is equally important because of the heavy cost of nitrogen. In wheat the new varieties have a greatly improved harvest index over the older varieties (Austin et al.[9]). However, in many pulses the plant assimilates a large amount of nitrogen but does not utilize it in grain development. Variability

TABLE 1.4. Species and cultivar difference in harvest index and nitrogen harvest index in wheat (*Triticum aestivum*) and chickpea (*Cicer arietinum*)

	Harvest index	Nitrogen harvest index
Wheat cultivars*		
Little Joss	36	71
Holdfast	36	69
Hobbit	48	72
Benoist 10483	48	73
Chickpea cultivars		
JG-62	23	46
M-109	30	51
M-119	35	69

* Based on Austin et al.[9] 1980.

in this character is now known (Table 1.4). For example, the nitrogen harvest index of the lower branches in pigeonpea is almost nil although they contain a considerable amount of nitrogen. It may be important to look for genetic differences in this character. This will be important if an improvement in protein content is not to result in a calorie penalty and vice versa.

Nitrogen fixation

Inorganic nitrogen is a key input in agriculture, both in developing and developed countries. Among the various energy inputs, nitrogen accounted for approximately 65 to 70% of the energy (Swaminathan[92]). Even in a developed country such as the United States, nitrogen is the third major energy input in crop production. Most of the grain crops contain 10–25% protein in grains and approximately 0.5 to 1.0% nitrogen in residual plant parts. It was, therefore, estimated that for wheat and chickpea, the amount of nitrogen harvested for a 6.0 tonnes grain yield, would be approximately 150 and 450 kg N ha^{-1}, respectively. Any effort to increase grain yield of cereals must involve greater input of nitrogen.

The possibility of improving nitrogen fixation, both in leguminous and non-leguminous crops is dealt with extensively in Chapter 8, the brief remarks that follow are from the point of view of plant breeders. There is now considerable evidence that both the host and *Rhizobium* exercise genetic control over nodulation (Riley[80]). There are several steps involved in establishing nodulation and most of these steps have been shown to be under genetic control. Since cereals lack in nitrogen fixing genes (nif), the transfer of such genes would be a prerequisite for any improvement. This could be a long-term objective, but for the near future improvement of nitrogen fixation in grain legumes and fodder legumes, the following objectives are important:

1. To improve the competitive ability of *Rhizobium* against those which inhabit the soils and cause ineffective nodulation. Some amount of success has been achieved in soybeans in this respect, which could possibly serve as a guide.
2. Bergersen[13] made calculations to show that there is, thermodynamically, a difference of only 5% excess energy requirement for nitrogen fixation as against nitrate assimilation. He gave no consideration to the energy requirements for nodule growth, transport of metabolites and other aspects. Many studies have now shown that photosynthate availability could limit nitrogen fixation (Sinha[82], Hardy & Havelka[45]). Therefore, while improving nodulation capacity, it must be kept in mind that increased nitrogen fixation could be realized only when provision for increased photosynthesis rate is made.
3. In most grain legumes nodules either disintegrate or become ineffective on the commencement of fruit setting (see Sinha[83]). Prolongation of nodule life after flowering should therefore be an important selection criterion.
4. It has now been shown that after harvest, both grain legumes and fodder legumes cause a legume effect on a subsequent crop which may be equivalent to 20 kg N ha^{-1} to 45 kg N ha^{-1} depending upon the crop species (De *et al.*[24]). This is important in a multiple cropping system where economy in nitrogen input is one of the objectives. It would be useful to determine variability in this respect among different leguminous species.
5. Relation between crop species and cultivars to non-symbiotic nitrogen fixation should be studied in greater detail. Is there any specificity in this non-symbiotic relationship based on leaching of specific compounds by the host?

Response to stress of mineral nutrients and water

In many crops, growth and yield are affected due to low level of some nutrients, particularly micronutrients or sometimes because of high levels causing toxicity. This whole area is considered at length in Chapter 4. In India, Agrawal and Sharma[2] carried out extensive studies for determining the susceptibility and toxicity limits of various crops to specific micronutrients. Varietal differences in these limits were found for zinc, boron, manganese and molybdenum in rice, wheat and maize.

Differences in tolerance to the aluminum toxicity of acid soils is known in many crops, particularly wheat (Foy[40]), and the character for tolerance has high heritability (Campbell and Lafever[21]). Rana *et al.*[77] have identified wheat varieties which are tolerant to sodic soils. They are now using them in a breeding programme involving agronomically superior varieties. These studies assume significance in the context of the growing problems of salinity and alkalinity in areas brought under irrigation. Thus, there are vast opportunities to develop new selection criteria to meet the shortages of inputs like nitrogen and phosphorus, and also to overcome the adverse effects arising from problems of soil stress factors.

Water is one of the key factors in crop production. Breeding for drought tolerance, a consequence of short supply of water, or water use efficiency could be a major objective in many areas. The limitations and opportunities in this respect have been discussed in several symposia (Reitz[79], Moss[66], Morgan[65], Steponkus et al.[88]), and are considered from a practical point of view in Chapter 6. Breeding for drought tolerance or water use efficiency needs to take into account agroclimatic parameters, particularly the total precipitation, its distribution and storage in the soil.

Besides the effect of water deficit, an equally important response of indeterminate crops such as pulses and oil-seeds, is the promotion of vegetative growth. Consequently when irrigation is applied, even when plants are flowering, there is an increase in dry matter production associated with a decrease in harvest index. Therefore no improvement in economic yield occurs. Selection against this behaviour could be advantageous for making these crops suitable

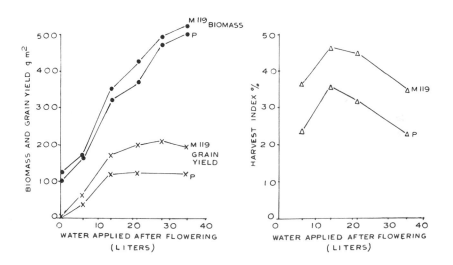

Fig. 1.3. Effect of water availability after flowering on biomass, grain yield and harvest index in chickpea, *Cicer arietinum*, (P is the parent, and M 119 a mutant).

for irrigated agriculture. That selection for such a character is possible has been observed in chickpea, *Cicer arietinum*, (Fig. 1.3). This appears to be one of the ways to improve harvest index in pulses, which are generally known to have a low harvest indices.

Differentiation and development in relation to sink

Growth, differentiation and development of plants which eventually lead to yield, are determined at different times during the life cycle of the plant. From the point of crop productivity the following phases are important:

1. Seed germination and seedling establishment
2. Vegetative growth
3. Flower initiation and development or tuber initiation and development
4. Grain development

The above phases are often clearly distinguishable and separable in most crops but some-times they overlap, depending on whether a crop is determinate or indeterminate in its growth habit. Physiological investigations over the years have attempted to relate the performance of these stages with yield. One of the stages which has been extensively studied is the post-anthesis period in cereals. It is, therefore, important to enumerate some important considerations:

1. The dry matter production in post-anthesis/flowering period is equal or sometimes even more if there are no constraints on water and nutrition.
2. The bulk of dry matter produced after anthesis, ranging from 60 to 80% can account for the grain yield.
3. Usually the uppermost leaf (the flag leaf in wheat and barley) or the upper leaves in sorg-hum and leaves above the cob in maize remain effective in photosynthesis during grain development (Asana[7], Fisher & Wilson[39]).
4. Dry matter production after anthesis is directly related to water availability in wheat (Passioura[73]).
5. Increase in the mean daily temperature adversely influences grain development and hence yield in wheat (Asana[8]). Similar observations have been recorded in other crops also.
6. The longer the duration of grain filling, the higher, usually, is the yield of a plant (Asana[8], de Wit[26]). This will be reflected in cereals in the duration between ear emergence and grain maturity.

From the above, it is clear that attainment of higher rate of dry matter production, for a longer duration after anthesis, would be helpful in increasing yield.

Flowering and the formation of 'sink'

Flowering time is important for the establishment of the 'sink' in most grain crops. Recent studies in various crop species indicate that flowering time is under genetic control, so that this character can be manipulated, along with the vegetative characters, to provide a mechanism for complementation between vegetative and reproductive structure. The following examples would serve to indicate the utility of such studies.

Wheat responds to both vernalization as well as day length (Evans et al.[37]). In some situ-ations it is profitable to obtain a photoinsensitive genotype, whereas for the regional or location specific adaptability, it may be necessary to have some degree of vernalization or photoperiod response. Klaimi and Qualset[55,56] observed that wheat cultivars exhibited various combina-tions of response to photoperiod and vernalization indicating that these two attributes are not associated in inheritance. Using diallel crosses between spring X winter types they showed that one or two duplicate genes governed these characters. The presence of minor genes and multiple alleles was not ruled out. Nevertheless, they concluded that a desirable pattern of heading could be obtained by manipulating genes for day length, or vernalization or both.

Bagnara *et al.*[10] reported inheritance of different developmental phases in *durum* wheats. They observed correlation between the time of initiation and the heading time. In sorghum also four major genes have been identified controlling flowering time (Quinby[75]). It appears that the time of initiation, the period between initiation and completion of panicle, and the period between panicle completion and emergence are regulated by different genes (Murfet[68]). This could provide an opportunity to select a genotype according to the specified needs of a location, depending upon the rainfall pattern, day length and temperature. Mishra[61] studied the time of initiation and the time of emergence in sorghum, using a mating system involving seven male parents and three female parents. Under field conditions, it was observed that F_1 behaviour followed the male parent. Thus the choice of male parent will be important in developing hybrids with the desirable flowering time.

Flowering behaviour in mung bean (*Vigna radiata*) has been studied in detail by Aggarwal & Poehlman[3] and Swindell and Poehlman[96]. A dominant or partially dominant gene for sensitivity to photoperiod was identified. The gene was expressed when the sensitive strain was grown in 16—14 hour photoperiod, but was not expressed when the populations were grown in a 12 hour photoperiod. An interesting point which emerged from their international mungbean nursery programme was the fact that the number of days to flower increased with the latitude, or at the same latitude by changing altitude. This clearly showed an interaction of photoperiod with temperature. Murfet[68] has reviewed the literature on genetic control of flowering in peas.

From the above and many other examples, it appears that the time of flowering can be manipulated in various crops. However, if we knew exactly the relationship of the preflowering phase to the reproductive and yielding capacity of a plant, the breeding programmes can be better planned. This should be one of the major objectives of research in this area in the coming years.

Yield components

In the final analysis, what is harvested are the grains, tubers or other economically useful parts. The following simple equation to describe yield was given by Ishizuka[48]:

Yield (of brown rice) — No. of spikelets per m^2 (No. of ears/m^2 X No. of spikelets/ear) X percentage of spikelets bearing a ripe grain X average grain weight (based on 1000 grains) X 1000.

This equation is based on the fact that a combination of grain number, grain weight and spike number is essential to raise yield levels in most grain crops. Several studies have shown that these components are negatively correlated and, therefore, are difficult to combine. Nevertheless, we know that despite having negative correlation, yield has been improved, both in wheat and rice. As stated earlier, an understanding of their development can be extremely useful in breaking these correlations. Appropriate agronomic practices which enable, through seed rate, to obtain the optimum number of spikes per square metre, will also help to offset the negative correlation observed at the individual plant level.

In sorghum and wheat considerable improvement has been achieved in yield components. It may, therefore, be worthwhile to analyse the reasons for this so that it could serve as a guide for the future. In India, Rao[78] made several crosses among indigenous inbred parents of sorghum in an effort to obtain yield heterosis, but with almost no success. When the indigenous male parents were crossed with Ck 60 a male sterile line from USA, a high degree of heterosis was

TABLE 1.5. Yield components of sorghum hybrids and their parents

Entry	Number of whorls/panicle	Branches per panicle	Number of grains/branch	Number of grains/panicle	1000-grain weight (g)	Grain weight/panicle (g)
msCK60B	9.3	54.8	24.10	1321	21.25	30.90
msCK60B × IS3691	11.0	89.5	27.40	2453	29.70	58.20
IS3691	13.0	90.3	13.35	1206	32.58	25.20
2219B × IS3691	12.0	94.3	23.29	2197	34.90	44.30
2219B	9.0	64.8	28.88	1872	17.63	35.30
CD at 5%	0.1	4.9	—	239	0.05	4.40

obtained. An analysis of yield components showed that the Indian parents had a large number of secondaries per panicle but had only a few grains per secondary. As against this, CK 60 had a fewer number of secondaries but a larger number of grains per secondary. The F_1 combined the character of the number of secondaries and the number of grains per secondary. The result was a larger panicle size (Table 1.5). It is likely that such a large panicle size developed only when there was sufficiently strong photosynthetically active source to support development. The study, however, points out that 'sink' size could be enlarged by a suitable choice of parents.

In wheat, there has been little progress in improving the yield potential to a level higher than that achieved in the early 1960's following the introduction of Norin dwarfing genes in a relatively photo-insensitive background. A major reason for this situation is that in *T.aestivum* the number of spikelets has not improved despite the fact that this character is under genetic control (Rahman *et al.*[76]). Increase in the number of spikelets in a uniculum wheat is associated with reduced grain weight. Thus increase occurred with a decrease in grain weight and the number of spikes. However, M.G. Joshi and colleagues (personal communication) at the IARI, New Delhi, adopted the technique of interspecific hybridization using *T.turqidum*, *T.durum*, *T.polonicum*, *T.spherococcum* and other species. This technique has resulted in generating a considerable amount of genetic variability in yield components. Particularly the number of spikelets has increased.

From the above, it will be clear that in a breeding programme a limit is reached when a gene pool responsible for yield components gets stabilized. At that stage hybridization within the same pool may not result in any further improvement. Component analysis and identification of a particular character in a distantly growing population or a related species may prove helpful in breaking negative correlations.

NEW TRENDS IN BREEDING AND SELECTION

(i) Individual plant vs population performance

A major shift in breeding work in recent years is the emphasis in selection programmes on population performance. It is usually conceded that the growth and development of an isolated individual plant becomes different in a population because of interplant competition. Some of the factors for which competition becomes operative are light, water, nutrients etc. Amongst these the erectophile types experience less competition than planophile types at the same plant population. Thus selection in early generations has to take into account the possibility of competitive effects in a population. Such studies have now been initiated (Tarhalkan & Rao[98]).

(ii) Ideotype concept

Among the various concepts and methods which helped in breeding improved crop plants, the ideotype concept has made a significant impact on the thinking of plant scientists. Jennings[50] made plant type an objective in plant breeding, taking into account leaf and stem characteristics. According to Donald[28] plant breeding has been based on defect elimination or selection for yield. According to him a valuable approach would be to breed for crop ideotypes, plants with model characteristics known to influence photosynthesis, growth and grain production. He defined ideotype as a biological model which is expected to perform or behave in a predictable manner within a defined environment. More specifically, a crop ideotype is a plant model which is expected to yield a greater quantity and/or quality of grain, oil or other useful produce when developed as a cultivar.

It is thus clear that a crop ideotype has to be based on genotype environment interaction. This necessitates physiological studies on the response of different plant processes to different environmental factors. It is only then that a model can be developed. On the basis of several studies, Donald[28] described a wheat ideotype. The ideotype consists of a short strong stem, few, small, erect leaves; a large ear (many florets per unit of dry matter of the tops); an erect ear; awns, and a single culm. Asana[7] also described a wheat ideotype suitable for unirrigated conditions. Apart from the morphological design postulated by Donald[28], a significant point was made that a successful crop plant would be of low competitive ability relative to its mass and of high efficiency relative to its environmental resources. Indeed, it has been shown that high yielding cultivars of barley and rice are suppressed or eliminated in mixtures (Suneson[89], Jennings and De Jesus[51]). This concept is of vital importance for breeding crops for monoculture and mixed croppings, since the plant will have to experience different kinds of competition under these two situations.

More recently, several efforts have been made to propose ideotypes of various crop plants. In most instances a generalized ideotype has been suggested for a crop. Obviously this is not adequate in view of the fact that crops experience different environmental effects. Secondly, what applies to a cereal which has determinate growth, may not apply to an indeterminate plant. It is necessary to understand as to which parts or processes in a plant limit the yielding capacity. Only then could a meaningful ideotype be suggested. Bhargava *et al.*[15] have proposed an ideotype of *Brassica sp.* which, with high inputs could yield up to 6 tonnes ha^{-1} or more.

It is true that hardly any effort has been made to breed for an ideotype, which ultimately may or may not be the most suitable. Nevertheless, the concept has triggered a thinking process and in combination with gap analysis (Fig. 1.4) as suggested by Swaminathan[93] provides a scientific orientation to a breeding programme.

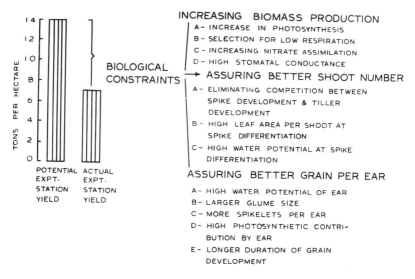

Fig. 1.4. Constraints which require attention for bridging the gap between potential and actual experimental yields in wheat.

(iii) Breeding methodology for determinate vs indeterminate crops

Historically the breeding methodology was developed with crops such as wheat, maize, sugarbeet, clover, etc. In these crops selection for yield is usually based on the yield of an individual plant in well spaced populations. Individual selections are multiplied to develop enough seed material for testing in a population. Where the yield of the shoot contributes significantly, this method remains useful because even in a population up to a certain limit the yield of the mother shoot is retained. Interestingly, the same methodology is usually utilized for selection in indeterminate crops such as chickpeas, pigeonpeas, mustard, etc. In these crops, an individual plant could yield well in a spaced population because of the fruiting on tertiary branches. However, such high yielding selections fail to retain their performance in a population because of reduced contributions of late order branches. Obviously, the same methodology would not suit the determinate and indeterminate plants. The ideotype concept would have possibly even greater relevance to indeterminate crops. The plant breeder would have to visualise the performance of an isolated plant when placed in a population to make his selection more effective.

(iv) Exploitation of hybrid vigour

Hybrid vigour has been extensively utilized in improving crop plants (see Wittwer[104], Sinha and Khanna[84]). Efforts to identify suitable parents through laboratory technique have not been successful (Riley[80]). There is no doubt that this phenomenon would continue to help in improving crop plants, but the following questions are important:

1. Can heterosis be fixed?
2. What will be the performance of F_1 hybrids produced through protoplast fusion of parents characterized by good combining ability?

The first question has attracted the attention of many workers and maybe some answer will emerge. Murty & Rao[69] have tried to introduce genetically controlled apomixis in *Sorghum* cultivars in order to fix heterosis. However, the second question has not been tested even in *Solanaceous* species where protoplast isolation, fusion, differentiation and development have already been achieved. In species such as *Lycopersicum esculentum, Solanum melanogena, Nicotiana sp.* heterotic hybrids have been obtained. At the same time protoplast fusion technique has also been developed. Therefore, it is time that the comparisons of F_1 hybrids obtained through sexual reproduction and protoplast fusion are made. This might help not only in understanding the phenomenon of heterosis, but also provide a tool for determining specific combining ability in laboratory conditions.

In addition, it is important to characterise the nature of heterosis at the level of different components. This might help in identification of suitable parents in a shorter time.

(v) Breeding for farming systems

The history of plant breeding objectives and methodology show that the improvement of an individual crop was the main concern in the past. It continues to be so in many parts of the world, particularly the temperate regions where the effective growth period is restricted to six months or less. This area is represented by Europe, North America, USSR, Japan, Argentina, Oceania, etc. On the other hand, in tropical regions there are vast areas where the limitations of water and high temperature are major constraints and limit cropping to one season or sometimes to none. However, there are vast areas stretched around the equator in South America, Africa and Asia where both temperature and water availability are favourable factors and make it possible to grow crops all the year round. This is done either through the use of field crops or by growing perennial crops. In some of these areas, the following practices have been in vogue from time immemorial:

1. Multiple cropping
2. Intercropping or mixed cropping
3. Sole perennial crops
4. Intercropping annual and perennial crops.

These practices are briefly described so that the breeding objectives can be appreciated and defined.

(a) *MULTIPLE CROPPING*

In many parts of South East Asia, Africa, the Indian sub-continent and other equatorial regions the mean daily temperature remains around $20°C$ the precipitation is 100 cm or more and solar radiation is in the range of 350 cal cm^2 day^{-1} to 700 cal cm^2 day^{-1}. Therefore, both the radiation and temperature regime are conducive to growing plants. In India, depending upon the agroclimate some of the following cropping patterns are practiced in a yearly cycle (see Swaminathan[95]):

1. Rice — mung bean
2. Sorghum — pigeonpea
3. Maize —wheat
4. Rice — wheat
5. Maize — potato — wheat
6. Maize — potato — wheat —mung bean
7. Rice — rice — rice
8. Rice — rice
9. Pigeon pea — wheat
10. Sorghum and cowpea — wheat
11. Maize and cowpea — wheat
12. Mung bean — mustard.

From the above, it would be clear that various crop combinations are practiced or are possible. The reasons for this are as follows:

1. Monsoon rains start from the middle of June and depending upon the region continue up to the end of September. Therefore, a period of 90 to 100 days is available when a crop can be grown without irrigation or with only one irrigation.
2. During the period of rains, the maximum temperature is between $33-38°C$ and the minimum between $25-30°C$. Therefore, the conditions are suitable for a C_4 plant or a plant like rice.
3. The rainy season recharges the aquifer with water in the various river basins and plains. There is also stored water in large dams which is utilized for irrigation.
4. The post-rainy season is characterized by lower mean daily temperature, reduced day length and low evapotranspiration. These conditions are conducive to growing C_3 crops such as wheat, barley and chickpea.
5. In some coastal regions there is a second monsoon and temperature remains favourable.

It will be obvious from the above analysis that multiple cropping is an ideal mechanism for maximising per day productivity by making effective use of natural resources such as radiations and water at suitable temperature. Therefore, the prime requirement would be to tailor crops and varieties to suit different agroclimatic needs. This necessitates cultivation of three or sometimes even four crops in a year. Accordingly, the first requirement for breeding crops for such situations would be to shorten the life span of a variety. Period-fixed, rather than season bound, varieties will be needed for developing efficient multiple cropping systems.

Along with the shortening of life span, an essential requirement will be the improving of dry matter production, through obtaining complete light interception as soon as possible. This means that the plant will have to have fast leaf area development. Usually a high specific leaf weight is associated with high photosynthesis rate. Accordingly while selecting for a higher leaf area production, it would be important to obtain it through an increased number of leaves rather than the expansion of leaves, as the latter may bring down the photosynthesis rate.

The shortening of life span also means that the plant should have a shorter juvenile phase, but no substantial reduction in the period of reproductive growth and grain development phase. Some examples will illustrate the point of view clearly.

In *Sorghum* old varieties took 130 days or more from sowing to harvesting, but the grain filling occurred in the last 40–45 days. This was partly due to declining temperatures. The new hybrids of *Sorghum* take about 95 days from sowing to harvesting while having higher grain yield potential than the older varieties. One main factor which caused this change is the early initiation of panicle differentiation which occurs in 26 to 28 days from sowing. Similar examples can be found in rice and other monsoon season crops.

Among the winter crops, the life span of old varieties of wheat and chickpeas used to be 160 days or more, while the sowing had to be done in late October or early November. This did not provide enough time for land preparation and other operations between a monsoon season crop and the winter season crop. The result was that only one crop could be grown which would be either a monsoon crop, or a winter season crop. In recent years, not only the duration of winter season crop was reduced but the sowing could be delayed by as much as one month.

Some of the objectives and parameters of breeding for multiple cropping system can be summarised as below:

1. Shortening the life cycle of crops. This would need a good understanding of the genetic control of various phases of plant growth.
2. Faster initial growth and dry matter accumulation.
3. Change in the date of sowing without adversely influencing the various phases of plant growth.
4. Adaptation to temperature change.
5. Complementary rather than allelopathic effects of the preceding crop on the succeeding crop.

(b) BREEDING FOR MIXED CROPPING OR INTERCROPPING

Mixed cropping or intercropping is extensively practiced in several parts of Asia, Africa and Latin America. Aiyer[4] prepared a comprehensive review of this practice in India wherein he considered the advantages and disadvantages of this system. Several questions in respect of the role of legumes in intercropping were raised. Many of these questions are relevant even today and have yet to be answered. Willey[102] reviewed the work on intercropping and attempted to analyse the scientific basis of this practice. Despite the importance of this practice in crop production, crop stability and adaptability to adverse situations, such as pests and diseases and low water availability, the fact remains that no serious effort has been made to breed plant types specifically for intercropping. Often the crops and varieties selected for monoculture performance have been included in intercropping experiments.

It may be worthwhile to restate the kinds of mixed cropping classified by Aiyer[4] and then to analyse the breeding objectives. Forms of mixed cropping are:

1. MIXTURE OF PERMANENT CROPS:
(a) Fruit orchards such as coconuts, mangoes, jack fruit guavas, oranges etc. The crown of trees mingle and develop canopies resulting in two or more tiers at irregular heights.
(b) Coffee plantations. Coffee bushes supporting vines of pepper. Alternatively plants of different heights such as coconut, arecanut, cloves, cocoa, coffee and even tea bushes.
(c) Areca palm with betel vines or areca, cardamoms and pepper.

2. MIXING PERMANENT AND TEMPORARY CROPS

The temporary crops grown mixed with permanent crops are in the nature of:

(a) Nurse crop

(b) Catch crop

(c) Green manure crop

3. MIXED CROPPING WITH FODDER GRASSES

Mixed cropping of legumes and cereals or other non-legumes:

(a) Monsoon season crops: Sorghum and pigeonpeas

(b) Winter season crops: Wheat and chick pea, barley and peas, etc.

The advantages and disadvantages of various mixtures in respect of soil moisture extraction, nutrient use and pest sensitivity have to be studied. Indeed, there is a relatively small amount of data on soil moisture profiles, nutrient changes and other aspects to fully describe the utility of this cropping pattern. However, the broad objectives will be the following:

1. Utilization of land space to the fullest extent possible in tropical regions
2. Making effective use of available moisture
3. Making effective use of legumes, either as fodder or grain crops, in improving the efficiency of the cropping system
4. Mixed cropping as a means of protection against disease and pests
5. Stability in production in dry regions.

A recent review by Willey[102] appreciates the value of mixed cropping in respect of many of the above objectives. In some instances it appears that the conclusions cannot be definitive in the absence of quantified data. Willey[102] has suggested various methods for evaluating the efficiency of mixed cropping or the sole crop performance.

Some of the following examples can illustrate the breeding objectives:

1. MIXED CROPPING IN PLANTATION CROPS

In coconut and arecanut orchards there is differential transmission of light through the plantation canopies. Nair et al.[70] studied light transmission on the ground at different hours of the day. The transmitted light varied from 1000 lux to 30,000 lux in different seasons. The light response curves for cocoa, pineapple, ginger and other crops indicate that this amount of light could be enough for those crops. Accordingly, it has been observed that coconut, cocoa and pineapple constitute a good multistoreyed multiplecropping pattern at the Central Plantation Research Institute, Kasargod, India. However, it is now important to improve this pattern by understanding the following:

1. The influence of different plant types of coconut on transmission of light on the ground.
2. Genotypic differences in light saturation in cocoa, pineapple, ginger, etc.
3. Nutritional and water requirements of different canopies so as to maximize total production.
4. Efforts to minimize competition during the development of economically important plant organs.

2. MIXED CROPPING IN DRYLAND

Drylands are the areas where the amount and distribution of rainfall is uncertain. In most places the duration of the rainy period is confined to nearly three months or so. Depending upon the rainfall this could be adequate for one crop. When the rainfall is more than adequate for one crop but insufficient to raise two successive crops, then one long duration and another a short duration crop are grown to take full advantage of the available moisture. The examples of this situation are the mixed cropping of sorghum and pigeonpea, pearl millet and pigeonpea, cotton and mung bean, castor and pigeonpea.

In addition, the dryland crops are raised on stored moisture in winter when only a small amount of rain is expected. Since the time of rainfall is uncertain, it influences plant development in relation to the time of rain. Usually a determinate and an indeterminate crop is used in a mixture. For example, wheat and chickpea, barley and peas, barley and chickpea, etc.

It would be obvious from the above discussion that water availability is an important factor influencing mixed cropping practice. So far only such varieties which performed well in monoculture have been used in mixed cropping. This probably may not be the best answer. In a recent experiment, Willey & Rao[103] employed seventeen genotypes of pigeonpeas in a 1:1 mixture of one hybrid of sorghum. There was no significant reduction in the yield of sorghum as compared to the sole crop of sorghum, but the yield of pigeon peas varied from 463 to 885 kg ha^{-1}. This indicated clearly that some specific genotypes of pigeon peas were more suitable than others. Tarhalkan and Rao[98] also tested the performance of two pigeonpea varieties in mixtures with CSH-6 sorghum. It was shown that the yield of sorghum was unaffected but the variety HY 3-A of pigeonpea yielded more than HX-2.

It would thus be advisable that breeding for mixed cropping has to be developed as a programme in most cases independent of monoculture breeding.

(vi) Breeding for small farms vs large farms

Small farms are often a characteristic of the developing countries, while large farms are common in the developed countries. Although broadly speaking foodgrains, fiber and industrial crops are produced in both kinds of farms, they differ in details. For example, human labour is available easily in the developing countries and a farmer usually grows crops to meet the requirements of his family. As against this a farmer in a developed country has very limited labour and he works as a commercial enterprise. Therefore, the objectives in breeding of the developing countries and the developed countries differ considerably:

(a) DEVELOPING COUNTRIES:

The following should be the main objectives:

1. Increase in productivity of land per unit area per year. This would involve multiple relay and inter-cropping.
2. Resistance to disease and pests in a farming system.
3. Breeding for grain production as well as straw or other plant residues as animal feed. This is important because animal power would continue to play a major role in agriculture economy of the developing countries.
4. Breeding for optimizing returns from a suitable blend of monetary and non-monetary inputs.
5. Breeding for suitability for incorporation in contingency plans of crop production in drought and flood prone areas.

(b) DEVELOPED COUNTRIES:
1. Breeding for maximizing output in relation to human labour because labour is scarce.
2. Breeding for synchrony in maturity and harvesting so as to fit in with 'superfarm' management.
3. Suitability for mechanized harvesting, processing and transport.
4. Improving efficiency of energy output to energy input.

The above mentioned instances are only some of the differences in requirements of the developing and developed countries. They, however, would differ much more depending upon a specific situation since, even in some of the developing countries and in countries where land is socially owned, super-farm production technology is being adopted.

(vii) Increasing the market competitiveness of commercial crops

Natural products like cotton, jute and rubber, face severe competition from synthetics. After cost escalation of petroleum products in recent years, natural products are again attracting attention. It is, however, necessary for plant breeders to be aware of market requirements and preferences and to help in incorporating to the greatest extent possible the characteristics for which the synthetic products are sought after by the consumer. The work done in India by cotton breeders and technologists indicates that there is much scope for successful research in this field.

It is now well recognised that loss in mechanical properties of easy-care finished fabric is to a great extent due to the inherent non-homogenous structure of cotton fibres and non-uniformity of distribution of cross links. There is also large variation in the convolution angle, spiral angle, and X-ray angle. In view of this, Indian cottons offer excellent genetic material for a detailed study of the behaviour of different varieties of cotton in their response to chemical finishing treatments.

In the work at the Cotton Technological Research Laboratory, Bombay, a large number of varieties of cultivated cottons were screened in yarn form after being given a standard cross-linking treatment with dimethylodihydroxyethylene urea (DMDHEU). There is a wide variation in the toughness retention within *Gossypium hirsutum* cottons after cross-linking treatment. For instance, the variety 66BH5/91 has toughness retention of 73% whereas the variety H.14 exhibited poor toughness retention of about 17%. Similarly, in the case of *G.arboreum* and *G.herbaceum* cottons, the varieties such as Sanjay and Jayadhar have exhibited very high toughness retention, whereas the variety AK235 has very poor toughness retention.

Eight varieties of cotton which had shown promise while screening in yarn form were further screened in fabric form for their easy-care properties. It was noted that four of them: Hybrid-4, Diqvijay, Deviraj and Sanjay have high toughness retention and with the exception of Hybrid-4, also have good wash-wear qualities.

The above studies have demonstrated that varietal response exists in cotton to finishing treatment and that cottons with high toughness character, more uniform fine structure, more circular cross-sectional shape, etc., would be more suited for easy-care treatments. It is of interest that some of the *G.arboreum* and *G.herbaceum* cottons are inherently suited for cross-linking treatment. It would, therefore, be desirable for the present day cotton breeder to introduce a new parameter in his breeding programme, viz., response to easy-care finishing. Similar studies are needed by botanists in all crops which are used as industrial raw material so that consumer preferences can be satisfied.

LOOKING AHEAD

It is now well established that rice bran contains considerable amounts of silicon. Efforts have been made to use rice bran for obtaining pure silicon which could be used in photovoltaic cells to harness solar energy. However, the extent of variability for this character is not known among the rice cultivars. After identifying a genotype having higher silicon content, it might be useful to incorporate this character through breeding in agronomically desirable cultivars. This could provide a cheaper source of silicon cells for direct sunlight use.

It is now clear that carbon dioxide concentration is increasing in the atmosphere. This might lead to changes in temperature and precipitation. Consequently, adaptability to both temperature and precipitation would become important. Although a plant breeder selects for micro-environmental conditions prevailing at a given time, it is important that a beginning is made for breeding for various weather probabilities. Detailed specifications can be developed for plant breeders on the basis of weather models and their implications on growth and development constructed by production physiologists. Early warning on CO_2 concentration should be followed by timely action in the field of crop breeding. Countries with facilities for creating different weather combinations under glass house conditions particularly, should start screening genotypes suitable for different growing conditions. The International Board for Plant Genetic Resources should initiate research on screening germplasm for growing conditions which may prevail in the twenty-first century.

Parameters based on economics, ecology, energy availability and employment requirements will dominate plant breeding programmes during the remaining part of the twentieth century. Plant breeders working in countries with small farms may have to tailor their selection criteria to the socio-economic and agro-ecological conditions prevailing in their countries. Fortunately many of the developing countries characterised by small farms have also abundant sunshine throughout the year. The period of maximum insolation in these countries, however, generally coincides with periods of minimum precipitation. Developing countries in the tropics and sub-tropics are now making major investments in irrigation. Therefore, there are opportunities for a 'cafeteria approach' in terms of breeding varieties for different management systems.

The most important change in outlook, however, relates to looking at breeding problems as a total system not only in terms of crops grown in rotation but also in terms of every part of the plant. As mentioned earlier, rice husk is today a good source of solar grade silicon for use in photovoltaic systems. Therefore, depending upon the state of development of post-harvest technology, selection criteria should include the potential value of every part of the plant. A multi-disciplinary approach to plant breeding will be even more important in times ahead, than it has been in the past.

REFERENCES

1. Abrol, Y.P. and Nair, T.V.R. Uptake and assimilation of nitrogen and its relationship to grain and protein yield. 113–131. *Nitrogen Assimilation and Crop Productivity.* S.P. Sen, Y.P. Abrol and S.K. Sinha (eds.), 1978.
2. Agarwala, S.C. and Sharma, C.P. *Recognising micronutrient disorders of crop plants on the basis of visible symptoms and plant analysis.* Botany Department, Lucknow University, Lucknow, 1979.
3. Aggarwal, V.D. and Poehlman, J.M. Effects of photoperiod and temperature on flowering in mung bean (*Vigna radiata(L)* Wikzek) *Euphytica* 26, 207–219, 1977.

4. Aiyer, A.K.Y.N. Mixed Cropping in India. *Indian J. Agri. Sci.* **29**, 439–543, 1949.
5. Allard, R.W. *Principles of Plant Breeding.* John Wiley & Sons Inc. New York & London, 1960.
6. Anonymous. Tolerance to adverse soils. *The International Rice Research Institute, Annual Report for 1977.* 113–129, 1978.
7. Asana, R.D. Growth habits of crops of non-irrigated areas. Important characteristics of plant types. *Indian Fmg.* **18**, 25–27, 1968.
8. Asana, R.D. *Physiological Approaches to Breeding of drought resistant Crops.* Indian Council of Agricultural Research, New Delhi, 1975.
9. Austin, R.B., Bingham, J., Blackwell, R.D., Evans, L.T., Ford, M.A., Morgan, C.L. and Taylor, M. Genetic improvement in winter wheat yields since 1900 and associated physiological changes. *J. Agri. Sci.* **94**, 675–689, 1980.
10. Bagnara, D., Bozzini, A. and Mugnozza, G.T.S. Inheritance of the duration of developmental phases of the biological cycle of *Triticum durum* Desf. *Genetica Agraria* **25**, 31–55, 1971.
11. Bartholomew, D.P. and Kadzimin, S.B. Pineapple. *Ecophysiology of tropical Crops.* Paulo de T. Alvim and T.T. Kozlowski (eds.), 113–156, 1977.
12. Begg, J.E. and Turner, N.C. Crop water deficits. *Advan. Agron.* **28**, 161–217, 1976.
13. Bergersen, F.J. The central reactions of nitrogen fixation. *Plant and Soil, special vol.* 511–524, 1971.
14. Bhagsari, A.S. and Brown, R.H. Photosynthesis in peanut (*Arachis*) genotypes. *Peanut Sci.* **3**, 1–5.
15. Bhargava, S.C., Tomar, D.P.S. and Sinha, S.K. Physiological basis of productivity in *Brassica* ecotypes. *Proc. National Seminar on Research and Production strategies in Oilseed Crops.* IARI. New Delhi, 1979.
16. Bhatia, C.R. and Rabson, R. Bioenergetic considerations in cereal breeding for protein improvement. *Science.* **194**, 1418–1419, 1976.
17. Bressani, R. Legumes in human diets and how they might be improved. *Nutritional improvement of food legumes by breeding.* 15–42. PAG & FAO, Rome, 1973.
18. Brewbaker, J.L. *Agricultural Genetics.* Prentice-Hall Inc. New Jersey, 1964.
19. Buringh, P., Van Heemst, H.D. and Staring, G.J. *Computation of Absolute Maximum Food Production of the World.* Agricultural University, Wageningen, The Netherlands, 1975.
20. Calvin, M. *Hydrocarbons via Photosynthesis.* ACS Rubber Division Meeting, San Francisco, California, 1976.
21. Campbell, L.G. and Lafever, H.N. *Heritability and gene effects for Aluminium tolerance in wheat.* Proc. 5th Intern. Wheat Genetics Symposium. New Delhi, 963–977. S. Ramanujam (ed.), 1979.
22. Croy, L.I. and Hageman, R.H. Relationship of nitrate reductase activity to grain protein production in wheat. *Crop Sci.* **10**, 280–285, 1970.
23. Dalling, M.J. and Loyn, R.H. Level of activity of nitrate reductase at the seedling stage as a predictor of grain nitrogen yield in wheat (*Triticum aestivum L.*) *Aust. J. Agri. Res.*
24. De R. Role of legumes in intercropping systems. *Proc. Nuclear Techniques in the development of management practices for multiple cropping systems.* IAEA, 1980.
25. De Vries, C.A., Ferwerda, J.D. and Flach, M. Choice of Food Crops in relation to actual and potential production in the tropics. *Neth. J. Agri. Sci.* **15**, 241–248, 1967.
26. De Wit, C.T., van Laar, H.H. and van Keulen, H. Physiological potential of Crop production. *Plant Breeding Perspectives,* 47–82. J. Sneep and A.J.T. Hendriksen (eds.).
27. Deckard, E.L., Lambert, R.J. and Hageman, R.H. Nitrate reductase activity in Corn leaves as related to yields of grain and grain protein. *Crop Sci.* **13**, 343–350, 1973.
28. Donald, C.M. The breeding of Crop ideotypes. *Euphytica.* **17**, 385–403, 1968.
29. Dornhoff, G.M. and Shibles, R.M. Varietal differences in net photosynthesis of soybean leaves. *Crop Sci.* **10**, 42–45, 1970.
30. Duncan, W.G., McCloud, D.E., McGraw, R.L. and Boote, K.J. Physiological aspects of peanut yield improvement. *Crop Sci.* **18**, 1015–1020, 1978.
31. Dunstone, R.L. and Evans, L.T. The Role of changes in cell size in the evolution of wheat. *Aust. J. Plant Physiol.* **1**, 157–165.
32. Ehleringer, J. Implications of quantum yield differences on the distribution of C_3 and C_4 grasses. *Oecologia,* **31**, 255–267, 1978.
33. Ehleringer, J. and Bjorkman, O. Pubescence and leaf spectra characteristics in a desert shrub, (*Encelia farinosa*) *Oecologia,* **31**, 1978.

34. Eilrich, G.L. and Hageman, R.H. Nitrate reductase activity and its relationship to vegetative and grain nitrogen in wheat (*Triticum aestivum* L.). *Crop Sci.* 13, 59—66, 1973.

35. Elliot, F.C. *Plant Breeding and Cytogenetics.* McGraw Hill Book Co. Inc. New York, 1958.

36. Evans, L.T. and Dunstone, R.L. Some physiological aspects of evolution in Wheat Aust. *J. Biol. Sci.* 23, 725—741, 1970.

37. Evans, L.T., Wardlaw, I.F. and Fischer, R.A. Wheat. *Crop Physiology.* 101—149. L.T. Evans (ed.). Cambridge University Press, 1975.

38. FAO. Land resources for populations of the future. *Report of the second FAO/UNFPA Expert Consultation.* FAO. Rome, 1980.

39. Fisher, K.S. and Wilson, G.L. Studies on grain production in *Sorghum vulgare* II. Sites responsible for grain dry matter production during post-anthesis period. *Aust. J. Agri. Res.* 22, 37—47, 1971.

40. Foy, C.D. Differential aluminium and manganese tolerance of plant species and varieties in acid soils. *Cien. Cult.* 28, 150—155, Sao Paulo, 1976.

41. Frankel, O.H. The dynamics of plant breeding. *Proc. XII International Cong. Genetics.* Vol. 3, 309—325. C. Oshima (ed.).

42. Grafius, J.E. Heterosis in barley. *Agron. J.* 51, 551—554, 1959.

43. Hageman, R.H., Leng, E.R. and Dudley, J.W. A biochemical approach to corn breeding. *Advan. Agron.* 19, 45—86, 1967.

44. Hamblin, J. and Donald, C.M. The relationship between plant form competitive ability and grain yield in a barley cross. *Euphytica* 23, 535—542, 1974.

45. Hardy, R.W.F. and Havelka, U.D. Photosynthate as a major factor limiting nitrogen fixation by field-grown legumes with emphasis on soybean. *Symbiotic nitrogen fixation in plants.* P.S. Nutman (ed.), 421—439, 1975.

46. Hayes, H.K., Immer, F.R. and Smith, D.C. *Methods of plant breeding.* McGraw Hill Book Co., New York, 1955.

47. Hsiao, T.C. Plant responses to water stress. *Ann. Rev. Plant Physiol.* 24, 519—570, 1973.

48. Ishizuka, Y. Engineering for higher yields. *Physiological aspects of crop yield.* J.D. Eastin, F.A. Haskins, C.Y. Sullivan and C.H.M. Van Bavel (eds.), 15—25. American Society of Agronomy, Madison, USA, 1969.

49. Jain, H.K. and Kulshrestha, V.P. Dwarfing genes and breeding for yield in bread wheat. *Z. Pflanzenzuchtg.* 76, 102—112, 1976.

50. Jennings, P.R. Plant type as a rice breeding objective. *Crop Sci.* 4, 13—15, 1964.

51. Jennings, P.R. and De Jesus, J. Studies on competition in rice I. Competition in mixtures of varieties. *Evolution,* 22, 119—124, 1968.

52. Johnson, V.A., Schmidt, J.W. and Mattern, P.J. Cereal breeding for better protein impact. *Econ. Bot.* 22, 16—25, 1968.

53. Khan, M.A. and Tsunoda, S. Evolutionary trends in leaf photosynthesis and related leaf characters among cultivated wheat species and its wild relatives. *Japan J. Breeding,* 20, 133—140, 1970.

54. Kishitani, S., Takano, Y. and Tsunoda, S. Optimum leaf-area and nitrogen content of single leaves for maximizing the photosynthesis rate of leaf. Canopies: A simulation in rice. *Japan J. Breeding,* 22, 1—10, 1972.

55. Klaimi, Y.Y. and Qualset, C.O. Genetics of heading time in Wheat (*Triticum aestivum* L.). I. The influence of photoperiod response. *Genetics* 74, 139—156, 1973.

56. Klaimi, Y.Y. and Qualset, C.O. Genetics of heading time in wheat (*Triticum aestivum* L.). II. The inheritance of vernalization response. *Genetics* 76, 119—133, 1974.

57. Loomis, R.S. and Williams, W.A. Maximum crop productivity: an estimate. *Crop Sci.* 3, 67—72, 1963.

58. Lorimer, G.H., Andrews, T.J. and Tolbert, N.E. Ribulose diphosphate oxygenase II. Further proof of reaction products and mechanism of action. *Biochemistry* 12, 18—23, 1973.

59. Manabe, S. and Wetherald, R.T. On the distribution of climate change resulting from an increase in CO_2 content of the atmosphere. *J. Atmos. Sci.* 37, 99—118, 1980.

60. McCashin, B.G. and Canvin, D.T. Photosynthetic and respiratory characteristics of mutants of *Hordeum vulgare* L. *Plant Physiol.* 64, 354—360, 1979.

61. Mishra, S.P. Selection for heterosis and combining ability in grain *Sorghum bicolor* L. Moench. *Ph.D. Thesis,* IARI, New Delhi, 1979.

62. Mishra, S.P., Sinha, S.K. and Rao, N.G.P. Genetic analysis of nitrate reductase in relation to yield heterosis in Sorghum. *Z. Pflanzenzuchtg.* **85**, 1980.

63. Mitra, R. and Bhatia, C.R. Bioenergetic consideration in the improvement of oil content and quality in oil-seed Crops. *Theor. Appl. Genetic* **54**, 41–47, 1979.

64. Mooney, H.A., Bjorkman, O. and Berry, J.A. Photosynthetic adaptations to high temperatures. *Environmental Physiology of Desert Organisms.* Neil F. Hadley (ed.). Halsted Press. 138–151, 1975.

65. Morgan, J.M. Differences in adaptation to water stress within crop species. *Adaptation of plants to water and high temperature stress.* N.C. Turner and P.J. Kramer (ed.), John Wiley & Sons, New York. 369–382, 1980.

66. Moss, D.N. Plant modification for more efficient water use: The challenge. *Plant modification for more efficient water use.* J.F. Stone (ed.), Elsevier Scientific Publishing Co. Amsterdam, 311–320, 1975.

67. Murata, Y. Physiological responses to nitrogen in plants. *Physiological aspects of crop yield.* J.D. Eastin, F.A. Haskins, C.Y. Sullivan and C.H.M. Van Bavel (eds.). Amer. Soc. of Agronomy. 235–259, 1969.

68. Murfet, I.C. Environmental interaction and the genetics of flowering. *Ann. Rev. Plant Physiol.* **28**, 253–278, 1977.

69. Murty, V.R. and Rao, N.G.P. Apomixis in breeding grain sorghums. *Sorghum in Seventies.* N.G.P. Rao and L.R. House (ed.). Oxford & IBH Publishing Co., New Delhi, 517–523.

70. Nair, P.K.R., Varma, R., Nelliat, E.V. and Bavappa, K.V.A. Beneficial effects of crop combinations of coconut and cocoa. *Indian J. Agric. Sci.* **45**, 165–171, 1975.

71. National Academy of Sciences. Firewood Crops: Shrubs and tree species for energy production, 1980.

72. Osmond, C.B. Pathways and function photorespiration. *Biological Application of Solar Energy.* A. Gnanam, S. Krishnaswamy and J.S. Kahn (eds.), The McMillian Co. of India Ltd. 26–28, 1980.

73. Passioura, J.B. Physiology of grain yield in wheat growing on stored water. *Aust. J. Plant Physiol.* **3**, 559–565, 1976.

74. Penning de Vries, F.W.T., Brunsting, A.H.M. and Van Laar, H.H. Products, requirements and efficiency of biosynthesis quantitative approach. *J. Theor. Biol.* **45**, 339–377, 1974.

75. Quinby, J.R. Leaf and panicle size of sorghum parents and hybrids. *Crop Sci.* **10**, 251–253, 1970.

76. Rahman, M.S., Halloran, G.M. and Wilson, J.H. Genetic control of spikelet number per ear with particular reference to rate of spikelet initiation in hexaploid wheat, *Triticum aestivum* L. Thell. *Euphytica*, **27**, 69–74, 1978.

77. Rana, R.S., Joshi, Y.C., Qadar Ali and Singh, K.N. Adaptive characteristics of salt tolerant genotypes. *Annual Report*, Central Soil Salinity Research Institute, Karnal, India. 56–60, 1976.

78. Rao, N.G.P. Genetic analysis of some Exotic Indian crosses in Sorghum II. Combining ability and components of genetic variation. *Indian J. Genet.* **30**, 362–376, 1970.

79. Reitz, L.P. Breeding for more efficient water use – Is it real or a mirage? *Agri. Meterol.* **14**, 3–11, 1974.

80. Riley, R. Methods for the future. *Plant breeding perspectives.* J. Sneep and A.J.T. Hendriksen (eds.), 321–369, 1979.

81. Sinclair, T.R. and de Wit, C.T. Photosynthate and nitrogen requirements for seed production by various crops. *Science* **189**, 565–567, 1975.

82. Sinha, S.K. Yield of grain legumes: Problems and prospects. *Indian J. Genet.* **34A**, 988–994, 1973.

83. Sinha, S.K. Food Legumes: Distribution, adaptability and biology of yield. *FAO Plant Production paper.* FAO, Rome, 1977.

84. Sinha, S.K. and Khanna, R. Physiological, biochemical and genetic basis of heterosis. *Advan. Agron.* **27**, 123–174, 1975.

85. Sinha, S.K. and Nicholas, D.J.D. Nitrate reductase in relation to drought. *Physiology and biochemistry of drought resistance.* L.G. Paleg and D. Aspinall (eds.). Academic Press. Sydney, 1981.

86. Sinha, S.K. and Swaminathan, M.S. The absolute maximum food production potential in India: An estimate. *Current Sci.* **48**, 425–429, 1979.

87. Sinha, S.K., Aggarwal, P.K., Chaturvedi, G.S., Koundal, K.R. and Khanna-Chopra, R. A comparison of physiological and yield characters in old and new wheat varieties. *J. Agr. Sci.* Cambridge Univ. Press. 1981.

88. Steponkus, P.L., Cutler, J.M. and O'Toole, J.C. Adaptation to water stress in rice. *Adaptation of plants to water and high temperature stress.* N.C. Turner and P.J. Kramer (eds.). John Wiley & Sons, New York, 401–418, 1980.

89. Suneson, C.A. Survival of four barley varieties in a mixture. *J. Amer. Soc. Agron.* **41**, 459–461, 1949.

90. Swaminathan, M.S. Mutation breeding. *Proc. XII Intern. Congr. Genetics.* C. Oshima (ed.), 3, 327–347, 1969.

91. Swaminathan, M.S. Basic research needed for further improvement of pulse crops in Southeast Asia. *Nutritional Improvement of Food legumes by breeding.* Max Milner (ed.). Protein Advisory Group of the UN Systems, UN. New York. 61–68, 1973.

92. Swaminathan, M.S. Indian Agriculture at the Crossroads. *Presidential address.* Indian Society of Agricultural Economics, New Delhi, 1977.

93. Swaminathan, M.S. Opportunities and problems in the developing countries. *Plant breeding perspectives.* J. Sneep and A.J.T. Hendriksen (eds.). 293–320, 1979.

94. Swaminathan, M.S. Recent trends in crop improvement in India. *Proc. 5th Int. Wheat Genetics Symp.* S. Ramanujam (ed.), 1267–1298, 1978.

95. Swaminathan, M.S. Past, present and future trends in tropical agriculture. *Perspectives in World Agriculture.* Agriculture Bureau, London, 1–47, 1980.

96. Swindell, R.E. and Poehlman, J.M. Inheritance of photoperiod response in mungbean (*Viona radiata* (L) Wilczek). *Euphytica.* 1978.

97. Takano, Y. and Tsunoda, S. Curvilinear regression of the leaf photosynthesis rate on leaf nitrogen content among strains of *Oryza* species. *Japan J. Breed.* 21, 69–76, 1971.

98. Tarhalkan, P.P. and Rao, N.G.P. Genotype density interactions and development of an optimum sorghum – pigeonpea inter-cropping system. *Proc. of the Symposium on Intercropping of pulses.* IARI, New Delhi. 1978.

99. Wallace, D.J., Ozburn, J.L. and Munger, H.M. Physiological genetics of crop yield. *Advan. Agron.* 24, 92–146, 1972.

100. Warner, R.L., Hageman, R.H., Dudley, J.W. and Lambert, R.J. Inheritance of nitrate reductase activity in *Zea mays. Proc. Nat. Acad. Sci.* 62, 785–792, 1969.

101. Wilson, D. and Cooper, J.P. Assimilation rate and growth of *Lolium* populations in the glasshouse in contrasting light intensities. *Ann. Bot.* 33, 951–965, 1969.

102. Willey, R.W. Intercropping – its importance and research needs. *Field Crop Abstracts.* 32, 1–10, 73–85, 1979.

103. Willey, R.W. and Rao, P. Growth studies in Sorghum pigeonpea intercropping with particular emphasis on canopy development and light interception. *Proc. International Intercropping Workshop.* ICRISAT, Hyderabad, 1979.

104. Wittwer, S.H. Global aspects of food production. Research and technology needs for the 21st Century. *Global Aspects of Food Production.* M.S. Swaminathan and S.K. Sinha (eds.). Academic Press, London. 1981.

105. Zelitch, I. Photosynthesis, photorespiration and plant productivity. Academic Press. 1971.

106. Zwartz, J.A. and Hautvast, J.G.A. Diagnosis of current world nutrition problems. *Plant breeding perspectives.* J. Sneep and A.J.T. Hendriksen (eds.). 28–46, 1979.

CHAPTER 2

STRATEGY OF BREEDING FOR DISEASE RESISTANCE

R.R. Nelson

Department of Plant Pathology, Pennsylvania State University,
University Park, PA 16802, U.S.A.

INTRODUCTION

It has been nearly seventy-five years since man became aware that the control of plant diseases could be achieved in a dramatic and spectacular fashion through hereditary means. The discovery by Biffen, that resistance in wheat to stripe rust was conditioned by a single gene, provided a firm scientific basis for exploiting the inherent resistance present in most crop species to a multitude of causal agents. The fact that the gene for stripe rust resistance was inherited independently of other parental traits, contributed to the conclusion that resistance could be readily and easily combined with other characteristics desired in commercial varieties. Thus, a new strategy was added to man's meager arsenal of schemes to combat the many organisms capable of inducing plant diseases.

The first seventy-five years of resistance breeding have been highlighted by an array of spectacular accomplishments and a series of incredible disasters[5]. The effective longevity of some resistance genes is astonishing given the acknowledged genetic potential for variation within most of our mutable and treacherous plant pathogens. On the other hand, the rapid and sometimes predictable demise of certain resistance genes is a graphic reminder that the purposeful manipulation of hereditary material often has made variation a most fortuitous trait for some plant pathogens.

There should be little need to document the fact that disease resistance will continue to be the singularly most important disease control strategy available to man, although it does have some acknowledged limitations.[5] The use of single-gene resistance has been the dominating genetic means of disease control and well it should have, given the general effectiveness and relative ease of incorporation into acceptable plant types. The most challenging issue for the future focuses on those array of diseases that cannot be controlled by the conventional use of single resistance genes. Fortunately for man, most species of cultivated plants or their wild relatives possess adequate resistance to most of their parasites. The question that confronts us concerns the best utilization of that available resistance. Most certainly some non-conventional schemes must be devised and assessed. Perhaps at the root of the matter is the need to re-examine some of our traditional genetic assumptions and beliefs. The laws of Mendelian genetics and the laws derived therefrom are not carved of granite, current dogma notwithstanding. One of the intents of this chapter is to objectively and logically challenge certain of our genetic assumptions and offer some new ones in the hope of re-shaping some of our future breeding strategies for disease resistance.

Of course, the main objective of this chapter is to explore the various breeding strategies available to us and to consider the factors that ultimately dictate our choice of a particular approach. As national and international breeding programs continue to develop genetic materials with the capability of wider geographical adaptation, differences among pathogen populations and climatic regimes among different geographical areas will probably dictate that no single genetic strategy will be universally successful. Available genetic options must be scrutinized carefully and placed in perspective with other considerations.

TWO KINDS OF DISEASE RESISTANCE

Man has found it convenient to classify disease resistance into two major types. The host either resists the establishment of a successful parasitic relationship by restricting the infection site and the infection process, or it resists the subsequent colonization and reproduction of the parasite following successful infection. Resistance is considered in this context as an active, dynamic response of the host to a parasite and, thus, excludes such passive phenomena as immunity, klenducity, or disease escape. **Resistance to infection** is referred to by such terms as hypersensitivity, race-specific resistance, non-uniform resistance, vertical resistance, or major gene resistance. **Resistance to subsequent colonization and reproduction** is characterized variably as partial resistance, slow rusting, field resistance, generalized resistance, durable resistance, race-nonspecific resistance, uniform resistance, horizontal resistance, minor gene resistance, and a series of other terms too numerous to list. The fact that resistance is relative, places the two kinds of resistance on a continuous scale. The difference between a pinpoint fleck reaction and a small necrotic lesion is a relative indication of host response to a parasite. The difference between a small necrotic lesion and a somewhat larger one is just as relative. Hypersensitivity at the cellular level, for example, technically can be considered as a resistance to growth. The point is that two major kinds of resistance phenomena are evidenced in host reactions towards parasites.

Cultivars reacting differentially to races of a pathogen are considered to possess genes for race-specific resistance. Race-specific resistance genes confer effective resistance, often hypersensitive in nature, against a certain race or races, but are equally ineffective against other races.

It is considered to be literally a 'feast or famine' type of resistance. Placed in an epidemiological framework, race-specific resistance affects the extent of disease onset by reducing the amount of effective initial inoculum. Races lacking virulence genes to overcome race-specific resistance genes are disqualified from epidemic encounters. Race-specific resistance genes are considered to be of no value when confronted by pathogen populations with the virulence genes to match them. From a genetic standpoint, race-specific resistance is usually, but not necessarily, conditioned by a single gene, and a major gene at that, according to current acceptance.

Race-nonspecific resistance is considered to retain some degree of effectiveness regardless of racial exposure. It would be naive, however, to assume or expect that nonspecific resistance would be uniformly effective against all races or sub-populations of races in light of the known ability of racial populations to differ in virulence and parasitic fitness.

Epidemiologically, nonspecific resistance is rate-reducing in nature; it affects the rate of disease increase (measured in one way by the apparent infection rate) and, thus, the terminal amount of disease. Reduction in the amount and rate of disease increase is accomplished by host mechanisms that restrict the number of successful infections, extend the latent period (the time required from infection to the subsequent production of inoculum), restrict the amount of tissue that is colonized from a single infection site, and reduce the amount and duration of inoculum production or sporulation.

From a genetic standpoint, nonspecific resistance is usually, but not necessarily, conditioned by several genes and minor genes at that according to current acceptance.

My case for the two kinds of resistance has been made in a simplistic fashion by design. Naturally, all forms of resistance which allow some level of sporulation are not race-nonspecific. Race-specific resistance does occur in which sporulation is not inhibited entirely. In a similar vein, hypersensitive infection sites are known to increase in size and sporulate on more mature plants when the host's resistance mechanism appears to erode. It is further evident that pre-infection factors play a role in some forms of rate-reducing resistance. Thus, rate-reducing resistance is not just a resistance to growth and sporulation; there are many examples in which nonspecific resistance can be characterized partially by the presence of fewer lesions. Thus, I have no quarrel with any exceptions that might exist to my simplistic treatment of the two kinds of disease resistance in plants.

BREEDING STRATEGIES USING SINGLE RESISTANCE GENES

In all probability, the use of single, so called 'major', genes will continue to be the most used genetic strategy to control plant diseases. Excluding varietal development through the process of selection from existing and diverse plant genotypes, and considering only those cases in which man has intentionally manipulated genetic material, the vast majority of all resistant varieties developed in the last seventy-five years have relied on the use of single genes. Single gene resistance has been remarkably successful, even though the periodic and dramatic breakdown of that resistance over the years has created a tendency for some to label that approach as risky and suspect. It is my intent to summarily dismiss the subject of the traditional single gene resistance by stating that, if it is usable for a particular disease in a particular geographical area, by all means use it. It is as simple as that.

The obvious alternative to genetic homogeneity, that suspicious trademark of widely used race-specific resistance, is genetic diversity, the heir apparent to solving some of our more tenacious plant disease problems; or so the argument goes. A later section of this chapter will document nature's decision to diversify its plant species.

Other than the traditional use of single gene resistance, other genetic strategies have been proposed for the non-traditional use of single genes to create some novel forms of genetic diversity. The major proposals include multilines and gene development. Each merits some discussion.

Multilines

Multilines are mechanical mixtures of near-isogenic lines each containing a different single race-specific resistance gene. The breeding approach to multiline development is the traditional backcrossing method involving a well adapted and acceptable recurrent parent and donor lines each containing a different and desired race-specific gene. Growers purchase and plant Certified Seed and harvest seed that cannot be resold or used as Certified Seed, thus retaining the genetic integrity of the multiline. The number of component lines used in a multiline is not fixed, but rather is influenced by the number of available component lines and the epidemiological capacities of the pathogen in question. Component lines can be mixed in equal or unequal proportions.

Multiline theory states that multilines restrict the terminal amount of disease by reducing the amount of effective initial inoculum X_0 *sensu* Vanderplank[14] available for disease onset and by retarding the rate of subsequent disease increase *r sensu* Vanderplank[14]. It is said to be a powerful tool that utilizes the combined virtues of race-specific and nonspecific resistance. The reduction of an effective X_0 assumes that the initial inoculum is comprised of different pathogen genotypes, some of which lack the virulence gene(s) to overcome any of the specific genes and are thus summarily dismissed from epidemic participation. Inoculum of pathogen genotypes possessing one or more virulence genes to match the specific resistance genes in one or more of the component lines is also dismissed from epidemic involvement should it fail to fall on plants possessing the resistance genes that can be overcome. Successful establishment results only from inoculum that falls on a compatible plant whose resistance gene is matched by the necessary virulence gene. Thus, in short, initial establishment of the pathogen is a matter of chance probabilities.

At some point in time following initial disease onset, successful infection sites generate new inoculum to initiate the next disease cycle. As the new inoculum departs from the infection site in search of a palatable plant, chance probabilities dictate that only a portion of that inoculum will eventually find one; that which does not is dismissed from further combat. The proportion of unsuccessful infections will be reflected in a reduction in the rate of potential disease increase, thus mimicking the forte of non-specific resistance.

The theoretical assumption that multilines effectively mimic the driving forces behind non-specific resistance seems, on the surface, to be impeccably logical, although I rather suspect that some provisions may need to be applied if the assumption is to be of universal validity. To my way of thinking, all forms of rate-reducing resistance should have some practical and useful merit and, as such, a plant genotype reducing the potential terminal disease from 100% to 98% is not meritorious. The assumption that multilines mimic some practical measure of rate-reducing resistance itself assumes that some portion of the multiline has become susceptible because of the presence of a pathogen genotype with one or more matching virulence genes, i.e. multilines could not reduce spread of disease if there were no disease. The question is what proportion of the multiline must remain race-specific resistant to permit the mimicking rate-reducing power of the multiline to have some practical effectiveness in managing disease to some acceptable level. Several investigators have set theoretical limits for susceptibility without ill-effects, although precise yield loss studies are a rarity. That limit will vary among diseases and among years which means that there is no fixed and safe limit of acceptable susceptibility. A given amount of disease might result in significant yield loss for one crop and none in another. Disease severities viz yield loss dictate the effectiveness of any genetic strategy designed to manage plant diseases. Whatever that limit might be would seem to dictate whether or not multilines can effectively practice the act of disease reduction.

The verdict on the potential usefulness of multilines to effectively manage variable pathogens is still in jury. It has been acclaimed to be 'conceptually innovative' or 'agriculturally conservative', 'a probable means to a permanent end' or a 'risky approach inviting the evolution of a super race' and, 'financially impractical and tedious' or 'the financial end to the single gene treadmill'. Be that as it may, it is a fact that the theoretical assumptions underpinning the multiline approach still remain theoretical assumptions or assumptions not adequately tested to deem their universal validity. Perhaps the singularly most important, yet unsubstantiated, assumption is that the deliberate incorporation of known susceptible germplasm as a part of a multiline would effectively stabilize the pathogen population and preclude the selection and increased frequency of a super race capable of devouring the total multiline. Should that assumption ultimately be shown to be valid and given the available resources and dictates, the multiline approach most certainly can be considered a viable and effective genetic scheme that speaks to genetic diversity.

Gene deployment

Some plant diseases are considered to spread over a wide geographical area in step-wise sequence. Stem rust of wheat in North America is one such disease. The rust fungus is said to spend its winters in the southern portion of the continental wheat belt, recuperating from its year's labors and regrouping for the next invasion to the north.[3] As it leaves its southern retreat, the stem rust population is said to be the epitome of genetic diversity. The gamut of diverse populations is said to range from weak, anaemic races that persist prudently only on the most susceptible of wheats to the bastions of super power who boast of having every virulence gene that evolution gave to the species. As predicted, the most fit best survive the early phases of the northward march, rapidly increasing their numbers and arriving in a deluge to devour their inadequately protected hosts in the northern plains and Canada.

Is there no hope for us at all? Proponents of gene deployment say there is.[5] They propose that the continental wheat belt be divided into different, but contiguous, zones and that a different race-specific resistance gene be used exclusively in one zone, i.e. gene A in Zone A, gene B in Zone B, etc. The number of zones, and thus the number of genes, is flexible, the final decision being influenced by the availability of genes and the magnitude of the problem.

Hold it a minute! If the rust population is that genetically diverse and if the super powers increase in frequency on the way north, what good would it do to have zones at all?; five zones means you have squandered five genes in a futile defense. Not so! You have neglected one point; you have neglected the concept of stabilizing selection as interpreted epidemiologically by Vanderplank[14]. The concept suggests that pathogen populations with increasing numbers of virulence genes will become increasingly less able to persist on simple host populations with little, if any, race-specific resistance. The gamut of virulence genes in the super races renders them unfit to survive on such simple hosts. So here is what you do: you plant a genetic loser in the south. As the rust populations begin their northern trek, they will run headlong into the southern loser; the super races which now are axiomatically unfit will fall by the way and, much to their surprise, the weak and the anemic races take over, doing virtually no damage but having that once-in-a-lifetime thrill of victory. Buffered with their new-found bravado, the weak races dominate the populations and continue northward to the zones where we have astutely positioned different and now effective single genes which easily defeat the weak races. So much for the thrill; they will never make it to Canada.

Varietal mixtures

Some discussion in recent years has focused on the potential strategy of varietal mixtures to manage or contain some highly variable and explosive plant pathogens. The theory behind the strategy is essentially akin to that of the multiline approach. Given a mixture of varieties each with a different racial resistance or level of resistance to a specific race or races, disease increase would be reduced in that portion of each successive generation of inoculum which would land on plants having an effective resistance gene capable of denying successful infection. The case of varietal mixture is well stated by Wolfe and Barrett[16].

At least a modicum of acceptance of Vanderplank's epidemiological concept of stabilizing selection is necessary to assume the effectiveness of varietal mixtures. Parasitic populations are said to be faced with a dilemma involving a trade-off between increasing virulence and decreasing fitness or more modest appetites and increased fitness. Wolfe and Barrett[16] do not contend that varietal mixtures are a permanent cure-all for all diseases. The rapidity and ease of creating different varietal mixtures to answer any unexpected (or expected) population shifts has merit.

There are some tangential questions that can be raised but need not be answered. Industry acceptance of mixed seed undoubtedly would be an issue. Are pure seed laws flexible enough to permit varietal mixtures? Are there means to determine in advance whether varietal mixtures would be effective against a specific pathogen? If regulatory and potential biological problems can be resolved, varietal mixtures may well be an additional genetic strategy.

Finally, it is acknowledged that some of the varieties used in any given mixture may not possess the traditional single gene, race specific resistance. This brief discussion of varietal mixtures is included in this section merely for convenience.

EPILOGUE

Some populations of plant parasitic fungi are more fit than other populations. This holds true for unique genotypes of particular races or for a species in general. Such a biological principle stands without challenge. It is of such truth that no lengthy discourse on the matter is necessary.

Why some populations are less fit than others merits some comment, although it is tempting, in a sense, to de-fuse the topic by invoking the impeccable assumption that some populations are less fit because they are less fit; that they are less fit because they lack certain vital genes for fitness. Some developments in recent years have placed certain complications in opting for such a convenient dismissal of the subject.

There is a school of thought that believes that nature tends to place some constraints on parasitic populations munching happily towards unbridled virulence.[11] Nature favors the more rational populations who conserve their energies for better endeavors; or so the argument goes. Teliologically, one could question the wisdom of carrying more weapons in the arsenal than are needed for any given confrontation or for any point in time; particularly if that were the only instance in which the weapon were used.

The same school of thought concedes that nature has no control, before the fact, over populations that accumulate more virulence genes than are necessary for survival and reproductive success. What nature does and can do, the school contends, is see to it that they are less fit to succeed than their less precocious relatives. Thus, the derivation of the axiom that populations with unnecessary genes for virulence are less fit to survive.

There are at least two theories regarding the ultimate fate of unnecessary genes of any kind. One theory presumes that genes that are not called upon to perform in a vital function will drop out of the population or remain at best in very low frequencies. The other theory, which I have come to support, contends that the issue is not whether a gene is unnecessary, but rather whether its presence in a genome is a detriment to the population carrying it. My support for the latter theory could be mounted in several ways. It is my belief, backed by no evidence whatsoever, that most genes activate whatever they activate for more than just a single function. For example, a gene contributing to the production of an enzyme that is necessary for pathogenesis also may contribute to the production of the same or a different enzyme for another purpose.

Two personal observations offer a modicum of support for my position. *Helminthosporium maydis* is a parasite that induces Southern leaf blight of corn. I have isolated *H. maydis* from small foliar leafspots occurring on species in 55 genera of the *Gramineae* in my wanderings throughout the world. The fungus appears to pose no pathological threat to species other than corn, although it has the genetic ammunition to parasitize other species in a modest way. A single gene for pathogenicity is all that is required for isolates of *H. maydis* to parasitize corn. Different genes are required to parasitize other gramineous species. One of our genetic studies, for example, revealed that isolates of *H. maydis* need thirteen different genes to be pathogenic to nine particular gramineous species. Through a series of genetic crosses and selection, a large number of isolates were developed each of which possessed from 1–13 of these pathogenicity genes. The question put to test was, "Are isolates of *H. maydis*, possessing all 13 genes carrying 12 useless genes with respect to their capacities to induce disease in corn?" The answer was a clear and resounding, **no**. Isolates carrying increasing numbers of the thirteen genes incited increasing amounts of disease (i.e. were more virulent) on corn than did isolates carrying fewer genes. The twelve genes other than the gene needed to parasitize corn were contributing to the intensity of the attack and thus to fitness.

That research gave me an explanation for my second general observation. Until then I was puzzled, albeit in a modest way, as to why I could frequently collect isolates of *Helminthosporium victoriae*, causal agent of Victoria blight of oats and champion of host-specific toxins, from other gramineous hosts from areas of the world where oat cultivars susceptible to the fungus had never been grown. There were collections of *H. victoriae* from *Agropyron* and *Hordeum* in Scotland, from *Agrostis* in Holland and from *Poa* in Switzerland near the tree line in the Swiss Alps. And, there was *Helminthosporium carbonum*, causal agent of a leaf spot of corn and other toxin producer, collected from a tuft of *Poa* just south of the Arctic Ocean. The genes in those two species that functioned in the parasitism of oats and corn were contributing in some way to the general parasitic fitness of the isolates in areas far removed from their favored hosts.

The matter of whether or not populations carrying unnecessary genes for virulence are less fit to succeed has become something more than a mere exercise in intellectual trivia. It has become a controversial issue in plant pathology, particularly among those who are concerned with managing plant diseases by means of disease resistance. The idea that populations with few (if any) virulence genes are more fit to succeed on hosts with few (if any) resistance genes than are populations with unnecessary virulence genes (in context with the same hosts) has penetrated our thinking on the potential value of gene deployment and on the genetic composition of multilines. Much has been said and written, pro and con, about the hypothesis of stabilizing selection.[14] Some, including myself,[7] have given up all hope of testing the hypothesis, since the built-in escape hatches preclude any valid test. If any hypothesis cannot be tested to determine its soundness, the hypothesis itself is suspect. Nonetheless, the idea of stabilizing selection is used by some who reflect on population dynamics and parasitic fitness. It has been theorized that within all parasitic species there are populations of super races that carry virtually every virulence gene the species has ever manufactured. It lies there in nature as a slumbering giant in minute frequency because its gamut of virulence genes has rendered it less fit to succeed on host populations that are no match for it on a gene for gene basis. There are those who believe that a substantial portion of a multiline must consist of known susceptible tissue so that stabilizing selection can protect the rest of the multiline from the ravages of a complex race. The proponents of deploying different resistance genes in different but contiguous regions believe that complex races will be stabilized against and would be unable to continue their travels to the next region.

Let us suppose, for the moment, that there are genes that function to induce disease and that there are other genes that condition overall fitness. It is really not an unreasonable tenet; in fact there is some evidence for it.[4] Let us further suppose that at least some of these different kinds of genes are inherited independently of one another. Again, not an unreasonable tenet; in fact there is some evidence of it, too. If these suppositions are true, as I believe them to be, then one might expect to find populations with few genes for virulence that are fit and other such populations that are unfit to succeed. One might also expect to find populations with many virulence genes, some or many of which may be unnecessary by our criteria, that are fit and other such populations that are unfit to succeed. There is no reason, to my way of thinking, why all of these combinations of virulence and fitness genes should not exist in natural populations at some point in time. If this is true, as I believe it is, then I am forced to reject the equation that simplicity equals fitness and return to the tenet that fit populations are fit because they are fit, pathogenic simplicity aside. So give some thought to these notions, young man. Perhaps one day you may want to monitor populations for their fitness and not just for their complexity.

To set the matter straight and in conclusion, I am not yet prepared to reject stabilizing selection as a valid concept of classical population genetics. Nor am I yet prepared to reject the belief that populations at the extremes of a normal distribution are, at least momentarily, less fit than populations about the mean. I am fully prepared, however, to reject the assertion that so-called unnecessary virulence genes render populations less fit. No one has ever provided me with evidence that any given virulence gene is unnecessary to a genome, let alone detrimental. In fact, I like to think that I have accumulated a little evidence to support the opposite contention.

History tells us that man should read the past before he writes his future. Logic and logical thought are essential to the development of new theories. Research is equally essential to the verification of new theories and should be undertaken before applying any new strategy whose success is largely dependent on the validity of the theory.

The success of gene deployment and multilines is irrevocably linked to the truth of stabilizing selection. There is evidence to support its validity and evidence to reject it. Any attempted use of gene deployment or multilines as a strategy of diversity must first ascertain whether more complex races of the pathogen in question will be cooperative enough to make it successful. I certainly hope that at least some of them will. We need every possible strategy that can be found.

NONSPECIFIC RESISTANCE AND GENETIC EQUILIBRIUM – THE NATURAL WAY OF LIFE

Race-nonspecific resistance, when viewed as a rate-reducing resistance, appears to be conditioned by several to many genes in most instances. Intensity usually is a quantitative trait. A quantitative trait is conditioned by many genes because one gene cannot accomplish the task. Each of the genes conditioning a quantitative trait contributes something to a collective venture that none can accomplish alone. Removing a single gene from a many-gene system should not drastically alter the expression of the trait. To what degree the trait may be altered would depend on the relative importance of the gene in the collective venture and the total number of genes controlling the trait. Greater fluctuations in the reaction of a host to different races should occur as more and more genes are removed. As all but one gene are removed, the host probably would react quite differently than it had before to the same races. It would react so differently, in fact, as to perhaps simulate a race-specific reaction to the races, i.e. resistant to some and susceptible to others.

A reverse of the sequence of events just described would, as I see it, represent the evolution of genes for disease resistance in plants. Co-epicenters, geographic areas in which both host and parasite have co-evolved, most accurately depict the story of the evolution of genes for virulence and resistance. The story is an incredible Darwinian classic. *Solanum demissum* and the late blight fungus in Mexico, *Tripsacum* and *Teosinte* and the northern leaf blight fungus in Mexico, and the old land races of rice and the rice blast fungus vividly depict the classic saga of the co-evolution of host and parasite. The relationship today between each of these host-parasite combinations is not sharply antagonistic. Each of the hosts sustains modest levels of disease, but disease is of no major concern to the host insofar as its ultimate reproductive success and pertinent survival are concerned. Both host and parasite had learned that the price of co-existence was preferable to the alternating thrill of victory and the agonies of defeat. A relaxation of selection pressure for both species highlighted their new relationship. Neither was or is today in serious jeopardy of extinction. They had become, in a sense, the 'odd couple' of the biological world.

Philosophical or teliological as these speculations may seem, they would seem to represent the sicentific explanation embodied in the co-evolution and co-existence of host and parasite in natural ecological niches. We can return to the beginning of their unique affair to learn how they arrived at their current status of equilibrate co-existence. The initial resistance of a host to a 'new' parasite probably was accomplished by a genetic change at a single locus. Plants with that resistance gene probably reacted to the parasite in a hypersensitive manner; probably like a modern-day cultivar whose resistance is conditioned by a single gene. Subsequently, new populations of the parasite evolved and strains pathogenic or virulent to the then resistant host were selected. The host, in turn, generated a new source of effective resistance; and so on. Such must have been the case, since both host and parasite still exist.

The process of co-evolution proceeded, perhaps stepwise in a gene-for-gene manner or perhaps in another way; it really doesn't matter. Host genotypes with fewer genes for resistance and pathogen genotypes with fewer genes for virulence probably were selected against and either dropped out of their respective populations or remained in low frequency. There can be little doubt but that the most fit to persist would dominate both populations. At any rate, it is difficult for me to imagine that all genotypes that existed throughout the process of co-evolution remain a part of the current populations of host and parasite. At various stages in evolution, all species generate some deleterious genes and genotypes containing them would be at a great selective disadvantage.

How long the process of co-evolution has occurred and what stage of co-evolution currently exists are not at all important. By the time the host and parasite had reached a stage of acceptable co-existence, each had accumulated a substantial number of resistance genes or virulence genes. The fact that they now exist in relative harmony indicates that the last resistance gene incorporated into the host genome did not confer a massive or hypersensitive response against the pathogen. Nor did the last virulence gene pose a serious threat to the host. There was and is no immediate selection pressure on either adversary. Each had found a mutual safety in numbers of genes for resistance or virulence. Genes that once functioned separately with only temporary success in the earlier stages of co-evolution now function collectively with a more permanent success.

If the assumption is correct that the long process of co-evolution resulted in the ultimate accumulation of many resistance and virulence genes in the host and parasite, genetic probabilities suggest that subsequent changes in the future course of their co-evolution will be largely subtle ones. At least it seems unlikely that either opponent will enjoy the massive superiority exhibited in the earlier stages of their co-evolution. Genetic changes in the parasitic populations may result in subtle improvements in one or more of its fitness attributes. A change could occur, for example, that would reduce the latent period by a day or so or increase the parasite's ability to reproduce to some modest extent. Contrastingly, a small genetic improvement in the resistance of the host could extend the latent period by a day or more or curb the parasite's reproductive powers to some modest extent. The epidemiological consequences of either change could be to increase or decrease the number of generations of disease within a given year and, thus, condition the terminal amount of disease.

SOME THOUGHTS ON GENES FOR DISEASE RESISTANCE

Man has not only found it convenient to characterize disease resistance into two major types, but he has also found it convenient to conclude that the two types of disease resistance are conditioned by different kinds of genes. The implications, for example, in the terms 'major genes' and 'minor genes' are reasonably obvious. If major gene resistance is the same as race-specific resistance and minor gene resistance is the same as nonspecific resistance, the two different kinds of resistance must be conditioned by different genes; or so the argument goes.

Several years ago, I proposed the concept that genes for race-specific and nonspecific resistance are the **same** genes.[6] The concept since has been developed in greater detail.[9,10] The two major kinds of disease resistance, my argument contended, are not indications of the action of different genes, but rather are expressions of different actions of the same genes in different genetic backgrounds. There are, in fact, no major genes and minor genes. There are only genes for disease resistance. Certainly, different genes may confer varying levels of expression of effectiveness and the expression of resistance genes may be influenced or modified by other resistance genes in different genetic backgrounds; background is the key to gene action.

The implication of the concept is that we recognize and characterize genes as being 'major' when they are separate and race-specific and characterize the same genes as 'minor' when the host response suggests a nonspecific or rate-reducing type of resistance.

I again submit that genes function one way when they are separate and another way when they are together. Again, the *Solanum demissum* — potato late blight story has relevance. Niederhauser[12] stated, "In Mexico we were forced to concentrate on the multigenic field resistance when it was found that no tuber-bearing *Solanum* species was immune or hypersensitive to the pathogen when exposed in the field". He also stated, "Field resistance is defined here as the resistance exhibited by a plant to all races capable of causing on it more than a hypersensitive reaction". And, finally, field resistance "is characterized by slow-spreading lesions in which sporulation is sparse. The lesions per plant are fewer, and tend to be on older, lower leaves". Niederhauser[12] apparently was discussing a resistance identical to what is now considered to be a rate-reducing type of resistance.

Man attempted to make use of the impressive resistance demonstrated by *S. demissum* to a multitude of races of *Phytophthora infestans* in the Toluca Valley of Mexico. Although *S. demissum* displayed a rate-reducing resistance, genes were extracted singly from the species and a race-specific or hypersensitive resistance was the result. Thus, the birth of the R-genes which were considered major genes because man has been scientifically and philosophically schooled to react with major favor to genes eliciting such a dramatic host response. It is difficult for me to comprehend how man can extract a major gene from a host genotype expressing minor gene resistance.

The first R-gene, R_1, soon fell prey to a race of *P. infestans* and man went back to *S. demissum* and extracted another R-gene, R_2. That gene broke down and man went back for another. And so on. One gene could not contain the highly variable late blight fungus. If man had wanted his cultivated potato, *S. tuberosum*, to behave like *S. demissum* against *P. infestans*, he should have taken all of the genes that protected *S. demissum* and put them back together again.

One final comment. If a race-specific gene is ineffective against a race, it is assumed not to function at all in resisting that race. Similarly, if five race-specific genes are all ineffective against a race, they are all presumed to be functionless against the race. It seems more reasonable to me that five singularly ineffective genes can be collectively functional and effective. If each contributes something in the way of resisting the parasite at some point after infection, the net, collective result logically should be a collective resistance to growth and/or reproduction.

Some intriguing facts arise from the literature. They can be summed up in one way as follows:

1.   Major genes are modified by minor genes and minor genes are modified by major genes.
2.   Major genes together exhibit additive effects and so do minor genes.
3.   Major genes and minor genes together express modifying and additive effects.
4.   Major genes mask minor genes and minor genes mask major genes.
5.   Major gene resistance is enhanced by minor genes and minor gene resistance is enhanced by major genes.

6. One dose of a major gene may confer susceptibility, whereas two, three, or four doses confer increasing levels of resistance.
7. Several genes collectively conditioning a rate-reducing resistance individually condition race-specific resistance and single genes controlling race-specific resistance collectively condition rate-reducing resistance.
8. Genes may be major in one background and minor in another.

It is evident that the long-fostered assumption that genes that illicit the hypersensitive response associated with race-specific resistance are major genes and those that do not are minor genes can no longer be accepted axiomatically. In truth we really do not know what major genes are and if we do not then why must we continue to characterize genes as being major or minor. Five race-specific genes, each conditioning a hypersensitive response to a single, different race but which are ineffective against the other four races are, in my opinion, no more major than five genes collectively conditioning an effective rate-reducing resistance against all of the five races.

There is, in fact, some evidence and considerable speculation that today's so-called minor genes controlling a quantitative trait are archaic 'major' genes that once conditioned a qualitative trait such as resistance against a particular genetic population of a pathogen. Although genetic changes rendered them ineffective in a qualitative sense, their presence in appropriate genetic backgrounds contributed to the overall ability of the host to resist the pathogen at some stage after successful infection. Contrary to this, when a race-specific resistance gene is overcome in modern varieties, the gene is usually removed from the scene, 'placed on the shelf' and replaced by another currently effective race-specific gene. So-called minor genes may appear to be minor only because we do not have access to a pathogen genotype that could identify them as major.

Natural populations of plants and their pathogens co-evolved to co-existence by reaching a genetic equilibrium based on the combined effects of many genes in each system. Man has upset that equilibrium for reasons well known to all of us. Many researchers are now suggesting that combining race-specific genes with race non-specific genes should provide long-lasting forms of disease resistance. My philosophy of breeding for disease resistance agrees in principle with their suggestions. So, "Go back young man and gather up your weary and defeated resistance genes of the past, take your currently successful genes, find some new ones if you can and build yourself a genetic pyramid. And for irony's sake, if not for mine, name your first cultivar 'Equilibrium'."

THE MECHANICS OF BREEDING FOR NONSPECIFIC RESISTANCE

Many plant breeders and plant pathologists have recognized the potential value of rate-reducing, nonspecific resistance in managing plant diseases at some economically acceptable threshold; they also have recognized the difficulties and problems of incorporating a multigenic resistance into an acceptable plant type. Their concern would seem justified given the proper assumption that the transfer of genes from rate-reducing donor lines to acceptable plant types would be sequential in nature. The selection within segregating populations for those plants carrying some of the resistance genes stored in the donor lines would be an extremely difficult task again given the proper assumption that the selected segregating plants would not express the same kind or degree of resistance associated with that of the donor lines. The concern for the judicious selection of plants within genetically diverse populations would be of similar magnitude were the breeding strategies to entail the pyramiding of single resistance genes into a single genetic background. The problems remain the same whether rate-reducing resistance is transferred sequentially from rate-reducing donor lines or constructed sequentially from the ground floor.

Although breeding for rate-reducing resistance remains a challenging task, it is neither as hopeless nor as difficult as currently perceived. Nature has left some clues for us; clues that we were unable to recognize until now. Techniques are now available for use in developing rate-reducing plant genotypes and they merit some discussion in this chapter.

Some forms of rate-reducing resistance express their resistance in part in the form of intermediate type or size of lesions, pustules, or whatever. Infections are considerably larger than those associated with the hypersensitive resistance, but are visibly smaller than those associated with a clearly susceptible reaction. Slow-blasting of rice, slow leaf-rusting of barley and wheat, and slow late-blighting of wild *Solanum* species are some examples of diseases which are curbed, in part, by a reduced size of the infection sites. Visual inspection of plants in segregating populations should identify those plants that have garnered some of the genes or combinations of genes necessary to a rate-reducing resistance. One, perhaps unnecessary, word of caution: do not select plants exhibiting a high level of resistance; they more than likely contain a single resistance gene which, by virtue of its solitude, reacts in a race-specific, hypersensitive manner.

A longer latent period often is a clue to rate-reducing resistance. A lengthening of the time required from the deposition of inoculum to the production of more inoculum should result in fewer generations of disease for any given season; **and** an epidemic is time!

Measuring the latent period for the mildew diseases and most of the rust diseases is a reasonably simple matter. The breaking of the rust sorus and the powdery appearance of a mildew lesion signal the end of the latent period. For diseases in which dew is required for sporulation, measuring latent period is not as straightforward. One acceptable and useful technique calls for the recording of the relative time required from an initial inoculation to the appearance of secondary lesions incited by inoculum produced in the primary lesions, i.e. secondary lesions cannot form until primary lesions have sporulated.

A plant genotype exhibiting a reduced number of infection sites, relative to a known susceptible genotype, probably is a candidate for rate-reducing resistance; fewer lesions, for example, equates to less disease. The tendency for the winter wheat variety Knox to sustain fewer mildew infections is one of the prime reasons that Knox exhibits slow-mildewing under natural and/or normal field conditions.

Selection of plants or genotypes with the ability to restrict lesion numbers can be discouraging in instances where massive, external inoculum from susceptible genotypes tends to mask the trait. For example, powdery mildew of winter wheat reached epidemic proportions in 1979 in many of our research and breeding plots at The Pennsylvania State University research farms. Inoculum was massive and it was omnipresent. In many of the plots, the variety Knox, an acknowledged slow-mildewing genotype, could not be distinguished from the susceptible variety, Chancellor. Mildew was present in unnatural and/or abnormal proportions, a circumstance not normally confronted by Knox in commercial production.

Selection for resistance that restricts lesion number can be accomplished within segregating populations or when many genotypes are tested in row trials, if certain parameters are followed. When the initial inoculum is deposited **uniformly** on all candidate plants, selection for restricted lesion number can be made by detecting differences in number of primary lesions, before secondary lesions appear. In the greenhouse, the use of settling towers enable the rapid evaluation of many candidate lines. Inoculum is introduced at the top of the tower and allowed to settle uniformly on the plants below. Coated glass slides or agar plates placed at regular intervals around the floor of the tower permit the accurate monitoring of the uniform distribution of inoculum. Settling towers are simple and inexpensive to construct and are an excellent tool for identifying the type of resistance that restricts lesion number. Greenhouse selection of candidate lines prior to field evaluation reduces the number of test lines and the cost of research.

The capacity of host genotypes to restrict the pathogens ability to reproduce inoculum may well be the singularly most important component of rate-reducing resistance. The impact of reduced sporulation on the rate and extent of disease increase is monumental. The winter wheat variety Redcoat has exhibited slow-mildewing tendencies for many years in Pennsylvania, although some erosion of its resistance is evident. Rouse[13] determined that the rate-reducing ability of Redcoat was accomplished for the most part by reduced sporulation, at any point in time or in terms of maximum cumulative sporulation. He concluded that the ability to limit pathogen reproduction may well be the prime component of rate-reducing resistance.

There are several rather simple, yet effective, methods of measuring sporulation of suspected candidate lines. The availability of these techniques will enable the investigator to recognize the more promising genotypes for inclusion in a breeding program. Comparison of the sporulation capacities of rate-reducing donor lines with early or advanced generation lines readily identifies the extent to which that trait has been sequentially transferred.

When the number of candidate lines to be tested for their potential rate-reducing capacities is of a workable magnitude or if field space is no major constraint, measuring the apparent infection rates (r) is a superb tool to detect rate-reducing or rate-limiting genotypes. Reduced infection rates reflect the sum total of all rate-reducing components and visually displays the extent of its effectiveness under field conditions.

The recently completed Ph.D. research of Dr. Reynaldo Villareal, under the guidance of myself and Dr. D.R. MacKenzie, is a classic example of an effort to detect and dissect rate-reducing resistance.[15] His working model was rice and the rice blast disease induced by the fungal pathogen *Pyricularia oryzae* Cav. Previous research and observations by scientists at the Institut de Recherches Agronomiques Tropicales et des Cultures Vivieres (IRAT), Ivory Coast, and the International Rice Research Institute (IRRI), Los Banos, Philippines, suggested that certain land varieties and other rice genotypes had some measure of rate-reducing resistance. Sixteen varieties, reported to be 'slow-blasting' or to have reduced apparent infection rates (r), were evaluated in 2 X 3 m microplots on the upland experimental farm of IRRI. Corn was planted around the rice plots to minimize interplot cross infection. When the rice plants were 24 days old, two pots of blasted rice seedlings, each pot containing about 20 seedlings and each seedling have about 75 sporulating blast lesions, were placed in the upwind corner of each plot. The plots were covered daily from 5 p.m. − 8 a.m. with plastic sheets to promote extended dew periods required for infection and sporulation. Disease severity and rate of disease increase were determined at 5-day intervals throughout the epidemic, initially by converting lesion number viz leaf area to severity and later by estimating severity on a percentage scale.

Nine of the 16 rice varieties expressed remarkable levels of rate-reducing resistance. Thirty-eight days after inoculation, the susceptible check varieties had sustained about 87% disease while the nine slow-blasting lines sustained disease levels of from less than one percent to 11%;

eight of the nine varieties had disease severities of less than three percent. Apparent infection rates or r-values of the susceptible check varieties ranged from 0.19—0.23, while r-values for the slow-blasting lines ranged from 0.02—0.11. The five varieties demonstrating the lowest r-values were evaluated in two additional field experiments with similar results.

Dr. Villareal then took the five best slow-blasting varieties into the IRRI phytotron to determine the components contributing to the rate-reducing resistance. To summarize very briefly, Villareal determined the following:

1. the susceptible check variety sustained a minimum of eight times more lesions than the slow-blasting lines;
2. the latent period ranged from 3.03 days for the susceptible variety to a maximum of 5.93 days for the best of the slow-blasting lines;
3. when maximum lesion size had been attained, the mean lesion size of the susceptible variety reached a maximum of 32 mm^2, whereas lesion size of the slow-blasting lines ranged from a minimum of 3 mm^2 to a maximum of 12 mm^2;
4. with an ingenious technique to determine the maximum cumulative sporulation for any given lesion, Villareal determined that lesions on the susceptible variety produced from 3—8 times more spores than lesions on the slow-blasting lines; and,
5. lesions on the susceptible variety continued to sporulate for up to 27 days, whereas the duration of sporulation ranged from 20—22 days for the slow-blasting lines.

Subsequent re-analysis of Villareal's data revealed that both maximum cumulative sporulation and maximum lesion size could be used singly to develop equations that predict, with great accuracy, the apparent infection rates for the slow-blasting varieties under field conditions. Essentially there would be no need to obtain r-values from micro-plot studies if values for sporulation capacities or lesion size could be obtained from greenhouse or growth chamber research.

The research of Villareal and additional research by Dr. MacKenzie and myself with support from the Rockefeller Foundation have prompted rice scientists at the International Center for Tropical Agriculture (CIAT), under the leadership of Dr. Peter Jennings, to attempt for the first time to develop a slow-blasting variety of upland rice. Should they be successful, the world will be witness to the second Green Revolution.

SOME THOUGHTS ON 'DEFEATED' RESISTANCE GENES

The periodic arrival of pathogen populations with virulence genes to overcome single, race-specific resistance genes has sent many such genes to the biological cemetery. They rest there with full honors having done their task gallantly if not indefinitely. It seems implicit that the tendency to bury defeated resistance genes is done with the presumption that they no longer are of any value at all; why else would they be buried.

My colleagues and I have been working with the winter wheat — powdery mildew system to evaluate the merits and demerits of certain genetic strategies to manage plant diseases at some acceptable level. Having access to the near-isogenic lines developed by Briggle[1] in a Chancellor background, each containing a single, different powdery mildew (Pm) resistance gene, we have bred two four-gene pyramids to test the theory that the additive effect of several defeated resistance genes would create an equilibrium with a pathogen genotype possessing all the virulence genes needed to match all of the resistance genes in the pyramid. The unbounded, if not biased, faith in the validity of my pyramid theory prompted us to assume that the pyramided resistance genes would create an equilibrate, slow-mildewing relationship with the pathogen. If such were the case, it would be necessary to evaluate each resistance gene and all possible two-gene combinations to ascertain what each gene may be contributing, if anything, to the successful collective venture. On the surface, it may seem naive to evaluate single, defeated genes since it is assumed that they are of no value when once defeated. Nonetheless, five near-isogenic lines with different Pm genes were compared to the susceptible recurrent parent, Chancellor, for their reaction to a pathogen isolate possessing all of the virulence genes needed to overcome the five Pm resistance genes. Three of the five isogenic lines, containing genes Pm3C, Pm4, or a gene identified as Michigan Amber, exhibited a dramatic residual effect expressed as reduced number of lesions and reduced sporulation when compared to Chancellor. These defeated genes were not totally overcome as one might have assumed.

Why did such a discovery remain undetected for so long? Defeated genes are diagnosed as being defeated based on a qualitative reaction. If a susceptible type reaction is observed on a genotype, it is assumed that the genotype is susceptible. The presumed all or nothing syndrome associated with race-specific genes dictates that such genes cannot have an intermediate reaction or some degree of residual effectiveness.

The residual effects of defeated resistance genes provide a beautiful and solid explanation to account for the fact that defeated genes were retained in wild populations, even after their defeat at some early point in the co-evolution of host and parasite. They were retained because they contributed something of value.

One heterozygous four-gene pyramid has been evaluated with a mildew isolate possessing all of the virulence genes needed to overcome the four resistance genes. The pyramid significantly and dramatically reduced lesion number and sporulation capacity when compared against several homozygous and heterozygous two-gene pyramids and the susceptible recurrent parent of the isolines. My theory is now documented that pyramiding 'major' genes gives the additive expression of 'minor' gene resistance.

The effects of gene dosage on the residual effects of defeated resistance genes

The F_1 of the crosses between each of the Pm isolines and the recurrent parent Chancellor were heterozygous for each of the Pm resistance genes. The availability of that F_1 seed permitted us to assess the effect of gene dosage on the residual effects of certain defeated powdery mildew resistance genes. Three-way comparisons among Chancellor, the homozygous isoline and the heterozygous F_1 lines revealed that the lines heterozygous for Pm genes 3c, 4, or Michigan Amber displayed approximately half of the residuality demonstrated by the isolines homozygous for those genes.

Vanderplank[14] states that race-specific, vertical resistance genes function against epidemic development of plant diseases by reducing the amount of effective initial inoculum (X_0) available for disease onset. Vanderplank[14] further states that vertical resistance genes have no influence on disease increase by races with virulence genes to match them. Mathematical analysis and modeling of plant disease epidemics are conducted with those assumptions of the role of vertical resistance genes. The dramatic effect on disease efficiency and sporulation by certain vertical Pm genes against races with virulence genes to overcome them clearly demonstrates that race-specific resistance also may have an effect on disease increase and can indeed influence the apparent infection rate (r), a trait heretofore attributed solely to genes for horizontal resistance.

Any researcher working with the genetic lines of winter wheat used in this study, but not knowing that the lines were isogenic for so-called 'major' Pm genes, would have concluded quite logically that the lines carrying genes Pm3c, Pm4 or MA possessed some level of rate-reducing or rate-limiting resistance. And yet, rate-reducing resistance, with rare exception, is considered to be under the control of so-called 'minor' genes or polygenes. The present research demonstrates that a gene may perform in a qualitative or a quantitative manner depending upon the genotype of the pathogen that confronts it. These results seem to lend support to the contention by Nelson that there are no major and minor genes for disease resistance, but only genes for disease resistance.

The results of the various aspects of the sporulation studies suggest that one benefit of combining the genes Pm4 and Michigan Amber in a breeding program might be to add the superior capacity of Pm4 to reduce total sporulation to the somewhat better ability of the Michigan Amber gene to resist the onset of sporulation.

The gene-for-gene concept, as proposed by Flor[2], states that for every resistance gene in the host there is a matching virulence gene in the pathogen. The present research in no way challenges the basic premise of the concept, but it may mandate that the gene-for-gene concept be restricted to a qualitative sense such as a black or white reaction type of resistant or susceptible and without reference to a potential quantitative or residual effect of a particular resistance gene.

It has been theorized that wild plant species and their pathogens have co-evolved to genetic equilibrium through the gradual and step-wise accumulation of resistance genes and virulence genes and that the resistance genes at some point in the co-evolutionary process conditioned race-specific reactions before being overcome by new and virulent strains of the pathogen. It seems logical now to speculate that the defeated genes were retained in host populations because their residual effects contributed something of biological value to the populations. It seems logical now to further speculate that defeated resistance genes should be retained in modern cultivars rather than be discarded for a currently effective gene.

The demonstrated residual effects of certain defeated Pm genes provide an opportunity to offer some speculative interpretations of some epidemiological and biological phenomena. For example, multilines with a given amount of susceptible tissue consistently sustain a percentage of disease which is something less than the percentage of susceptible tissue. That phenomenon could be attributed to the expression of the buffering effects of multilines if susceptible tissue were incorporated intentionally into the multiline. However, if a component line carrying a specific resistance gene were rendered susceptible by the advent of a pathogen genotype with a matching virulence gene, the less than expected disease severity may reflect the residual effects of the defeated gene, as well as any buffering effects that may accrue. Theoretically, the latter type of multiline should sustain less disease than a multiline with a known susceptible component line.

Although a statistically significant residual effect could not be demonstrated for Pm2, the differences between Pm2 and Chancellor were biologically obvious. It seems reasonable to speculate that the genotype of a pathogen should influence the capacity of a resistance gene to express residual effects and the extent of the residuality. The research of Martin and Ellingboe demonstrated that the powdery mildew resistance gene Pm4 interacted differently with different isolates of *E. graminis f. sp. tritici* with respect to the gene's ability to restrict the development of elongating secondary hyphae. Studies are currently in progress to examine the interaction of each Pm isoline with a series of isolates each possessing at least the virulence gene needed to match a particular Pm gene.

MULTIPLE EFFECTS OF SOME RESISTANCE GENES

Most cultivated and wild hosts are subject to potential attack by an awesome and substantial array of plant pathogens. And yet, most hosts appear to possess a suitable degree of resistance to most or all of their potential antagonists, i.e. few plant species have been rendered extinct at the hands of a pathogen. I have pondered for many years the question as to whether plant species could bear the genetic load of carrying different sets of resistance genes to combat each and every pathogen that might confront them. Might it be possible that some resistance genes might function in the battle against more than one pathogen? Can a host 'recognize' whether it is being attacked, for example, by a leaf rust pathogen, a stem rust pathogen, a stripe rust pathogen or a powdery mildew pathogen? All four of those pathogens are obligate parasitic and all exhibit the same style of attack.

Some of my recent research has generated a modest amount of evidence, perhaps somewhat circumstantial, that some resistance genes may indeed function against more than a single pathogen. Dr. E.L. Sharp and associates at Montana State University have been actively involved for many years in the detection and incorporation of resistance to stripe rust of wheat. They have found that the additive effects of what they call minor genes can confer acceptable resistance to stripe rust. Out of a cross of Itana X PI 178383, Sharp *et al.* obtained, among other things, four lines, one possessing no minor genes for stripe rust resistance, one carrying one minor rust gene, another carrying two minor rust genes and the fourth possessing three minor genes for stripe rust resistance. As far as I am aware, none of the breeding materials developed with minor gene resistance to stripe rust has been subject to testing for their reaction to powdery mildew of wheat. Dr. Sharp kindly provided me with seed of the four lines in question possessing from no genes to three minor genes for stripe rust resistance. The four lines were inoculated with a highly virulent isolate of the powdery mildew fungus; the results were of considerable interest to me. The wheat line carrying no minor genes for stripe rust resistance was quite susceptible to the mildew isolate; the line with one minor rust gene also was susceptible but to a visibly lesser extent; the line with two minor rust genes exhibited a substantial degree of resistance to mildew highlighted by a chlorotic appearance around the lesions; and, the line with three minor rust genes was highly resistant to mildew. My description of the reactions of the four lines has been stated in very general terms since additional research is needed to be more definitive. At any rate, I am now prepared to theorize that some resistance genes may indeed perform more than a single resistance function.

CONCLUDING REMARKS

My story is told for what it is worth. My sincere thanks to Editors Vose and Blixt for extending to me the opportunity to prepare this chapter. I admit to a certain bias in some of the issues that have been discussed; it is, however, a bias out of conviction rather than narrow-mindedness or at least I hope that is the case.

The next 75 years of resistance breeding will, no doubt, display an array of new strategies and methodology. Old problems and old scientists will be replaced by new problems and better scientists. There will be the continual need to re-examine old truths and replace them, if need be, with new ones. John Maynard Keynes put it aptly when he said "The difficulty lies, not in the new ideas, but in escaping from the old ones . . .". Plant scientists tend to be a conservative lot by nature. Some non-conservative thinking is clearly called for in the coming years as we attempt to meet our obligations to a hungry world. If this chapter whets even a single appetite, my efforts will have been worthwhile.

REFERENCES

1. Briggle, L.W. Near-isogenic lines of wheat with genes for resistance to *Erysiphe graminis* f. sp. *tritici*. *Crop Sci.* 9, 70–72, 1969.
2. Flor, H.H. Host-parasite interaction in flax rust – its genetics and other implications. *Phytopathology* 45, 680–685, 1955.
3. Hamilton, L.M. and Stakman, E.C. Time of stem rust appearance on wheat in the western Mississippi basin in relation to the development of epidemics from 1921 to 1962. *Phytopathology* 57, 609–614, 1967.
4. Hill, J.P. The heritability and genetic control of certain attributes of parasitic fitness of *Helminthosporium maydis* race T. *Ph.D. Thesis*. The Pennsylvania State University, Pa. 59 pp. , 1979.
5. Jensen, N.F. Intra-varietal diversification in oat breeding. *Agron. J.* 44, 30–34, 1952.
6. Nelson, R.R., MacKenzie, D.R. and Scheifele, G.L. Interaction of genes for pathogenicity and virulence in *Trichometasphaeria twicica* with different number of genes for vertical resistance. *Phytopathology* 60, 1250–1254, 1970.
7. Nelson, R.R. Stabilizing racial populations of plant pathogens by use of resistance genes. *J. Environ. Qual.* 1, 220–227, 1972.
8. Breeding plants for disease resistance: concepts and applications. Nelson, R.R. (ed.). Pennsylvania State Univ. Press, Pa. and London. 401 pp. 1973.
9. Nelson, R.R. Horizontal resistance in plants: concepts, controversies, and applications. *Proc. seminar on horizontal resistance to the blast disease of rice*. Galvez, G.E. (ed.). 1–20, Series CE-9. CIAT, Cali, Colombia. 1975.
10. Nelson, R.R. Genetics of horizontal resistance to plant diseases. *Ann. Rev. Phytopathology* 16, 359–378, 1978.
11. Nelson, R.R. The evolution of parasitic fitness. *Plant Disease*. Hoarsfall and Cowling (eds.), 23–46, Ch. II, Vol. 4. Academic Press, New York, 1979.
12. Niederhauser, J.S. Genetic studies on *Phytophthora infestans* and *Solanum* species in relation to late blight resistance in the potato. *Recent Adv. Bot.* 1, 491–497, 1961.
13. Rouse, D.I. Components of horizontal resistance to powdery mildew of wheat as related to disease progress and parasitic fitness. *Ph.D. Thesis*. The Pennsylvania State University, Pa. 92 pp. 1979.
14. Vanderplank, J.E. *Disease Resistance in Plants*. Academic Press, New York and London. 206 pp. 1968.
15. Villareal, R.L. The slow leaf blast infection in rice (*Oryza sativa L.*). *Ph.D. Thesis*. The Pennsylvania State University, Pa. 107 pp. 1980.
16. Wolfe, Martin S. and Barrett, John. Can we lead the pathogen astray? *Plant Disease* 64, 148–155, 1980.

CHAPTER 3

SEARCH FOR USEFUL PHYSIOLOGICAL AND BIOCHEMICAL TRAITS IN MAIZE

J.H. Sherrard, R.J. Lambert, M.J. Messmer, F.E. Below and
R.H. Hageman
Department of Agronomy, University of Illinois,
1102 S. Goodwin Avenue, Urbana, Illinois 61801, U.S.A.

Editors' note: There is still regrettably little information concerning genotypic effects of enzymes and other biochemical traits essential to plant growth. Professor Hageman and his group have done pioneer studies in this area, working with maize. Here they record some of their findings.

INTRODUCTION

Because grain yield is the product of metabolic events that are dictated by numerous physiological and biochemical processes, it seems logical that productivity could be enhanced by genetic manipulation of these processes. However, these processes are so complex (intertwined and regulated with provisions for alternate metabolic routes) that it is difficult to identify specific physiological or biochemical traits that can be directly related to changes in productivity.

Since grain yield is influenced by a number of physiological characteristics and biochemical processes operative within the plant it should be possible to manipulate these traits to produce higher yielding genotypes.

Changes found in physiological or biochemical traits of higher yielding newly released cultivars, relative to lower yielding cultivars, released in preceding years, provides one approach to identification of traits associated with yield. Another approach is to select arbitrarily the trait because of its prominent role in increasing grain production (e.g. dark respiration, photosynthesis, nitrogen metabolism).

Selection of cultivars or hybrids with higher grain yield, especially maize, may have been dictated in part by the plant's responsiveness to fertilizer-N. In a comparison of 19 commercial maize hybrids that were developed and released for farm production at intervals during the period 1939 through 1971, Duvick[15] found that when grown at a 'moderately high' fertility level, the more recently released hybrids were more productive. Welch[39] found that for recently developed maize hybrids, grain yields were decreased (46%) when the annual application of fertilizer N (180 kg/ha) to highly fertile central Illinois soil was omitted. Hageman[20] noted that the increases in maize yields in Illinois from 1950 to 1975 were associated with increased use of fertilizer-N, however percent grain N was unaffected. The newly developed semi-dwarf rice and wheat varieties that utilize greater amounts of N fertilizer without lodging, also produce greater yields of grain. These observations indicate that fertilizer-N has a direct effect on grain production. Increased assimilation of nitrogen could result in increased maize productivity through maximization of sink size, enhancement of grain development or by developing and maintaining an effective photosynthetic apparatus which should be conducive to vegetative development and sustained grain filling.

The potential for breeding for superior physiological or biochemical traits in crop plants is dependent on the genetic variability present in the species for the trait(s), the heritability of the trait, the development of procedures to accurately measure parameters that reflect these characteristics in the plant, and the positive association of the trait with yield. Because of the limited knowledge of the processes involved in achieving maximum grain yield, the identification of a trait causally related to grain yield is the most formidable barrier. It is envisioned that, selecting for appropriate physiological or biochemical trait(s) could result in higher grain yields than by selecting for grain yield alone.

IDENTIFICATION OF SELECTION TRAITS FOR INCREASING GRAIN YIELD

With respect to identification of changes in physiological or biochemical traits associated with improved productivity within a species, relatively few studies have been conducted. Duncan et al.[14] suggested that change in assimilate partitioning (improved harvest index) was responsible for increases in peanut yields over the past 40 years. From a study of old low yielding and new higher yielding soybean cultivars, Gay et al.[19] concluded that future yield improvement may be attained by lengthening the grain filling period. For some of the maize hybrids utilized by Duvick[15], increased field canopy photosynthesis was associated with the newer released higher yielding hybrids (personal communication, D.B. Peters, USDA–ARS, Urbana, IL).

Wilson[41] has found that selection for decreased dark respiration in conjunction with selection for long leaves of erect habit in *Lolium perenne* was associated with an annual increase of 30% in digestible organic matter.

With respect to photosynthesis, several investigators have shown marked differences in photosynthetic rate per unit leaf area between varieties of several crop plants (Dornhoff and Shibles[11]; Evans[17]; Moss and Musgrove[27]). Heritability of light saturated photosynthetic rate was high for *Lolium perenne* (Wilson and Cooper[40]) and maize (Moss and Musgrave[27]). In one or two studies with soybeans, there was a tendency for higher photosynthetic rates in the higher yielding varieties, however even in these studies there were major exceptions.[17] Although selection for divergent photosynthetic rates in maize over five generations was successful, no correlative response in yield was established.[27] Crosbie *et al.*[8] found that selection for carbon dioxide exchange rates resulted in an actual advance of about 5% per cycle with realized heritability of 0.26 to 0.33. One cycle of selection for higher exchange rates did not affect plant height, tillering or grain yield. Except for the report of Harrison *et al.*[22] that canopy apparent photosynthesis of soybeans was significantly correlated with seed yield in a number of soybean lines derived from two crosses, there is little evidence that photosynthesis, especially when expressed as rates, is useful as a selection criterion for grain yield.

Fakorede and Mock[18] reported that recurrent selection for grain yield had no effect on an estimate of carbon dioxide-exchange rate/ha for the BSSS(R) X BXCB1(R) population cross. They concluded that increases in grain yield resulted from increases in translocation efficiency and in duration of transport of assimilate to the grain. In another study, Allison and Watson[1] found the stalk to contain excess capacity for supplying dry matter to the grain during the grain filling period. A number of similar reports led Tollenaar[38] to conclude that sink capacity rather than source capacity commonly limits grain yield of maize grown in regions unaffected by a short growing season. Source limitations appear to exist in short growing season environments.[24] Tollenaar[38] suggested that where sink limitation exists, improved grain yield could be achieved by selecting for factors which influence photosynthate supply to the ear during the flowering period. Alternatively, N rather than carbohydrate supply could be the factor limiting the development of sink capacity.

The availability of N at various stages of plant growth should affect the potential grain yield of cereal crops. However, the effects of N are confounded by interactions with other nutrients and metabolic systems, especially carbon metabolism. Further complications are imposed by metabolic variation among genotypes and environmental effects. As a result it has been difficult to determine the optimum amount of N that should be available to the various plant parts at various stages of development to ensure that N is not the factor limiting grain yield. The following portion of this article will review a limited number of the studies concerned with the role of N in grain yield and describe attempts to identify possible selection traits involving N metabolism for improving grain yield in maize.

Some insight into the relative importance of N availability during vegetative and reproductive growth phases is provided by evaluation of the importance of sink or source in limiting grain yield. Studies conducted by Earley *et al.*[16] in Illinois and by Prine[31] in Florida showed that the level of irradiation during flowering had more influence on grain yield of maize than during the grain filling period. Light energy could influence N-metabolism because light reduced ferredoxin is required for the assimilation of nitrite as well as for the generation of NADPH for the fixation of CO_2 by green leaves.[2]

The following considerations emphasize the major role of N in plant growth and development. Kernel initiates, pollen and developing embryo all have relatively high N/C ratios. Grain production is sharply reduced by adverse environmental conditions that occur during the onset and early phases of reproduction. Such stresses could affect many components related to normal plant development, however, several studies indicate that N-metabolism is more sensitive to such adverse conditions than carbohydrate metabolism. Nitrate reductase is more sensitive to decreases in leaf water potential than either photosynthesis or respiration and nitrate reductase is more sensitive to elevated temperatures than either phosphoenolpyruvic or ribulose 1,5-diphosphate carboxylase.[20] Nitrogen metabolism in maize was more adversely affected by decreased irradiation than carbohydrate metabolism.[26] A study of the rate of change in leaf, stalk and ear carbohydrates and reduced N contents of the maize hybrid Mo17 X B73, showed little or no remobilization of carbohydrate from the vegetation throughout ear development.[3] In contrast there was a rapid depletion of reduced N from the leaves and stalks that was initiated concurrent with the linear phase of ear growth. The early and extensive loss of reduced N from the leaves may indicate that the capacity to supply reduced N to the ear is inadequate. In addition, this loss of N from the leaves could be the cause of the diminished rate of photosynthesis commonly noted between anthesis or shortly after until maturity.

An adequate N supply is considered to be essential in the establishment of maximum sink capacity and in maintaining photosynthetic capacity in rice.[29] However, Murata and Matsushima[29] suggested that excessive use of N is undesirable since it can result in excessive leaf area expansion which is inversely correlated with photosynthetic capacity. Thomas and Thorne[37] found that large applications of fertilizer N to spring wheat at planting resulted in larger plants with increased leaf area but did not increase yield. In contrast, Hucklesby et al.[23] increased grain yield and grain protein in three cultivars of winter wheat with increased fertilizer N application on three dates during late spring. Application of supplemental fertilizer N to maize during the reproductive stage resulted in an increase in grain protein production[10] indicating that the rate of supply of reduced N is limiting grain N accumulation under 'normal' farming conditions. The data of Hucklesby et al.[23] indicate it is the unavailability of nitrate to and in the plant, during the grain filling period, that could be responsible for the constancy of the grain protein percentages commonly observed. The preceding results concur with an earlier investigation of Ingle et al.[25] in which changes were measured in the composition of developing maize kernels. They indicated that endosperm cell division was essentially completed within 28 days following pollination. An increase in protein content of the endosperm following this stage was attributed to production of specialized storage protein.

The initial experiments by Zieserl et al.[42] indicated that level of leaf nitrate reductase activity (NRA) might serve as a simple selection criterion relating N metabolism and yield of grain and grain protein. This view was supported by the work of Brunetti and Hageman[6] who found a highly significant correlation between NRA and actual accumulation of nitrogen by wheat seedlings. However, in subsequent experiments correlations between NRA and either grain yield or grain N among diverse genotypes have not been consitent. Using six maize hybrids, Deckard et al.[10] found significant but low correlations between seasonal canopy NRA and grain yield and plant and grain reduced N content. The low correlations indicate that measured NRA does not accurately reflect the *in situ* assimilation of nitrate among diverse genotypes and/or that factors other than the level of NRA are involved in grain yield and grain protein. Subsequent work provides some insight as to why NRA may not be a good measure of *in situ* reduction. Shaner and Boyer[36] have shown that it is the flux of nitrate rather than the nitrate content of the leaf that regulates the level of NRA in the maize leaf. Later, Reed and Hageman[34,35] noted that a given flux rate of nitrate will not result in the same amount of NRA among maize genotypes. From their work with two wheat varieties Rao et al.[32] indicated that absorption and translocation of nitrate as well as the level of NRA effected in situ accumulation of reduced N. They concluded that no single identifiable trait can be used as a criterion in selecting wheat genotypes for efficient N utilization. Because of these interactions of nitrate availability and level of NRA, it is suggested that changes in levels of reduced N of the entire plant with time may provide a more accurate estimate of nitrate assimilation than measurement of NRA. Among wheat genotypes Dalling et al.[9] found that variations in efficiency of remobilization of vegetative N to the grain precluded a close relationship between NRA and accumulation of N in the grain. They also established that the remobilization of vegetative N to the grain was one of the complicating factors. In maize, ear removal depressed both photosynthesis and nitrate reduction but had little effect on N remobilization, indicating that remobilization was controlled independently of the ear sink.[7] The concurrent onset of proteolysis, loss of leaf reduced N, amino N and chlorophyll for both the eared and earless plants in this study indicate that factors other than N supply could also affect ear development and plant senescence.

With bean plants (Phaseolus vulgaris L.) Neyra et al.[30] observed seasonal patterns of nitrogenase activity to be maximal at flowering and decline rapidly thereafter. In contrast, the observed NRA of the leaf canopy increased markedly after flowering. Soil application of fertilizer N at this time caused a marked increase in leaf NRA and a near doubling of bean yields. Thus, it appears that in some but not all instances, there is a causal relationship between NRA and accumulation of N in the grain.

The increase in in vivo leaf NRA in response to added NO_3^- that occurs at most stages throughout the growth cycle of the maize plant, but particularly during the reproductive phase, also implies the leaf has the potential to assimilate more NO_3^- in situ, with a resultant increase in grain protein, if NO_3^- is available to the leaf blades. An increased NO_3^- supply is difficult to achieve during grain development because there is a progressive decrease in ability of the plant to absorb NO_3^- from the soil. The decreased availability of NO_3^- could result from root senescence, a diminishing rate of NO_3^- uptake, recycling of NO_3^- from the plant to the soil or a decrease in soil NO_3^- supply or availability. This problem might be overcome by the application of suitable forms of N directly to the plant foliage.

Muruli and Paulsen[28] used mass selection with half-sib families to investigate the feasibility of improving N use efficiency in maize. The objective was to increase yields by selecting from a Mex-Mix population grown at low and high levels of N (0 and 200 kg/ha). Synthetic lines developed from the highest yielding families grown at the low level (N-efficient) and high level of N (N-inefficient) were then tested for yield at 0, 50, 100 and 200 kg N/ha. At low levels of N the N-efficient selection outyielded both the original population and the N-inefficient selection but was not responsive to high levels of N. The N-inefficient selection yielded most at high levels of N. It was concluded that N use efficiency could be improved through selection. The relationship between yield and several biochemical and physiological traits of N metabolism was also examined at both silking and maturity. Of these traits, the quantity of N accumulated by the plant (stover plus grain) between silking and maturity was the most important in distinguishing between the ten highest yielding lines in each selection and the means of the original population.

Reed et al.[33] found that two maize genotypes low in NR terminated NO_3^- absorption earlier than two high NR genotypes. In addition, the low NR genotypes had higher proteolytic activities which increased earlier, and had a higher percentage of grain N and harvest indices

TABLE 3.1. Phenotypic correlation values between various physiological traits and mature grain components and total leaf dry weight at anthesis for ten maize genotypes. The traits were measured on samples
representative of all leaves or stalk (sheaths excluded) at seven days postanthesis. Plants were grown
under field conditions on the University of Illinois South Farm in 1979. The genotypes were the
three strains, Illinois high protein (IHP), Illinois low protein (ILP) and Illinois reverse low
protein (IRLP); the inbreds Mo17 and B73; and the hybrids B73 $\times$ IHP, B73 $\times$ ILP,
B73 $\times$ IRLP, B73 $\times$ Mo17 and B84 $\times$ Mo17.

Trait	Mature grain		All leaves
	Dry wt. g/plant	Protein, %	Dry wt. g/plant
		r	
Dry wt. (all leaves, g/plant)	+0.85**	−0.12	+1.00
Dry wt. (stalk, g/plant)	+0.94**	−0.42	+0.87**
Reduced N (all leaves, g/plant)	+0.67*	+0.08	+0.97**
Reduced N (stalk, %)	−0.81**	+0.90**	−0.43
Reduced N (all leaves, %)	−0.40	+0.88**	−0.29
Crude grain protein (mature, %)	−0.59	+1.00	−0.12

* Significant at the 5% level, n = 10.
** Significant at the 1% level, n = 10.

than the high NR genotypes. They concluded that leaf proteolytic activity was more closely
related to the accumulation of grain N than leaf NRA. In a study of five maize genotypes
Below et al.[4] found the remobilization of both vegetative dry weight and N contributed to ear
development but with the N contributing to a greater degree. Although the rate of N accumulation in the developing ear was not related to the rate of dry matter accumulation, these data
do not preclude the possibility that dry matter accumulation is dependent upon movement of
N to the developing grain. Excessive remobilization could result in an increased percentage of
grain N either directly, or indirectly by decreasing the rate of assimilate supply through simultaneous degradation of the leaf photosynthetic apparatus.

Below[3] correlated numerous potential selection traits with components of grain yield for
10 maize genotypes grown under field conditions. The genotypes used were uniquely divergent
in grain yield and grain protein and included inbreds and hybrids. The traits were measured at
seven days post anthesis. The more pertinent and significant relationships are shown in Table
3.1. A comparison of the correlation values obtained for the hybrids and inbreds compared
separately (n = 5) indicate that the overall relationship was not excessively influenced by the
pooling of the data. For leaf dry weight versus grain yield the hybrids had a r value of +0.84
and the inbreds a value of +0.53. For leaf reduced N content the values were: hybrids +0.61;
inbreds +0.07 (+0.86 if the extremely low yielding IHP was omitted). For leaf reduced N concentration, the values were: hybrid −0.83, inbreds −0.56. For stalk reduced N concentration
the values were: hybrids −0.96**; inbreds −0.81. In all cases the trends were similar and
diverged only in degree of correlation with the hybrids having higher values.

TABLE 3.2. Phenotypic correlation values between various physiological traits and mature grain components and leaf blade dry weight at anthesis for twenty S_1 maize families. The leaf traits were measured on the first leaf, above the ear leaf at anthesis (A) and the first leaf below the ear leaf at 30 days after anthesis (A + 30).

Trait	Mature grain		Leaf blade (A)
	Dry wt. g/plant	Protein, %	dry wt. g/plant
		r	
Leaf Blade			
Dry wt. (A, g)	−0.37	+ 0.20	+ 1.00
Specific dry wt. (A, g/area)	−0.44*	+ 0.58**	+ 0.30
Fresh wt. (A, g)	−0.46*	+ 0.20	+ 0.86**
Reduced N (A, mg)	−0.24	+ 0.22	+ 0.86**
Reduced N (A, %)	+ 0.52*	−0.41	−0.24
Midrib			
NO_3^- (A + 30, %)	+ 0.16	−0.61**	−0.02
Grain			
Protein (mature, %)	−0.59**	+ 1.00	+ 0.20

* Significant at the 5% level, n = 20.
** Significant at the 1% level, n = 20.

These correlation values (Table 3.1) are applicable only to the material evaluated. Different relationships may be obtained when these traits are applied to a different plant population or when the same population is grown in a different environment.

The following discussion illustrates these points. In an independent second study the parental population was an Illinois version of the super stiff-stalk synthetic maize. The population [20 S_1 families (ears)] was selected on basis of leaf blade nitrate reductase activity (NRA, g^{-1} fresh wt.) of individual seedlings at 40 days after planting. For both experiments similar correlations were found only between leaf weight and leaf N content and between percentage grain crude protein and grain yield. In contrast with the data of Table 3.1 data of Table 3.2 show negative correlations between leaf weight or leaf N content and grain yield. Leaf reduced N concentration was positively associated with yield (Table 3.2) and conversely (Table 3.1). By comparison with data of several other experiments, the negative correlation between leaf weight and grain yield is atypical, thus it is likely that environment as well as plant population could have affected the divergent result. The environment was adverse for maize production in 1980 (University of Illinois weather records). Although the cause of the inverse relationship between leaf dry weight and grain yield (Table 3.2) is unknown, this relationship would affect the other leaf parameters. Nevertheless, the data of both experiments indicate a relationship between leaf reduced N at anthesis, however, a positive relationship was obtained only when leaf N was expressed as content in the first experiment and as concentration in the second study.

The primary objective of the study of the 20 S_1 families (Table 3.2) was to test the usefulness of seedling leaf NRA as a selection trait. The 20 S_1 families were obtained by recurrent divergent phenotypic selection from 320 plants. Ten ears (families) were ultimately selected from the 10 selfed plants that had the greatest seedling level of leaf NRA (high NRA group). Ten ears were also selected from the 10 plants with the smallest level of leaf NRA (low NRA group). In 1979, the mean seedling NRA for each group was 38.0 and 15.4 (in relative units g^{-1} fresh weight). Five replicate plots were planted with kernels from each of the 20 selected ears in 1980 and the seedlings evaluated for leaf NRA and numerous other vegetative traits. The mean NRA values for the high group was significantly higher at 40 days after planting (26.1 vs 17.2) and at anthesis (16.3 vs 13.0) but not at anthesis plus 30 days (5.5 vs 5.2) for the high NRA group than for the low NRA group, respectively. For the correlative response traits shown in Table 3.2 there were no significant differences between the two groups.

Collectively, these data illustrate some of the difficulties encountered in the identification of useful physiological or biochemical selection traits. These problems indicate that the successful use of such traits is dependent upon a detailed understanding of the metabolism of the plant in relation to grain yield as a function of genotype and environment. Such information should also be useful in the application of genetic-engineered changes in enzymes and metabolic systems. Because enhanced use of fertilizer N has, under most conditions, greatly increased productivity of cereal crops, an understanding of N metabolism and its interaction with other metabolic systems is deemed desirable. In particular, there is need to understand the N requirements for development and maintenance of an effective photosynthetic apparatus, maximum sink initiation and development, effective pollination and a high rate and duration of grain fill among diverse plant populations grown under different environmental conditions.

SELECTION OF PLANT POPULATION BY PHYSIOLOGICAL AND BIOCHEMICAL TRAITS

Two projects will be described which have utilized physiological and biochemical traits associated with N metabolism as the selection criterion in standard breeding programs, in an attempt to develop a superior plant population.

A divergent phenotypic recurrent selection project in a maize synthetic was initiated in 1974 at the University of Illinois.[13] In this procedure, individual plants from a diverse population were sampled. These plants were selfed and kernels bulked from the ears of plants with the desirable trait(s). The plants regenerated from this seed were randomly intermated and the grain harvested and designated as a completed cycle. Successive cycles were generated by repeating the selection process. Stiff-stalk synthetic was used as the parental material because several superior inbreds had been developed from this population. The sole selection criterion was leaf blade *in vivo* (plus NO_3^-) NRA of upper canopy leaves at 10 and 20 days after anthesis. The NRA was measured as described by Hageman and Hucklesby.[21] Leaf blade NRA was chosen as the selection criterion for the following reasons. An increased level of NRA was taken as an approximation of an increased amount of reduced N made available to the plant (based on data available in 1974) during the critical stage of grain development. The increased amount of newly reduced N should increase the amounts available for development of sink capacity, increased grain protein and maintenance of the photosynthetic apparatus. It was visualized that high NRA postanthesis would be positively associated with prolonged leaf duration. The divergent low selection would be the converse of these changes. NRA at the two sampling dates was integrated and modified to reflect the level of NRA and the rate of change in activity between the two dates. Selection in the high strain was for plants with a high level of NRA at both sampling dates and which maintain activity between the sampling dates. Selection in the low strain was for low NRA and a decrease in activity between dates. In each cycle of selection

TABLE 3.3. Direct and correlative response to divergent selection for leaf nitrate reductase activity. Selection was made from an Illinois version of RSSSC super stiff-stalk synthetic maize. Results are the averages of data obtained in 1978 (one experiment) and 1979 (two experiments) conducted on the University of Illinois farm

	NRA		Stover dry weight plant^{-1} at maturity		Grain yield	
Cycle	High	Low	High	Low	High	Low
			——— g ———		——— q/ha ———	
0	31	31	188	188	73	73
1	31	25	190	193	74	81
2	32	20	211	198	74	80
3	34	19	208	189	73	76
4	37	17	202	203	73	82
LSD 0.05	4	2	12	–	–	5

NRA was expressed as modified integrated area and computed as follows: (Integrated area) (Slope + 1)

$$\left(10 \; \frac{(NRA_{10} + NRA_{20})}{2} \right) \left(\frac{(NRA_{20} - NRA_{10})}{10} + 1 \right)$$

where NRA_{10} and NRA_{20} represents the nitrate reductase activity (μmol NO_2^- g f.wt.$^{-1}$ h; *in vivo* + NO_3^-) at 10 and 20 days postanthesis, respectively. 10 represents the 10 day interval between sampling and 1 is a constant.

200 self-pollinated plants from each strain were assayed for leaf blade NRA. The modified integrated NRA value (MIA) was used to select the highest or lowest 40 plants in each strain. During 1978 and 1979 the available cycles were grown in replicated trials at Urbana, Illinois to determine the responses of selection for NRA and the correlated response for certain other traits. The data for four cycles of selection for high and low leaf blade NRA during grain filling are presented in Table 3.3. Selection for leaf blade NRA in the high strain increased NRA by

16 percent from cycle 0 to cycle 4 with most of the progress occurring between cycle 2 and cycle 4. The low NRA strain had a continuous response (45% decrease) to selection from cycle 0 to cycle 4. The reasons for the difference in response of the two NRA strains to selection is not clear. However, selection theory dictates that the rate of response to selection is a function of the genetic variance. Since the maximum genetic variance at a locus occurs between gene frequencies of 0.4 to 0.6, a synthetic variety that had a large number of loci with alleles for high NR levels at frequencies above 0.6 would have less genetic variance and respond less to selection. Selection for low NRA in a strain described above would increase the frequency of alleles for low NRA into the 0.4 to 0.6 range which would increase the genetic variance and response to selection for low NRA. Since the stiff stalk synthetic strain used in this study had undergone several generations of selection for grain yield there could have been a correlated response for high NRA which may have resulted in gene frequencies for high NRA alleles to be above the 0.4 to 0.6 range.

With respect to the correlative responses, the high but not the low NRA strain showed a significant linear increase (b value of 4.6 g) in stover dry weight. The largest increase in stover weight in the high strain was noted for cycle 2 and cycle 3. The low strain showed little change in stover dry weight until cycle 4 when an increase occurred. The change in stover dry weight of the high NRA strain was not reflected in increased grain yields of the high strain (Table 3.3). The high NRA strain showed no change in grain yield. In contrast, the low NRA strain had a significant correlated response for grain yield. There was an 11 percent increase in grain yield in the low NRA strain from cycle 0 to cycle 1 with another small increase for cycle 4.

The concept that the high and low NRA strains would diverge in leaf area duration was not substantiated by visual evaluation of senescence (retention of chlorophyll) or by grain moisture at maturity.

In four cycles of selection there was a significant correlative response in percent grain crude protein to selection for high NRA when grown at 224 kg/ha of supplemental N in both 1978 and 1979 (Table 3.4). The low NRA selections showed little change in percent grain protein during four cycles of selection. In contrast, when the two NRA selections were grown in a low N environment (112 kg/ha) in 1979 there was no significant response in grain protein regardless of strain or cycle of selection (Table 3.4). The plants were grown on a field that had been in continuous maize production for 10 years previous to 1979 and had received 112 kg/ha of N annually. The uptake of N by the plants and heavy summer rainfall could have depleted the soil nitrate by midseason in 1979. Based on the work of Deckard et al.[10], a low supply of soil-N, especially after anthesis, could have been the reason for the lack of a correlated response for grain protein content of the high NRA selection.

The increase in grain protein content in response to selection for high NRA did not show the classical inverse relationship between grain yield and percent grain protein. The selection for low leaf NRA did not change percentage grain protein but did change grain yield. In the low NRA selections, the lack of change in percentage of grain protein indicates that the relative proportions of dry matter (starch) to reduce N accumulated by the grain was not altered by the decreased amounts of NRA in the leaves during grain fill. In conjunction with newly reduced N, the remobilization of N from the vegetation during grain fill was sufficient to maintain a relatively constant ratio of reduced N to dry weight in the grain. In terms of total protein per ha, the low NRA selections produced as much or more grain protein as the high NRA selections. In the high NRA selections, the increased leaf NRA during grain fill enhanced the percentage of grain protein. Whether this indicates that reduction of nitrate in the leaves during grain fill interferes with the photosynthetic capacity is open to conjecture.

The second experimental approach involved use of mass selection with half-sib families. Selection of the families was based on the characteristics of several plants generated from an individual ear. Families for selection in the following cycle are generated by randomly intermating several families having the required trait(s). Successive cycles are generated by the same procedure. This procedure has advantages for determining which N parameters should be used

TABLE 3.4. Correlative response of grain crude protein with divergent selection for leaf nitrate reductase activity. Other details as for Table 3.1.

Cycle	224 kg N/ha				112 kg N/ha	
	1978–1979*		1979		1979	
	High	Low	High	Low	High	Low
				—%—		
0	9.5	9.5	9.2	9.2	9.2	9.2
1	10.1	9.5	10.1	9.3	9.0	9.0
2	9.9	9.8	9.5	9.4	9.1	8.7
3	9.7	9.7	9.3	9.5	9.1	9.3
4	10.0	9.7	9.6	9.3	9.4	8.9
5	–	–	9.7	9.4	9.3	9.2
LSD 0.05	0.4	–	0.6	–	–	–

* Mean value for percent grain protein.

as selection criteria. The method allows for the analysis of a number of characters since some of the plants from each family can be destructively sampled and others reserved for grain yield. In contrast, the phenotypic recurrent divergent selection procedure does not allow for destructive plant sampling. In addition, the half-sib family procedure can be used to obtain estimates of the genetic variance and genetic correlations among traits. These estimates allow prediction of the response to selection and the information generated can be used in the selection of desirable traits. Selection theory dictates that for a trait to be an aid in selecting for grain yield the trait must (a) have a higher heritability than grain yield and (b) have a high genetic correlation (0.7 to 0.9) with grain yield. Successful selection increases the probability that improved inbreds with the desired expression of traits can be developed from the population.

The half-sib family procedure with modifications has been effective in improving grain yield of maize populations with increases of 2.4 to 5.3 percent observed per cycle (Bshadur[5]; Dudley et al.[12]). The objectives of the program were to identify various physiological and biochemical traits that exhibited desirable heritability and correlation characteristics and to test these traits for their suitability in selection for enhanced grain yields.

TABLE 3.5. Genetic and phenotypic correlations between several physiological or biochemical traits and grain yield for 100 half-sib maize families. The traits and yield were measured for the families derived from an Illinois version of RSSSC super stiff-stalk maize synthetic. Plants were grown in 1980 on the University of Illinois farm

Trait and measurement date	Correlation (w standard error)	
	Genotypic	Phenotypic
	──────────── r ────────────	
At anthesis		
Dry weight (all leaves, g plant^{-1})	+ 0.52 ± 0.20	+ 0.16 ± 11
Reduced N (all leaves, g plant^{-1})	+ 0.70 ± 0.19	+ 0.20 ± 0.10
N, all forms (total plant, g plant^{-1})	+ 0.88 ± 0.19	+ 0.16 ± 0.09
At maturity		
Reduced N (plant + grain, g plant^{-1})	+ 0.79 ± 0.28	+ 0.77 ± 0.04
NO_3^- (total plant, g plant^{-1})	+ 0.85 ± 0.19	+ 0.23 ± 0.10
Crude protein (grain, %)	−0.41 ± 0.19	−0.14 ± 0.10
Harvest index (dry weight)	+ 1.24 ± 0.21	+ 0.41 ± 0.08
10 days before anthesis		
Reduced N (leaf blade*, %)	+ 0.80 ± 0.20	+0.25 ± 0.10
15 days postanthesis		
Width (individual leaf,* cm)	−0.41 ± 0.20	−0.06 ± 0.10
Nitrate reductase activity (leaf blade,* units g^{-1} fresh wt.)	+ 0.62 ± 0.21	+ 0.38 ± 0.09
Height (plant, cm)	+ 0.75 ± 0.22	+ 0.37 ± 0.09

*   The individual leaf sampled was the one below the ear leaf.

Kernels from each of 100 ears (families) were planted in duplicate plots in blocks on the University of Illinois farm in 1980. The source material (an Illinois version of RSSSC) was obtained by three cycles of selection for high grain and protein yield by the modified-ear-to-row method of breeding. Nitrogen was supplied at the rate of 224 kg/ha.

From these plots, whole plants were harvested at 40 days after planting, at anthesis and at physiological maturity (black layer formation). The harvested plants were subdivided into leaves and stalk (including sheaths), and dried, weighed and assayed for nitrate and reduced N. The leaf below the ear was also removed from other plants at 10 days prior to anthesis and 15

TABLE 3.6. Means, range in means, heritabilities* and genetic coefficients of variability for several physiological and biochemical traits for 100 half-sib maize families. Other details as for Table 3.5.

Trait and measurement date	Mean	Range	h^2*	G.C.V.
			———%———	
At anthesis				
Dry weight (all leaves, g plant^{-1})	54	64 — 41	60 ± 11	6.1
Reduced N (all leaves, g plant^{-1})	1.5	1.1 — 1.8.8	44 ± 12	5.9
N, all forms (total plant, g plant^{-1})	2.7	2.0 — 3.6	39 ± 19	5.7
At maturity				
Reduced N (plant + grain, g plant^{-1})	3.8	3.8 — 5.8	31 ± 15	6.2
$NO_3^- - N$ (total plant, mg plant^{-1})	174	76 — 284	34 ± 15	17.1
Crude protein (grain, %)	11.9	10.5 — 13.4	38 ± 17	2.7
Yield (grain, g plant^{-1})	116	73 — 161	31 ±13	8.4
Harvest index (dry weight)	0.46	0.46 — 0.54	37 ± 12	6.1
10 days before anthesis				
Reduced N, (leaf blade,** %)	3.5	3.1 — 4.0	31 ± 10	2.1
15 days postanthesis				
Width (individual leaf,** cm)	11.2	9.8 — 13.1	68 ± 16	4.2
Nitrate reductase activity (leaf blade,** units g^{-1} fresh wt.)	42	29 — 60	46 ± 3	9.1
Height (plant, cm)	224	205 — 247	67 ± 11	2.9

* Heritability (h^2) in the narrow sense.
** The individual leaf sampled was the one below the ear leaf.

and 40 days following anthesis. These leaves were subdivided into lamina and midrib and processed and analyzed as the other samples. In addition NRA (*in vivo* + NO_3^-) was measured on portions of the lamina of the individual leaf. Leaf area per plant and plant height were determined shortly after anthesis. Grain yield and harvest index (stover) were determined at physiological maturity.

From these various measurements, estimates of genetic and phenotypic correlations between the traits and grain yield were calculated. Some of the pertinent and significant values are presented (Table 3.5). The genotypic correlation for reduced N content of all leaves at anthesis

(+ 0.7) falls within the guidelines for a useful selection trait. In contrast with the data for 10 diverse genotypes (Below[3]), the reduced N content of all leaves had a slightly higher phenotypic correlation than leaf dry weight. The correlation value for NRA is below the theoretical level of a useful selection trait and is lower than the correlation value for leaf reduced N content. This is consistent with the current view that the *in situ* assimilation of nitrate is a function of the availability of nitrate as well as the level of the enzyme. Although the correlations involving nitrate as a trait are high (+ 0.85 and +0.88, genotypic), the use of nitrate as a selection trait is considered questionable. This opinion is based on the knowledge that environmental conditions have a pronounced effect on the availability and assimilation of nitrate. The negative genetic correlation (−0.41) for leaf width is consistent with the use of long narrow leaves as a selection trait in a forage grass (Wilson[41]). The generally low phenotypic correlations (Table 3.5) reflect the environmental effects on this population.

Significant amounts of additive genetic variance as measured by the heritability (h^2) value were found for a number of the traits measured (Table 3.6). The h^2 values for the traits listed are all high enough to indicate that progress from selection could be expected. These data have to be interpreted with some caution because of the lack of an estimate of the genotype X environment interaction. The lack of this estimate inflates the h^2 value, however, additional cycles of selection will allow for additional estimates and comparison of the observed with the predicted selection response. The genetic coefficient of variability indicates that most of these traits have about the same relative amount of genetic variation (4—10%). In general, traits measured at maturity (harvest index for stover, percent grain protein and grain yield) had lower h^2 values (31—37%) than the other traits (42—68%). The range in family mean values (Table 3.6) also indicates considerable diversity in this population for the traits measured.

Based on the genetic correlation and heritability estimate, reduced-N at anthesis was chosen as the selection criterion. Using the appropriate prediction equations the gain in grain yield using yield as the sole selection criterion should be 6 percent per cycle. Using the selection criteria of yield and reduced N at anthesis the predicted gain should be 8 percent.

For maize, published data can be interpreted to indicate that N availability may impose a greater limitation on grain yield than does the supply of photosynthate. However, the interaction between these two components of yield should not be ignored in identifying selection traits. The metabolic N requirement is important in developing and maintaining an effective photosynthetic apparatus for achieving a high rate and duration of grain filling. N is also important in initiating and developing the ear sink. Availability of high levels of N to the plant roots shortly after anthesis usually has little influence on increasing yield, however such late applications can be expected to increase grain protein, especially when expressed on a percent basis. This indicates that deposition of carbohydrate and nitrogen are independent processes. Additional information on the metabolism of carbon and nitrogen and their interactions should allow further progress to be made toward increasing yield by selection for appropriate physiological or biochemical criteria.

ACKNOWLEDGEMENT

The work was supported in part by a grant from the Science and Education Administration of the US Department of Agriculture and Grants No. 5901-0410-8-0144-0 and 59-2171-1-1-705-0 from the Competitive Grants Office, a grant from Pioneer Hi-Bred International, Inc. and Allied Chemicals.

REFERENCES

1. Allison, J.C.S. and Watson, D.J. The production and distribution of dry matter in maize after flowering. *Ann. Bot.* **30**, 365–381, 1966.
2. Beevers, L. and Hageman, R.H. Nitrate and nitrite reduction. *The Biochemistry of Plants.* Vol. 5, 115–168. Miflin, B.J. (ed.). Academic Press, New York. 1980.
3. Below, F.E. Nitrogen metabolism as related to productivity in ten maize (*Zea mays* L.) genotypes diverse for grain yield and grain protein. *M.S. Thesis.* University of Illinois, Urbana, 1981.
4. Below, F.E., Christensen, L.E., Reed, A.J. and Hageman, R.H. Availability of reduced N and carbohydrates for ear development of maize (*Zea mays* L.). *Plant Physiol.*, (in press). 1981.
5. Bshadur, K. Progress from modified ear-to-row selection in two populations of maize. *Ph.D. Thesis.* University of Nebraska, Lincoln, 1974.
6. Brunetti, N. and Hageman, R.H. Comparison of *in vivo* and *in vitro* assays of nitrate reductase in wheat (*Triticum aestivum* L.) seedlings. *Plant Physiol.* **58**, 583–587, 1976.
7. Christensen, L.E., Below, F.E. and Hageman, R.H. The effects of ear removal on senescence and metabolism of maize (*Zea mays* L.). *Plant Physiol.*, (in press). 1981.
8. Crosbie, T.M., Pearce, R.B. and Mock, J.J. Selection for high CO_2 exchange rate among inbred lines of maize. *Crop Sci.* **21**, 629–631, 1981.
9. Dalling, M.J., Halloran, G.M. and Wilson, J.H. The relation between nitrate reductase activity and grain nitrogen productivity in wheat. *Aust. J. Agric. Res.* **26**, 1–10, 1975.
10. Deckard, E.J., Lambert, R.J. and Hageman, R.H. Nitrate reductase activity in corn leaves as related to yield of grain and grain protein. *Crop Sci.* **13**, 343–350, 1973.
11. Dornhoff, G.M. and Shibles, R.M. Varietal differences in net photosynthesis of soybean leaves. *Crop Sci.* **10**, 42–45, 1970.
12. Dudley, J.W., Alexander, D.E. and Lambert, R.J. Gene improvement of modified protein maize. CIMMYT – *Purdue Int. Symp. on Protein Quality in Maize. Mexico, 1972.* 1974.
13. Dunand, R.T. Divergent phenotypic recurrent selection for nitrate reductase activity and correlated responses in maize. *Ph.D. Thesis.* University of Illinois – Urbana-Champaign. 1980.
14. Duncan, W.G., McCloud, D.E., McGraw, R.L. and Boote, K.J. Physiological aspects of peanut yield improvement. *Crop Sci.* **18**, 1015–1020, 1978.
15. Duvick, D.V. Genetic rates of gain in hybrid maize yields during the past 40 years. *Maydica* **22**, 187–196, 1977.
16. Earley, E.B., McIlrath, W.O., Seif, R.D. and Hageman, R.H. Effect of shade applied at different stages in corn (*Zea mays* L.) production. *Crop Sci.* **7**, 151–156, 1967.
17. Evans, L.T. The physiological basis of crop yield. Crop Physiology: Some Case Histories. Evans, L.T. (ed.). Cambridge University Press. London. 327–355. 1975.
18. Fakorede, M.A.B. and Mock, J.J. Changes in morphological traits associated with recurrent selection for grain yield in maize. *Euphytica* **27**, 397–409, 1978.
19. Gay, S., Egli, D.B. and Reicosky, D.A. Physiological aspects of yield improvement in soybeans. *Agron. J.* **72**, 387–391, 1980.
20. Hageman, R.H. Integration of nitrogen assimilation in relation to yield. *Nitrogen Assimilation of Plants.* Hewitt, E.J. and Cutting, C.V. (eds.). 591–611. Academic Press, New York. 1979.
21. Hageman, R.H. and Hucklesby, D.P. Nitrate reductase from higher plants. *Methods in Enzymology.* San Pietro, A. (ed.). 23, 491–503. Academic Press, New York. 1971.
22. Harrison, S.A., Boerma, H.R. and Ashley, D.A. Heritability of canopy-apparent photosynthesis and its relationship to seed yield in soybeans. *Crop. Sci.* **21**, 222–226, 1981.
23. Hucklesby, D.P., Brown, C.M., Howell, S.E. and Hageman, R.H. Late spring applications of nitrogen for efficient utilization and enhanced production of grain and grain protein of wheat. *Agron. J.* **63**, 274–276, 1971.
24. Hume, D.J. and Campbell, D.K. Accumulation and translocation of soluble solids in corn stalks. *Can. J. Plant Sci.* **52**, 363–368, 1972.
25. Ingle, J., Beitz, D. and Hageman, R.H. Changes in composition during development and maturation of maize seeds. *Plant Physiol.* **40**, 835–839, 1965.
26. Knipmeyer, J.W., Hageman, R.H., Earley, E.B. and Seif, R.D. Effect of light intensity on certain metabolites of the corn plant (*Zea mays* L.). *Crop Sci.* **2**, 1–5, 1962.

27. Moss, D.N. and Musgrave, R.B. Photosynthesis and Crop Production. *Adv. in Agron.* **23**, 317—336, 1971.
28. Muruli, B.I. and Paulsen, G.M. Improvement of nitrogen use efficiency and its relationship to other traits in maize. *Maydica* **26**, 63—73, 1981.
29. Murata, Y. and Matsushima, S. Rice. L.T. Evans (ed.). *Crop Physiology: Some Case Histories.* 73—99. Cambridge University Press, London, 1975.
30. Neyra, C.A., Franco, A.A. and Pereira, J.C. Seasonal patterns of nitrate reductase and nitrogenase activities in *Phaseolus vulgaris* (L.). *Plant Physiol.* **63**, 421—424, 1977.
31. Prine, G.M. Critical period for ear development among different ear-types of maize. *Soil Crop Sci. Soc. Fla. Proc.* **33**, 27—30, 1973.
32. Rao, K.P., Rains, D.W., Qualset, C.O. and Huffaker, R.C. Nitrogen nutrition and grain protein in two spring wheat genotypes differing in nitrate reductase activity. *Crop Sci.* **17**, 283—286, 1977.
33. Reed, A.J., Below, F.E. and Hageman, R.H. Grain protein accumulation and the relationship between leaf nitrate reductase and protease activities during grain development in maize (*Zea mays* L.). *Plant Physiol.* **66**, 164—170, 1980.
34. Reed, A.J. and Hageman, R.H. The relationship between nitrate uptake, flux and reduction and the accumulation of reduced N in maize (*Zea mays* L.). I. Genotypic variation. *Plant Physiol.* **66**, 1179—1183, 1980a.
35. Reed, A.J. and Hageman, R.H. The relationship between nitrate uptake, flux and reduction and the accumulation of reduced N in maize (*Zea mays* L.). II. The effect of nutrient nitrate concentration. *Plant Physiol.* **66**, 1184—1189, 1980b.
36. Shaner, D.L. and Boyer, J.S. Nitrate reductase activity in maize (*Zea mays* L.) leaves. I. Regulation by nitrate flux. *Plant Physiol.* **58**, 499—504, 1976.
37. Thomas, S.M. and Thorne, G.E. Effect of nitrogen fertilizer on photosynthesis and ribulose 1,5-diphosphate carboxylase activity in spring wheat in the field. *J. Exp. Bot.* **26**, 43—54, 1975.
38. Tollenaar, M. Sink-source relationships during reproductive development in maize. A review. *Maydica* **22**, 49—75, 1977.
39. Welch, L.F. Nitrogen use and behaviour in crop production. Agric. Exp. Stn., College of Agric. Urbana, IL. *Bull. 761*, 31—33, 1979.
40. Wilson, D. and Cooper, J.P. Diallel analysis of photosynthetic rate and related leaf characters among contrasting genotypes of *Lolium perenne. Heredity* **24**, 633—649, 1969.
41. Wilson, D. Plant design and the physiological limitations to production from temperate forages. XIII Intl. Bot. Congr. Abstracts 206. 1981.
42. Zieserl, J.F., Rivenbank, W.L. and Hageman, R.H. Nitrate reductase activity, protein content and yield of four maize hybrids of varying plant populations. *Crop Sci.* **3**, 27—32, 1963.

CHAPTER 4

EFFECTS OF GENETIC FACTORS ON NUTRITIONAL REQUIREMENTS OF PLANTS

P. B. Vose

IAEA Project, Centro de Energia Nuclear na Agricultura,
University of Sao Paulo, CP96 Piracicaba, 13.400 SP, Brazil

INTRODUCTION

Although soil and fertilizer science and practice try to ameliorate unfavourable soil conditions there are frequently soil conditions which represent a continuing problem over large areas. Soil acidity (Al and Mn) toxicity, salinity, toxicities due to B, Cu, Fe, etc. and deficiencies of Mn, Fe, Cu, B, Mo and Zn are examples. On such soils appropriate fertilizer and cultural practices will help to alleviate the condition, but there is always a tendency to revert to the unimproved situation, and an appropriate choice of crop variety is essential. There is now sufficient evidence to indicate that a high measure of selection for adaptation to such conditions is possible, as is specifically indicated in later sections. The possibilities of selecting for efficient use of nutritional elements even under conditions of high fertility is also being recognized.

The recognition that varieties of crop plants differ in their response to nutrients and soil conditions, has lead to work directed towards elucidating the factors involved. If we know better what the factors are and the range of response, then we have the possibility of developing a given variety to suit some particular growing conditions. In practice of course, plant breeders have always tried to do this by purely empirical methods of selecting for yield, growth type, disease resistance and quality attributes. The increasing work that has accumulated on many yield factors and the related biochemistry, has brought nearer the possibility of selection for specific conditions.

Of the early citations, Bourcet[1] recorded differential uptake of I by plants, Mooers[2] noted that genotypes vary in their response to levels of soil fertility and Hoffer and Carr[3] found differences in the uptake of Al by maize and related it to the incidence of root rot. Probably Gregory and Crowther[4] were the first to investigate consciously the interrelationship of genetic factors and nutrition when they showed that barley varieties responded differentially to the major elements, and drew attention to the possibility of developing varieties suited to defined soil types.

A certain amount of practical selection for nutritional variation as a means of obtaining suitably adapted varieties, took place before the agronomists and breeders involved understood what they were actually selecting. Thus before it was known that 'whiptail' of cauliflower was due to Mo deficiency suggestions were made to use resistant varieties (Clayton[5], Wessels[6] and Magee[7]), in effect varieties we would now consider to be especially efficient in their use of Mo. Similarly in Brazil, conscious selection and breeding of wheat varieties for acid soils has been done since 1925, before it was known that the *crestamento* or 'burning' of wheat leaves was due to aluminium toxicity (Beckman[8], de Silva[9]). This undoubtedly represents both the earliest and longest sustained successful effort to breed for a specific nutritional requirement.

Despite such early examples it is a fact that soil science and agronomy during the last fifty years have concentrated on changing the soil environment to suit the crop, rather than attempting to breed better adapted crop varieties. This has been due partly to the compartmental nature of much agricultural research, partly to the challenge that changing the soil environment represented, and partly (quoting an earlier review[10]) because 'the breeder of any crop plant must take so many factors into account, e.g. yield, disease resistance, growth habit, agronomic adaptability, resistance to lodging etc., and commercial factors such as earliness, quality, size, uniformity and suitability for packing and transport, that there is little inducement to consider an additional factor such as nutritional efficiency, unless forced to do so by extreme requirements'.

When a number of reviews concerned with genetic factors in plant nutrition or ion uptake appeared in the early Sixties (Myers[11], Epstein[12], Gerloff[13], Vose[10], Epstein and Jefferies[14]) it was possible to attempt a complete coverage of the literature, but in 1982 this is barely practical and in the meantime a number of symposia have also been published (IAEA/FAO, 1966[15], Klimashevsky, 1974[16] and Wright, 1976[17]). More recent contributions have emphasised genetic specificity of ion content and concentration (Saric[404]), and the interaction of soil fertility and crop genotype (Wegozyn *et al.*[405]). The earlier articles certainly generated some momentum, but the primary cause of the great increase in interest has been that the oil crisis and the World economic situation in general are causing a complete re-appraisal in regard to the cost of fertilizers and soil amelioration of the poorer types of marginal soils. This is particularly true in regard to the less developed countries where the cost of fertilizer has always been the major production cost.

There is too, a growing acceptance of the fact that increased food for the less developed countries is going to have to be produced in those countries, and that it is going to be necessary to increase production on large areas of poor soils with such problems as acidity, salinity, iron deficiency and many microelement difficulties. Thus the International Rice Research Institute is undertaking an extensive screening[18,19,20] of rice varieties for adaptation to problem soils with aluminium and manganese toxicities, salinity, iron and zinc deficiencies. Similarly CIAT (Centro Internacional de Agricultura Tropical) is now routinely screening[21,22] rice, corn, sorghum, peanuts and *Phaseolus* for tolerance to aluminium toxicity to improve adaptation to the large areas of infertile soils of the humid and sub-humid tropics.

The mechanisms of nutritional variation comprise a number of integrated processes including absorption, translocation, assimilation, detoxication and metabolic factors such as enzyme activity, together with root morphology, size, and the factors which determine the distribution of dry matter between root and shoot. Undoubtedly many nutritional variants are very complex but for convenience four main types of varietal response have been previously recognized[10,23]: (i) differential yield response and (ii) differential nutrient uptake and translocation, (iii) varying requirement for specific elements, and (iv) the differential resistance to toxicity, such as due to Al, Mo, heavy metals and salinity.

Thus we have 'responding' and 'non-responding' varieties and genotypes according to the degree of response to increased nutrients. There are varieties which are 'efficient' or 'non-efficient' converters of nutrients into dry matter, some being specifically efficient when grown at lower levels of a particular nutrient. Varieties may be 'efficient' or 'inefficient' for uptake or translocation. Genotypes may be 'accumulators', 'highly acquisitive' or 'non-accumulators' of certain elements. Varieties may have 'high' or 'low' requirements for certain microelements. In the case of toxicity response we recognize 'tolerant' or 'resistant' genotypes compared with 'sensitive', 'non-tolerant' or 'susceptible' types. Terminology is quite important to avoid confusion. In particular, use of the term 'efficiency' needs to be specifically defined as it may refer to uptake, response, metabolism or dry matter production.

While much recent work has concentrated in the identification and selection of cultivars able to withstand mineral stresses in problem soils, it is equally important to recognize that cultivars which are efficient in their use of the major nutrients N and P, and of certain microelements, could be at least of equal value at the very highest levels of soil fertility and fertilization, through the more efficient use of fertilizers.

GENETICS OF NUTRITIONAL VARIATION

To make use of any nutritional character in a crop improvement programme requires an initial adequate range of variation of the character, particularly in the direction in which improvement will be sought, to make selection appear promising. Given this, it is necessary to confirm to what degree the character is heritable, and the mode of gene action. Whether for example it is dominant or recessive, simple or multigenic, additive or non-additive.

Although Weiss[24] investigated the genetics of iron utilization in the soybean as long ago as 1943 we still know comparatively little about the detailed genetics of nutritional variation. Some progress has taken place and although many investigations are inconclusive in general it appears that when a nutritional character is likely to be advantageous, e.g. tolerance of a toxicity or efficient utilization of a specific element, it tends to exhibit some degree of dominance.

Lyness[25] compared varieties of maize and found differential yield response to P, and a difference in P-absorbing capacity. When inbred P-inefficient and P-responsive lines were crossed the gene for P-inefficiency behaved as a recessive in the F_1 generation. Also working with maize Harvey[26] studied the absorption and utilization of N by maize inbreds and hybrids.

Differential nutritional response was due to inherent differences in genetic constitution. The transmissibility of the nutritional complexes present in the inbred lines was demonstrated by the response of hybrids to nutritional treatments. It was suggested that the mode of inheritance was either partial dominance of the genetic complex for efficient utilization of ammonium-N, or else dominance divided between factors for efficiency and inefficiency. Similar differences were found when comparing tomato varieties at high and low levels of N, P and K. In particular it was found that the differential response to K supply was inherited but that the genetic complex for efficient utilization of limited K was complicated. F_1 hybrids did not show complete dominance.

Differences in the uptake of radioactive P by two inbred maize lines and their hybrids was reported by Rabideau et al.[27]. Varieties sometimes show differential response to high levels of phosphate, some being more sensitive than others. This is particularly the case with the soybean. From crosses of phosphate tolerant and phosphate sensitive plants Bernard and Howell[118] found that a pair of alleles at a single locus determined the response. Thus NpNp plants were not sensitive to high phosphate levels but npnp plants showed high sensitivity. Intermediate response was given by heterozygote Npnp.

Gorsline et al. in several publications[28,29,30] showed with maize that there was differential accumulation of Ca, Mg and K in the ear leaves of single crosses and inbreds. It was found that differential accumulation was highly inherited under an essentially gene additive scheme for Ca and Mg, but K had a more complicated mode of inheritance, including non-additive gene action. In the case of Al and Mn content the differences were found to be heritable, but apparently separate genetic factors appeared to control the concentration in the ear leaf or in the grain independently. In the case of the ear leaf it was suggested that Al and Fe were controlled by three genes with major effects, while Mn was under the control of at least two genes with major effects, plus one or more additional genes.

The genetic background to a number of instances of toxicity tolerance, such Al-tolerance, is now better known than when Beckman[31] first studied it in wheat. He found that tolerance was dominant, but that there were intermediate types in F_1 populations. In F_2, of a cross of susceptible X tolerant, he recorded 462 tolerant, 191 intermediate and 159 susceptible plants. In some barley populations Al tolerance is controlled by a major dominant gene (Reid[32]) and this also appears to be the case in some wheat varieties. However, Lafever[33] has reported that genetic control can be more complicated in other wheat populations, where there are indications that there are two or more major genes, plus modifier genes involved. This was also essentially found by Kerridge and Kronsted[34], who although they determined a single dominant gene in a cross between a tolerant and an Al-sensitive variety, theorised that as the moderately tolerant parent was not as tolerant as some other varieties, then it was likely that several modifying genes were involved in addition to the dominant one. The tolerance of hexaploid wheats to Al toxicity has been ascribed to the D genome by Slootmaker[35], and confirmed by Prestes et al.[36] who found the Al tolerance factor associated with chromosome 5D.

Rhue[37] has shown in corn that when F_1 hybrids are obtained by crossing two inbred lines of different Al tolerance, then they exhibit tolerance equal to or greater than that of the more tolerant parent, indicating tolerance to be dominant. F_2 and backcross populations showed only two distinct classes of tolerance when screened over a wide range of Al stress, with segregating ratios of tolerant:sensitive plants of 3:1 and 1:1. Progeny from selfed plants were also segregated into distinct tolerance classes. It was concluded that Al tolerance in corn is controlled at a single locus by multiple alleles, as evidenced by the wide range of tolerance as well as the existence of distinct classes of tolerance in S_1, F_2 and backcross populations. There was no evidence for cytoplasm effects on Al tolerance. Vose[38] selected ryegrass, *Lolium perenne* L., for Al tolerance through four mass selection generations, and each generation required progressively increased concentration of Al for adequate screening, clearly indicating the dominant nature of the tolerance.

Work on manganese toxicity by Ouellette and Dessureaux[39] compared parents with progeny from a diallel cross of five plants. With 25 ppm Mn the yield of the parent was reduced by 80% but that of the crosses only by 38%. Considerable improvement in tolerance was obtained after only two generations of selection and cross breeding. Wilkins[50] working on lead tolerance in *Festuca ovina* found that in most cases high tolerance was dominant over both medium tolerance and non-tolerance. It appeared that there was not a simple major gene with two alleles, but the difficulty of establishing with accuracy the tolerance of an individual plant made it impossible to establish how many alleles were involved, or indeed whether they were located at more than one locus, to an extent that the system could be called polygenic.

Despite the fact that it has been known for a long time that there are considerable species and variety differences in salt tolerance of plants, very little work seems to have been done on the inheritance of tolerance. In one study with soybeans Abel[274] compared tolerant and sensitive varieties, and found that tolerance was related to reduced translocation of chloride to the leaves. Crosses of tolerant and sensitive material indicated that tolerance was controlled by a single locus, Ncl being dominant for tolerance while ncl was the recessive with high chloride translocation and salt sensitivity.

Weiss[24] first demonstrated that when soybean varieties were grown in nutrient solution low in Fe they showed differential performance and marked variation in chlorosis and much work has subsequently been carried out on this by Brown[40] and co-workers. Weiss showed on the basis of F_2^2, F_3 and backcross populations of crosses between efficient and inefficient genotypes, that Fe utilization was determined by a single gene. The performance of F_1 plants from efficient X inefficient crosses established the complete dominance of the Fe allele, and the absence of maternal inheritance. More recently Fehr[401] has indicated that resistance to iron chlorosis in soybeans is controlled by many genes and can be considered a quantitative character, even though the major gene Fe may be primarily responsible for iron use efficiency. Bell et al.[51,52] found that plants homozygous for the yellow stripe maize mutant (ys_1) were unable to utilize ferric Fe, but heterozygous plants were able to use both ferrous and ferric Fe. In sorghum there is apparently no clear cut dominance of Fe-efficiency[290,402], while *Phaseolus* also apparently exhibits quantitative inheritance[403], it being hypothesised that the Fe deficiency trait is controlled primarily by two major gene pairs.

The genetics of a number of plant requirements for specific elements are known. Pope and Munger[43] found that magnesium deficiency induced chlorosis in celery, *Apium graveolens*, was determined by a single gene. Crosses of the susceptible variety Utah 10B with resistant varieties showed that Utah 10B inherits recessively by simple 3:1 Mendelian ratio the tendency to become Mg deficient. The same workers found[44] that the celery variety S-48-54-1 was susceptible to boron deficiency, being determined by double recessive alleles at a single locus. The situation for boron deficiency in beet appears to be different as it was suggested[45] that the genetic system was complicated in view of the very wide array of tolerance shown by different varieties.

In view of the range of possible mechanisms that can be responsible for genotypic variation in nutrition it is not surprising that different genetic systems may function for the same element in different plant species. Continuing with boron, it was demonstrated by Wall and

Andrus[46] that the mutant known as 'little stem' in tomatoes, *Lycopersicum esculentum*, was due to a single recessive gene, and that plants having this condition had a lower boron content in the leaves. Raising the boron concentration in the nutrient medium made it possible to eliminate the deficiency, showing clearly that the genetically controlled little stem condition was responsible for reducing transport up the stem.

As Gabelman[47] has pointed out, much of our existing knowledge of the potential inheritance of nutritional variation has come from the study of atypical response to nutritional level, e.g. as shown by the appearance of deficiency symptoms in especially sensitive varieties when more tolerant varieties remain apparently unaffected.

Such studies have been comparatively easy because it has been simple to classify progeny, and while valuable in establishing the principle of nutrient variations and providing us with some insight into the possible modes of inheritance, we are still lacking broadly based data on the inheritance of factors for efficient utilization of the major elements N, P and K. A start has been made, largely by the Wisconsin group.

A study[48] was made of the inheritance of potassium utilization in strains of snapbeans, *Phaseolus vulgaris*, grown with low K level. Strains were classed into efficient and inefficient utilizers of K, based on the severity of appearance of deficiency symptoms. Results suggested that control was by a single pair of alleles possibly with modifier genes, and that efficiency was recessive. In contrast, similar studies reported[47] for tomatoes indicate that in this species the trait for tolerance of low K is quantitatively inherited, as had essentially been found by Harvey[26].

The earlier suggestions[26] of complicated inheritance of efficiency for nitrogen utilization have also been found by O'Sullivan *et al.*[49] with tomatoes with both dominance and additive gene effects. At a more specific level Warner *et al.*[53] found that the level of nitrate reductase in two inbred lines of corn, *Zea mais*, was controlled by two loci. The heritable nature of nitrate reductase level in corn lines was further considered by Hageman[54], and is brought up to date in Chapter 3. Gallagher *et al.*[407] have claimed major gene control of NRA activity in wheat in comparisons of the high yielding UC44-111 (high NRA) and Anza (low NRA) and progenies. Apparently UC44-111 has a single dominant gene (N*ra*) which accounted for most of the variation in NRA.

VARIATION IN RESPONSE TO MAJOR NUTRIENTS

(i) Nitrogen

It was suggested[10] that one of the most significant advances would be to improve the efficiency of nitrogen utilization. Not too much progress has been made, but it is now becoming clear that the apparent rather complicated inheritance of N-utilization efficiency can almost certainly be ascribed to the fact that there are at least four major variables discernible in the overall process: 1. uptake of nitrate (and/or ammonium to a lesser extent), 2. the level and activity of nitrate reductase, 3. the size of the storage pool of nitrate, which is in turn related to nitrate reductase activity, 4. the ability to mobilise and translocate nitrogen from leaves and other parts to the developing grain. It is thus almost inevitable that any general study of nitrogen utilization efficiency and its inheritance will indicate a complex situation, incapable of resolution, because we are dealing with the end result of a number of processes, each of which will almost certainly be under separate genetic control. Future work will have to emphasize the efficiency and control of specific limiting processes.

Differential growth response to N nutrition has been recognized for many crops, e.g. maize[25,26,55,75], barley[4], oats[81], tomatoes[26,49], rice[76,77], ryegrass[57,78,79,80], and apple rootstocks[82], so it is probably a general phenomenon.

Historically, Harvey[26] studied the absorption and utilization of N by maize inbreds and hybrids, and the response to different levels of N, K and P by tomato. He found that in maize grown in solution culture there was a differential response to ammonium and nitrate N. There was also a differential response in per cent dry matter, those lines which did equally well on ammonium and nitrate N having the higher DM content with ammonium nutrient. Hoener and De Turk[55] found differential growth response of high- and low-protein maize varieties when grown at different N levels in nutrition cultures. The low-protein variety yielded more at the lowest level of N but with high N the high-protein variety yielded much more. The high-protein line absorbed and assimilated more nitrate, while the low-protein line accumulated only 49, 48 and 71% of nitrogen, nitrate and dry matter as the more effective line. They suggested that the differential growth response of maize varieties to different N levels might be due to the high-protein plants having a greater enzymatic capacity for reducing nitrate than the low-protein plants. Similarly Burkholder and McVeigh[75] found differential growth of lines and hybrids of maize particularly in response to high levels of N. All hybrids exceeded either or both parent lines in DM production.

Much later, Hageman et al.[56] were able to show that the maize hybrid Hy2 X 0h7 had a consistently higher level of nitrate reductase activity than the hybrid WF9 X C103. The nitrate reductase activity was positively correlated with water soluble protein content and negatively with nitrate content. As it was known that the hybrid Hy2 X 0h7 consistently outyielded WF9 X C103 it was suggested that differences in N metabolism were the critical factors determining yield.

Subsequently, most studies on limiting processes have concentrated on nitrate uptake or accumulation, and related nitrate reduction activity. Vose and Breese[57] found that ryegrass genotypes which utilized N inefficiently in terms of dry matter production had higher nitrate content and lower α-amino and amide nitrogen than did efficient genotypes. In the case of excised wheat roots, differences in the uptake of ammonium-N by different genotypes were demonstrated by Picciurro et al.[58] and also for nitrate by Brunetti et al.[59]. A radiation induced *Triticum durum* mutant had lesser efficiency for NH_4 uptake than the parent Capelli, and this was reflected in straw N content. Bread wheat had a much lower ammonium uptake efficiency than *durum* wheats[58]. Differences were found in nitrate uptake by the roots of two barley varieties[60]. Large varietal differences, as much as 100 per cent, have been found for nitrate accumulation in spinach by Maynard and Barker[61] and Sistrunk and Cash[62]. Dobrunov and Sarsenbaev[96] found differences in the absorption and assimilation of nitrate and ammonium ions by rice varieties. Kuban 3 absorbed and utilized more nitrate in the formation of amide and protein than Dubovsky 129, although the latter had greater uptake capacity per unit weight and was more resistant to the inhibiting effect of high concentrations.

In maize, Haroon[63] found five inbred and seven F_1 hybrids showed variation in nitrate concentration in stem bases. Thus W64A had highest nitrate accumulation in stems, as also did the F_1 hybrids with this line as a parent. In general, nitrate concentrations in the stem base of F_1 hybrids were intermediate between those of parent inbreds. Similarly, Chevalier and Schrader[64] found that inbreds of maize differed significantly as a group. Consistently W64A accumulated most while A632 accumulated least. Significant differences were observed between individual lines. Absorption studies followed essentially the same pattern.[65]

Neyra and Hageman[66] found a close association between nitrate accumulated by excised maize roots and their *in vitro* nitrate reductase activity. On the other hand Jackson[67] has found that although nitrate uptake was fairly similar in three corn varieties, the proportion accumulated in the roots, translocated and reduced varied substantially. These more recent results are not inconsistent with the earliest findings[55] in maize that high-protein plants contained more ammonia and amide than did low-protein plants, suggesting that they had greater enzymatic capacity for reducing nitrates than had low-protein plants. However, the difficulty has been emphasised[408] of developing simple physiological or biochemical screening criteria to identify at the seedling stage, genotypes having superior potential for accumulation of reduced-N.

Instances of variation in nitrate reductase level could in fact be an indirect reflection of capacity to take up and assimilate molybdenum, as nitrate reductase activity is dependent on Mo as an electron carrier.[98] Moreover it is well known that nitrate accumulates in plants suffering from Mo deficiency.[99,100]

A somewhat different situation was found in wheat by Seth *et al.*[72] who studied N utilization in high- and low-protein wheat varieties in the field and in solution culture. It was found that protein content of the vegetative parts of the low-protein wheat was as high as in the case of the high-protein varieties. The differences in protein content arose in the kernels, and we can now recognize this as a N mobilization and transport phenomenon. The situation is less complicated in vegetative material. Thus Vose and Breese[57] found in ryegrass, *Lolium perenne*, that different genotypes varied in their efficiency of N utilization, some tending towards high dry matter yield and relatively low N content (termed efficient), or lower yielding types with high N content, regarded as basically inefficient converters of N. Using 'tiller plants' they were able to show that the ranking of genotypes was the same whether grown at low or high N levels. However the efficient genotypes showed an increased yield relative to the inefficient genotype when grown under high N nutrition as opposed to low N nutrition.

Differential response to high and low N nutrition seems to be of general significance, e.g. Antonovics *et al.*[80] reported populations of *Lolium perenne* adapted to very high levels of N growing on cliff tops frequented by large numbers of sea birds, and genotypes adapted to low N established in poor pasture. Similarly, O'Sullivan *et al.*[49] found with vegetative growth of tomatoes that two classes of plants could be recognized, i.e. strain 62 which responded much more to added increments of N ('responder') and strain 34 which was a 'non-responder'. In beans, *Phaseolus vulgaris*, a study[101] of over a hundred varieties indicated differences in the efficiency of N conversion into grain and leaves. With grain crops the work of Huffaker and Rains[68] emphasizes the need not only to consider nitrogen uptake and assimilation but the mobilization of stored nitrogen. It was found that the breakdown and retranslocation of protein from the foliage to grain during seed filling were critical physiological markers for grain protein content. Anza and US44-111 showed marked differences in nitrate content of leaves. Although Anza contained considerably more nitrate, it absorbed less nitrate over a two hour period in short term uptake studies.[69] Nitrate reductase showed consistently higher levels of the enzyme in UC44-111 than in Anza. Thus the genotype with low nitrate reductase leaf activity had higher levels of tissue nitrate.[70]

The lower reductive potential of Anza for absorbed nitrate was reflected in lower percentage of grain protein. However, although UC44-111 had a much greater capacity than Anza to respond to N fertilizer and reductivity assimilate nitrogen, the per cent protein in the grain was not much greater because of the much lower efficiency of UC44-111 in remobilizing and translocating the reduced nitrogen to the grain.[71]

This was essentially the same situation found by Seth et al.[72], Johnson et al.[73] and McNeal et al.[74] where differences in high- and low-protein wheat varieties are due to genetic control of N re-export and translocation. Following heading, high protein lines increase more rapidly in per cent nitrogen in the grain than low protein lines due partly to more efficient trans-location of N. Mikesell and Paulsen[83] showed that there was little difference in the trans-location of amino acids between high- and low-protein lines, but that the main difference was that grain protein in low-protein lines depended on N already present in lower leaves at anthesis, but that high-protein varieties needed not only high translocation efficiency from lower leaves but also continued assimilation of N by the flag leaf after anthesis. It therefore appears that in grain crops we are seeking a combination of high NO_3 reductive capacity com-bined with efficient remobilization of straw protein to grain protein. Nitrate reductase is only one factor in utilization of fertilizer nitrogen and translocation of remobilized vegetative nitrogen to the developing grain is at least as, if not more, important.

Most studies of variation in nitrogen utilization efficiency have ignored the related assimila-tion and translocation of carbohydrates. Stoy[102] pointed out that although there were many cases of differences in rates of photosynthesis between species and varieties there was no clear cut relationship found between photosynthesis activity and yield performance. Thus for cereal crops, as in the case of nitrogen, primary assimilation of CO_2 appears in general to be of lesser importance than the efficiency with which the assimilates are translocated. Assimilate distribution studies have made most progress in wheat in which they have been related to leaf type[103,104], growth stage[105,106] and contrasting yield varieties[107,108,109]. There is now a great need to link up varietal studies on nitrogen and carbon assimilation efficiency.

In one study[112] where this was partly done it was found that maize genotypes responded differently in terms of CO_2 assimilation, temperature, light intensity and N-nutrition. An early attempt was made by Yamada[77], comparing N-nutrition of *indica* and *japonica* rice varieties. *Japonica* varieties gave a striking response in terms of grain yield, with increasing fertilization. The *indica* variety on the other hand was characteristic in giving the highest yield of all the varieties without fertilizer, but showed no response to higher fertility levels. The DM pro-duction of the *indica* variety was greater than that of the *japonica* types, but whereas the latter utilized about 50% of the DM in grain production, the *indica* variety utilized less than 40%. With increasing fertilization the grain/straw ratio of the *indica* variety decreased so that the proportion fell to 25%. It was suggested that the distribution of DM between grain and straw was determined by varietal differences in N and carbohydrate metabolism. Thus in the *indica* type N absorption is high and its assimilation requires a large proportion of the available photo-synthates. This leads to the production of cell-wall components and vigorous vegetative growth. In contrast, within the *japonica* types response to fertilization takes one of two main forms, either increase in panicle number or increase in panicle size.

Asana et al.[111] noted in a study with wheat that the two most 'efficient' varieties, as measured by the conversion of nitrogen into dry matter and grain, were not the highest yield-ing. They pointed out that developmental aspects of growth, such as tillering, also had to be taken into account. Frey[81] had earlier found that ten oat varieties responded to nitrogen differentially through yield components. Thus Maxim produced high relative grain yield in-creases because both number of heads per plant and number of seeds per head were stimulated, whereas Clintland responded only in number of seeds per head, while Bond showed little response to N fertilization in any of the yield components.

Although Harvey[26] studied the differential response of maize lines to ammonium and nitrate, and differences have been found in the uptake of ammonium-N by the excised roots of different genotypes of wheat,[58] nearly all attention has been devoted to the nitrogen reduction step of nitrogen assimilation. This is logical inasmuch as in fertile soils the activity of nitrifying bacteria normally results in nitrate being the dominant form of nitrogen available to the plant.

However, the suggestion[84] that only in very unusual circumstances does the NH_4 ion constitute a major source of nitrogen for the crop, should not be taken too literally. In most fertile soils some NH_4 will be available to the plant, while in acid soils ammonium fertilizer may stay unchanged in form for two or three weeks. Obvious instances of ammonium feeding are when grassland is heavily fertilized with ammonium sulphate, and the normal situation in rice soils.[85,86] Growth rates are reported[89] to be increased when some ammonium is present in addition to nitrate, moreover the use of nitrification inhibitors will increase the amount of NH_4 in soil.[88] As it is now known that NO_3^- and NH_4^+ ions are utilized equally well by plants, with certain differences in uptake and assimilation, it makes it all the more important that the ammonium incorporation step should receive as much attention as the reduction of nitrate, as it is in any event a following step in nitrogen assimilation following nitrate reduction. Unlike nitrate, the ammonium status of plants cannot be determined by simple analysis, partly because of the difficulty of achieving an accurate estimation but principally because free ammonium in plants is toxic at levels which can be easily determined. Attempts have therefore been made to estimate the level of ammonium nutrition by the carboxylate/amide ratio,[87] or the Cation-Anion/Organic-N ratio.[89] Varietal differences reflected by these ratios do not appear to have been reported yet.

The earlier idea[90,91,92] that ammonia is assimilated by its direct incorporation into α-ketoglutaric acid to form glutamic acid, in the presence of glutamic dehydrogenase, has been confirmed by studies with ^{15}N.[93,94] In this scheme, other amino-acids and aspartic acid will be formed from glutamic acid through transamination. It follows that if the amination of α-ketoglutaric acid through the mediation of L-glutamate dehydrogenase (GDH) is the principal means of NH_4^+ ion assimilation then the comparative level and activity of this enzyme between genotypes could be an indicator for relative efficiency of nitrogen utilization. Like nitrate reductase, GDH is an inducible enzyme, with its level being influenced by ammonia level.[95,96] As NH_4^+ ions are normally taken up by plants more rapidly than NO_3^- ions, it is possible that the amination step is rarely, if ever, rate-limiting, but it should be considered to be so until proven otherwise. In carbon-3 plants it may also be necessary to consider the role of the glutamine synthetase (GS)/glutamate synthase (GOGAT) mediated nitrogen metabolism system, of which photorespiration is now believed to be an integral part.[409,410,411]

Nobody appears to have followed up the finding of Iwata and Baba (quoted in 76) that in the case of rice, normally an ammonium feeder, the ability of roots to oxidise α-naphthylamine varies with the degree of response to N fertilization. Low response varieties showed higher α-naphthylamine oxidation activity than high response varieties but the activity decreased after mid season. This was correlated with the fact that under high N low response rice varieties absorb more N during early growth than do high response varieties, the pattern being reversed in the later growth stages.[97]

A good deal more attention needs to be paid to the ammonium uptake and assimilation processes in plants in general, and also in varieties.

(ii) Phosphorus

Variety effects relating to phosphorus seem to take three distinct forms: phosphorus efficiency in relation to growth, i.e. the ability to respond well to phosphorus fertilization; tolerance of high levels of phosphorus nutrition; and differential P-accumulation. These responses are not necessarily mutually exclusive, in fact the evidence rather points to the fact that they are linked.

De Turk et al.[112], Smith[113] and Lyness[25] pioneered studies with phosphorus. De Turk et al.[112] found marked differences in the response of two first-generation maize crosses to phosphate fertilization. The responder had earlier maturity and increased yield. It appeared that the non-responsive cross could better absorb phosphate from a limited supply but if the supply was adequate it failed to maintain high absorption. Non-responsive crosses also had a higher concentration of inorganic P during vegetative growth. Smith[113], using sand solution culture and sand/soil mixtures, found marked differential response of maize inbreds and hybrids to low P. Hybrids tended to approach the more efficient parent. This was found with all cultural techniques, and the responses were the same whether measured as vegetative weight or ear weight. It was concluded that the ability to grow relatively well on a low level of P is inherited, and it is dominant over lack of such ability, as found also by Lyness.[25]

Lyness[25] found both a differential yield response and a difference in P-absorbing capacity when he grew 21 varieties of maize at different levels of P in sand culture and nutrient solution. One variety which had a high P requirement apparently had only a limited capacity for P absorption from low P concentrations. Both Lyness[25] and Smith[113] found that there was also a direct correlation between P efficiency and the number of secondary roots in relation to primary roots, the latter finding that the root type of the efficient parent was dominant in hybrids.

Lyness[25] did appear to obtain basic differences in capacity for P utilization and response and this has been confirmed by more recent work with beans, *Phaseolus vulgaris*, in which Whiteaker et al.[114] screened 54 different lines for efficiency of P utilization in nutrient culture with very low levels of P. Lines were selected to represent extremes in response to phosphorus stress and were classified as efficient, moderately efficient and inefficient based on dry weight production, when related to mg dry weight yield per mg of P in the tissue. These values varied from 380 to 671 mg. At high levels of P two types of response were noted. Some lines showed a growth increase as solution P was increased, but others showed no growth response. Lines which were most efficient at stress P levels were not always responders. Under P stress, net photosynthesis per unit of P was higher in an efficient than in an inefficient line. These experiments clearly emphasized that differences in P utilization are related to the participation of P in metabolic processes, and not only due to absorption.

The variable behaviour of *Phaseolus vulgaris* under low phosphorus conditions ('P-stress') was further studied by Lindgren et al.[115], who used excised roots to evaluate the differences in P absorption by 59 lines. P-absorption rates varied greatly between lines and relative values between efficient and inefficient lines remained constant with plant age. Heritability estimates derived from parent-offspring regression in families X inefficient lines were estimated as 40%. The uptake of P by excised roots did not predict P uptake and translocation in intact plants. However, as this estimate was only done over a 24 hour period and the data relates to whole shoots this might not reflect the true situation, as the ^{32}P was likely to be preferentially translocated to the youngest developing leaves.

Relative lack of response to fertilizer by soybeans has interested agronomists for a long time, moreover, sensitivity to phosphorus could be a problem with band fertilization under intensive culture. It has been suggested that this lack of response is due to the soybean having been cultured in China over a long period on poor soils, so that natural selection has been away from responding genotypes. Moreover, US commercial soybean varieties are said to be based on no more than 30 introduced Oriental genotypes. Even so, Howell[116] reported significant soybean varietal differences in yield response over a range of P treatments, and Howell and Bernard[117] were able to classify 44 varieties on the basis of sensitivity to high fertilizer P, the later genetical studies having been referred to.[118] It was apparent from the studies that some soybean varieties showed distinct sensitivity to high P levels, and among these varieties Chief was classified as tolerant of high P and Lincoln was sensitive, the latter appearing to be the genetic origin of the P-sensitive response in other varieties. Howell and Foote[119] found that the difference in P-sensitivity of these varieties was located primarily in the roots, so that Lincoln soybeans took up more phosphorus than did Chief from low-N and high-P solutions although the difference was greater with high P.

Howell and Bernard's work[117] was carried out in solution culture but results carried out in the field by Dunphy et al.[120,121] confirmed the response pattern of varieties and showed that the varieties responding more to fertilizer were usually the higher yielding varieties.

Similar variation in sensitivity to high P-levels is also shown by *Phaseolus vulgaris*. Rice[122] found that tolerance of *Phaseolus* to high P nutrition was inherited maternally, with efficiency associated with large-seeded varieties.

Inability to respond to P, due to adaptation to low fertility, was also demonstrated by Jowett[129] for a Pb-tolerant population of the grass *Agrostis tenuis*, in contrast to a 'normal' pasture population which had not been subject to extreme selection pressure towards low fertility survival. Similarly, Snaydon and Bradshaw[149] found differences in the response of natural populations of *Trifolium repens* to phosphate. In contrast, where there has been long selection under high fertility nutritional variability may be effectively eliminated. Thus Vose[150] compared twelve commercial varieties of perennial ryegrass for efficiency of phosphate uptake from soil and obtained virtually no difference with the varieties selected, when using a technique which eliminated differential root growth as a factor.

The question arises as to the relationship of P to other elements and the mechanism of variety-fertility interaction. Peterson[123] compared four lines of beans with promising P-response characteristics with two standard varieties and four levels of phosphorus. Seed yield increases were obtained with increasing P, response being due to seed set rather than seed size increase, and occurred at rates where the P concentration in the leaves exceeded levels usually regarded as optimum. It was concluded that improved N supply for the plants and not improved P nutrition by itself was responsible for the yield increases at the highest P fertilizer rates. Although some increased nodulation was due to P, nodulation did not appear to be solely responsible for the increased N supply. This suggests that increased P enhanced root growth and hence a greater soil volume was exploited. Hughes[124] similarly found a close connection existed between yield and leaf N, leaf N and leaf P, and leaf N and pods per plant when he compared soybean 18 varieties over two years at different fertilizer levels.

Paulsen and Rotimi[161] examined whether varietal differences in uptake of P could be affecting varietal performance through an indirect effect on micronutrient uptake, in this case Zn. They grew Chief (tolerant) and Lincoln (sensitive) varieties of soybeans at different P and Zn levels in solution culture. Both varieties were susceptible to Zn deficiency induced by high phosphorus levels, and high P level decreased growth of the P-sensitive variety more than the

P-tolerant variety, but Zn was decreased in both varieties equally. Added Zn overcame the effects of high P on the P-tolerant variety Chief, but not on the sensitive variety. Failure of Lincoln soybeans to respond to increased Zn supply suggested that high P had an additional effect besides inducing Zn deficiency. Interaction of P with Zn decreased Zn in the leaves, and the causal effect appeared to originate in the roots.

Differences in phosphorus uptake, accumulation and content have been similarly recognized in maize, although in this robust species visible sensitivity effects as in soybeans, are not recognized. Rabideau et al.[27] studied the absorption of ^{32}P by two inbred lines of maize and their hybrids and found that the hybrid with the largest root system absorbed the greatest amount of P. One line and its corresponding hybrids consistently absorbed more ^{32}P than did other lines, and this was associated with more rapid early development. Ferreyra[138] was able to demonstrate significant differences in the uptake of P by excised roots of two maize varieties and a kinetic analysis of the results indicated differences in carrier concentration in the roots. Other workers have shown substantial differences in the accumulation of P by maize[29,30,125,126,127]. In particular, Baker et al.[126] selected lines of inbred maize for high and low P accumulation when grown in soil. Thus lines Pa36 and Pa32 accumulated more P than the lines PaW703 and WH. Clark and Brown[131] determined differences in uptake of P from nutrient solution by Pa36 and WH. They found that the accumulator Pa36 took up and accumulated higher amounts of P from low P concentrations than WH when either grown separately or together with WH in the same container. Experiments with Al as a competing ion showed that Pa36 was also better able to accumulate P than WH. Following Woolhouse's[132] suggestion that increased P uptake under stress conditions may be due to increased phosphatase activity of the roots, the level of this enzyme was assayed. Phosphatase activities of Pa36 roots were higher than those of WH roots, and phosphatase activity increased with decreasing nutrient P.

This work emphasizes the need to define terminology, in particular to define 'efficiency'. It seems quite wrong to apply loosely the term 'phosphorus efficient' to lines that accumulate higher concentrations of phosphorus than other lines, when grown at a given level of phosphorus. They can be more correctly described as 'P-accumulators' or 'efficient P-accumulators'. Clark and Brown's[131] data show that in 3 out of 4 of the P-nutrient levels WH yielded more than Pa36, and that when the total amount of P taken up is calculated (as opposed to %P) both varieties have the same except at the highest level of P. Consequently in terms of dry matter produced per unit of P content, WH is more efficient than Pa36 at three out of the four levels! In general terms, the accumulation of elements by plants seems to be evidence of dry matter conversion inefficiency, rather than characterising capacity for nutrient response. For example in a comparison[130] of various clones of white clover it was found that concentration of P in high- and low-yielding clones showed that high-yielding clones had the lowest per cent P and the lowest yielding clones the highest P. Of course from the animal feeding point of view the P content may be very important. Thus Seay and Henson[145] found in red clover that variation in yield of P amongst 30 clones showed differences as great as fivefold at the first cut. Accumulation in plants is often evidence of a major limiting factor to growth, e.g. minor element deficiency. Point is given to this observation by the report[133] that Pa36 grew very poorly on a low-P soil compared with WH. This was found to be its poor capacity for either uptake or translocation of Mo, in contrast to WH.

That phosphorus nutrition is very closely bound up with both the availability and utilization of microelements was further demonstrated by Brown et al.[134,135,136] and Clark et al.[137]

with sorghum varieties. Twelve genotypes were grown on both an acid soil (pH 4.3) high in available Al, the same soil limed to pH 5.2 which made Cu marginally available, and another alkaline soil (pH 7.5) with low availability of Fe. Genotypes response varied widely. Under low-P or Al-toxicity conditions the most efficient P-absorbing genotypes took up more P than the less effective absorbers and grew normally, while plants which were ineffective absorbers developed P-deficiency symptoms. But on alkaline soil although those genotypes efficient for P-absroption took up more P than did the less effective genotypes this induced Fe chlorosis which did not appear in genotypes less capable of absorbing P. Thus a highly efficient P-absorbing line which is well adapted to growth on acid soils may be highly susceptible to iron deficiency when grown on alkaline soils. Under Cu-deficiency stress one of the genotypes developed Cu deficiency symptoms accompanied by high P accumulation in the leaves.

ROOT MORPHOLOGY

A good deal of the work on varietal differences in phosphorus nutrition has not satisfactorily established whether the causal effect is basically metabolic or whether it is due to differences in size and type of roots and rooting patterns. This applies to many elements but it is particularly relevant to P where root interception and contact, plus diffusion, are the important mechanisms of supply, and probably also with Fe, Mn and Mo[139,140,141,142,143,144]. Rabideau's et al.[27] work relating ^{32}P uptake to size of root system has already been mentioned. Smith[113] too found the P efficiency in the lines with which he worked was directly related to the high ratio of secondary to primary roots. In perennial ryegrass Troughton[146] found that the uptake of ^{32}P by branched roots can be as much as 30% greater than by unbranched roots. Schenk and Barber[406] have found that P-acquisition in maize can be increased by development and selection of hybrids with more fibrous root systems.

Nye[147] has emphasized that when a nutrient is at a low concentration in the medium, root surface area is most likely to be the limiting factor in ion absorption. Low P concentration is a frequent situation. Lindgren et al.[115] found that the rate of P absorption by beans from low concentrations was negatively correlated with the weight of the excised roots of the different lines. They ascribed the increased rate of P absorption associated with decreased root surface area to basic differences in P-uptake capacity at a metabolic level. Comparing the root systems of three varieties of sugarcane Evans[148] found that there was not necessarily any relationship between the absorbing surface and the total length or weight of roots although there were large differences in the total absorbing surface of different varieties. Raper and Barber[151,152] found that although the soybean variety Harosoy 63 had a more extensive root system than that of Aoda, the capacity for nutrient absorption at high nutrient levels was twice as great in Aoda as Harosoy 63. However at low nutrient levels the capacity of each variety to absorb nutrients was about the same, so that Harosoy 63 would have an advantage because of its more extensive root system.

Investigations of root systems as a contributory factor in varietal differences in nutritional response have not received as much attention as their importance warrants, although an extensive rooting system can be valuable to take advantage of the largest possible volume of soil both for nutrients and water. It is essential that the investigator should determine at an early stage whether supposed nutritional variation is due to metabolism and physiology or whether it is the result of gross differences in morphological root pattern and growth. The writer[160] grew in solution culture three ryegrass strains which had given widely different yields when grown as spaced plants in the field. In culture the relative order of yield remained the same as in the field, but the actual difference in yield was very small, although the highest yielding variety had the most roots. It was concluded that the larger root system of the highest yielding variety enabled it to take advantage of the soil volume when spaced widely in the field, but it had no superior metabolic efficiency. Root effects will be especially marked when plants are grown singly or with wide spacing, but possibly of less importance for cereal crops or pasture in which plants are grown close together. For these, increased depth of rooting for greater drought resistance may be significant.

Haahr[153] found in barley that the variety Carlsberg II was able to take up a larger amount of water from deeper soil layers than a mutant which had a much shallower root system. Wheat has been shown[154,155,157] to have considerable varietal differences in root depth and form, which may have significance for cultivation, fertilization and drought resistance. For example, on zinc-deficient soils in the Punjab, India it was found[156] that when a deep rooted wheat variety like Kalyan Sona was grown in rotation with a deep rooted cotton crop, zinc deficiency was aggravated. The deficiency problem is much less when a wheat variety with spreading lateral roots is used. Klimashevsky[16] and others[159,158] have stressed the work being carried out in the USSR on varietal differences in rooting pattern. Kirichenko et al.[158] have carried out long term experiments with spring and winter wheat, barley, sunflower and sugar-beet, and are among the few having data on the root pattern of progenies, as a result of conscious breeding. Substantial differences have been found in the growth pattern and physiological activity of roots of different millet varieties.[159]

(iii) Potassium, Calcium and Magnesium

There are very many references to varietal differences in content of the major cations, but not as many as might be expected in relation to response or requirement for K, Ca or Mg, although some of these were quite early. Thus Hoffer[162] found that selfed lines and crosses of six lines of maize grown on clay and loam soils showed differences in absorption of K. Hybrids had a higher K content than the selfed lines. Harvey[26] found significant differential growth of tomato lines in response to K, and this differential response was inherited. Hybrids showed as striking responses as the parents. Tisdale and Dick[164] reported that varieties of cotton varied with respect to potassium requirements for normal development and for withstanding *Fusarium* wilt attack. The differential response appeared to be associated with groups of varieties' degree of resistance to attack.

More recent work with tomato (Lingle and Lorenz[165]) has shown that varieties bred for high production and dense planting rates can make high demands on the K absorption and transport mechanism. Thus high yielding varieties VF-145 and VF13L, bred for mechanical harvesting, are very liable to K deficiency and this is accentuated by the short growing season. O'Sullivan et al.[49], Gerloff[166] and Gabelman[47] have described wide differences in the degree of response of tomato lines to K. Thus inefficient strain 94 produced only half as much dry matter (0.95 g per 5 mg K per plant) as the efficient strain 98 (1.97 g per 5 mg K). However, when a more adequate amount of K was given (200 mg K available per plant) strain 98 yielded about 25% less than strain 94 (6.48 compared with 8.26 g). Clearly strain 94 has a requirement for higher substrate K.

The efficiency of potassium utilization in strains of *Phaseolus vulgaris* has been studied by Shea et al.[48,167] Plants were grown at stress (5 ppm) and high (200 ppm) levels of K. Differential response to low K nutrition appeared to be associated with efficiency in K utilization at the metabolic level. It was found that the unusual capacity of some strains to produce normal growth on low K was not due to greater seed size or root systems. The variations in efficiency of potassium utilization were not associated with higher levels of potassium in efficient plants. Efficiency of potassium utilization did not appear to be all associated with substitution of sodium for potassium. In contrast, such association was found by Baker[168] working with red beet, *Beta vulgaris*. In this species the cultivated types derive from primitive forms of saline habitat, and respond to Na as a fertilizer. It was found that the effect of adding Na to low-K cultures improved the yields of both K-efficient and K-inefficient red beet 2.7–3.7 times. When Na was given there were no longer yield differences between K-efficient and K-inefficient lines, as apparently Na could substitute for K. With soybeans Dunphy *et al.*[120] found that varieties which exhibited the most severe potassium deficiency were not usually the ones that showed greatest yield response to K-fertilizer.

The problem of interrelationship of plant content substrate level of element and degree of response is also presented by calcium. De Turk[163] reported that the inbred maize line Tr required five times the concentration of Ca (50 ppm) for normal growth, compared with only 50 ppm needed by Reid yellow dent. Snaydon[169] found that the micro-distribution of white clover, *Trifolium repens* within 10 m^2 was closely correlated with the calcium content of the soil, although not related to the pH, while Bradshaw and Snaydon[170] found that an upland population of *Trifolium repens* grew better at a low level of calcium than did a population from chalk downland. They later recorded[171] a marked differential response to calcium with the grass species *Festuca ovina* and that some genotypes were apparently adapted to low Ca. It was further found[172] that when populations of *Trifolium repens* from acid and calcareous soils were grown with different levels of Ca in nutrient solution (lowest level, 4 ppm), the plants from acid low-Ca habitat grew better with low Ca than plants from calcareous soils. With high Ca levels the calcareous genotypes grew better than those from acid soils. Similarly, Clarkson[173] compared two closely related species of the grass *Agrostis, stolonifera* common in high Ca soils and *setacea* which occurs in acid low Ca soils. When grown at increasing levels of Ca (with pH 4.5) it was found that *stolonifera* accumulated greater amounts of Ca and grew better correspondingly, while *setacea*, adapted to low Ca soil, did not respond to increasing Ca above 0.25 mM. It seemed likely that its uptake mechanism was saturated.

Robinson[130] found significant differences between clones of white clover for both Ca% and total Ca content, but no relationship was found between Ca content of the plant and response to liming. Big differences in Ca content of white clover, *Trifolium repens*, were found by Vose and Koontz[174] and Vose and Jones[175]. The latter grew 11 commercial varieties in three soils differing in Ca-content and pH (5.1, 6.2, 7.2). On the low-Ca soil Ca-contents ranged from 0.86 to 3.10% and on the high-Ca soil from 1.45 to 3.8%. Of two highest relative yielding varieties on the low-Ca soil, one had 0.89% Ca and the other 3.0% Ca content. Three varieties yielded most on the high Ca soil and all varieties yielded the least on low Ca soil. Certain varieties having their highest Ca content coupled with high Mn content on the lowest Ca soil, suggested that manganese toxicity was more the limiting factor for these varieties than Ca deficiency. Three distinct nutritional types were recognized: Szolnok which showed considerable capacity to resist Mn-toxicity if sufficient Ca was present; Dutch which showed great sensitivity to Mn regardless of Ca level; and S.100 Nomark which showed the least response to differences in Ca level.

Calcium content has been directly related to performance in some instances, e.g. Lineberry and Burkhart[176] found that the strawberry variety Klondike showed Ca-deficiency symptoms within the normal concentration range of the variety Blakemore. Clark and Brown[177] grew maize inbreds on soils known to induce Ca and Mg deficiency and were able to distinguish 36 lines that were Ca efficient (Ca content 0.29—0.93%) and 44 lines that were Mg efficient (Mg content 0.033—0.119%), using the parameters of Ca and Mg concentration, dry matter yield and degree of Ca or Mg deficiency. Clark[178] reported that the line Oh43 grew better and produced more dry matter compared with A251, at both high and low Ca. Oh43 could produce the same amount of dry matter at about 25% of the amount of Ca in solution as A251 and had fewer deficiency or toxicity symptoms. This indicates the complexity of the Ca-response pattern. Also working with maize, Bruetsch and Estes[179] found comparatively small differences in the Ca and Mg accumulation by 12 commercial hybrids, but that those with higher Ca content tended to be later maturing. Rivard and Bandel[180] analysed 120 maize hybrids and found a Ca range of 0.36—0.65% in ear leaves and 0.23—0.34% Mg, and concluded that the differences were not sufficiently great to affect the interpretation of plant analyses.

Little work has been done on varietal differences in Ca utilization at the metabolic level, although Brown[181] studied the differential uptake of Fe and Ca by maize, and found that there was differential distribution of Ca to newly emerging leaves and Ca-deficiency appeared if there was insufficient Cu present. Although giving Cu prevented apparent Ca deficiency, too high a level caused Fe deficiency symptoms. Greenleaf and Adams[182] noted that the Ca-deficiency condition in tomatoes known as 'blossom-end rot' was under genetic control through Ca metabolism.

Maume and Dulac[183,184] found varietal differences in N, P, K, Ca and Mg content of wheat in 1934, but it was almost thirty years before pressure of the need to determine especially Ca and Sr uptake in relation to nuclear fission fall-out stimulated a great deal of work in this area. Thus the Pennsylvania group of mainly Thomas, Baker, Gorsline, Ragland and Bradford worked with maize, and in a whole series of papers[28,29,30,41,125,185,186,187,188,189,190] recorded the differential accumulation of Ca, Mg, K, Sr, P, Cu, B, Zn, Mn, Al and Fe. Such differential accumulation is highly inherited and has been referred to in a previous section. Although additive gene effects were not observed[29] for most mineral elements it was found[185] that Ca and Sr were mainly controlled by the same gene acting in an additive way. As one might expect, correlation between Sr and Ca was very high. Differences in element ratios supported the conclusion[25] that plants have different capacities to absorb selectively elements like Mg and Ca. Previous work[186] had suggested that the mechanisms of uptake of different elements were independent, and under separate genetic control.

The Minnesota group of primarily Myers, Rasmusson, L.H. Smith and Kleese have worked extensively on Ca and Sr accumulation in wheat and barley[191,192,193,194,195,196,197] and soybeans[198,199]. It was found that varieties differ in their capacity to absorb Ca and Sr and that some varieties show greater capacity to discriminate against ^{89}Sr in the grain than do others. The reciprocal grafting experiments of Kleese[199] with soybean indicated that in this species the control of differences is in the shoots rather than the roots. As in maize the heritability for Ca and Sr accumulation are great enough to make it practical to select within barley[42,194] and later work suggested that Ca content was easiest to select.[197]

The contents of cations and the balance between them may be of importance from nutritional aspects in specific cases, but in the overall situation we are primarily interested in efficiency of production and the internal plant factors which relate to efficiency of conversion. A comparatively small proportion of the work has related differences in concentration to yield. Plants appear to have far less capacity to withstand Mg-deficiency stress, compared with other cations, presumably because of its essentiality for chlorophyl and enzymic transphosphorylations. With Mg, varietal differences in efficiency of utilization and performance are therefore closely connected. Wallace[200] found that the limiting value for magnesium deficiency in the apple, 'Laxton's Superb', to be quite different from that in other varieties. Jones et al.[201] found that under Mg-stress conditions the tomato variety Potentate had lower Mg concentration than two other varieties, and also had more severe symptoms. We have already noted Pope and Munger's[43] genetic work on Mg chlorosis in celery. Differential susceptibility to Mg-deficiency of other celery varieties has been found.[202] Vose and Griffiths[203] recorded that of two oat varieties, one showed Mg-deficiency symptoms both earlier and more pronounced than the other, and the difference was subsequently reflected in the differential effect of Mg deficiency on yield.

Sayre[204,205] carried out extensive analyses of the mineral content of maize in Ohio, finding varietal differences in most elements, but especially Ca and Mg. He found the variety WF9 to be Mg-efficient and Oh40B to be Mg-inefficient. Foy and Barber[206] worked with these varieties and found higher Mg in the nodes and stems of the Ohio variety Oh40B, and attributed the difference between the inbreds for Mg-accumulation to immobilization of Mg in stems and nodes. Although the C.E.C. of Oh40B was slightly higher than the Indiana WF9 it was not sufficient to account for the observed differences. Further work by Schauble and Barber[207] was unable to find any differences between the two inbreds for nodal morphology or Mg-protein complexes, and it was suggested that higher accumulation of Mg by Oh40B nodes was due to a defective transport system. Selecting varieties from earlier work,[177] Clark[208] grew B57 (Mg efficient) and Oh40B (inefficient) maize varieties at different levels of Mg. B57 produced more dry matter and developed fewer Mg-deficiency symptoms, but above 1.2 mM Oh40B dry matter yields were greater than those of B57. To produce comparable dry matter yields at low Mg levels Oh40B required about ten times more Mg than B57. Whatever the Mg level in solution, B57 tops contained about twice the Mg concentration of Oh40B, but Oh40B roots had higher Mg concentrations than B57 at low Mg. It was suggested that Oh40B had the ability to take up as much Mg as the efficient B57 but has inability to translocate Mg to the top. High Ca also affected Oh40B yields more than B57.

The mechanisms of varietal differences in the cation concentration of plant leaves are probably quite varied. Root size and morphology will clearly play a role as discussed earlier. There is adequate evidence of variation in uptake and transport of ions, e.g. see reviews by Myers[11], Epstein and Jefferies[14], Läuchli[128]. Pinkas and Smith[209] found differential strontium accumulation in two barley varieties under metabolic control of the transport into the xylem of the roots. In an ion uptake study with six varieties of barley, Jefferies and Epstein[214] found that the variety Chevron had a potassium uptake mechanism subject to competition from Ca and Mg. Growth of this variety is restricted to soils of low Ca and Mg status. Millaway and Schmid[210] investigated genetic control of Rb-absorption by excised maize roots. Picciurro and Brunetti[211] found differences in the uptake of ^{22}Na by excised roots of different varieties of tomato. However, differences in uptake by roots may not reflect differences in concentration in the leaves. Thus Vose[23] reported although the oat variety Letoria contains in the mature leaves only half the potassium content of many oat varieties, short term studies of potassium uptake with ^{42}K showed that the K-uptake of Letoria seedlings was much higher at this stage. Clearly there was also a transport factor involved in the older plants.

CATION EXCHANGE CAPACITY

Varietal differences in cation exchange capacity are quite common but are not necessarily related to cation content. Butler *et al.*[212,217] found that the C.E.C. of seven varieties of rye-grass varied by a factor of 1.5, and obtained significant correlations in the levels of polyvalent cations, Al, Ti and Fe. Vose[213] found that the total cation content and K content of twelve perennial ryegrass varieties were highly correlated ($> + 0.8$) with differences in their C.E.C. The proportion of K to Ca was not related to variations in C.E.C. The range of C.E.C. 16.4–19.8 mequ/100 g was similar to that found by Mouat[215] for six clones of ryegrass (17.2–20.9 mequ/100 g). Examination of twenty-three varieties of *Phaseolus vulgaris* (Vose, un-published), showed a wide C.E.C. range of 19.0–29.9 mequ/100 g dwt, total cation range of 214–287 mequ/100 g, K content of 39–75 mequ/100 g, Ca content 125–185 mequ/100 g and Mg content of 27–38 mequ/100 g. However, there was no correlation between C.E.C. and total cations or any single cation. Some varieties with the lowest C.E.C. were known to be the most tolerant of acid soils, in line with previous studies[38,218,219,220,221]. Whether there is any correlation between C.E.C. and cation content probably depends on soil characteristics. In a little noticed reference Browne[216] reported that five varieties of sugarcane grown on four soils absorbed widely differing amounts of K depending on the soil involved. A variety which contained the highest when grown on an organic soil contained the lowest percent on the other soils.

STOCK-SCION EFFECTS

We have noted that Mg content of leaves can be determined by translocation factors in the stem, and references in later sections indicate that one manner in which plants overcome potential toxicity situations, due for example, to manganese or salinity, is to retain the ion in the roots. Moreover, susceptibility of certain varieties to various microelement deficiencies seems to be associated with retention of the element in the roots and/or poor translocation. Considerable influence of the effect of rootstock on the response to and accumulation and concentration of many elements has been noted for a wide range of crops, including sunflower, Jerusalem artichoke and walnut[221], citrus[225,226,227,228,229,233,235], apples[223,231,234,236], chlorine in avocado[230], salt accumulation in almonds[229], and composition and phosphorus response in the grape vine[22,232,237]. Most stock/scion effects are due to the stock, but in at least a number of cases, e.g. boron accumulation[221] and Na content of citrus[227] the effect is due to the scion. Such extensive data for the effects of stock and scion in various perennial crops suggest that translocation effects (or defects) are much more common than appreciated. They have largely been missed in annual plants as grafting is not normally carried out except experimentally, as for example with soybeans[199], sunflower[221] and tomatoes[243]. This suggests an area for future research.

VARIETAL EFFECTS OF SECONDARY ELEMENTS

Nitrogen, P, K, Ca and Mg are major elements normally the subject of routine fertilization but many secondary elements may be limiting. Over the years, practical experience and some experimentation has shown cultivators and agronomists that certain plant varieties do badly on particular soils, even showing toxicity or deficiency symptoms, while others grow quite well. The elements associated with such effects are summarized in Table 4.1.

Until recently such observations of tolerance or susceptibility had been largely haphazard, and the general and obvious reaction was simply to avoid growing the susceptible variety. We

TABLE 4.1. Elements associated with common major deficiency and toxicity effects in plants

Element	Effect	Notes
Fe	deficiency (chlorosis)	alkaline soils (toxicity on acid Ferralsols and Acrisols)
Mn	deficiency and toxicity	toxicity occurs on acid soils
Al	toxicity	major toxicity on acid soils, never known to be deficient
B	deficiency and toxicity	toxicity occurs on alkaline soils
Cu	deficiency	excess copper may cause Fe chlorosis
Mo	deficiency	excess Mo content harmful to animals may occur on certain alkaline soils
Se	none in plants	excess accumulation harmful to animals
Zn	deficiency and toxicity	Cu, Zn and Pb toxicity may result from contaminated mine or other waste
(P)		(excess P may help to induce Zn or Cu deficiency)
Na + Cl	toxicity from excess salinity	

now realize that we can do better than that, and for every obvious case of minor element deficiency there are probably many more going undetected but causing less than optimum yields, as Foy[255] has pointed out for example in the case of Al-toxicity. This implies that all new varieties should be consciously screened for their 'nutritional adaptability' much more rigorously than in the past.

Valuable variability has been shown in the field, primarily under cultivated conditions, but also in natural habitats. In most cases bred varieties which have subsequently turned out to be expecially resistant to some toxicity or deficiency have done so purely fortuitously. It is however possible with such knowledge either to grow the variety under the particular conditions to which it is best suited, or to use the variety as a source of genes of the character for incorporation in a breeding and selection programme. The choice of crop varieties resistant to boron or saline toxicities is common-place in farming areas with these problems. Conversely a variety which is specifically bred for a number of important characteristics can be lacking an essential nutritional requirement, e.g. the garden beet variety 'Good For All' developed for canning had to be abandoned because of its extreme susceptibility to boron deficiency.[257] Plant breeding stations with their well-fertilized soils can easily miss such deficiencies in their material in the early stages of selection. A practical example of this is de Mooy's[262] observation that when new 'improved' soybean varieties were introduced in calcareous soils in Iowa they developed Fe chlorosis as they had never been tested for Fe-deficiency.

Selecting plants from natural habitats has the advantage of widening the potential gene pool and under certain circumstances can offer a high degree of natural selection for the desired character, as has been demonstrated in the case of resistance to heavy metal toxicities. Nevertheless this procedure tends to be very limited in application, as the ecotypes obtained usually do not have the other desired agronomic features normally derived through generations of breeding and selection. Exceptionally, forage grass ecotypes taken from natural habitats may have in large degree suitable growth habit, but even they may not have the capacity for good seed production which is so essential to a modern variety. There are nevertheless special cases where only tolerance and persistence are required, and growth habit and yield are of less account; for example in developing genotypes of grasses suitable for reclaiming derelict industrial sites and mine waste tips[269,270,271]. For such use even good seed production may not be essential if the grass is of creeping habit, e.g. *Agrostis* and *Festuca spp.* and can reproduce vegetatively.

(1) Screening methods

Screening of existing and obsolescent varieties also has a role, because it is from these that plant breeders select much of their parent material for developing improved cultivars. Therefore, if nutritional defects are known they can be avoided. Moreover the development of extremely high yielding varieties is going to place greater stress on the plant's capacity to acquire, transfer and utilize micronutrients. Some steps are now being taken in this direction. Following the work of Walker et al.[257], Kelly and Gabelman[45,260] developed methods for screening red beet *Beta vulgaris* for B deficiency susceptibility. Randhawa[238] has described routine methods of screening for micronutrient deficiency susceptibility in India, in sand culture, pots and in the field. Agarwala et al.[239,240] have used sand culture to screen a great number of crops for Zn, Mn, Cu, Mo and B deficiency susceptibility. Brown and Jones have described screening soybeans[241], cotton[242] and sorghum[136] for Fe, Zn and Cu stress and resistance to Al and Mn toxicities. Ponnamperuma[20] has described methods for mass screening of rice varieties for salinity tolerance, zinc deficiency, iron tolerance, and P-requirement, primarily using soils suitably modified.

Randhawa[238] reports that field tests are preferred and that pot tests may not indicate the same relative order of susceptibility. This is readily understandable as a variety with a much branched root system but poor inherent uptake capacity might appear much better in the field than when grown in pots where root growth would be unimportant. Brown[244,245] has described screening plants for efficiency using very low P and Fe levels.

Mn and Al are major factors in the soil acidity complex[259] and much current effort is being put into developing screening methods[17] to select tolerant plants although there is some early work, as indicated in Table 4.2.

Variation in salt tolerance has been known for a long time, e.g. Hayward and Wadleigh[261] cite many references to varietal differences, including cotton, alfalfa, almonds, oats, wheat, barley and maize. At that time they suggested that although many varietal differences were established, further screening and testing for salt tolerance on a genetic basis was required. Most of the early data was used by farmers and agronomists to select from existing varieties and it has been quite recently that Epstein's[263,264,265] work has demonstrated that selection and breeding of extremely salt tolerant cultivars is practicable. The problems of screening for salt tolerance have been discussed by Epstein[263] and Nieman and Shannon[266].

The very specialized area of selection and adaptation to extreme toxicities of heavy metals, such as Pb, Zn and Cu which result from mine and other contamination has been almost entirely the work of Bradshaw and his co-workers described in a series of papers, e.g.[267,268,269, 270,271]. The methods used for screening and selection of grasses have recently been discussed[271], mainly employing soil contaminated to different extent. The practicality of the

TABLE 4.2. Summary of early screening work for identifying tolerance to aluminium and manganese toxicities associated with acid soils

Crop(s)	Tolerance	Reference
Alfalfa	low pH	Schander[246]
Alfalfa	Mn-toxicity	Dessureaux[247]
Barley	Al-toxicity	Reid et al.[248,249]
Barley	low pH	Stølen[258]
Maize	Al-toxicity	Rhue and Grogan[250]
Rice	Al-toxicity	Howeler and Cadavid[251]
Rice, maize, peanuts, sorghum, cow-pea, cassava, *Phaseolus vulgaris*	Al-toxicity, + Ca deficiency	Spain[21]
Ryegrass	Al and Mn toxicity	Vose[38]
Wheat and barley	Al and Mn toxicity	Neenan[256]
Wheat	Al-toxicity	de Silva[9]
Wheat	Al-toxicity	Moore[252]
Wheat	Al-toxicity	Campbell and Lafever[253]
Wheat, barley, maize, rice, sorghum, soybeans	Al-toxicity	Konzak et al.[254]
General	Al and Mn toxicity	Foy[255]

methods is attested to by the production of Pb/zn tolerant *Festuca rubra* var. 'Merlin', a Pb/Zn tolerant *Agrostis tenuis* var. 'Goginan' for acidic wastes, and an *Agrostis tenuis* cultiva 'Parys' for neutral and acid Cu-contaminated areas.

(i) *PHYSIO-GENETIC BASIS FOR SCREENING METHODS*

The practical screening and selection methods that are now possible have a theoretical base on a large amount of work that has been carried out on the nature of varietal differences in response to Fe, Mn, Al, B, Cu, Mo, Zn, NaCl and Pb.

The earlier work, most of which is still relevant, was summarized by Vose[10], the more recent work on variation in response to micronutrients by Brown et al.[243], and the present situation on tolerance to aluminium toxicity by Foy[255,272]. The extensive work on Fe-deficiency stress, a model of the type of work that needs to be done for all micronutrients, has been summarized by Brown.[273]

Table 4.3, which follows, attempts to collate some of the relevant data now available to us on varietal response and effects, and differences in varietal physiology for Fe, Mn, B, Cu, Mo, Pb, Zn and NaCl.

The differential response of genotypes to deficiency or excess of the micro-elements can be generally attributed to differences in absorption, translocation and what might loosely be defined as 'cell factors'. Recognizing absorption and translocation differences is relatively easy, but the ability of certain genotypes apparently to function effectively with a lesser content of an element under deficiency conditions, or with higher levels in the case of toxicity situations, is a good deal more complex and at present little understood.

TABLE 4.3. Some known differences in plant genotype response, effects and physiology,
for Fe, Mn, B, Cu, Mo, Pb, Zn and NaCl

Crop(s)	Nature of Research	Reference
IRON		
soybean	physiology and inheritance of Fe-utilization (classic!)	Weiss[24]
lupins and soybeans	susceptibility to development of Fe-chlorosis related to enzymatic activity	Brown and Hendricks[275]
maize	plants homozygous for yellow stripe mutant ys_1 unable to use ferric iron	Bell et al.[51]
soybeans	genotypic effect of rootstock on Fe-chlorosis	Brown et al.[276]
maize	Indiana Wf-9 corn absorbed and translocated Fe more efficiently than Ohio 40-B	Foy and Barber[206]
soybeans	susceptibility to chlorosis associated with poorer capacity to absorb Fe from Fe-chelate	Brown and Tiffin[277]
soybeans	tolerance to chlorosis associated with much greater root reductive capacity	Brown et al.[278]
maize	effect of ys_1 locus on uptake and utilization of Fe	Bell et al.[279]
rice	Fe-uptake by varietal effects of excised rice roots associated with kinetic differences. Pebifun greater than Siam-29	Shim and Vose[280]
soybeans	utilization efficiency of Fe by two genotypes and isolines	Brown et al.[281]
maize	genotypic differences in Fe and Ca uptake distribution. Insufficient Cu gave Cs deficiency, too much gave Fe-deficiency	Brown[181]
citrus	Fe-uptake related to rootstock	Wallihan and Garber[282]
soybeans	phosphate decreased ^{14}C incorporation into organic acids much more in Fe-inefficient than Fe-efficient genotypes	Brown[283]
maize	iron uptake related to genotype	Brown and Bell[284]
maize	mechanism of differential response of Wf-9 and Ohio 40B genotypes to Fe	Odurukwe and Maynard[285]
citrus (grapefruit)	differential effect of sixteen rootstocks on Fe-chlorosis	Wutscher et al.[286]
tomato	Fe inefficient tomato mutant, with controlling mechanism in the root	Brown et al.[287]
soybeans	roots of iron efficient Hawkeye release more reductants than iron inefficient PI	Ambler et al.[288]
soybeans	suggested reductants released by roots are phenolic compounds which assist in releasing Fe from chelates	Brown and Ambler[287]
sorghum	efficient and inefficient Fe-utilizing lines and F_1 hybrids. Fe-efficiency related to maintenance of low plant P concentrations	Mikesell et al.[290]

Crop(s)	Nature of Research	Reference
maize	exudate concentration of Fe in efficient Wf-9 seven times that of inefficient ys_1/ys_1. The latter lacks efficient mechanism for moving Fe from root cortex to xylem	Clark et al.[291]
tomato	comparison of Fe-efficient and inefficient lines suggested Fe enters plant primarily by lateral roots	Brown and Ambler[292]
sorghum	Fe efficient lines release more reductant from roots and contain less P than inefficient	Brown and Jones[293]
general	determining 'iron efficiency'	Brown and Jones[245]
tomatoes	Fe-stress increased nitrate reductase activity in Fe-efficient (calcicolous) plants but not in Fe-inefficient (calcifuge) plants	Brown and Jones[294]
weeping lovegrass *Eragrostis curvula*	strain differences in susceptibility to Fe-related chlorosis due to inhibited Fe-metabolism rather than reduced Fe-uptake	Foy et al.[295]
chickpea	varietal susceptibility to Fe-deficiency characterized by reduced catalase, peroxidase and starch phosphorylase activity	Agarwala et al.[296]
general	genetically controlled chemical factors in Fe-stress response	Brown[273]
Phaseolus	inheritance of Fe-deficiency	Coyne et al.[403]
rice, chickpea, pigeon pea, jute, peanut	genotypic differences in Fe-uptake and utilization	Kannan[415]
general	factors affecting uptake	Olsen and Brown[413]
barley	differential response to Fe-stress	Fleming and Foy[416]
sorghum	Fe use efficiency in grain sorghum hybrids	Esty et al.[402]
sorghum	variability in genotypic tolerance to low Fe	Williams et al.[417]
MANGANESE		
peas	varietal differences in susceptibility to 'Marsh spot' (Mn-deficiency) disease	Ovinge[297]
cereals	variation in susceptibility to Mn-deficiency	Gallagher and Walsh[298]
peas	differences in susceptibility not accompanied by corresponding differences in Mn-content	Walsh and Cullinan[299]
rice	Mn conc. in leaves of Pebifun was four times Siam-29, but at toxic levels Siam-29 more resistant due to less absorption	Lockard[300]
oats	records Avon, a variety exceptionally resistant to Mn deficiency	Toms[301]
general	notes variation in response to deficiency and excess	Millikan[302]
oats	Mn-deficiency susceptible Star had high-leaf, low-root Mn, and resistant S.171 had low-leaf, high-root Mn	Vose and Griffiths[203,303]

Crop(s)	Nature of Research	Reference
oats	varieties resistant to Mn deficiencies characterized by higher Mn uptake by leaf discs than susceptible varieties	Vose[303]
oats	varietal differences in uptake and distribution of Mn and relationship to 'pools'	Munns *et al.*[305,306]
oats	Mn can be re-distributed under conditions of deficiency; preferentially to youngest leaf and grain	Vose[307]
pasture legumes	variation in response to Mn-excess	Andrew and Hegarty[308]
wheat, oats, barley	differing sensitivity to Mn-deficiency	Nyborg[309]
oats	variation in response to Mn-stress	Brown and Jones[310]
soybeans	varietal response to Mn and Fe toxicities; Mn tolerance control in plant top	Brown and Jones[311]

BORON

sunflower, Jerusalem artichoke, walnut, citrus	concentration and accumulation dependent on rootstock (classic!)	Eaton and Blair[221]
grape vine	varietal difference in B requirement	Scott[312,313]
garden beet	varietal difference in susceptibility to deficiency	Walker[257]
citrus	variation in B content with rootstock	Haas[314]
celery	inheritance of susceptibility to deficiency determined by single gene difference	Pope and Munger[44]
tomato	brittle stem, B-deficiency	Andrus[315]
tomato	inheritance of B-response	Wall and Andrus[46]
tomato	variation in B transport	Brown and Jones[316]
tomato	genetic control of B uptake	Brown and Ambler[317]
tomato	genetics of boron transport	Wann and Hills[318]

COPPER

cereals	Cu content in grain	Greaves and Anderson[319]
oats	grain content and Cu requirement of varieties	Rademacher[320,321]
various	copper-phosphorus induced Fe-chlorosis	Brown *et al.*[322]
pasture legumes	differences in ability to take up Cu from Cu-deficient soil or solution	Andrew and Thorne[323]
wheat, oats, barley	sensitivity to Cu-deficiency	Smilde and Henkens[324]
Agrostis stolonifera	evolution of Cu tolerance	Wu *et al.*[326]
cereals	differential response of genotypes to Cu	Nambiar[327]
various species	Cu tolerance	Gartside and McNeilly[325]

MOLYBDENUM

cauliflower	Mo control of 'whiptail'; varieties	Waring *et al.*[328]
brassicas	varietal susceptibility to deficiency	Neenan and Goodman[329]
legumes	different Mo requirements	Andrew and Milligan[330]
maize	deficiency symptoms	Dios Vidal and Broyer[331]

Crop(s)	Nature of Research	Reference
soybean	variation in progeny response	Harris et al.[332]
cauliflowers	response of cultivars to Mo	Chipman et al.[333]
maize	differential response of two inbreds to Mo stress	Brown and Clark[334]

LEAD

Festuca ovina	lead tolerance	Wilkins[50,335]
Agrostis tenuis	adaptation for tolerance	Jowett[336,337]
Festuca ovina	genetics of Pb tolerance	Urquhart[338]

ZINC

soybeans	different tolerance to Zn-excess, Hudson Manchu tolerated 10 times, concentration tolerated by susceptible vars. (classic!)	Earley[339]
oats	Zn requirement: Mulga oats much higher requirement than Algerian	Williams and Moore[340]
Agrostis tenuis	Zn toxicity tolerance	Bradshaw[341]
maize	variation in Zn content of grain	Massey and Loeffel[342]
soybeans	variety Sanilac more susceptible to deficiency than Saginaw	Ellis[390]
maize	variation in Zn-P interaction. Zn efficiency genetically controlled inbreds and crosses	Halim et al.[343]
soybeans	variation in Zn-P interaction	Paulsen and Rotimi[161]
Phaseolus vulgaris 'navy bean'	differential susceptibility to Zn-deficiency	Ambler and Brown[344]
Phaseolus vulgaris 'navy bean'	Tolerance to excessive Zn-levels; Saginaw tolerated higher levels than Sanilac	Polson and Adams[345]
wheat	differential tolerance to Zn deficiency; response to Zn application not related to varietal susceptibility	Sharma et al.[346]
Anthoxanthum odoratum	genetics of Zn-tolerance	Gartside and McNeilly[347]
Agrostis tinuis	Zn-tolerance	Gartside and McNeilly[348]
wheat	varietal performance on Zn-deficient soils; dwarf variety WG377 exceptionally tolerant of deficiency	Sharma and Sharma[349]
rice	relative varietal susceptibility to Zn-deficiency. 34 varieties showed wide differences	Agarwala et al.[350]

SODIUM CHLORIDE

strawberry clover *Trifolium sp.*	in sand culture Nebraska more resistant than Colorado or Idaho strains	Gauch and Magistad[351]
lettuce	differential salt tolerance of six varieties	Ayers et al.[352]
green beans *Phaseolus sp.*	differential salt tolerance of six varieties	Bernstein and Ayers[353]
barley and wheat	variation in tolerance when grown in salinized soil	Ayers et al.[354]
citrus	Na content of citrus influenced by the scion variety	Cooper et al.[227]

Crop(s)	Nature of Research	Reference
carrots	differential salt tolerance of five varieties	Bernstein and Ayers[355]
almonds and stone-fruits	salt accumulation was affected by rootstock	Bernstein et al.[229]
cotton	individual genotypes varied in resistance to increased salinity	Strogonov et al.[356]
Typha sp.	clones showed differential salt tolerance, hybrid clones intermediate tolerance	McMillan[357]
various	specifically variation in salt tolerance at germination	Wahhab et al.[358]
grape	differential tolerance of four varieties grown in sand culture, with different leaf C1 before injury apparent	Ehlig[359]
Agropyron	differential tolerance within twenty-five strains. Agropyron elongatum distinctly superior Na-tolerance	Dewey[360]
rice	laboratory technique for testing salinity tolerance	Sakai and Rodrigo[361]
rice	technique for determination and relative tolerance of salinity at germination and seedling stage	Pearson[362,363] Peason et al.[364]
crested wheatgrass Agropyron desertorum	breeding and selection for salinity tolerance	Dewey[365]
avocado	uptake of chloride affected by rootstock	Embleton et al.[230]
barley	Na content of salt-susceptible varieties increased, but not salt-tolerant. Susceptible had higher Na and Cl, but lower K	Greenway[366]
citrus	chloride uptake influenced by rootstock variety	Hewitt and Furr[233]
soybeans	capacity for chloride exclusion inherited and related to reduced translocation of chloride to the leaves	Abel[274]
grape	rootstock affects chloride accumulation by leaves	Bernstein et al.[367]
sugarcane	differential variety response to saline media	Tanimoto[368]
Agropyron elongatum, tall wheat grass and Agropyron intermedium	species differ in NaCl tolerance, intermedium being much more susceptible, associated with low root Ca	Elzam and Epstein[369,370]
rice	screening techniques and differential varietal response to salinity	Shafi et al.[371] Janardhan[372] Barakat et al.[373] Janardhan and Murty[374] Datta[375] Purohit and Tripathi[376] Bari et al.[377] Ponnamperuma[20]
rice	breeding for salinity tolerance based on crosses between Thomas 347 and Magnolia	Akbar and Yabuno[378]
tobacco cells, Nicotiana	developing in tissue culture lines of cells tolerant to NaCl	Nabors et al.[379]

Crop(s)	Nature of Research	Reference
Nicotiana sylvestris *Capsicum annuum*	development of NaCl resistant lines of cells	Dix and Street[380]
tomato, *Lycopersicum esculentum* *Lycopersicum cheesmanii*	*cheesmanii* (Galapagos) ecotypes much more salt tolerant, and Na appeared to have very low toxicity for this ecotype	Rush and Epstein[264]
barley	selection for NaCl tolerant genotypes; and field testing with sea water irrigation	Epstein and Norlyn[265] Epstein[263]

The various factors are best known for iron deficiency, which has received the most attention, due to its importance in Ca and P-induced chlorosis and chlorotic conditions induced by excess use of Cu-based fungicidal sprays. Differences in iron transport are the major factors separating Fe-efficient and Fe-inefficient plants. Brown[273] has emphasized that it is the ability of Fe-efficient plants positively to respond to iron deficiency situations that distinguishes them from inefficient genotypes. An iron-inefficient plant may have adequate iron in its roots but be showing visible stress from lack of iron in its tops. In the same situation the iron efficient plant can make iron available for transport and utilization by the leaves.

It appears that iron absorption and transport is mediated by the release of hydrogen-ions by the root, thus lowering the pH of the root zone, favouring Fe^{3+} solubility and reduction of Fe^{3+} to Fe^{2+}. The reduction of Fe^{3+} to Fe^{2+} is carried out by chemically undefined reductants that are released by the roots, making it possible for the Fe^{2+} to enter the root.[413] It is likely that Fe^{2+} is kept reduced by reductant in the root. Subsequent transport to the tops appears to be through a route whereby Fe^{2+} is oxidised to Fe^{3+} (no detectable Fe^{2+} is found in the xylem), chelated by citrate, and Fe^{3+}-citrate is translocated in the xylem.

Genotypes which respond to iron stress apparently have the capacity to increase or maintain hydrogen ion excretion from the roots; excrete reducing compounds from the roots; increase the rate of reduction of Fe^{3+} to Fe^{2+} at the root surface; and increase the level of citrate in the root sap. This general pattern of iron stress response was initially worked out for the well known soybean varieties Hawkeye (efficient) and PI-54619-5-1 (inefficient)[276,277,278,281], but has been confirmed by work with maize[181,284,291] and tomato[287,292]. It is this total stress response which characterizes Fe-efficient genotypes.

In every other instance the mechanism of efficiency has been much less investigated, and the situation with manganese for example is very confused. It could very well be that root reductive capacity is also a governing factor in Mn uptake, as in the case of Fe. Lockard[300] found that Mn concentration in the leaves of the rice variety Pebifun was four times that of Siam-29 and that at toxic levels of Mn then Siam-29 was more resistant due to lower absorption. Shim and Vose[280] chose their experimental material for excised root studies of Fe uptake on the basis of this work. They found that Pebifun also showed much greater uptake of Fe than did

Siam-29, even at the basic level of short term studies on excised roots, clearly pointing to similarities in the uptake of Fe and Mn by rice.

Similarly, the greater tolerance of Siam-29 to high levels of Mn can also be explained by relatively low capacity for Mn uptake of this variety. Support for the root having a controlling influence comes from the work of Munns *et al.*[306] who found that root manganese was held in three 'pools', including non-labile and labile fractions, the latter being considered to supply Mn to the shoot. Varieties showed considerable variation in the size of the pools, particularly the non-labile pool.

Vose and Griffiths[203,303] recorded that Mn-deficiency tolerant oat variety S.171 had low leaf but high root-Mn concentration, while the deficiency susceptible Star had high leaf-Mn content and low root content. Clearly lack of Mn in the leaves was not related to susceptibility. Walsh and Cullinan[299] had earlier found that differences in susceptibility of peas to Mn-deficiency bore no relationship to Mn-content. This suggests possibly that deficiency tolerant varieties have a lower inherent need for Mn in metabolic processes.

Some control by leaves of Mn-response cannot be ruled out. Thus, Vose[303] compared the uptake of Mn by vacuum infiltrated leaf discs and found that a deficiency-tolerant variety had much greater short term uptake capacity than a Mn deficiency-susceptible variety, while varieties of intermediate tolerance had median uptake. Brown and Jones[311] believed that Mn toxicity tolerance appeared to be controlled in the plant tops. It might be that there is more than one mechanism responsible for Mn-response, whether to deficiency or toxicity.

With boron there seems little doubt that leaf accumulation is primarily influenced by the capacity of the root to take up boron, and susceptibility to B deficiency is a reflection of low root uptake efficiency and consequent translocation. The early reciprocal grafting experiments of Eaton and Blair[221] showed that the stock predominantly affected leaf content of sunflower, citrus and walnut, although with the latter crop B absorption was not wholly governed by the rootstock, but there was some scion effect too. Thus the variety Eureka accumulated more in the leaves than Payne when grown on the identical stock. In tomatoes Wall and Andrus[46] and Brown and Jones[316] found that root factors determined B transport.

In the case of Mo, Zn and Cu there is really too little evidence to be able to assert with any confidence mechanisms for tolerance or susceptibility to deficiency stress. The ability of certain tolerant varieties to grow on low-Mo soils has been ascribed to a basic inherent ability to take up more Mo than do susceptible varieties[333,334]. However, susceptibility might well be connected with the capacity of susceptible varieties to take up P more efficiently, this resulting either in competition with Mo uptake, or the Mo binding with P, thus making less Mo available for metabolic processes. Brown and Clark[334] found that the maize variety Pa36, efficient for P uptake, was particularly susceptible to low Mo.

A possible phosphorus interaction seems particularly likely with Zn[161,343]. The soybean variety Sanilac is more susceptible to Zn deficiency than Saginaw[390] and on low-Zn soil developed much more severe symptoms while accumulating twice as much P in the tops as did the latter variety[344] although Zn contents were not greatly different. P-interaction is clearly not the whole story as Saginaw has been shown also to tolerate excessive levels of Zn better than Sanilac, so it is unusually effective at both high and low Zn levels.[345]

Cu-deficiency tolerant varieties of both wheat[319] and oats[320,321] appear to have greater capacity to accumulate higher amounts of Cu under deficiency conditions than have susceptible varieties, and not merely to have the means of functioning at lower inherent cell-Cu content. With Cu too, there appears to be a close connection with P. Thus Brown[391] found that high P accentuated the effect of Cu deficiency in wheat, leading to suppressed calcium translocation, the effect of P being greater with the variety 'Atlas 66' than Monon.

Present evidence suggests that the mechanisms of toxicity tolerance are complex and inter-related. Thus Foy[272] has described many effects of Al toxicity and the response of tolerant and susceptible varieties and has concluded that the exact physiological mechanisms of toler-ance are debatable and are possibly different in different plants. We would conclude that they are controlled by different genes. This is probably the case for most toxicity tolerance, but we know less about Cu, Zn and Pb because less work has been done.

Some common features can be discerned in tolerant genotypes which can withstand excess cations. These are: (i) an inherent capacity — mechanism unknown — of the cells of tolerant plants to function with higher than normal amounts of the toxic elements; (ii) exclusion (of Al[392,393]) at the plasmalemma; (iii) a mechanism whereby the metal is either kept in non-ionic form and/or is bound onto the cell wall, thus reducing the effective active concentration in the cell and keeping the metal from reaching metabolic sites and permitting P-uptake and metabolism; and (iv) an ability to function with smaller amounts of P and possibly Ca.

A few examples must suffice: all Al tolerant varieties possess attributes (i), (iii) and (iv), possibly (ii) as well, but this has not been sufficiently widely researched to be certain of its universality. Mn tolerance appears in major part to depend on the plants functioning with higher than normal levels, as found by Neenan[256] for wheat and Foy et al.[394] for cotton, but reduced absorption could also be a factor.

Copper and zinc tolerance depend primarily on inactivation of the metal and adaptation to low levels of Ca and P. Thus Turner and Gregory[395] found there were big differences in the extent of binding of Zn in roots, particularly in the cell wall fraction, and Peterson[396] found that Zn was especially bound in the pectate fraction of the root cell wall. Inactivation was similarly found to be the primary means of Cu tolerance in Agrostis tenuis[397], while in the Cu-tolerant Zambian shrub Becium homblei, Reilly[398] found that Cu was inactivated in non-ionic form perhaps as a protein complex.

The capacity to function at low fertility levels, especially of Ca and P seems to be a uni-versal feature of Cu, Zn and Pb tolerant plants[267,268]. Jowett[399] found that a lead tolerant population of Agrostis tenuis was adapted to low levels of Ca and P, while Khan[269,400] found that copper and zinc tolerant plants of the same species were adapted to low levels of phosphorus. It is interesting to note that although heavy metal tolerance has been shown to be primarily metal-specific[270], heavy metal tolerant plants tend also to show a greater general tolerance than do non-tolerant genotypes. Possibly this is related to their inherent capacity for growth at low fertility levels.

Although we are still a long way from knowing the full mechanisms that make plants tolerant of saline conditions, a number of factors concerned with salinity tolerance can be recognized. As in the case of microelements we are primarily concerned with differential up-take, transport and 'toxicity response'. It is clear, though, that tolerance is not always due to the same factors, and a system which appears operational in one genus may appear to have relatively little importance in another. Frequently tolerance appears to depend on regulating ion content, in particular excluding either Na or Cl from the leaves. In some cases it appears that a relatively high concentration of either one or the other is possible, but a high concentra-tion of both Na and Cl often seems typical of a susceptible genotype[261,366,389].

There are exceptions to this, and we find exceptionally salt tolerant plants such as the mangrove, Avicenna marina[381], tolerating very high levels of NaCl, as does the wild tomato ecotype *Lycopersicum cheesmanii*[264]. Moreover, even at a cellular level both in intact plants and in tissue cultures[379,380], some NaCl tolerant genotypes do appear to have the capacity to tolerate higher osmotic pressures and higher NaCl content than do others.[386] It is not known how some genotypes can have substantially higher content of Na and Cl than others, without apparent toxicity. This enhanced, but at present indefinable, toxicity resistance is undoubtedly a factor in many instances of tolerance. However, it appears that intact plants are generally more tolerant than tissue cultures, and much data point to Na or Cl exclusion from the leaves, and the ability to regulate the ion content of the plant or at least the upper parts as a major factor.

Sodium exclusion has been most widely studied, and operates in two ways: discrimination against Na uptake by roots in favour of K, and discrimination against Na transport by stems. It is generally accepted[214,364,389] that ion uptake by roots is mediated by highly selective mechanisms, one operating at low concentrations such as might be found in typical soils (mechanism 1), and a second mechanism operational at much higher concentrations (mechanism 2), which could be typical of a fertilizer band or the conditions in a highly saline soil. It has been known for a long time that the uptake of K and Na at low concentrations was carried out by a selective site with a high affinity for K, but with little ability to take up Na in the presence of K, while at higher concentrations mechanism 2, which normally has an equal affinity for Na, is of increased importance[384,385]. At comparatively low salt concentrations mechanism 1 provides a means whereby the plant can accumulate K even when that ion is present in much lower relative concentration than Na. In effect there is positive discrimination in favour of K as opposed to Na.

In many situations where salinity is a problem, high concentrations of cations are inevitably found. It is of interest then that in the mangrove *Avicenna marina* mechanism 2, predominant at high concentrations, has been found to resemble the normal mechanism 1, both in selective preference for potassium and resistance to interference by sodium. In this case potassium absorption at high concentration is predominantly via the type 2 mechanism. *A. marina* appears to be the only example so far recorded in which both mechanisms 1 and 2 show preferential affinity for potassium.[381] How far we can accept such an extreme case of salt tolerance adaptation as relevant to 'normal' economic crops is difficult to say, but in fact hardly any salt tolerant genotypes have been screened for cation uptake characteristics, so the phenomenon might be more widespread than we at present know. There is also the possibility that ion uptake characteristics could be altered by mutation breeding.[412]

In the case of salt tolerant *Agropyron cristatum*[369,370,388] the type 2 mechanism contributed much more to the total rate of absorption of chloride and alkali cations than in the case of a less tolerant ecotype. We have in both cases, mechanisms of osmotic adjustment, which while providing for the efficient uptake of potassium also have the effect through solute absorption of maintaining internal osmotic pressures higher than that of the external medium, thus preventing osmotic loss of water.

Poor stem transport of Na relative to K is well known,[382] and this discrimination against Na being transported to the leaves is clearly a major factor in the salt tolerance of some varieties. Species differences in the ability to translocate Na to plant tops are long established.[383] In citrus it was found that Na content was influenced by the scion variety.[227]

There seems little direct evidence of discrimination against Cl uptake by roots, but there are certainly wide differences in the ultimate Cl transport to leaves, tolerance being related to reduced translocation. In citrus[233], grape[367] and avocado[230] leaf content of Cl is influenced by the rootstock. In one of the few studies of the inheritance of salt tolerance Abel[274] showed that in soybeans tolerance was related to reduced chloride transport controlled by a single locus. The recessive had high chloride translocation and enhanced salt sensitivity.

It is apparent that with variation in a variety of possible mechanisms of salt tolerance, considerable selection and breeding for increased salt tolerance should be possible. Dewey[365] carried out breeding for increased salinity tolerance in crested wheatgrass and Akbar and Yabuno[378] have made some progress in rice, a crop in which a great deal of screening work is at present being done (see Table 4.3). Probably Epstein's selection experiments[263,265] with barley, based largely on 'Composite Cross XXI'[389] together with the associated sea-water irrigation experiments have most clearly demonstrated the practicability of consciously breeding for substantially increased salt tolerance in economic crops.

In general, it is clear that considerable modification and selection for improved nutritional characteristics is practicable, even if we still do not know as much about the genetics of many nutritional attributes as we would like. Especially in the case of the micro-elements a fair degree of selection is possible, both for efficiency of utilization and against toxicities, even with existing material. The possibility should not be overlooked of using mutation breeding techniques to obtain 'nutritional mutants' in otherwise useful varieties.[412] It is comparatively common for a plant with an induced mutation to retain largely unchanged its other significant agronomic attributes.

It is at present impossible to say how far selection for very specific nutritional characters can be carried out. It seems likely that extreme selection for one character might alter the balance of the others, giving rise to a new problem. This will probably be less likely in the case of microelements. Clearly there will be a limit to how far one can select for such a fundamental trait as the efficiency of N utilization at a cellular level, but probably there is great selection potential for efficiency of re-mobilization and translocation of N-assimilates.

The increasing information that is becoming available on efficient nutritional genotypes makes it clear that conscious selection for nutritional attributes of crop plants has passed beyond the curiosity stage, and has already contributed to the development of released varieties tolerant to Al-toxicity (for acid soils) and to Fe-deficiency. A recent international symposium[414] clearly indicated that positive selection and breeding for ability to withstand Fe-deficiency was now the most practicable way of overcoming this widespread problem on alkaline soils. In the future we can expect to see more salinity-tolerant varieties, and varieties more efficient in their utilization of the major elements.

REFERENCES

Note: references are listed primarily according to the section in which they first appear, but many are relevant to more than one section. Recent work is presented in the important reference 418.

General

1. Bourcet, P. Sur l'aborption de l'iode par les végétaux. C.R. Acad. Sci., Paris **129**, 768–770, 1899.
2. Mooers, C.A. Varieties of corn and their adaptability to different soils. *Univ. Tenn. Agr. Exp. Sta. Bull.* **126**, 1922.
3. Hoffer, G.N. and Carr, R.H. Accumulation of aluminium and iron compounds in corn plants and its probable relation to root rots. *J. Agric. Res.* **33**, 801–824, 1923.

4. Gregory, F.G. and Crowther, F. A physical study of varietal differences in plants. Part I. A study of the comparative yields of barley varieties and different manuring. *Ann. Bot.* **42**, 757—770, 1928.

5. Clayton, E.E. Investigations of cauliflower diseases on Long Island. *NY Agr. Exp. Sta. Bull.* **506**, 1—15, 1924.

6. Wessels, P.H. Soil acidity studies with potatoes, cauliflower and other vegetables on Long Island. *Cornell Univ. Agr. Exp. Sta. Bull.* **536**, 1—42, 1932.

7. Magee, C.J. Whiptail disease of cauliflower can almost be eliminated by liming. *Agr. Gaz. N.S.N.* **46**, 911—914, 1933.

8. Beckman, I. Technical Communication. Cultivation and Breeding of Wheat (*Triticum vulgare*) in the South of Brazil (in Portuguese). IX Int. Genetics Meeting, Bellagio (1953) (English version in reference 16).

9. de Silva, A.R. Application of the plant genetic approach to wheat culture in Brazil. *Proc. Workshop, Beltsville, 22—23 November 1976.* Cornell Univ. Agr. Exp. Sta. Spec. Pub., Ithaca, New York, 1977.

10. Vose, P.B. Varietal differences in plant nutrition. *Herbage abstracts* **33**, 1—12, 1963.

11. Myers, W.M. Genetic control of physiological processes: a consideration of differential ion uptake by plants. *Symposium of Radioisotopes in the Biosphere.* R.S. Caldecott and L.A. Snyder (eds.). University of Minnesota, Minneapolis, 201—226, 1960.

12. Epstein, E. Selective ion transport in plants and its genetic control. *Desalination Research Conference.* Nat. Res. Counc. Pub. **942**, 284—298, 1963.

13. Gerloff, G.C. Comparative mineral nutrition of plants. *Ann. Rev. Plant Physiol.* **14**, 107—124, 1963.

14. Epstein, E. and Jefferies, R.L. The genetic basis of selective ion transport in plants. *Ann. Rev. Plant Physiol.* **15**, 169—184, 1964.

15. IAEA/FAO Isotopes in Plant Nutrition and Physiology. *Proc. Symp. Vienna, September 1966.* IAEA, Vienna, 1967.

16. Klimashevsky, E.L. (ed.). *Variety and Nutrition* (in Russian, with English summaries), pp.284, Siberian Branch USSR Acad. Sci., Irkutsk, 1974.

17. Wright, M.J. (ed.). Plant Adaptation to Mineral Stress in Problem Soils. *Proc. Workshop, Beltsville, 22—23 November 1976.* Cornell Univ. Agr. Exp. Sta. Special Pub., Ithaca, New York, 1977.

18. International Rice Research Institute. *Ann. Rept. 1965.* Los Baños, Philippines, 1966 and subsequently.

19. International Rice Research Institute. *Ann. Rept. 1975*, Los Baños, Philippines, 1976.

20. Ponnamperuma, F.N. Screening rice for tolerance to mineral stresses. *Proc. Workshop, Beltsville, 22—23 November 1976*, 341—353. Cornell Univ. Agr. Exp. Sta. Spec. Pub., Ithaca, New York, 1977.

21. Spain, J.M. Field studies on tolerance of plant species and cultivars to acid soil conditions in Colombia. *Proc. Workshop, Beltsville, 22—23 November 1976*, 213—222. Cornell Univ. Agr. Exp. Sta. Spec. Pub., Ithaca, New York, 1977.

22. Centro Internacional de Agricultura Tropical, CIAT, *Ann. Rept. 1976*, CIAT, Cali, 1977.

23. Vose, P.B. The concept, application and investigation of nutritional variation within crop species. *Isotopes in Plant Nutrition and Physiology. Proc. Symp. Vienna 1966*, 539—548. IAEA, Vienna, 1967.

Genetics and Inheritance

24. Weiss, M.G. Inheritance and physiology of efficiency in iron utilization in soybeans. *Genetics* **28**, 253—268, 1943.

25. Lyness, A.S. Varietal differences in the phosphorus feeding capacity of plants. *Plant Physiol.* **11**, 665—688, 1936.

26. Harvey, P.H. Hereditary variation in plant nutrition. *Genetics* **24**, 437—461, 1939.

27. Rabideau, G.S., Whaley, W.G. and Heimsch, C. The absorption and distribution of radioactive phosphorus in two maize inbreds and their hybrids. *Amer. J. Bot.* **37** 93—99, 1950.

28. Gorsline, G.W., Ragland, J.L. and Thomas, W.I. Evidence for inheritance of differential accumulation of calcium, magnesium and potassium by maize. *Crop Sci.* **1**, 155—156, 1961.

29. Gorsline, G.W., Thomas, W.I. and Baker, D.E. Inheritance of P, K, Mg, Cu, B, Zn, Mn, Al and Fe concentrations by corn, *Zea mais* L., leaves and grain. *Crop Sci.* **4**, 207–210, 1964.

30. Gorsline, G.W., Thomas, W.I. and Baker, D.E. Major gene inheritance of Sr—Ca, Mg, K, P, Zn, Cu, B, Al—Fe, and Mn concentration in corn, *Zea mais* L.. *Penn. State Univ. Bull.* **746**, 1968.

31. Beckman, I. Cultivation and breeding of wheat in the south of Brasil (in Portuguese). *Agronomia Sulriograndense* **1**, 64–72, 1954.

32. Reid, D.A. Genetic control of reaction to aluminium in winter barley. *Proc. 2nd Int. Barley Genet. Symp.*, 409–413. Washington State University Press, 1970.

33. Lafever, H.N., Campbell, L.G. and Foy, C.D. Differential response of wheat cultivars to aluminium. *Agron. J.* **69**, 563–568, 1977.

34. Kerridge, P.C. and Kronsted, W.E. Evidence of genetic resistance to aluminium toxicity in wheat, *Triticum aestvum. Agron. J.* **60**, 710–711, 1968.

35. Slootmaker, L.A.J. Tolerance to high soil acidity in wheat-related species, rye and triticale. *Euphytica* **23**, 505–513, 1932.

36. Prestes, A.M., Konzak, C.F. and Hendrix, J.W. quoted by Reid, D.A. Aluminium and manganese toxicities in the cereal grains. *Proc. Workshop Plant Adaptation to Mineral Stress in Problem Soils, Beltsville, 22–23 November 1974*, 55–64. Cornell University, 1977.

37. Rhue, D. Genetic control of Al tolerance in corn, *Zea mais. Agron. J.*, 1978.

38. Vose, P.B. and Randall, P.J. Resistance to aluminium and manganese toxicities in plants related to variety and cation-exchange capacity. *Nature*, **196**, 85–86, 1962.

39. Ouellette, G.J. and Dessureaux, L. *Canad. J. Plant Sci.* **38**, 206–214, 1958.

40. Brown, J.C. Genetically controlled factors involved in absorption and transport of ions by plants. *Advances in Chemistry*, **162**, *Bio-inorganic Chemistry* II, 43–103, 1977.

41. Naismith, R.W., Johnson, M.N. and Thomas, W.I. Genetic control of relative calcium, phosphorus and manganese accumulation on chromosome 9 in maize. *Crop Sci.* **14**, 945–949, 1974.

42. Fick, G.N. and Rasmusson, A.C. Heritability of Sr-89 and Ca-45 accumulation in barley seedlings. *Crop Sci.* **7**, 315–317, 1967.

43. Pope, D.T. and Munger, H.M. Heredity and nutrition in relation to magnesium deficiency chlorosis in celery. *Proc. Amer. Soc. Hort. Sci.* **61**, 472–480, 1953.

44. Pope, D.T. and Munger, H.M. The inheritance of susceptibility to boron deficiency in celery. *Proc. Amer. Hort. Sci.* **61**, 481–486, 1953.

45. Kelly, J.F. and Gabelman, W.H. Variability in the tolerance of red beet varieties and strains to boron deficiency. *Proc. Amer. Soc. Hort. Sci.* **76**, 409–415, 1960.

46. Wall, J.R. and Andrus, C.F. The inheritance and physiology of boron response in the tomato. *Am. J. Bot.* **49**, 758–762, 1962.

47. Gabelman, W.H. Genetic potentials in Nitrogen, Phosphorus and Potassium efficiency. *Proc. Workshop Plant Adaptation to Mineral Stress in Problem Soils. Beltsville, 22–23 November 1976*, 205–212. Cornell Univ. Agr. Exp. Sta. Spec. Pub., Ithaca, New York, 1977.

48. Shea, P.E., Gabelman, W.H. and Gerloff, G.C. The inheritance of efficiency in potassium utilization in strains of snap beans, *Phaseolus vulgaris. Proc. Amer. Soc. Hort. Sci.* **91**, 286–293, 1967.

49. O'Sullivan, J., Gabelman, W.H. and Gerloff, G.C. Variations in efficiency of nitrogen utilization in tomatoes, *Lycopersicum esculentum*, grown under nitrogen stress. *J. Amer. Soc. Hort. Sci.* **99**, 543–547, 1974.

50. Wilkins, D.A. The measurement and genetical analysis of lead tolerance in *Festuca ovina. Scottish Plant Breeding Station Rept.*, 85–98, 1960.

51. Bell, W.D., Bogorad, L. and McIlrath, W.J. Response of yellow-stripe maize mutant (ys_1) to ferrous and ferric iron. *Bot. Gaz.* **120**, 36–39, 1958.

52. Bell, W.D., Bogorad, L. and McIlrath, W.J. Response of yellow-stripe in maize. I. Effect of ys_1 locus on uptake and utilization of iron. *Bot. Gaz.* **124**, 1–8, 1962.

53. Warner, R.L., Hageman, R.H., Dudley, J.W. and Lambert, R.J. Inheritance of nitrate reductase activity in *Zea mais. Proc. Nat. Acad. Sci.* **62**, 785–792, 1969.

54. Hageman, R.H. *Univ. Ill. Coll. Agric. Spec. Pub.* **36**, 1975.

Nitrogen

55. Hoener, I.R. and De Turk, E.E. The absorption and utilization of nitrate nitrogen during vegetative growth by Illinois high and Illinois low protein corn. *J. Amer. Soc. Agron.* **30**, 232, 1938.

56. Hageman, R.H., Flesher, D. and Gitter, A. Diurnal variation and other light effects influencing the activity of nitrate reduction and nitrogen metabolism in corn. *Crop Sci.* **1**, 201, 1961.

57. Vose, P.B. and Breese, E.L. Genetic variation in the utilization of nitrogen by ryegrass species, *Lolium perenne* and *L. multiflorum*. *Ann Bot.* **28**, 251, 1964.

58. Picciurro, G., Ferrandi, L., Bonifoti, R. and Bracciocurti, G. Uptake of ^{15}N-labelled NH_4^+ in excised roots of a *durum* wheat mutant line compared with *durum* and bread wheat. *Isotopes in Plant Nutrition and Physiology Proc. IAEA Symp.*, Vienna, 5–9 September 1966, 511, IAEA, 1967.

59. Brunetti, N., Picciurro, G. and Boniforti, R. Atti Simp. *Int. Agrochim.* **8**, 239, 1971.

60. Smith, F.A. *New Phytol.* **77**, 769, 1973.

61. Maynard, D.M. and Barker, A.V. *Comm. Soil Sci. and Plant Anal.* **2**, 461–470, 1971.

62. Sistrunk, W.A. and Cash, J.N. *J. Amer. Soc. Hort. Sci.* **100**, 307–309, 1975.

63. Haroon, M. *Ph.D. Thesis.* Univ. of Wisconsin, Diss. *Abstr.* **36**, 331B, 147 pp., 1975.

64. Chevalier, P. and Schrader, L.E. *Crop Sci.* **17**, 1977.

65. Schrader, L.E. Uptake, accumulation, assimilation and transport of nitrogen in high plants. *Nitrogen in the Environment.* Vol. 2, Nielsen, D.R. and MacDonald, J.G. (eds.), 101–141, Academic Press, New York, 1978.

66. Neyra, C.A. and Hageman, R.H. *Plant Physiol.* **56**, 692, 1975.

67. Jackson, W.A. Nitrate acquisition and assimilation by high plants: processes in the root system. *Nitrogen in the Environment.* Vol. 2, Nielsen, D.R. and MacDonald, J.G. (eds.), 45–48, Academic Press, New York, 1978.

68. Huffaker, R.C. and Rains, D.W. Factors influencing nitrate acquisition by plants; assimilation and fate of reduced nitrogen. *Nitrogen in the Environment.* Vol. 2, Nielsen, D.R. and MacDonald, J.G. (eds.), 1–43, Academic Press, New York, 1978.

69. Rao, K.P., Rains, D.W., Qualset, C.O. and Huffaker, R.C. *Crop Sci.*, 1978.

70. Gallagher, L.W. Genetic variation in nitrate reductase activity in two wheat genotypes (*Triticum aestivum*) and its relation to nitrogen assimilation. *Ph.D. Thesis.* Univ. Calif. Davis, 1976.

71. Qualset, C.O. *et al. Unpublished data.* Cited by Huffaker, R.C. and Rains, D.W., ref. 68.

72. Seth, J., Herbert, T.T. and Middleton, G.K. Nitrogen utilization in high and low protein wheat varieties. *Agron. J.* **52**, 207–209, 1960.

73. Johnson, V.A., Mattern, P.J. and Schmidt, J.W. Nitrogen relations during spring growth in varieties of *Triticum aestivum*. *Crop Sci.* **7**, 664–667, 1967.

74. McNeal, F.H., Berg, M.A. and Watson, C.A. Nitrogen and dry matter in five spring wheat varieties at successive stages of development. *Agron. J.* **58**, 605–608, 1966.

75. Burkholder, P.R. and McVeigh, I. Growth and differentiation of maize in relation to nitrogen. *Amer. J. Bot.* **27**, 414–424, 1940.

76. Baba, I. Mechanism of response to heavy manuring in rice varieties. *Internat. Rice Comm. News Letter*, FAO, **10**, No. 4, 9–16, 1961.

77. Yamada, N. The nature of fertilizer response in *japonica* and *indica* rice varieties. *Internat. Rice Comm. News Letter*, FAO, **8**, 14–19, 1959.

78. Holmes, W. and MacLusky, D.S. The intensive production of herbage for crop drying. Part 6. A study of the effect of intensive N fertilizer treatments on species and strains of grass. *J. Agric. Sci.* **46**, 267–286, 1955.

79. Hunt, I.V. Comparative productivity of 27 strains of grasses in the West of Scotland. Part 1. Total production for three years. *Brit. Grassl. Soc.* **11**, 49–55, 1956.

80. Antonovics, J., Lovett, J. and Bradshaw, A.D. *Isotopes in Plant Nutrition and Phsyiology.* Proc. IAEA Symposium, Vienna 5–9 September 1966, 549, 1967.

81. Frey, K.J. Yield components in oats. II. The effects of nitrogen fertilization. *Agron. J.* **51**, 605–608, 1959.

82. Ruck, H.C. and Bolas, B.D. Studies in the comparative physiology of apple rootstocks. I. The effect of nitrogen on the growth and assimilation of Malling apple rootstocks. *Ann. Bot.* Lond. **20**. 57–68, 1956.

83. Mikesell, M.E. and Paulsen, G.M. Nitrogen translocation and the role of individual leaves in protein accumulation in wheat grain. *Crop Sci.* 11, 919–922, 1971.

84. McKee, H.S. *Encyclopaedia Plant Physiol.* 8, 564, 1958.

85. Merzari, A.H. and Broeshart, H. *Isotope Studies on the Nitrogen Chain. Proc. Symp. IAEA, Vienna, 1967,* 79, IAEA, 1968.

86. Broeshart, H. *Nitrogen-155 in Soil-Plant Studies. Proceedings of Research Coordination Meeting, Sofia, December 1969,* 47, IAEA, Vienna, 1971.

87. Reisenauer, H.M. Absorption and utilization of ammonium nitrogen by plants. *Nitrogen in the Environment, Vol. 2,* Nielsen, D.R. and MacDonald, J.G. (eds.), 157–170, Academic Press, New York, 1978.

88. Hiatt, A.J. Critique of absorption and utilization of ammonium nitrogen by plants. *Nitrogen in the Environment, Vol. 2,* Nielsen, D.R. and MacDonald, J.G. (eds.), 191–199, Academic Press, New York, 1978.

89. Van Egmond, F. Nitrogen nutritional aspects of the ionic balance of plants. *Nitrogen in the Environment, Vol. 2,* Nielsen, D.R. and MacDonald, J.G. (eds.), 171–189, Acad. Press, New York, 1978.

90. Chibnall, A.C. *Protein metabolism in the Plant.* Yale University Press, 1939.

91. Mothes, K. *Planta,* 30, 726, 1940.

92. Archibald, R.M. *Chem. Rev.* 37, 726, 1945.

93. Kumazawa, K. *Final Res. Rept. Mimeo* 1–25, IAEA Laboratory, Seibersdorf, Vienna, 1969.

94. Birecka, H. *Final Res. Rept. Mimeo* 1–15, IAEA Laboratory, Seibersdorf, Vienna, 1969.

95. Joy, K.W. *Plant Physiol.* 44, 849, 1969.

96. Dobrunov, L.G. and Sarsenbaev, B. Variety differences in nitrogenous nutrition of rice. *Variety and Nutrition,* 159–168, Klimoshevsky, E.L. (ed.), Siberian Inst. Pl. Phus. Biochem. Irkutsk, 1974 (in Russian, English summary).

97. Takahashi, Y., Iwata, I. and Baba, I. Studies on the varietal adaptability for heavy manuring in rice. I. Varietal differences in nitrogen and carbohydrate metabolism affected by different supply of nitrogen. *Proc. Crop. Sci. Soc. Japan* 28, 22–24, 1959.

98. Nicholas, D.J.D. and Nason, A. Mechanism of action of nitrate reductase from *Neurospora. J. Bio. Chem.* 211, 183–197, 1954.

99. Hewitt, E.J. and Jones, E.W. The production of molybdenum deficiency in plants in sand culture with special reference to tomato and *Brassica* crops. *J. Pomol.* 23, 254–262, 1947.

100. Mulder, E.G. Importance of molybdenum in the nitrogen metabolism of microorganisms and higher plants. *Plant and Soil* 1, 94–119, 1948.

101. Amaral, F. de Araujo Lopes do Eficiencia de Utilicação de Nitrogenio, Fósforo e Potássio de 104 Variedades de Feijoeiro, *Phaseolus vulgaris. Ph.D. Thesis,* ESALQ, Piracicaba, University of São Paulo, 1975.

102. Stoy, V. Use of tracer techniques to study yield components in seed crops. *Tracer Techniques for Plant Breeding. Proc. Panel Vienna 1974,* 43–55, IAEA, Vienna, 1975.

103. Austin, R.B. and Longden, P.C. A rapid method for the measurement of rates of photosynthesis using $^{14}CO_2$. *Ann. Bot.* 31, 245, 1969.

104. Austin, R.B., Ford, M.A. and Edrich, J.A. *Photosynthesis, translocation and grain filling in wheat.* Rep. Pl. Breed. Inst. Cambridge, 1973, 159, 1974.

105. Stoy, V. Photosynthesis, respiration, and carbohydrate accumulation in spring wheat in relation to yield. *Physiol. Plant Suppl. IV,* 1, 1965.

106. Stoy, V. The translocation of ^{14}C-labelled photosynthetic products from the leaf to the ear in wheat. *Physiol. Plant* 16, 851, 1963.

107. Lupton, F.G.H. The analysis of grain yield of wheat from measurements of photosynthesis and translocation in the field. *Ann. Appl. Biol.* 64, 363, 1969.

108. Lupton, F.G.H. Further experiments in photosynthesis and translocation in wheat. *Ann. Appl. Biol.* 71, 69, 1972.

109. Ruckenbauer, P. Yielding ability and translocation pattern of $^{14}CO_2$ labelled assimilated in contrasting wheat varieties. *Trans. 3rd Symposium of Accumulation and Translocation of Nutrients and Regulators, Warsaw, 14–18 May, 1973,* 124, 1973.

110. Istatkov, St., Mladenova, Y. Investigations with isotopes in the heterosis phenomenon in maize. *Nitrogen-15 in Soil-Plant Studies, Research Coordination Meeting, Sofia,* 227–238, IAEA, 1971.

111. Asana, R.D., Ramaiah, P.K. and Rao, M.V.K. The uptake of nitrogen, phosphorus and potassium by three cultivars of wheat in relation to growth and development. *Indian J. Plant Physiol.* **9** **2**, 85–107, 1968.

Phosphorus

112. De Turk, E.E., Holbert, J.R. and Howk, B.W. Chemical transformations of phosphorus in the growing corn plant, with results on two first generation crosses. *J. Agric. Res.* **46**, 121–141, 1933.
113. Smith, S.N. Response of inbred lines and crosses in maize to variations of N and P supplied as nutrients. *J. Amer. Soc. Agron.* **26**, 785–804, 1934.
114. Whiteaker, D., Gerloff, G.C., Gabelman, W.H. and Lindgren, D. Intraspecific Differences in Growth of Beans at Stress Levels of Phosphorus. *J. Amer. Soc. Hort. Sci.* **101**, 472–475, 1976.
115. Lindgren, D.T., Gableman, W.H. and Gerloff, G.C. Variability of phosphorus uptake and translocation in *Phaseolus vulgaris* under phosphorus stress. *J. Amer. Hort. Sci.* **102**, 674–677, 1977.
116. Howell, R.W. Phosphorus nutrition of soybeans. *Plant Physiol.* **29**, 477–483, 1954.
117. Howell, R.W. and Bernard, R.L. Phosphorus response of soybean varieties. *Crop Sci.* **1**, 311–313, 1961.
118. Bernard, R.L. and Howell, R.W. Inheritance of phosphorus sensitivity in soybeans. *Crop Sci.* **4**, 298–299, 1964.
119. Howell, R.W. and Foote, B.D. Phosphorus tolerance and sensitivity of soybeans as related to uptake and translocation. *Plant Physiol.* **39**, 610–613, 1964.
120. Dunphy, E.J., Kurtz, L.T. and Howell, R.W. Responses of Different Lines of Soybeans to High Levels of Phosphorus and Potassium Fertility. *Soil Sci. Soc. Amer. Proc.* **32**, 383–385, 1968.
121. Dunphy, E.J., Meissen, M.E., Fulcher, C.E. and Kurtz, L.T. Tolerance of soybean varieties to high levels of phosphorus fertilizer. *Soil Sci. Soc. Proc.* **30**, 233–236, 1966.
122. Rice, R. Physiology and inheritance of differential growth response under high phosphorus levels among different lines of beans, *Phaseolus vulgaris*. *Ph.D. Thesis*, Univ. Wisconsin, Madison, *Diss. Abstr.* **35** **11**, 5220B.
123. Peterson, G.A. *Ph.D. Thesis*, Iowa State University, 1967.
124. Hughes, J.L. *Ph.D. Thesis*, University Nebraska, 1971.
125. Baker, D.E., Bradford, R.R. and Thomas, W.I. Accumulation of Ca, Sr, Mg, P and Zn by genotypes of corn, *Zea mais*, under different soil fertility levels. *Isotopes in plant nutrition and physiology*, 465–477, IAEA, Vienna, Austria, 1967.
126. Baker, D.E., Jarrell, A.E., Marshall, L.E. and Thomas, W.I. Phosphorus uptake from soil by corn hybrids selected for high and low phosphorus accumulation. *Agron. J.* **62**, 103–106, 1970.
127. Baker, D.E., Wooding, F.J. and Johnson, M.W. Chemical element accumulation by populations of corn, *Zea mais*, selected for high and low accumulation of P. *Agron. J.* **63**, 404–406, 1971.
128. Läuchli, A. Genotypic variation in transport, 372–393. *Encyclopedia of Plant Physiology*. V. Lüttge and M.G. Pitman (eds.), New Series, **2**, part B. Springer-Verlag, Berlin, 1976.
129. Jowett, D. Adaptation of a lead-tolerant population of *Agrostis tenuis* to low soil fertility. *Nature*, Lond. **184**, 43, 1959.
130. Robinson, R.R. Mineral content of various clones of white clover when grown on different soils. *J. Amer. Soc. Agron.* **34**, 933–939, 1942.
131. Clark, R.B. and Brown, J.C. Differential phosphorus uptake by phosphorus-stressed corn inbred. *Crop Sci.* **14**, 505–508, 1974.
132. Woolhouse, H.N. Differences in the properties of the acid phosphatases of plant roots and the significance in the evolution of edaphic ecotypes. *Ecological aspects of the mineral nutrition of plants*. I.H. Ronson (ed.), 357–380, *Brit. Ecol. Soc. Symp.* **9**, Blackwell, Oxford.
133. Clark, R.B. and Brown, J.C. Corn lines differ in mineral efficiency. *Ohio Report* **60** (5):83–86, 1975.
134. Brown, J.C., Clark, R.B. and Jones, W.E. Efficient and inefficient use of phosphorus by sorghum. *Soil Sci. Soc. Amer. J.* **41**, 747–750, 1977.
135. Brown, J.C. and Jones, W.E. Phosphorus efficiency as related to iron inefficiency in sorghum. *Agron. J.* **67**, 468–472, 1975.
136. Brown, J.C. and Jones, W.E. Fitting plants nutritionally to soils. III. Sorghum. *Agron. J.* **69**, 410–414, 1977.

137. Clark, R.B., Maranville, J.W. and Ross, W.M. *Differential phosphorus efficiency in sorghum. Abstract 77—843, Abstracts Series,* Nebraska Agricultural Experiment Station, 1977.

138. Ferreyra Hernandez, F.F. Absorcão de Fósforo em Raizes Destacadas de Milho (*Zea mais* L.): Diferencas entre Hibridos e Linhagens. *Ph.D. Thesis,* ESALQ, Piracicaba, University of São Paulo, 1978.

139. Barber, S.A. The role of root interception, mass-flow and diffusion in regulating the uptake of ions by plants from soil. *IAEA Tech. Rept.* No. 65, 39—45, IAEA, Vienna, 1966.

140. Barber, S.A., Walker, J.M. and Vasey, E.H. Mechanisms for the movement of plant nutrients from the soil and fertilizer to the plant root. *Agric. Food Chem.* 11, 204, 1963.

141. Jenny, H. *Growth in Living Systems.* Zarrow (ed.), Basic Books Inc., 1961.

142. Lavy, L.T. and Barber, S.A. Movement of molybdenum in the soil and its effect on availability to the plant. *Soil Sci. Soc. Amer. Proc.* 28, 93, 1964.

143. Olsen, S.R. Phosphorus diffusion to plant roots. *Plant Nutrient Supply and Movement. IAEA Tech. Rept.* 48, 130—139, 1965.

144. Passioura, J.A. *Plant and Soil* 18, 225, 1963.

145. Seay, W.A. and Henson, L. Variability in nutrient uptake and yield of clonally propagated Kenland red clover. *Agron. J.* 50, 165—168, 1958.

146. Troughton, A. *Report on the work carried out during the tenure of a national research fellowship, New Zealand, July 1958—July 1959,* 1959.

147. Nye, P.H. The relation between the radius of a root and its nutrient absorbing power (α): some theoretical considerations. *J. Expt. Bot.* 24, 783—786, 1973.

148. Evans, H. Studies on the absorbing surface of sugar-cane root systems. I. Method of study with some preliminary results. *Ann. Bot.* 2, 159—182, 1938.

149. Snaydon, R.W. and Bradshaw, A.D. Differences between natural populations of *Trifolium repens* in response to mineral elements. I. Phosphate. *J. Expt. Bot.* 13, 422—434, 1962.

150. Vose, P.B. *Unpublished,* 1960.

151. Raper, C.D. and Barber, S.A. Rooting systems of soybeans. I. Differences in root morphology among varieties. *Agron. J.* 62, 581—584, 1970.

152. Raper, C.D. and Barber, S.A. Rooting systems of soybeans. II. Physiological effectiveness as nutrient absorption surfaces. *Agron. J.* 62, 585—588, 1970.

153. Haahr, V. Nuclear methods for detecting root activity. *Tracer Techniques for Plant Breeding. Panel Proc. 1974,* 57—63, IAEA, Vienna, 1975.

154. Newbould, P., Ellis, F.B., Barnes, B.T., Howse, K.R. and Lupton, F.G.H. Intervarietal comparison of the root systems of winter wheat. Agricultural Research Council, Letcombe Laboratory, *Ann. Rept. 1969,* 38, 1970.

155. Subbiah, B.V., Katyal, J.C., Narasimham, R.L. and Dakshinamurti, C. Preliminary investigations on root distribution of high yielding wheat varieties. *Int. J. Appl. Radiat. Isotopes* 19, 385, 1968.

156. Swaminathan, M.S. The role of nuclear techniques in agricultural research in developing countries. *Proc. 4th Int. Conf. Peaceful Uses of Atomic Energy, Geneva, 1971,* 12, 3—32, IAEA, Vienna, 1972.

157. Vedrov, N.G. Root system of cereal crops as an object of breeding, 139—144. *Variety and Nutrition,* Klimashevsky, E.L. (ed.), Siberian Inst. Pl. Phys. Biochem. Irkutsk, 1974 (in Russian, English summary).

158. Kirichenko, F.G., Danilichuk, P.V. and Kostenko, A.I. On method of plant breeding according to the vigour of root growth, 93—104. *Variety and Nutrition,* Klimashevsky, E.L. (ed.), Siberian Inst. Pl. Phys. Biochem. Irkutsk, 1974 (in Russian, English summary).

159. Vasilchuk, N.S. and Kumakov, V.A. Development and physiological activity of the roots of different millet varieties, 145—152. *Variety and Nutrition,* Klimashevsky, E.L. (ed.), Siberian Inst. Pl. Phys. Biochem. Irkutsk, 1974 (in Russian, English summary).

160. Vose, P.B. Nutritional response and shoot/root ratio as factors in the composition and yield of genotypes of perennial ryegrass, *Lolium perenne. Ann. Bot.* 26, 425—437, 1962.

161. Paulsen, G.M. and Rotimi, O.A. Phosphorus-zinc interaction in two soybean varieties differing in sensitivity to phosphorus nutrition. *Soil Sci. Soc. Amer. Proc.* 32, 73—76, 1968.

Calcium, Potassium and Magnesium

162. Hoffer, G.N. Some differences in the functioning of selfed lines of corn under varying nutritional conditions. *J. Amer. Soc. Agron.* **18**, 322–334, 1926.

163. De Turk, E.E. Plant Nutrient Deficiency Symptoms. Physiological Basis. *Ind. Eng. Chem.* **33**, 648–653, 1941.

164. Tisdale, H.B. and Dick, J.B. Cotton wilt in Alabama as affected by potassium supplements and as related to varietal behaviour and other important agronomic problems. *J. Am. Soc. Agron.* **34**, 405–426, 1942.

165. Lingle, J.C. and Lorenz, O.A. Potassium nutrition of tomatoes. *J. Amer. Soc. Hort. Sci.* **94**, 679–683, 1969.

166. Gerloff, G.C. Plant efficiencies in the use of nitrogen, phosphorus and potassium. *Proc. Workshop, Beltsville, 22–23 November 1976*, Wright, M.J. (ed.), 161–173. Cornell Univ. Agr. Exp. Sta. Spec. Pub., Ithaca, New York, 1977.

167. Shea, P.F., Gerloff, G.C. and Gabelman, W.H. Differing efficiencies of potassium utilization in strains of snap beans, *Phaseolus vulgaris* L. *Plant and Soil* **28**, 337–346, 1968.

168. Baker, L. Inheritance and basis for efficiency of potassium utilization in the red beet, *Beta vulgaris* L. *Ph.D. Thesis*, Univ. of Wisconsin, Madison (Diss. Abstr. 29, 494–495B), 1968.

169. Snaydon, R.W. Micro-distribution of *Trifolium repens* L. and its relation to soil factors. *J. Ecol.* **50**, 133–143, 1962a.

170. Bradshaw, A.D. and Snaydon, R.W. Population differentiation within plant species in response to soil factors. *Nature* **183**, 129–130, 1959.

171. Snaydon, R.W. and Bradshaw, A.D. Differential response to calcium within the species *Festuca ovina* L. *New Phytol.* **60**, 219–224, 1961.

172. Snaydon, R.W. and Bradshaw, A.D. Differences between natural populations of *Trifolium repens* L. in response to mineral nutrients. II. Calcium, magnesium and potassium. *J. Appl. Ecol.* **6**, 185–202, 1969.

173. Clarkson, D.T. Calcium uptake by calcicole and calcifuge species in the genus *Agrostis* L. *J. Ecol.* **53**, 427–435, 1965b.

174. Vose, P.B. and Koontz, H.V. The uptake of strontium and calcium from soils by grasses and legumes and the possible significance in relation to Sr-90 fallout. *Hilgardia* **29**, 575–585, 1960.

175. Vose, P.B. and Jones, D.G. The interaction of manganese and calcium on nodulation and growth in varieties of *Trifolium repens*. *Plant Soil* **18**, 372–385, 1963.

176. Lineberry, R.A. and Burkhart, L. Nutrient deficiencies in the strawberry leaf and fruit. *Plant Physiol.* **18**, 324–333, 1943.

177. Clark, R.B. and Brown, J.C. Differential mineral uptake in maize inbreds. *Commun. Soil Sci. Plant Anal.* **5**, 213–227, 1974a.

178. Clark, R.B. Plant efficiencies in the use of calcium, manganese and molybdenum. *Proc. Workshop, Beltsville, 22–23 November 1976*, Wright, M.J. (ed.), 195–191. Cornell Univ. Agr. Exp. Sta. Spec. Pub., Ithaca, New York, 1977.

179. Bruetsch, T.E. and Estes, G.O. Genotype variation in nutrient uptake efficiency in corn. *Agron. J.* **68**, 521–523, 1976.

180. Rivard, C.E. and Bandel, V.A. Effect of variety on nutrient composition of field corn. *Commun. Soil Sci. Plant Anal.* **5**, 229–242, 1974.

181. Brown, J.C. Differential uptake of Fe and Ca by two corn genotypes. *Soil Sci.* **103**, 331–338, 1967.

182. Greenleaf, W.H. and Adams, F. Genetic control of blossom-end rot disease in tomatoes through calcium metabolism. *J. Am. Soc. Hort. Sci.* **94**, 248–250, 1969.

183. Maume, L. and Dulac, J. Différences variétales, dans l'absorption de l'azote, de l'acide phosphorique et de la potasse par des blés ayant atteint use meme époque physiologique dans un meme milieu. *C.R. Acad. Sci.* Paris **198**, 149–202, 1934.

184. Maume, L. and Dulac, J. Absorption de N, P_2O_5, K_2O, CaO, MgO par différentes variétés de blé observés la meme année dans un meme milieu. *Ann. Ec. Agric.* Montpellier, **23**, 96–103, 1934.

185. Gorsline, G.W., Thomas, W.I. and Baker, D.E. Relationship of strontium-calcium within corn. *Crop Sci.* **4**, 154–156, 1964.

186. Baker, D.E., Thomas, W.I. and Gorsline, G.W. Differential accumulation of strontium, calcium and other elements by corn *Zea mais* L. under greenhouse and field conditions. *Agron. J.* 56, 352–355, 1964.

187. Gorsline, G.W., Baker, D.E. and Thomas, W.I. Accumulation of eleven elements by field corn *Zea mais* L. *Pennsylvania State Agric. Exp. Sta. Bull.* 725, 1965.

188. Bradford, R.R., Baker, D.E. and Thomas, W.I. Effect of soil treatments on chemical element accumulation of four corn hybrids. *Agron. J.* 58, 614–617, 1966.

189. Thomas, W.I. and Baker, D.E. Application of biochemical genetics in quality improvement and plant nutrition. II. Studies in the inheritance of chemical element accumulation in corn *Zea mais* L. *Qual. Plant Mat. Veget.* 13, 98–104, 1966.

190. Craig, W.E. and Thomas, W.I. Prediction of chemical accumulation in maize double-cross hybrids from single-cross data. *Crop Sci.* 10, 609–610, 1970.

191. Rasmusson, D.C., Smith, L.H. and Myers, W.M. Effect of genotype on accumulation of strontium-89 in barley and wheat. *Crop Sci.* 3, 34–37, 1963.

192. Smith, L.H., Rasmusson, D.C. and Myers, W.M. Influence of genotype upon relationship of strontium-89 to calcium in grain of barley and wheat. *Crop Sci.* 3, 386–389, 1963.

193. Rasmusson, D.C., Smith, L.H. and Kleese, R.A. Inheritance of Sr-89 accumulation in wheat and barley. *Crop Sci.* 4, 586–589, 1964.

194. Young, W. and Rasmusson, D.C. Variety differences in strontium and calcium accumulation in seedlings of barley. *Agron. J.* 58, 481–483, 1966.

195. Rasmusson, D.C. and Kleese, R.A. Isogenic analysis of strontium-89 accumulation in barley. *Crop Sci.* 7, 617–619, 1967.

196. Kleese, R.A., Rasmusson, D.C. and Smith, L.H. Genetic and environmental variation in mineral element accumulation in barley, wheat and soybeans. *Crop Sci.* 8, 591–593, 1968.

197. Rasmusson, D.C., Hester, A.J., Fick, G.N. and Byrne, I. Breeding for mineral content in wheat and barley. *Crop Sci.* 11, 623–626, 1971.

198. Kleese, R.A. Relative importance of stem and root in determining genotype differences in Sr-89 and Ca-45 accumulation in soybeans *Glycine max* L. *Crop Sci.* 7, 53–55, 1967.

199. Kleese, R.A. Scion control of genotypic differences in Sr and Ca. Accumulation in soybeans under field conditions. *Crop Sci.* 8, 128–219, 1968.

200. Wallace, T. Chemical investigations relating to magnesium deficiency of fruit trees. *J. Pomol.* 18, 145–160, 1940.

201. Jones, J.O., Nicholas, D.J.D. and Wallace, T. Experiment on the control of magnesium deficiency in greenhouse tomatoes. *Progress report I. Rep. Agric. Hort. Res. Sta. Bristol for 1943*, 48–53, 1944.

202. Johnson, K.E.E., Davis, J.F. and Benne, E.J. Occurrence and control of magnesium-deficiency symptoms in some common varieties of celery. *Soil Sci.* 91, 203–207, 1961.

203. Vose, P.B. and Griffiths, D.J. Varietal differences in the manganese and magnesium nutrition of oats, Paper presented to British-Dutch-Scandinavian Biology Meeting, SEB, Amsterdam, April, 1968.

204. Sayre, J.D. Mineral nutrition of corn. *Corn and Corn Improvement.* Sprague, G.F. (ed.), 293–294. Academic Press, New York, 1955.

205. Sayre, J.D. Magnesium needs of inbred corn lines, 17, *69th Ann. Rept. Ohio Agric. Exp. Sta. Bull.* 705, 1951.

206. Foy, C.D. and Barber, S.A. Magnesium absorption and utilization by two inbred lines of corn. *Soil Sci. Am. Proc.* 22, 57–62, 1958.

207. Schauble, C.E. and Barber, S.A. Magnesium immobility in the nodes of certain corn inbreds. *Agron. J.* 50, 651–653, 1958.

208. Clark, R.B. Differential magnesium efficiency in corn inbreds. I. Dry-matter yields and mineral element composition. *Soil. Sci. Soc. Am. Proc.* 39, 488–491, 1975.

209. Pinkas, L.L.H. and Smith, L.H. Physiological basis of differential strontium accumulation in two barley genotypes. *Plant Physiol.* 41, 1471–1475, 1966.

210. Millaway, R.M. and Schmid, W.E. Genetic control of rubidium absorption by excised corn roots: a preliminary survey using several inbred varieties. *Transact. Illinois State Acad. Sci.* 60, 250–258, 1967.

211. Picciurro, G. and Brunetti, N. Assorbimento del Sodio (Na-22) in radici escisse di alcune varietà di *Lycopersicum esculentum. Agrochimica* 13, 347–357, 1969.

212. Butler, G.W. and Johns, A.T. Some aspects of the chemical composition of pasture herbage in relation to animal production in New Zealand. *J. Aust. Inst. Agric. Sci.* 27, 123–133, 1961.
213. Vose, P.B. The cation content of perennial ryegrass *Lolium perenne* in relation to intraspecific variability and nitrogen/potassium interaction. *Plant and Soil* 19, 49–64, 1963.
214. Epstein, E. Dual pattern of ion absorption by plant cells and by plants. *Nature* 212, 1324–1327, 1966.
215. Mouat, M.C.H. Genetic variation in root cation-exchange capacity of ryegrass. *Plant Soil* 16, 263–265, 1962.
216. Browne, C.A. "Soils and Men" *U.S. Dept. Agr. Yearbook*, pp. 777–806. US Govt. Printing Office, Washington, D.C.
217. Butler, G.W., Barclay, P.C. and Glenday, A.C. Genetic and environmental differences in the mineral composition of ryegrass herbage. *Plant and Soil* 16, 214–228, 1962.
218. Drake, M., Vengris, I. and Colby, W.G. Cation exchange capacity of plant roots. *Soil Science* 72, 139–147, 1951.
219. McLean, E.O., Adams, D. and Franklin, R.S. Cation exchange capacity of plant roots as related to nitrogen contents. *Proc. Soil Sci. Soc. Amer.* 20, 345–347, 1956.
220. Knight, A.H., Crooke, W.M. and Inkson, R.H.E. Cation exchange capacities of tissues of higher and lower plants and their related uronic acid contents. *Nature* 192, 142–143, 1961.
221. Eaton, F.M. and Blair, G.Y. Accumulation of boron by reciprocally grafted plants. *Plant Phys.* 10, 411–424, 1935.
222. Lagatu, H. and Maume, L. Influence de la nature du porte-greffe sur le mode d'alimentation NPK de la vigne greffe. *C.R. Acad. Sci.* Paris 209, 281–284, 1939.
223. Pearse, H.L. Water culture studies with apple trees. II. The seasonal absorption of nitrogen and potassium by Cox's orange pippin on Malling rootstocks. IX and XII, *J. Pomol.* 17, 344–361, 1940.
224. Sinclair, W.E. and Bartholomew, E.T. Effects of rootstock and environment on the composition of oranges and grapefruit. *Hilgardia* 16, 125–176, 1944.
225. Haas, A.R. Effect of the rootstock on the composition of citrus trees and fruit. *Plant Phys.* 23, 309–330, 1948.
226. Smith, P.F., Reuther, W. and Specht, A.W. The influence of rootstock on the mineral composition of Valencia orange leaves. *Plant Physiol.* 24, 455–461, 1949.
227. Cooper, W.C., Gorton, B.S. and Olson, E.O. Ionic accumulation in citrus as influenced by rootstock and scion and concentration of salts and boron in the substrate. *Plant Physiol.* 27, 191–203, 1952.
228. Wallace, A. and Smith, R.L. Rootstock influence on the potassium, calcium, magnesium nutrition of citrus. *Better Crops with Plant Food* 39, 9–14, 1955.
229. Bernstein, L., Brown, J.W. and Hayward, H.E. The influence of rootstock on growth and salt accumulation in stone-fruit trees and almonds. *Proc. Am. Soc. Hort. Sci.* 68, 86–95, 1956.
230. Embleton, T.W., Matsumura, M., Storey, W.B. and Garber, M.J. Chlorine and other elements in avocado leaves as influenced by rootstock. *Proc. Am. Soc. Hort. Sci.* 80, 230–236, 1962.
231. Awad, M.M. and Kenworthy, A.L. Clonal rootstock, scion variety and time of sampling influences in apple leaf composition. *Proc. Am. Soc. Hort. Sci.* 83, 68–73, 1963.
232. Cook, J.A. and Lider, L.A. Mineral composition of bloomtime grape petiole in relation to rootstock and scion variety behaviour. *Proc. Am. Soc. Hort. Sci.* 84, 243–254, 1964.
233. Hewitt, A.A. and Furr, J.R. Uptake and loss of chloride from seedlings of selected *Citrus* rootstock varieties. *Proc. Am. Soc. Hort. Sci.* 86, 194–200, 1965.
234. Sistrunk, J.W. and Campbell, R.W. Calcium content differences in various apple cultivars as affected by rootstock. *Proc. Am. Soc. Hort. Sci.* 88, 38–40, 1966.
235. Wutscher, H.K., Olson, E.O., Shull, A.V. and Peynado, A. Leaf nutrient levels, chlorosis, and growth of young grapefruit trees on 16 rootstocks grown on calcareous soil. *J. Am. Soc. Hort. Sci.* 95, 259–261, 1970.
236. Roach, W.A. Distribution of molybdenum. *Nature*, Lond. 131, 202, 1933.
237. Mellado, L. Absorcion y Transporte de Fosforo en Plantas de Vid (*Vitis*) Injertadas. *Proceedings of Symposium on Isotopes and Radiation in Soil Plant Nutrition Studies. Ankara 1965.* 241, IAEA, Vienna, 1965.

Secondary elements: Screening techniques

238. Randhawa, N.S. Screening of crop varieties with respect to micronutrient stresses in India. *Proc. Workshop, Beltsville, 22—23 November 1976*, Wright, M.J. (ed.), 393—400. Cornell Univ. Agr. Exp. Sta. Spec. Pub., Ithaca, New York, 1977.

239. Agarwala, S.C., Sharma, C.P. and Sharma, P.N. Susceptibility of some high yielding varieties of wheat to deficiency of micronutrients in sand culture. *Proc. Int. Symp. Soil Fert. Evaluation, New Delhi* 1, 1047—1064, 1971.

240. Agarwala, S.C. and Sharma, C.D. Non-pathogenic disorders of crop plants, specially of high yielding varieties. *Current Trends in Plant Pathology*, S.P. Raychauduri and J.P. Verman (eds.), 40—52, 1974.

241. Brown, J.C. and Jones, W.E. Fitting plants nutritionally to soils. I. Soybeans. *Agron. J.* 69, 399—404, 1977.

242. Brown, J.C. and Jones, W.E. Fitting plants nutritionally to soils. II. Cotton. *Agron. J.* 69, 405—409, 1977.

243. Brown, J.C., Ambler, J.E., Chaney, R.L. and Foy, C.D. Differential responses of plant genotypes to micronutrients. *Micronutrients in Agriculture*, 389—418, *Soil Sci. Soc. Amer.*, Madison, 1972.

244. Brown, J.C. Screening plants for iron efficiency. *Proc. Workshop, Beltsville 22—23 November 1976*, Wright, M.J. (ed.), 355—358. Cornell Univ. Agr. Exp. Sta. Spec. Pub., Ithaca, New York, 1977.

245. Brown, J.C. and Jones, W.E. A technique to determine iron efficiency in plants. *Soil Sci. Soc. Amer. J.* 40, 398—405, 1976.

246. Schander, H. Bericht uber individuelle Unterschode ernahrungsphysiologischen Art bei verschiedene species und ihre experimentelle Erfassung. *Ber. Deutsch. Bot. Ges.* 57, 29, 1939.

247. Dessureaux, L. The selection of lucene for tolerance to manganese toxicity. *Nutrition of the Legumes*, 277—279. Hallsworth, E.G. (ed.), Butterworths, London, 1958, 1958b.

248. Reid, D.A., Fleming, A.L. and Foy, C.D. A method for determining aluminium response of barley in nutrient solution in comparison to response in Al-toxic soil. *Agron. J.* 63, 600—603, 1971.

249. Reid, D.A. Screening barley for aluminium tolerance. *Proc. Workshop, Beltsville 22—23 November 1976*, Wright, M.J. (ed.), 269—275. Cornell Univ. Agr. Exp. Sta. Spec. Pub., Ithaca, New York, 1977.

250. Rhue, R.D. and Grogan, C.O. Screening corn for aluminium tolerance. *Proc. Workshop, Beltsville 22—23 November 1976*, Wright, M.J. (ed.), 297—310. Cornell Univ. Agr. Exp. Sta. Spec. Pub., Ithaca, New York, 1977.

251. Howeler, R.H. and Cadavid, L.F. Screening of the cultivars for tolerance to Al-toxicity in nutrient solutions as compared with a field screening method. *Agron. J.* 68, 551—555, 1976.

252. Moore, D.P. Screening wheat for aluminium tolerance. *Proc. Workshop, Beltsville 22—23 November 1976*, Wright, M.J. (ed.), 287—295. Cornell Univ. Agr. Exp. Sta. Spec. Pub., Ithaca, New York, 1977.

253. Campbell, L.G. and Lafever, H.N. Correlation of Field and Nutrient Culture Technique of Screening Wheat for Aluminium Tolerance. *Proc. Workshop, Beltsville 22—23 November 1976*, Wright, M.J. (ed.), 277—286. Cornell Univ. Agr. Exp. Sta. Spec. Pub., Ithaca, New York, 1977.

254. Konzak, C.F., Polle, E. and Kittrick, J.A. Screening several crops for aluminium tolerance. *Proc. Workshop, Beltsville 22—23 November 1976*, Wright, M.J. (ed.), 311—327. Cornell Univ. Agr. Exp. Sta. Spec. Pub., Ithaca, New York, 1977.

255. Foy, C.D. General principles involved in screening plants for aluminium and manganese tolerance. *Proc. Workshop, Beltsville 22—23 November 1976*, Wright, M.J. (ed.), 255—267. Cornell Univ. Agric. Exp. Sta. Spec. Pub., Ithaca, New York, 1977.

256. Neenan, M. The effects of soil acidity on the growth of cereals with particular reference to the differential reaction of varieties thereto. *Plant and Soil XII* No. 4, 324—338, June 1960.

257. Walker, J.C., Jolivette, J.P. and Hare, W.W. Varietal susceptibility in garden beet to boron deficiency. *Soil Sci.* 59, 461—464, 1945.

258 Stølen, O. Investigations on the tolerance of barley varieties to high hydrogen-ion concentration in the soil. *Royal Vet. and Agric. Coll. Yearbook* (Copenhagen) 81—107, 1965.

259. Vlamis, J. Acid soil infertility as related to soil-solution and solid-phase effects. *Soil Sci.* 75, 383—394, 1953.

260. Kelly, J.F. and Gabelman, W.H.   Variability in the mineral composition of red beet *Beta vulgaris* varieties in relation to boron nutrition. *Proc. Amer. Soc. Hort. Sci.* 76, 416–424, 1960.

261. Hayward, H.E. and Wadleigh, C.H.   Plant growth in saline and alkali soils. *Advanc. Agron.* 1, 1–38, 1949.

262. de Mooy, C.J.   *Iron deficiency in soybeans. What can be done about it?* Pm-531, Cooperative Extension Service, Iowa State University, Ames, IA 500010, 1972.

263. Epstein, E.   Genetic potentials for solving problems of soil mineral stress: adaptation of crops to salinity. *Proc. Workshop, Beltsville 22–23 November 1976*, Wright, M.J. (ed.), 73–82. Cornell Univ. Agr. Exp. Sta. Spec. Pub., Ithaca, New York, 1977.

264. Rush, D.W. and Epstein, E.   Genotype responses to salinity. *Plant Physiol.* 57, 162–166, 1976.

265. Epstein, E. and Norlyn, J.D.   Seawater-based crop production: a feasibility study. *Science* 147, 249–251, 1977.

266. Nieman, R.H. and Shannon, M.C.   Screening plants for salinity tolerance. *Proc. Workshop, Beltsville 22–23 November 1976*, Wright, M.J. (ed.), 359–367. Cornell Univ. Agr. Exp. Sta. Spec. Pub., Ithaca, New York, 1977.

267. Turner, R.G.   Heavy metal tolerance in plants. *Ecological Aspects of the Mineral Nutrition of Plants.* I.H. Rorison (ed.), Blackwell Scientific Publications, Oxford and Edinburgh, 399–410, 1969.

268. Antonovics, J., Bradshaw, A.D. and Turner, R.G.   Heavy metal tolerance in plants. *Adv. Ecol. Res.* 7, 1–85, 1971.

269. Walley, K., Khan, M.S.I. and Bradshaw, A.D.   The potential for evolution of heavy metal tolerance in plants. I. Copper and zinc tolerance in *Agrostis tenuis. Heredity* 32, 309–319, 1974.

270. Bradshaw, A.D.   The evolution of heavy metal tolerance and its significance in land reclamation. *Heavy metals and the environment.* T.C. Hutchinson (ed.), Toronto Univ. Press, Toronto, Canada, 1976.

271. Humphreys, M.O. and Bradshaw, A.D.   Heavy metal toxicities. *Proc. Workshop, Beltsville 22–23 November 1976*, Wright, M.J. (ed.), 95–105. Cornell Univ. Agr. Exp. Sta. Spec. Pub., Ithaca, New York, 1977.

272. Foy, C.D.   Effect of nutrient deficiencies and toxicities in plants: acid soil toxicity. *Handbook of Nutrition and Food* M. Rechcigl (ed.), CRC Press Inc., Florida. In preparation. *Mimeo available.*

273. Brown, J.C.   Genetically controlled chemical factors involved in absorption and transport of iron by plants. *Advances in Chemistry Series, No. 162, Bioinorganic Chemistry II*, 93–103, 1977.

274. Abel, G.H.   Inheritance of the capacity for chloride inclusion and chloride exclusion by soybeans. *Crop. Sci.* 9, 697–698, 1969.

275. Brown, J.C. and Hendricks, S.B.   Enzymatic activity as indication of copper and iron deficiencies in plants. *Plant Physiol.* 27, 651–659, 1952.

276. Brown, J.C., Holmes, R.S. and Tiffin, L.O.   Iron chlorosis in soybeans as related to the genotype of rootstock. *Soil Sci.* 86, 75–82, 1958.

277. Brown, J.C. and Tiffin, L.O.   Iron chlorosis in soybeans as related to the genotype of rootstock. 2. A relationship between susceptibility to chlorosis and capacity to absorb iron from iron chelate. *Soil Sci.* 89, 8–15, 1960.

278. Brown, J.C., Holmes, R.S. and Tiffin, L.O.   Iron chlorosis in soybeans as related to the genotype of rootstock. 3. Chlorosis susceptibility and reductive capacity at the root. *Soil Sci.* 91(2), 127–132, 1961.

279. Bell, W.D., Bogorad, I. and McIllrath, W.J.   Yellow-stripe phenotype in maize. *Bot. Gaz.* (Chicago) 124, 1–8, 1962.

280. Shim, S.C. and Vose, P.B.   Varietal differences in the kinetics of iron uptake by excised rice roots. *J. Expt. Bot.* 16, 216–232, 1965.

281. Brown, J.C., Weber, C.R. and Caldwell, B.E.   Efficient and inefficient use of iron by two soybean genotypes and their isolines. *Agron. J.* 59, 459–462, 1967.

282. Wallihan, E.F. and Garber, M.J.   Iron uptake by two *Citrus* rootstock species in relation to soil moisture and CaCO$_3$. *Agron. J.* 60, 50–52, 1968.

283. Brown, J.C.   Iron chlorosis in soybeans as related to the genotype of rootstock. 5. Differential distribution of photosynthetic C-14 as affected by phosphate and iron. *Soil Sci.* 105, 159–165, 1968.

284. Brown, J.C. and Bell, W.D.   Iron uptake dependent upon genotype of corn. *Soil Sci. Soc. Amer. Proc.* 33, 99–101, 1969.

285. Odurukwe, S.O. and Maynard, D.N. Mechanism of the differential response of Ind. Wf 9 and Ohio 40 B corn seedlings to iron nutrition. *Agron. J.* **61**, 694—697, 1969.

286. Wutscher, H.K., Olson, E.O., Shull, A.V. and Peynado, A. Leaf nutrient levels, chlorosis, and growth of young grapefruit trees on 16 rootstalks grown on calcareous soil. *J. Amer. Soc. Hort. Sci.* **95**, 259—261, 1970.

287. Brown, J.C., Chaney, R.L. and Ambler, J.E. A new tomato mutant inefficient in the transport of iron. *Physiol. Plant.* **25**, 48—53, 1971.

288. Ambler, J.E., Brown, J.C. and Gauch, H.G. Sites of iron reduction in soybean plants. *Agron. J.* **63**, 95—97, 1971.

289. Brown, J.C. and Ambler, J.E. "Reductants" released by roots of Fe-deficient soybeans. *Agron. J.* **65**, 311—314, 1973.

290. Mikesell, M.E., Paulsen, G.M., Ellis, Jr. R. and Casady, A.J. Iron utilization by efficient and inefficient sorghum lines. *Agron. J.* **65**, 77—80, 1973.

291. Clark, R.B., Tiffin, L.O. and Brown, J.C. Organic acids and iron translocation in maize genotypes. *Plant Physiol.* **52**, 147—150, 1973.

292. Brown, J.C. and Ambler, J.E. Iron-stress response in tomato *Lycopersicon esculentum* Mill., I. Sites of Fe reduction, absorption, and transport. *Physiol. Plant.* **31**, 221—224, 1974.

293. Brown, J.C. and Jones, W.E. Phosphorus efficiency as related to iron (in) efficiency in sorghum. *Agron. J.* **67**, 468—472, 1975.

294. Brown, J.C. and Jones, W.E. Nitrate reductase activity in calcifugous and calcicolous tomatoes as affected by iron stress. *Physiol. Plant* **38**, 273—277, 1976.

295. Foy, C.D., Voight, P.W. and Schwartz, J.W. Differential susceptibility of weeping lovegrass strains to an iron-related chlorosis on calcareous soils. *Agron. J.* **69**, 491—496, 1977.

296. Agarwala, S.C., Mehrotra, S.C., Sharma, C.P. and Bisht, S.S. Effect of iron deficiency on growth, chlorophyll, oxygen uptake, tissue iron and activity of certain enzymes in three varieties of chickpea. *J. Ind. Bot. Soc.* **56**, 1977.

297. Ovinge, A. Het optreden van kwade harten in Schokkers in Zeeland in 1934. *Landbouwk*, Tijdschr. **47**, 315—383, 1935.

298. Gallagher, P.H. and Walsh, T. The susceptibility of cereal varieties to manganese deficiency. *J. Agr. Sci.* **33**, 197—203, 1943.

299. Walsh, T. and Cullinan, S.J. Investigation on marsh spot disease in peas. *Min. Proc. R. Irish Acad. B.*, **50**, 279—285, 1945.

300. Lockard, R.G. Mineral nutrition of the rice plant in Malaya. *Bull. 108*, Dest. Agric. Federation of Malaya. Kuala Lumpur, 148, 1959.

301. Toms, W.J. Avon — an oat variety resistant to Mn deficiency. *J. Dept. Agric. W. Aust.* **8**(5), 523, 1959.

302. Millikan, C.R. Plant varieties and species in relation to the occurrence of deficiencies and excesses of certain nutrient elements. *J. Aust. Inst. Agric. Sci.* **27**, 220—233, 1961.

303. Vose, P.B. and Griffiths, D.J. Manganese and magnesium in the grey speck syndrome of oats. *Nature* **191**, 299—300, 1961.

304. Vose, P.B. Manganese requirement in relation to photosynthesis in *Avena*. *Phyton* **19**, 133—140, 1962.

305. Munns, D.N., Johnson, C.M. and Jackson, L. Uptake and distribution of manganese in oat plants. 1. Varietal variations. *Plant Soil* **19**, 115—126, 1963.

306. Munns, D.N., Jacobson, L. and Johnson, C.M. Uptake and distribution of manganese in oat plants. 2. A kinetic model. *Plant and Soil* **19**, 193—204, 1963 N.

307. Vose, P.B. The translocation and redistribution of manganese in *Avena*. *J. Exp. Bot.* **14**, 448—457, 1963.

308. Andrew, C.S. and Hegarty, M.P. Comparative responses to manganese excess of eight tropical and four temperate pasture legume species. *Aust. J. Agric. Res.* **20**, 687—696, 1969.

309. Nyborg, M. Sensitivity to manganese deficiency of different cultivars of wheat, oats, and barley. *Can. J. Plant Sci.* **50**, 198—200, 1970.

310. Brown, J.C. and Jones, W.E. Differential response of oats to manganese stress. *Agron. J.* **65**, 624—626, 1974.

311. Brown, J.C. and Jones, W.E. Manganese and iron toxicities dependent on soybean variety. *Comm. in Soil Sci. and Plant Analysis* **8**, 1—15, 1977.

312. Scott, L.E.   An instance of boron deficiency in the grape under field conditions. *Proc. Amer. Soc. Hort. Sci.* **38**, 375—378, 1941.
313. Scott, L.E.   Boron nutrition of the grape. *Soil Sci.* **57**, 55—65, 1944.
314. Haas, A.R.C.   Boron content of citrus trees grown on various rootstocks. *Soil Sci.* **59**, 465—479, 1945.
315. Andrus, C.F.   Brittle stem, an apparently new sublethal gene in tomato. *Tomato Genetics Coop. Rep.* **5**, 5, 1955.
316. Brown, J.C. and Jones, W.E.   Differential transport of boron in tomato *Lycopersicon esculentum* Mill., *Physiol. Plant.* **25**, 279—282, 1971.
317. Brown, J.C. and Ambler, J.E.   Genetic control of uptake and a role of boron in tomato. *Soil Sci. Soc. Am. Proc.* **37**, 63—66, 1973.
318. Wann, E.V. and Hills, W.A.   The genetics of boron and iron transport in tomato. *J. Hered.* **64**, 370—371, 1973.
319. Greaves, J.E. and Anderson, A.   Influence of soil and variety on the copper content of grains. *J. Nut.* **11**, 111—118, 1936.
320. Rademacher, B.   Kupfergehalt, Kupferbedarf und Kupferaneignungsvermögen verschiedener Hafersorten als Grundlage für die Züchtung gegen die Heidermoorkrankheit widerstandsfähiger Sorten. *Z. Pfl. Krankh.* **47**, 545—560, 1937.
321. Rademacher, B.   Uber der Veranderungen des Kupfergehaltes, den Verlau der Kupferaufnahme und den Kupferentzug beim Hafer. *Bodenk. Pflanzenernahr* **19**, 80—108, 1940.
322. Brown, J.C., Holmes, R.S. and Specht, A.W.   Iron, the limiting element in a chlorosis. II. Copperphosphorus induced chlorosis dependent upon plant species and varieties. *Plant Physiol.* **30**, 457—462, 1955.
323. Andrew, C.S. and Thorne, P.M.   Comparative responses to copper of some tropical and temperate pasture legumes. *Aust. J. Agr. Res.* **13**, 821—835, 1962.
324. Smilde, K.W. and Henkens, C.H.   Sensitivity to copper deficiency of different cereals. *Neth. J. Agr. Sci.* **15**, 249—258, 1967.
325. Gartside, D.W. and McNeilly, T.   The potential for evolution of heavy metal tolerance in plants. II. Copper tolerance in normal populations of different plant species. *Heredity* **32**, 335—348, 1974b.
326. Wu, L., Bradshaw, A.D. and Thurman, D.A.   The potential for evolution of heavy metal tolerance in plants. III. The rapid evolution of copper tolerance in *Agrostis stolonifera*. *Heredity* **34**, 165—187, 1975.
327. Nambiar, E.K.S.   Genetic differences in the copper nutrition of cereals. I. Differential responses of genotypes of copper. *Aust. J. Agric. Res.* **27**, 453—463, 1976.
328. Waring, E.J., Wilson, R.D. and Shirlow, N.S.   Whiptail of cauliflower. Control by the use of ammonium molybdate and sodium molybdate. *Agr. Gaz.* N.S.W. **59**, 625—630, 1948.
329. Neenan, M. and Goodman, O.G.   Varietal susceptibility of brassicas to molybdenum deficiency. *Nature* **174**, 792—793, 1954.
330. Andrew, W.D. and Milligan, R.T.   Different molybdenum requirements of Medics and subterranean clover on a red brown soil at Wagga Wagga, New South Wales. *J. Aust. Inst. Agr. Sci.* **20**, 123—124, 1954.
331. Dios Vidal, R. and Broyer, T.C.   Deficiency symptoms and essentiality of molybdenum in corn hybrids. *Agrochimia* **9**, 273—284, 1965.
332. Harris, H.B., Parker, M.B. and Johnson, B.J.   Influence of molybdenum content of soybean seed and other factors associated with seed source on progeny response to applied molybdenum. *Agron. J.* **57**, 397—399, 1965.
333. Chipman, E.W., Mackay, D.C., Gupta, H.C. and Cannon, H.B.   Response of cauliflower cultivars to molybdenum deficiency. *Can. J. Plant Sci.* **50**, 164—168, 1970.
334. Brown, J.C. and Clark, R.B.   Differential response of two maize inbreds to molybdenum stress. *Soil Sci. Soc. Am. Proc.* **38**, 331—333, 1974.
335. Wilkins, D.A.   A technique for the measurement of lead tolerance in plants. *Nature* **180**, (4575), 37, 1957.
336. Jowett, D.   Adaptation of a lead-tolerant population of *Agrostis tenuis* to low soil fertility. *Nature* **184**, 43, 1959.
337. Jowett, D.   Population studies on lead-tolerant Agrostis tenuis. *Evolution* **18**, 70—80, 1964.
338. Urquhart, C.   Genetics of lead tolerance in *Festuca ovina*. *Heredity* **26**, 19—33, 1971.

339. Earley, E.B.   Minor element studies with soybeans. I. Varietal reaction to concentration of zinc in excess of the nutritional requirement. *J. Amer. Soc. Agron.* **35**, 1012–1023, 1943.

340. Williams, C.H. and Moore, C.W.E.   The effects of stage of growth on the copper, zinc, manganese and molybdenum contents of Algerian oats grown on thirteen soils. *Aus. J. Agr. Res.* **3**, 343–361, 1952.

341. Bradshaw, A.D.   Populations of *Agrostis tenuis* resistant to lead and zinc poisoning. *Nature* **169**, 1098, 1952.

342. Massey, H.F. and Loeffel, F.A.   Factors in interstrain variation in zinc content of maize *Zea mays* L. kernels. *Agron. J.* **59**, 214–217, 1967.

343. Halim, A.H., Wasson, C.F. and Ellis, Jr. R.   Zinc deficiency symptoms and zinc and phosphorus interactions in several strains of corn *Zea mays* L. *Agron. J.* **60**, 267–271, 1968.

344. Ambler, J.E. and Brown, J.C.   Cause of differential susceptibility to zinc deficiency in two varieties of navy beans *Phaseolus vulgaris* L. *Agron. J.* **61**, 41–43, 1969.

345. Polson, D.E. and Adams, M.W.   Differential response of navy beans *Phaseolus vulgaris* to zinc. I. Differential growth and elemental composition at excessive Zn levels. *Agron. J.* **62**, 557–560, 1970.

346. Sharma, C.P., Agarwala, S.C., Sharma, P.N. and Ahmad, S.   Performance of eight high yielding wheat varieties in some zinc deficient soils of Uttar Pradesh and their response to zinc amendments in pot culture. *J. Indian Soc. Soil Sci.* **19**, 93–100, 1971.

347. Gartside, D.W. and McNeilly, T.   Genetic studies in heavy metal tolerant plants. I. Genetics of zinc tolerance in *Anthoxanthum odoratum. Heredity* **32**, 287–297, 1974.

348. Gartside, D.W. and McNeilly, T.   Genetic studies in heavy metal tolerant plants. II. Zinc tolerance in *Agrostis tenuis. Heredity* **33**, 306–308, 1974.

349. Sharma, C.P. and Sharma, P.N.   Comparative growth and zinc uptake of some high yielding varieties of wheat in Tanai soils of Uttar Pradesh with and without zinc amendment. *Geophytology* **6**, 267–271, 1976.

350. Agarwala, S.C., Chatterjee, C. and Sharma, C.P.   *Relative susceptibility of some high yielding rice varieties to zinc deficiency.* Geobios, 1977.

351. Gauch, H.G. and Magistad, D.C.   Growth of strawberry clover varieties and of alfalfa and ladivo clover as affected by salt. *J. Amer. Soc. Agron.* **35**, 871–880, 1943.

352. Ayers, A.D., Wadleigh, C.H. and Bernstein, L.   Salt tolerance of six varieties of lettuce. *Proc. Am. Soc. Hort. Sci.* **57**, 237–242, 1951.

353. Bernstein, L. and Ayers, A.D.   Salt tolerance of six varieties of green beans. *Proc. Am. Soc. Hort. Sci.* **57**, 243–248, 1951.

354. Ayers, A.D., Brown, J.W. and Wadleigh, C.H.   Salt tolerance of barley and wheat in soil plots receiving several salinization regimes. *Agron. J.* **44**, 307–310, 1952.

355. Bernstein, L. and Ayers, A.D.   Salt tolerance of five varieties of carrots. *Proc. Am. Soc. Hort. Sci.* **61**, 360–366, 1953.

356. Strogonov, B.P., Ivanitskaya, E.F. and Chernudeva, I.P.   Effect of high concentrations of salts on plants. *Fiziol. R. T.* **3**, 319–327, 1956.

357. McMillan, C.   Salt tolerance within a *Typha* population. *Amer. J. Bot.* **46**(7), 521–526, 1959.

358. Wahhab, A., Jabbar, A. and Muhammad, F.   Salt tolerance of various varieties of agricultural crops at germination stage. *Pak. J. Sci. Res.* **11**(2), 71–80, 1959.

359. Ehlig, C.F.   Effect of salinity on four varieties of table grapes grown in sand culture. *Proc. Amer. Soc. Hort. Sci.* **76**, 323–335, 1960.

360. Dewey, D.R.   Salt tolerance of twenty-five strains of *Agropyron. Agron. J.* **52**, 631–635, 1960.

361. Sakai, K. and Rodrigo, M.   Studies on a laboratory method of testing salinity resistance in rice varieties. *Trop. Agriculturist* **116**, 179–184, 1960.

362. Pearson, G.A.   The salt tolerance of rice. *Int. Rice Comm. Newsletter* **10**(1), 1–4, 1961.

363. Pearson, G.A.   A technique for determining salt tolerance of rice. *Int. Rice Comm. Newsletter* **10**, 5–7, 1961.

364. Pearson, G.A., Ayers, A.D. and Eberhard, D.L.   Relative salt tolerance of rice during germination and early seedling development. *Soil Sci.* **102**, 151–156, 1966.

365. Dewey, D.R.   Breeding crested wheatgrass for salt tolerance. *Crop Sci.* **2**, 403–407, 1962.

366. Greenway, H. Plant response to saline substrates. I. Growth and ion uptake of several varieties of *Hordeum* during and after sodium chloride treatment. *Aust. J. Biol. Sci.* **15** (1), 16–38, 1962.

367. Bernstein, L., Ehlig, C.F. and Clark, R.A. Effect of grape rootstocks on chloride accumulation in leaves. *J. Am. Soc. Hort. Sci.* **94**, 584–590, 1969.

368. Tanimoto, T.T. Differential physiological response of sugarcane varieties to osmotic pressures of saline media. *Crop Sci.* **9**, 683–688, 1969.

369. Elzam, O.E. and Epstein, E. Salt relations of two grass species differing in salt tolerance. I. Growth and salt content at different salt concentrations. *Agrochimica* **13**, 187–195, 1969a.

370. Elzam, O.E. and Epstein, E. Salt relations of two grass species differing in salt tolerance. II. Kinetics of the absorption of K, Na, and Cl by their excised roots. *Agrochimica* **13**, 196–206, 1969b.

371. Shafi, M., Majid, A. and Ahmad, M. Some preliminary studies on salt tolerance of rice varieties. *West Pak. J. of Agric. Res.* **8**(2), 117–123, 1970.

372. Janardhan, K.V. Note on modified method for screening rice varieties for tolerance to saline conditions. *Indian J. Agric. Sci.* **41**, 504–507, 1971.

373. Barakat, M.A., Khalid, M.M. and Atia, M.H. Effect of salinity on the germination of 17 rice varieties. *Agric. Res. Rev.* Cairo, UAR, 49(2), 219–224, 1971.

374. Janardhan, K.V. and Murty, K.S. Studies in salt tolerance in rice. III. Relative salt tolerance of some local and high yielding rice varieties. *Oryza* **9**(1), 24–34, 1972.

375. Datta, S.K. A study of salt tolerance of 12 varieties of rice. *Curr. Sci.* **41**, 456–457, 1972.

376. Purohit, D.C. and Tripathi, R.S. Performance of some salt tolerant paddy varieties in Chambal commanded area to Rajasthan. *Oryza* **9**(2), 19–20, 1972.

377. Bari, G., Hamid, A. and Awan, M.A. Effect of salinity on germination and seedling growth of rice varieties. *Int. Rice Comm. Newsletter* **22**(3), 32–36, 1973.

378. Akbar, M. and Yabuno, T. Breeding of saline-resistant varieties for rice. III. Response of F_1 hybrids to salinity in reciprocal crosses between Thomas 347 and Magnolia. *Jpn. J. Breed.* **25**(4), 215–220, 1975.

379. Nabors, M.W., Daniels, A., Nadolny, L. and Brown, C. Sodium chloride tolerant lines of tobacco cells. *Plant Sci. Letters* **4**, 155–159, 1975.

380. Dix, P.J. and Street, H.E. Sodium chloride resistant cultured cell lines from *Nicotiana sylvestris* and *Capsicum annuum*. *Plant Sci. Lett.* **5**, 231–237, 1975.

381. Rains, D.W. and Epstein, E. Preferential absorption of potassium by leaf tissue of the mangrove, *Aricenna marina*, an aspect of halophytic competence in coping with salt. *Aust. J. Biol. Sci.* **20**, 847–857, 1967.

382. Wallace, A., Hemaidan, N. and Sufi, S.M. Sodium translocation in bush beans. *Soil Sci.* **100**, 331–334, 1965.

383. Huffaker, R.C. and Wallace, A. Sodium absorption by different plant species at different potassium levels. *Soil Sci.* **87**(3), 130–134, 1959.

384. Epstein, E. and Hagen, C.E. A kinetic study of the absorption of alkali cations by barley roots. *Plant Physiol.* **27**, 457–474, 1952.

385. Fried, M. and Noggle, J.C. Multiple site uptake of individual cations by roots as affected by hydrogen ion. *Plant Physiol.* **33**, 139–144, 1958.

386. Repp, G.I., McAllister, D.R. and Wiebe, H.H. Salt resistance of protoplasm as a test for the salt tolerance of agricultural plants. *Agron. J.* **51**, 311–314, 1959.

387. Udovenko, G.V. Reasons of different responses of plant varieties and species to saline soils. 219–224. *Variety and Nutrition*, Klimashevsky, E.L. (ed.), Siberian Inst. Pl. Phys. Biochem., Irkutsk, 1974 (In Russian).

388. Epstein, E. Mineral metabolism of halophytes. *Ecological Aspects of the mineral nutrition of plants.* British Ecol. Soc. Symp. No. 9, 345–355, Blackwell, Oxford, 1969.

389. Suneson, C.A. and Wiebe, G.A. *Crop Sci.* **2**, 347, 1962.

390. Ellis, B.G. Response and susceptibility. *Zinc deficiency – a symposium. Crop Soils* **18**(1), 10–13, 1965.

391. Brown, J.C. Calcium movement in barley and wheat as affected by copper and phosphorus. *Agron. J.* **57**, 617–621, 1965.

392. Henning, S.J. Aluminum toxicity in the primary meristems of wheat roots. *Ph.D. Thesis.* Dept. Soil Sci., Oregon State Univ., Corvallis, Oregon, 1975.

393. Rhue, R.D. The time-concentration interaction of Al toxicity in wheat root meristems. *Ph.D. Thesis.* Dept. Soil Sci., Oregon State Univ., Corvallis, Oregon, 1976.

394. Foy, C.D., Fleming, A.L., Armiger, W.H. Differential tolerance of cotton varieties to excess manganese. *Agron. J.* **61**, 690—694, 1969.

395. Turner, R.G. and Gregory, R.P.G. The use of radioisotopes to investigate heavy metal tolerance in plants. *Isotopes in Plant Nutrition and Physiology*, 493—509, IAEA, Vienna, 1967.

396. Peterson, P.J. The distribution of ^{65}Zn in *Agrostis tenuis* and *A. stolonifera* tissues. *J. Expt. Bot.* **20**, 863—875, 1969.

397. Turner, R.G. The sub-cellular distribution of zinc and copper within the roots of metal-tolerant cloves of *Agrostis tenuis*. *New Phytologist* **69**, 725—731, 1969.

398. Reilly, C. The uptake and accumulation of copper by *Becium homblei* (De Wild) Duvig and Plancke. *New Phytol.* **68**, 1081—1087.

399. Jowett, D. Adaptation of a lead-tolerant population of *Agrostis tenuis* to low soil fertility. *Nature* **184**, 43, 1959.

400. Khan, M.S.I. The process of evolution of heavy metal tolerance in *Agrostis tenuis* and other grasses. *M.Sc. Thesis*, University of Wales, 1969.

401. Fehr, W.R. Control of iron-deficiency chlorosis in soybeans by plant breeding. *Proc. Int. Symp. Iron Nutrition and Interactions in Plants*. Provo, Utah, 12—14 August, *Abstracts*, 1981.

402. Esty, J.C., Onken, A.B., Hossner, L.R. and Matheson, R. Iron use efficiency in grain sorghum hybrids and parental lines. *Agron. J.* **72**, 589—592, 1980.

403. Coyne, D.P., Korban, S.S., Knudsen, D. and Clark, R.B. Inheritance of iron deficiency in crosses of bean, *Phaseolus vulgaris*, cultivars. *Proc. Int. Symp. Iron Nutrition and Interactions in Plants*. Provo, Utah, 12—14 August, *Abstracts*, 1981.

404. Saric̓, M.R. Genetic specificity in relation to plant mineral nutrition. *J. Plant Nutrition* **3**, 743—766, 1981.

405. Wegozyn, V.A., Hill, R.R. Jr. and Baker, D.E. Soil Fertility — Crop Genotype Associations and Interactions. *J. Plant Nutrition* **2**, 607—627, 1980.

406. Schenk, M.K. and Barber, S.A. Root characteristics of corn genotypes as related to P-uptake. *Agron. J.* **71**, 921—924, 1979.

407. Gallagher, L.W., Soliman, K.M., Qualset, C.O., Huffacker, R.C. and Rains, D.W. Major gene control of nitrate reductase activity in common wheat. *Crop Sci.* **20**, 717—720, 1980.

408. Reed, A.J. and Hageman, R.H. Relationship between nitrate uptake, flux, and reduction, and the accumulation of reduced nitrogen in maize *Zea mays* L., I. Genotypic variation, II. Effect of nutrient nitrate concentration. *Plant Physiol.* **66**, 1179—1183, 1980.

409. Keys, A.J., Bird, I.F., Cornelius, M.J., Lea, P.J., Wallsgrove, R.M. and Miflin, B.J. Photorespiratory nitrogen cycle. *Nature* **275**, 741—743, 1978.

410. Wallsgrove, R.M., Keys, A.J., Bird, I.F., Cornelius, M.J., Lea, P.J. and Miflin, B.J. The location of glutamine synthetase in leaf cells and its role in the re-assimilation of ammonia released in photorespiration. *J. Expt. Bot.* **31**, 883, 1980.

411. Woo, K.C., Berry, J.A. and Turner, G.L. Release and refixation of ammonia during photorespiration. *Carnegie Inst. Yearbook*, **77**, 240—245, 1978.

412. Vose, P.B. Potential use of induced mutants in crop plant physiology studies. *Int. Symp. Induced Mutations as a Tool for Crop Plant Improvement*. IAEA, SM-251/44, IAEA, 159—181, Vienna, 1981.

413. Olsen, R.A. and Brown, J.C. Factors related to ison uptake by dicotyledonous and monocotyledonous plants. I. pH reductant, II. The reduction of Fe^{3+} as influenced by roots and inhibitors, III. Competition between roots and external factors for Fe. *J. Plant Nutrition* **2**, 629, 1980.

414. Proceedings Int. Symp. Iron Nutrition and Interactions in Plants, Provo, Utah, 12—14 August 1981, *J. Plant Nutrition*, **5**, 229—1002, 1982.

415. Kannan, S. Genotypic differences in iron uptake and utilization in some crop varieties. *Proc. Int. Symp. Iron Nutrition and Interactions in Plants*, Provo, Utah, 12—14 August, *Abstracts*, 1981.

416. Fleming, A.L. and Foy, C.D. Differential response of barley varieties to Fe stress. *Proc. Int. Symp. Iron Nutrition and Interactions in Plants*, Provo, Utah, 12—14 August, *Abstracts*, 1981.

417. Williams, E.P., Clark, R.R., Yusuf, Y., Ross, W.M. and Maranville, J.W. Variability of sorghum genotypes to tolerate low iron. *Proc. Int. Symp. Iron Nutrition and Interactions in Plants*, Provo, Utah, 12—14 August, *Abstracts*, 1981.

418. *Proc. Int. Symp. Genetic Specificity of Mineral Nutrition of Plants*, Belgrade, 30 August—2 September, 1982, Saric, M.R. (ed.) 386 pp. Serbian Acad. Sci. and Arts, Belgrade, 1982 (Also Special Vol. 72 *Plant and Soil*, 1983).

CHAPTER 5

BREEDING AND SELECTION FOR RESISTANCE TO LOW TEMPERATURE

C. Stushnoff
Department of Horticultural Science

B. Fowler
Crop Development Centre

Anita Brûelé-Babel
Department of Crop Science

*University of Saskatchewan,
Saskatoon, Saskatchewan, S7N 0W0, Canada*

INTRODUCTION

Essentially all members of the plant kingdom, with the possible exception of those found in warm tropical regions, are vulnerable to low temperature stress at some stage of their life cycle. Over large regions of the earth's surface low temperatures are of such magnitude that only taxa with special adaptive features grow naturally and only a few have been developed by mankind as cultivated crops to provide food, fibre, aesthetic beauty and shelter.

Several strategies have been identified in plants which enable their survival, not only through the course of normal seasonal changes typical of north and south temperate regions, but also through severe and unpredictable fluctuations over the course of a few days or even hours in

115

extreme cases. Plant breeders must be concerned with the nature of these processes in their quest to produce new cultivars which are resistant to low temperature stress.

Perhaps one of the most highly developed and successful strategies to cope with cold stress and to maximize adaptation to a short season in the plant kingdom, is the summer annual with the seed providing an ideal means of avoiding cold injury. Most cereals, vegetables, and many forage and seed legumes are grown as summer annuals and although winter survival is not a problem, specific threshold temperatures for germination, growth, and development to attain maturity are critical concerns.

A number of very important food crops of tropical or sub-tropical origin such as tomatoes, rice, and corn can suffer from exposure to temperatures well above freezing ($0°C$ to $+20°C$) resulting in reduced growth and yields from chilling injury. Temperatures below $0°C$ can cause extensive damage to most agronomic and horticultural crops particularly in the tender seedling stage early in the season or late in the season as fruits approach maturity at the end of the growing season.

Winter annuals such as winter wheat and fall rye provide significant agronomic advantages where they can be grown without serious losses from low temperature injury. Even though consistent production in northern regions often depends on survival from adequate snow cover and refined management techniques the prospects of obtaining significant increases in yield are very good if a significant break-through, by means of breeding hardy cultivars, could be achieved.

Herbaceous perennials, also representing significant agronomic and horticultural crops, employ protection of the meristematic growing points through proximity to the soil surface and snow cover as a strategy for survival in cold regions. This group generally possess capability for early season growth responses as an adaptive advantage. Some plants in this group such as rhubarb and many grasses have very low heat requirements to initiate spring growth thus having selective advantage for early growth in regions with cool moist environments.

Trees and to a lesser degree shrubs and vines are exposed to seasonal and diurnal temperature variation throughout their life cycle. Reproductive and structural meristematic tissues are unceasingly vulnerable to the vagaries of nature. Relatively recent elucidation of some mechanisms controlling resistance to low temperatures have helped explain both the limits of natural vegetation for a number of hardwood tree species and fruit species as well as the basis for resistance in hardy cultivars. Deciduous woody plants are characterized by complex series of phenophases associated with environmental triggering stimuli to achieve growth and development, and rest and dormancy at the appropriate time of year in sequence and in anticipation of severe temperature changes. Many conifers but only a few broadleaf evergreens possess extreme cold resistance, as evidenced by their natural distribution to the polar regions and high altitude mountain forests. Essentially all woody plants cultivated by mankind whether for their fruits, lumber or aesthetic beauty, from the polar regions to the tropics can suffer from low temperature injury.

It is evident that about the only common denominator in the study of cold hardiness is low temperature and even that factor is highly unpredictable in severity and timing from year to year. Plants have evolved various strategies enabling them to survive and flourish in spite of cold stress; the mission for the plant breeder and the cold stress physiologist is to unravel these complexities and direct a selection program to enhance stress resistance in the crop of concern. In this chapter we will attempt to provide an overview of theoretical mechanisms, selection techniques and approaches to breeding for cold hardiness with a variety of herbaceous and wood plant species used as examples (Table 5.1).

TABLE 5.1. Examples of cold stress injury in various taxa and tissues from +10° to −196°C

DESCRIPTION OF EVENT AND MECHANISM OF RESISTANCE

EXAMPLES OF PLANTS AND STRUCTURES INVOLVED

°F	°C		
+50	+10	**Chilling Injury** - Membrane phase transition; reduced fluidity affects enzymes, metabolism, ion leakage. - Restricted germination, growth, and reproduction.	**Tropical and Sub Tropical Origin** - Fruits: Bananas, Papaya, Passiflora, etc. - Warm Season Vegetables: Tomatoes, Cucumbers, Peppers, etc. - Monocots: Corn, Rice
+32	0	Ice and water in equilibrium.	
+23	−2 −4 −5 −6	**Spring Frost** - Blossoms, germinating seedlings, newly emerged shoot growth. - Ice crystallization following nucleation of supercooled water. - Bacterial nucleators, *Pseudomonas, Erwinia.*	**Fruit Crops in Bloom or Pre Bloom** Strawberries, Citrus, Peaches, Apricots, Cherries, Plums, Pears, Apples. **Grape Shoots** **Potatoes at High Elevations** **Spring Seeded Grains and Vegetables** **Vegetable Transplants.**
+14	−8 −10	**Extracellular Bulk Water Freezes** in hardy plant tissues − No injury in most cases.	
+10.4	−12	**Winter Injury to Herbaceous Crowns and Tree Roots** - Equilibrium freezing (tolerance) - Dehydration stress - Increasing cell sap concentration - Supercooling possible in certain tissues.	**Winter Cereals (Crown Temps)** Oats (−13°), Barley (−15°), Wheat (−21°) Rye (−30°) **Strawberries** (−12°) **Overwintering Vegetables for Seed** Cabbage (−12°) **Grasses and Forages** - Kentucky bluegrass (−25°) - Alfalfa (−25°)
−4	−20 −24		
		Winter Injury to Dormant Flower Buds - Non equilibrium intracellular freezing of supercooled water and cell solutes (avoidance) - Usually rate dependent - Shift readily.	**Fruit Flower Buds** *Prunus*, Blueberry, Grape, Cloudberry **Ornamentals** - Azalea flowerbuds
		Winter Injury to Xylem Ray Parenchyma in Woody Plants - Non equilibrium, intracellular freezing of supercooled water and cell solutes - Usually rate independent - Limits northern distribution to - 40° isotherm.	**Hardwood and Fruit Trees** - Maple, Hickory, Ash, Oak, Elm - Apples, Pears, *Prunus.* **Seeds** - Lettuce seed (20 to 55% water)
−52.6	−38.1 −47	Homogeneous nucleation of pure water (10 μm di).	
−112	−80	**Winter Injury to Some Boreal Trees** - No supercooling, resistance by tolerance but exact mechanism is unknown.	*Salix caprea,* *Betula pubescens* **Dry Biological Dispersal Systems** - Seeds, Spores, Pollen - Some boreal species
−320	−196	**Ultimate Adaptation** - Low water content - Membrane conformation changes - Exact mechanism of tolerance is unknown.	*Salix pentandra, Cornus stolonifera* - Cryopreservation of vegetative meristems by artificial manipulation.

MECHANISMS OF COLD HARDINESS FOR PLANT
BREEDING PROGRAMS

It is not our intention to provide a complete review of all theories which have been put forth to explain cold hardiness, but rather to provide a conceptualized overview which we feel might be particularly useful to plant breeders (Table 5.1). For more complete theoretical details the reader is referred to Levitt[49], Li and Sakai[50], Lyons *et al.*[53], Mussell and Staples[57], Olien and Smith[61], and Weiser[87].

(i) Chilling Injury

Unlike most other types of low temperature injury, which have their greatest impact when plant systems are dormant or in transition to and from dormancy, sensitivity to chilling injury occurs from approximately $20°C$ to $0°C$. Chilling injury is manifested in a variety of physiological disruptions including germination, floral and fruit development, yield, and storage capability, all of which can have a very significant impact on productivity and end use of the product. Disruptions to normal development occur when chilling sensitive crops are exposed to some critical temperature which varies in level and duration for each crop. Yields can be reduced as a consequence of disruption to overall growth or as in the case of rice[57], severe yield losses can result from floral malfunction. Minor chilling stress can be reversible under non-stress temperatures and exposures of short duration to temperatures above the critical range can result in hardening which may reduce or eliminate injury during subsequent exposure to critical temperatures.

Considerable physiological research has been directed at metabolism and membrane function in relation to chilling injury and resistance. The reader is referred to Levitt[49], and Lyons *et al.*[53] for elaboration of metabolic disturbances; permeability changes; membrane fluidity, protein and lipid unsaturation; enzyme energy activation and cytological malfunctions, all of which have been implicated in chilling injury.

(ii) Breeding and selection for chilling injury

Vallejos[84] pointed out that although some breeding work has been attempted the results have generally been contradictory and not spectacular. For example, in the case of breeding for cold tolerance in seed germination; monogenic, polygenic, and maternal inheritance have all been reported. Vallejos[84] attributed the lack of progress to the fact that selection was largely confined to commercial cultivars which were selected for adaptation to mild climates, high yields and uniformity and should not be expected to produce striking variants as might exotic species-germplasm.

Tissue culture selection for chilling resistance has been used to obtain cell lines of *Nicotiana sylvestris* ($-3°C$) and *Capsicum annum* ($5°C$) but subsequent regeneration of callus from seedlings of the resistant lines was sensitive. This was attributed to either epigenetic control or segregation of chimeral plants in culture. Refinements to this promising approach are being pursued and it would seem that this approach plus the use of *in vitro* technique to accomplish somatic-genome hybridization should meet with success, particularly since many of the chilling sensitive species have been thoroughly researched in tissue culture Dix[11].

(iii) Frost tender plants

Plants which are sensitive to chilling injury are also classified as frost tender with little or no freezing resistance. Beans, cucurbits, corn, potatoes, and tomatoes will not resist exposure to more than $-2°C$ to $-3°C$ for very short periods of time and readily exhibit frost injury symptoms including a water soaked flaccid appearance with loss of turger followed by rapid drying upon exposure to warm temperatures. Ion leakage from cells is common following injury. Plants in this category are incapable of tolerating the formation of ice in their tissues whether extracellular or intracellular. There is recent evidence that the presence of certain bacteria, *Pseudomonas syringae* and *Erwinia herbicola* promote ice nucleation at warmer temperatures than in their absence and thus promote frost damage, for example on corn seedlings Arny et al.[4] No reports of progress to increase resistance to frost through breeding are known at this time.

(iv) Plants with some frost resistance

Many members of the *Cruciferae* and certain other herbaceous plants such as peas and spinach can tolerate several degrees of frost ($-3°C$ to $-5°C$). Many bulbs tolerate $-6°C$ to $-18°C$, while winter cereals, bluegrass and alfalfa survive exposure to $-20°C$ to $-30°C$. This group can tolerate the presence of extracellular ice in their tissues within specific limits, determined by genetic background and growing conditions leading to acclimation. In these plants water begins to freeze extracellularly at $-3°C$ to $-5°C$ and may extend to $-10°C$, depending among other factors on the freezing point depression from dissolved salts and organic compounds. The formation of extracellular ice creates a vapor pressure deficit promoting outward migration of intracellular water. As the temperature drops greater dehydration stress is created and if the rate and/or limit of temperature drop does not exceed the inherent capability of the cell system to resist this stress, equilibrium is reached with no injury. Freezing rates exceeding the rate of water loss to extracellular ice may induce supercooling or be followed by ice crystallization, intracellular freezing and death of individual cells. Levitt[49], suggested that prolonged exposure to sub-freezing temperatures may also cause damage from the attendant dehydration stress if equilibrium conditions of the systems are exceeded.

The most convincing recent research suggests that injury occurs during freezing at the moment of passing through a critical temperature, furthermore, the primary site of injury is concluded to be the plasma membrane Rajashekar et al.[72].

(v) Cold hardy plants

Very hardy plants, predominantly deciduous temperate woody plants also tolerate extracellular ice in their tissues, as in cortical stem tissues[76], and bud scales[37]. The lower limits of tolerance vary greatly with seasonal acclimation patterns, the rate and degree of temperature decline and the genetic capability of tissues to accommodate extracellular freezing and the accompanying dehydration stress.

Another distinct mechanism has been discovered in xylem ray parenchyma Quamme et al.[69], George et al.[24], and flower buds Graham and Mullin[26], of many woody plants. Freezing of floral primordia is avoided in these tissues by deep supercooling in the range of $-20°C$ to $-40°C$. For reasons which are not entirely understood, water which can nucleate and freeze as intracellular ice below critical temperatures, can remain liquid as low as $-47°C$ provided the limits of the system are not exceeded and nucleation does not occur.[78] When nucleation of this supercooled fraction does occur, intracellular ice forms suddenly and death occurs at a specifically defined point. Nucleation of this supercooled aqueous fraction results in a phase change which can be detected by calorimetric methods including differential scanning calorimetry and differential thermal analyses.

Differential thermal analyses studies in North America showed that a large number of eastern hardwood deciduous species which were characterized by the presence of low temperature deep supercooling exotherms during mid-winter acclimation were also found to be restricted

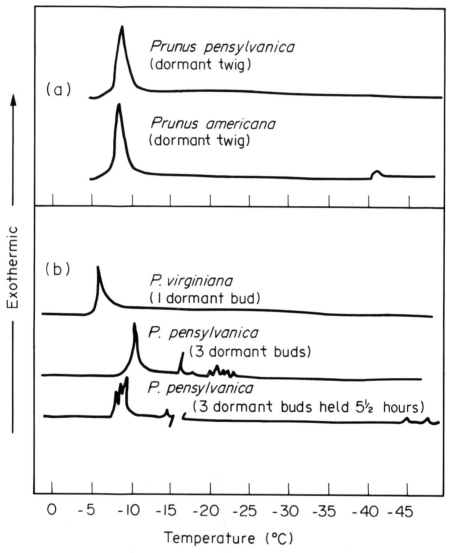

Fig. 5.1. Deep supercooling patterns, as determined by differential scanning calorimetry, found in dormant twigs and flower buds of *Prunus* species representing typical types of different freezing mechanisms. A. Hardy dormant twigs of *P. pensylvanica* with no low temperature exotherms and *P. americana* with low temperature exotherms typical of a less hardy species. B. Flower buds from the very hardy *P. viriginiana* with no low temperature exotherms compared to another very hardy species, *P. pensylvanica*, with low temperature exotherms which exhibit a very rapid hardening response following a short exposure to low temperature. This figure was synthesized from Burke and Stushnoff. (1979).

in their natural distribution zone to a northern boundary which coincided with a minus 40 isotherm[23]. Kaku and Iwaya[41] confirmed and elaborated this concept with woody plants in Japan. Apparently this low temperature supercooling phenomenon, found in the ray parenchyma of many species, serves not only as an avoidance mechanism for freezing but because supercooling is limited to approximately −40°C to −47°C it also provides a limit to the northern advancement of these species in a natural evolutionary sense.

A few species which have low temperature exotherms can also be found growing well north of the −40°C isotherm, as in the case of *Ulmus americana* and with this species recent research by Gusta[29] has shown that under natural conditions of extremely low temperature, in the vicinity of −40°C for two to three weeks, these exotherms may shift to temperatures as low as −55°C or may disappear entirely. Not only does this finding explain the exceptions which were brought up as a follow-up to the original study by George *et al.*[23]; but it also provides a promising opportunity for plant breeders to find special genetic variants, which do have a capability for a very low level of adaptation to cold conditions in spite of the presence of low temperature supercooling as an avoidance mechanism.

Of extreme interest and importance to plant breeders is the occurrence of deep supercooling in flower buds of a number of plant species. This phenomenon was first found in azalea flower buds by Graham and Mullin[26]. Further elaboration of deep supercooling in flower buds of azalea have been described by George *et al.*[24], Kaku *et al.*[42], and Sakai[76]. This phenomenon is of critical importance to fruit breeders and has been reported in flower buds of grape by Pierquet *et al.*[64], in blueberry by Biermann *et al.*[5], and in *Prunus* by Quamme[66], and Burke and Stushnoff[7]. The study by Burke and Stushnoff[7] used DSC to study deep supercooling in flower buds of many *Prunus* species, cultivars and hybrids and classified these taxa into three distinct groups (Fig. 5.1).

A number of taxa which had been recognized as having extreme cold hardiness in the flower buds had no low temperature exotherms; a second group had low temperature exotherms and had a history of extreme cold hardiness and the third group were those that had low temperature exotherms but were not capable of cold acclimation to the extreme levels of the former two groups. In this study it was also determined that the mechanism responsible for the extreme cold acclimation capability in the second group, even though low temperature exotherms were present, was that if flower buds of these taxa were exposed to short periods of sub-freezing temperatures the exotherms and the killing point shifted rapidly to much lower temperatures. This is perhaps the most significant point for breeding, in that it illustrates a genetic mechanism which is available for plant breeders to test readily and incorporate into progeny for development of cold resistance. The same phenomena has been reported for behaviour of extremely hardy species of grape *Vitis riparia* Pierquet *et al.*[64], and also in the case of some very hardy blueberry selections developed from a breeding program for cold hardiness Biermann *et al.*[5].

BREEDING AND SELECTION FOR RESISTANCE TO LOW TEMPERATURE STRESS IN HERBACEOUS CROPS

Winter survival of herbaceous crops is influenced by many variables such as soil heaving, ice encasement, desiccation, low temperature, flooding and disease. Even though the importance of these variables in winter survival has been recognized, it is often difficult to identify the primary variable responsible for winter damage. Also, because of the variability of the winter environment, these factors often interact to produce an extremely complex picture when plant growth resumes in the spring.

Attempts to unravel the mysteries of winter survival have lead to much confusion and, although many thousands of papers have been written on the subject, no unified theory has yet evolved to explain winter survival of any herbaceous crop. Consequently, one should not expect agreement among breeders as to the best approach to be used in breeding for improvement of this character. However recognition that several variables are involved, and that these variables interact, is the first step in the process of formulating a strategy for a breeding program aimed at improving winter survival. With these observations in mind the remainder of this discussion will be limited to a consideration of one aspect of winter survival, namely, resistance to low temperature stress. The bulk of the work on low temperature stress in herbaceous plants has been done on the winter cereals and, of these, winter wheat will be given primary consideration.

Development and maintenance of low temperature hardiness

Low temperature tolerance in plants is temperature induced and can easily be reversed. Because of the variable nature of this character and the great variability of winter environments, critical phases differ among species, areas and years. It is not our intent to review this subject for all situations. Rather, to emphasize the complex nature of this character, consideration will be given to the development and maintenance of low temperature tolerance using wheat as a model.

The shoot meristems of trees and shrubs generally overwinter above ground and, for this reason, are exposed to wide fluctuations in temperature. In contrast, the meristems of herbaceous plants are immersed in the soil which acts as a buffer to extremes in temperature. In the seedling stage, most can regenerate from undamaged crown tissue. Consequently, the temperature to which the crown is exposed is the critical factor in winter survival. The crown of the plant is usually located in the surface 5 cm. Therefore, it is the soil temperature at this depth which determines survival.

When exposed to cool fall temperatures, seedlings of cold tolerant cultivars have the ability to cold acclimate or 'harden off'[18,60,62,82]. This process usually does not start before soil temperatures fall below 10°C and is not completed until the crowns have been subjected to sub-freezing temperatures. Under field conditions cold acclimation is usually completed in four to six weeks. Cold acclimation in winter cereals can start shortly after germination.[3] However, establishment prior to the onset of conditions for cold acclimation is required for plants to achieve and maintain maximum cold tolerance under field conditions. Winter wheat plants should have a well developed crown and at least three leaves before fall freeze up. Exposure to warm temperatures during the hardening period will result in rapid dehardening.[30] However, at this stage, dehardened plants returned to conditions favoring acclimation, will quickly reharden.

Once plants have been cold acclimated a high level of hardiness can be maintained provided temperatures remain below freezing.[30,60] Prolonged exposure of hardened plants to temperatures slightly above freezing will reduce their low temperature tolerance.[60] Similarly, alternate freezing and thawing and extended exposure to high soil moisture or temperatures slightly above the killing temperature will reduce the plants low temperature tolerance.[32,60] Collectively, these factors usually result in a much reduced low temperature tolerance in plants by spring.

Exposure of plants to warm spring temperatures will eventually result in complete dehardening.[2,19,90] Once spring dehardening has been initiated, returning plants to conditions for cold acclimation will not reverse the dehardening process and loss of low temperature tolerance will result even if plants are maintained at temperatures below freezing.[31]

Selection for low temperature tolerance

One of the main objectives in most winter crop breeding programs is to insure that new cultivars have sufficient cold tolerance to survive the low temperature extremes of the environment for which the breeder is selecting. Therefore, the ideal situation for selection is field exposure at a uniform low temperature stress which approaches the extreme expected for the critical winter period of the area of concern. Survivors could then be selected with the confidence that they would provide the grower with the low temperature protection required for most winters. However, this optimum situation rarely, if ever, occurs, with the result that selection for low temperature tolerance can be a very frustrating experience.

The major frustration to plant breeders selecting within target production areas is the low frequency of test winters. This inability to control low temperature stress levels under field conditions has been long recognized.[10] Levitt[48] has estimated that test winters which produce differential winterkill occur as infrequently as once every 10 years. Reitz and Salmon[74] have reported a much higher frequency of differential winterkill for a number of winter wheat test sites in the American Great Plains area but they too had difficulty in obtaining an adequate test every year.

Further analyses of the frequency of test winters reveals that it is a combination of the cold hardiness potential of the cultivars under test and the stress levels experienced at the test site which determine whether or not differential winterkill occurs.[20] Therefore, the frequency of test winters for cultivars of a particular winter-hardiness class can be influenced by selecting sites which have a history of providing the desired levels of stress.[36,74] This can usually be accomplished by growing trials at or outside the margin of the production area for which the plant breeder is selecting.

When test winters are experienced, a second problem arises in that winterkilling within a trial is often irregular, resulting in a high experimental error and a greatly reduced ability to make meaningful selections.[54] Variability within trials can be caused by many factors other than genotypic differences. Differences in seeding depth, soil compaction, soil moisture, plant nutrition, ice encasement, snow cover, diseases and insects can all influence the level of winter survival. Among these variables, snow cover has the most dramatic influence on soil temperature.[21,43,77,90] This relationship combined with the difficulty in maintaining a uniform depth on test sites makes snow cover the most important source of experimental error that the plant breeder has to contend with. Ironically, it is the relationship between snow cover and soil temperature which also provides the plant breeder with the best opportunities for selecting for low temperature tolerance in herbaceous crops.

Through site selection and the control of snow cover, it is possible for the plant breeder to greatly increase the probability of differential winterkill at the temperature stress level of concern to a particular production area. On open fields the drifting of snow gives rise to small snow dunes which may, or may not, stabilize. When associated with low temperatures, this type of variable snow cover results in unpredictable and irregular winterkill patterns. In contrast, obstacles such as trees, fences, hills, roadways, etc., will cause a banking of snow and more predictable directional changes in stress levels, however, the patterns of changes are nearly as irregular as those found in open fields.[17] Our experience in this regard indicates that snow banking near a wind barrier or snow collection on the protected side of a hill usually results in a rate of change in the stress gradient which is too rapid to allow for the detection of small but important differences in low temperature tolerance.

We have had greater success in identifying small differences where snow depth is varied through the degree of field exposure, e.g. exposed summerfallow versus protected summer-fallow versus standing stubble. In this instance variability in stress level is detected and correct-ed for by the use of reference plots of known low temperature tolerance levels and the employ-ment of a moving average. Only plots with partial winterkill are considered and ratings are based on differences in percent winterkill among entries rather than on actual percent winter-kill of each plot. Utilizing this procedure a Field Survival Index (FSI) is calculated for each entry.[20] This method provides an objective measure of low temperature tolerance for each entry, reduces the experimental error associated with field trials, alleviates the difficulties arising from the absence of partial winterkill for a few entries in each trial, and allows for the pooling of results from different trials.

Because of the limitations inherent in field trials, there has been a continuing search for rapid and efficient methods for predicting low temperature tolerance. Most biochemical, physiological, and morphological characters change in the plant during cold acclimation and, based on these changes, a large number of prediction tests are possible. Most of the changes in these characters can be measured on plants grown and acclimated under both field conditions and controlled environments. The main disadvantages in utilizing field acclimated material is that the opportunity for evaluation usually only arises for one short period of each year and, during this period, there are often fluctuations in the level of low temperature tolerance. Use of controlled environments theoretically allows for more rigid control of conditions and greater flexibility in timing of experiments.

Ideally a screen for low temperature tolerance should be highly repeatable. It should be simple to conduct, highly correlated with low temperature ratings under field conditions, rapid and non-destructive and it should require only part of a single plant for analyses. With these criteria as a reference, Fowler et al.[22] evaluated the usefulness of 34 biochemical, physiological and morphological characters as predictors of winter survival for wheat. In spite of the criticism of field survival trials, the Field Survival Index (FSI) was found to be the most repeatable esti-mate considered. Therefore, if only the first three criteria on the list had to be met, i.e. high repeatability, correlation with field survival and simplicity, then clearly, field survival would be the best screen. If time is considered as a critical factor then many of the remaining screens become attractive alternatives. Among these, LT_{50} (temperature at which 50% of the popula-tion is killed) provided the best compromise. However, as was the case for all prediction tests, it provided at best a very coarse screen for field survival; experimental errors were too large to allow for the detection of small differences which are of practical concern to the plant breeder. Few of the characters considered could be measured without destroying the plant on which the measurement was made. This becomes an important consideration if one wishes to screen in segregating populations and be sure of saving the hardy individuals. Among the screens for which the plants measured could be saved, leaf water content and plant erectness gave the highest correlation with field survival. In addition, a combination of measures of leaf water content and plant erectness provided nearly as much information on field survival as did LT_{50}. High experimental errors associated with both these measurements were considered to be the main limitation to their use as screening tests.

Genetics of low temperature tolerance

Genetic studies have indicated that low temperature tolerance is a complex quantitative character which is strongly influenced by environment. Beyond this point, the search for a consensus on the mode of gene action controlling the expression of low temperature tolerance has been more difficult.

Cytoplasmic factors have been implicated in the control of low temperature tolerance.[9,52] However, agreement on this point is far from unanimous as several studies[44,81] have failed to show significant differences between reciprocals in F_1 hybrids.

Several researchers have reported on the cold hardiness levels of F_1 hybrids in wheat. In a number of instances the F_1 hybrids were hardier than the hardiest parent.[44,45,75] Others have observed that the hardiness levels of most F_1 hybrids were intermediate or tended toward the hardier parent.[14,52,75] In addition, Rosenquis[75] reported that three of twenty-one F_1 hybrids produced had hardiness levels similar to that of the less hardy parent. These results clearly support the concept that the control of low temperature hardiness is, indeed, complex.

Genetic analyses of later generations have done little to clarify this picture. Sutka[81] examined cold hardiness in a six-parent diallel and found significant effects for both general combining ability (GCA) and specific combining ability (SCA) indicating that both additive and non-additive gene action were important. A high GCA:SCA ratio demonstrated that there was a preponderance of additive variance. Erickson[14] also found that GCA was much greater than SCA (20.7 times), again indicating the importance of additive gene actions. Further evidence for additive gene action was presented by Law and Jenkins[47] when they observed that factors associated with three chromosomes of 'Cappelle Desprez' acted in an additive manner to influence cold tolerance. Analyses of another diallel cross suggested an additive-dominance system.[81] Dominant genes tended to act towards lower resistance, while recessive genes tended toward high resistance. Quisenberry[71] and Worzella[89] reported that under mild winter conditions, dominant genes control winter hardiness, while under more severe conditions, there was a lack of dominance.

Comparable modifying effects of environment have been reported in oats[39,55] and barley.[15] On a similar note, results from experiments with wheat by Gullord[27] and Gullord et al.[28] have been interpreted to suggest that, generally additive effects prevail under low moisture conditions while non-additive effects become more important under high moisture conditions. To this point, the evidence has been strongly in favor of additive control of low temperature tolerance under the more frigid environments. However, to make the picture even more interesting, Limin and Fowler[51] found that cold hardiness levels of wheat amphiploids produced from interspecific crosses depended mainly on the SCA of the parents involved.

Many cytogenetic studies utilizing various monosomic and substitution lines have been conducted to investigate the control of low temperature tolerance in wheat.[47,81] As a result of these studies, a total of 15 out of 21 chromosomes have been implicated as being involved in determining cold tolerance. Further inspection of these results indicates that the more tender cultivars tended to have fewer chromosomes influencing cold tolerance than did the more hardy cultivars.

It is clear from these observations that the inheritance of low temperature tolerance is extremely complex. Indications are that there are a large number of genes involved and that the mode of gene action is extremely variable. However, complex genetic interactions should not be too surprising when one considers that induction of the genetic system regulating low temperature tolerance also initiates a large number of biochemical, physiological and morphological changes in the plant. For similar reasons, it is probable that many genes have an indirect influence on the degree of expression of low temperature tolerance. Certainly the large reduction in cold tolerance resulting simply from doubling of chromosome number of rye[13] would suggest that a high degree of genetic balance or harmony is required for full expression of the genes directly controlling low temperature tolerance.

Breeding for improved low temperature tolerance

Progress from recent breeding efforts to improve the low temperature tolerance of herbaceous crops has not been very encouraging. The most concentrated effort has been with the cereal crops and even here the degree of success has been very limited. Three reasons can be identified as being mainly responsible for this lack of progress toward the production of super hardy cultivars.

(1) Exploitable genetic variability for low temperature tolerance appears to have been largely exhausted for most crops. This is certainly the case with wheat[20] and barley[25] where most evidence indicates that little more than small gains should be expected through conventional breeding programs. The only examples of transgressive segregation in these crops have come from crosses involving at least one parent with moderate or poor low temperature tolerance and the best selections from these populations have not surpassed hardy types already available. In both species, the search for improved low temperature tolerance has been expanded to include attempts at interspecific and intergeneric transfers. However, here too, the results have not been encouraging.[13,51] (2) The genetic control of low temperature tolerance is extremely complex. As with most quantitative characters, this usually means that the plant breeder has to work with small differences and large populations with the result that progress is slow. (3) Available selection methods are not accurate or precise enough to allow for the identification and selection of small but meaningful differences in low temperature tolerance. Field survival, with all its inherent problems, still remains the only screen precise enough to allow for identification of small differences which are important to the plant breeder.

Based on these observations, one has to be generally pessimistic that great strides can be made toward the production of super low temperature tolerant cultivars. Improvement which would allow for a large expansion of the present production areas for most crops are not going to be acquired easily. On the other hand, a great deal of improvement could be made in the low temperature tolerance of cultivars presently grown in most established production areas. Improved agronomic practices over the past 60 years have allowed most breeders to reduce their selection pressure for this character with the result that, for many production areas, there has been a reduction in low temperature tolerance of recent cultivar releases. A case in point has been the situation for winter wheat in the Western Canadian Prairies where 'Kharkov 22MC', released in 1912, remained the most cold tolerant cultivar available until the release of 'Norstar' in 1977.

Field studies conducted under the extreme winter conditions of Saskatchewan, Canada, have indicated that the genotype X environment interaction for field survival is small relative to the error associated with individual measurements.[22] This once again suggests that the main restriction on selection for low temperature tolerance is a poor ability to measure differences for this character. An inability to measure differences means that the probability of genetic gain, for low temperature tolerance, from single plant selections is poor. The usual method for improving the resolution in plant breeding trials is to grow more replicates. However, in early generation segregating populations or in bulk populations under selection this is not an option.

In this instance the safest approach is to apply mild selection pressure by utilizing one or more of the prediction tests that are highly correlated with field survival. At this stage more intensive pressures may be directed toward the selection for more simply controlled characters. In this manner, genotypes with poor low temperature tolerance can be eliminated thereby speeding up the selection program and allowing some control in population size until such time as critical field survival data is available. However, because of the high experimental error associated with low temperature prediction tests, selection pressure should only be increased slowly. Final selections among homozygous lines should be based on field survival data from several replicates.

BREEDING AND SELECTION FOR RESISTANCE TO LOW TEMPERATURE STRESS IN WOODY PLANTS

Woody plants are naturally distributed in great variety and abundance throughout most regions of the globe and have played a key role in providing food, shelter, and aesthetic beauty throughout the course of our short existance on this planet. Natural distribution patterns of most woody plants are often controlled by environmental temperature and as a consequence the success of our attempts to cultivate desirable species is also very much restrained by temperature stresses, particularly low temperature.

Fruit crop production is particularly vulnerable because a successful harvest depends not only on sound structural plants with functional water and nutrient transport systems plus photosynthetic ability, but also fully functional floral reproductive systems. Therefore the entire life cycle is exposed to low temperature stress greatly multiplying the problem of providing fully adapted cultivars. It is not surprising that these crops are grown in the most favorable climatic regions for each particular crop.

Total plant loss is devastating with most woody plants, in that a long lag time is required to replace a tree. Injury and loss of flower buds and flowers in bloom do not have a permanent affect but may cause catastrophic economic losses when they occur, which is usually a certainty in some major production regions throughout the world. The foregoing factors are good reasons to justify breeding for resistance to cold stress in these crops. In addition, breeding for cold resistance provides the opportunity to develop new industries where additional cold hardiness can be achieved, and indeed this has been the pattern of successful breeding programs as pioneering developments occurred in North America. Plant breeders have been successful in developing cold hardy cultivars of fruits.[38,67,79,91]

A plant breeder who is concerned with winter hardiness as an objective in a breeding program with a woody plant species will quickly recognize that many types of injury affecting various plant parts have been identified and reported in the literature. It will be necessary to identify and give appropriate priority to the most critical type of injury affecting the particular crop. A common basis of comparison in pioneer breeding programs was a rating of cultivar performance following a 'test winter'. A great deal of valuable information has been obtained from such observations but there are also a number of weaknesses in relying solely on 'test winters' for evaluation of woody plant material cold hardiness. 'Test winters' are usually infrequent and rarely similar in degree of severity and timing, whereas plant resistance to cold stress follows a definite seasonal response triggered by environmental stimuli such as day length and cold temperature. The changes which take place enabling a non-hardy plant to become resistant to winter cold temperatures are known as acclimation and the corresponding changes which occur in spring when hardiness is lost is known as deacclimation. Both of these responses are well covered in a review by Weiser.[87]

In some instances, for example very cold resistant crab apples, the cultivar may conform to a designation of hardy for all seasons, severity of stress and plant parts. In other cases, however, a cultivar may be lacking in only one or two critical components which would characterize it as being non-hardy for a particular region in use. It is evident that many growth processes can be dealt with more meaningfully in terms of breeding if the specific key element is recognized and the appropriate breeding program and selection technique developed for one factor at a time rather than trying to achieve an overall hardiness complex in one simultaneous effort. Intuitively, it would appear that a component approach might provide more rapid progress, and perhaps more important, a sharper focus on the procedure to achieve the desired end result. Several components have been identified which can be considered for breeding hardy fruit crops[79] and most of these are applicable to other woody plants as well; (1) time of initiation of acclimation, (2) rate of acclimation, (3) intensity of low temperature resistance, (4) capacity to resist loss of cold hardiness following moderation (above $0°C$) in mid-winter temperature, (5) capacity to regain cold hardiness if deacclimation occurs, (6) onset and rate of deacclimation.

Types of winter injury

(i) ROOT HARDINESS

is important in periods of sparse snow cover and deep frost penetration, particularly in young plants with a shallow poorly developed root system. Major problems have been introduced when non-hardy dwarfing apple rootstocks are grown in northern areas where they are not adapted to cold winter soil temperatures. Roots are generally much less hardy than above-ground structures. In the case of apples, twigs can survive $-45°C$ but even the hardiest known roots only to $-22°C$.

Root hardiness is a difficult problem because of sampling and because of the potential of undetectable latent injury which may show up in above-ground plant performance a year after the damage has occurred. Genetic variation for hardiness is present in apple rootstocks derived by both clonal and seed propagation and in seedling rootstock of pear, plum, cherry and peach for example.[38]

(ii) SOUTH-WEST INJURY

in the northern hemisphere, is common on trees with smooth, thin bark such as young apple, peach, cherry and mountain ash. Injury results from low temperature which follows a bright sunny day conducive to heating of the trunk from both incident and reflected radiation from the snow. The southwest side of the trunk may be up to $50°C$ warmer than the shaded side. Rapid temperature drop can result with changing weather patterns leading to injury manifested in cracking and drying of bark on the southwest side. Severe injury can lead to death during periods of water stress in the subsequent growing season and even moderate loss of bark results in loss of production in fruit trees and unsightly ornamental trees. This type of injury is very much cultivar related in apples but very little breeding has been directed towards solving this problem.

(iii) WINTER BURN

of conifers is similar to sunscald in the extent of differential warming and cooling in needles and can also be overcome by shading in winter as can sunscald.

(iv) *WINTER DESICCATION*

injury of evergreens may have symptoms similar to winter burn but is caused by severe dehydration stress because plant tissue moisture can not be replenished when soil and plant tissue
water is frozen. Some plants cannot tolerate a high degree of dehydration stress under conditions of low temperature, low humidity and strong winds.

(v) *BLACK HEART INJURY*

A common cause of injury to deciduous hardwoods and most fruit trees in which low temperature supercooling occurs is black heart injury. This is the result of xylem ray parenchyma cells
being killed followed by oxidative browning from poly-phenol oxidation. Death does not
usually result immediately or in some cases at all, but the tree is structurally weakened and
prone to secondary pathogenic invasion. In most cases of moderate injury new xylem is generated in the following growing season from more cold resistant cambial tissue which was un-
injured, and a central black heart wood core remains permanently in these trees. This problem
is of greatest concern to breeders in cold temperate regions particularly where temperatures
colder than $-40°C$ are likely to occur even infrequently.

(vi) *TIP DIE BACK*

is most common in young trees which have not acclimated early enough in fall to resist a rapid
temperature drop. This is often a problem to the nursery industry where rapid growth from the
vigorous fertilization and irrigation program may be desirable from a growth standpoint but
also interfere with rapid cold acclimation. The problem also occurs with terminal growth of
cultivars which are only marginally cold hardy especially if they have limited capacity to
acclimate early enough at the onset of winter.

(vii) *FLOWER BUD INJURY*

in *Prunus*[7,66], Rhododendron[23,42,76], some *Rubus*[73], *Vaccinium*[5] and *Vitis*[64] is caused by
nucleation of a supercooled aqueous fraction in the floral primordia. The lower limit of low
temperature supercooling was postulated to be approximately $-47°C$.[8] This phenomenon does
not occur in *Malus, Pyrus* or *Amelanchier*[40] flower buds even though it is present in the xylem
ray parenchyma of *Malus* and *Pyrus*.

In spite of the presence of this avoidance mechanism in flower buds, extremely cold resistant
genotypes are known to exist, because in these special genotypes the low temperature exotherms shift rapidly in response to sub-freezing temperatures. This shifting mechanism is
extremely useful in identifying hardy parental material for breeding programs and promises
to be of great benefit to plant breeders.

(viii) *KILLING OF VASCULAR CONNECTIVE TISSUE*

A closely related type of injury which affects the development of flower buds is the killing
of vascular connective tissue in flower buds. Spur bearing apples and cherries often exhibit this
type of injury wherein flowers fail to develop fully and dry up in spring. In *Rubus chamaemorus* browning of the vascular connective tissue clearly was the first point of injury followed
by death of the floral primordia.[73]

Artificial freezing tests

Researchers have used portable field freezing chambers and laboratory freezing chambers to standardize freezing programs for screening purposes. Although portable freezing chambers most closely approach natural conditions their extreme cost, bulkiness and inflexibility in terms of plant size accommodated, and plot design, make their use a questionable choice over laboratory freezing chambers. The most popular approach and one which has resulted in reliable data is to use laboratory freezing units which are equipped with accurate temperature control systems. These units can be cooled with liquid nitrogen[70] for precise control or much more economically by mechanical refrigeration.[33] The essential elements are a freezing chamber, an electronic process controller, a programmer, and a temperature recorder. The most common approach is to freeze at a predetermined rate and to remove samples at specified temperatures over a range likely to cover the response of the material. The range of temperatures selected will vary with the stage of acclimation throughout the test season.

Freezing and thawing rates are best determined experimentally, but in general full expression of hardiness is obtained by approaching equilibrium freezing.[49] Slow freezing rates 5°C/hour or less have given better separation than rapid rates with peach[86] and strawberry[35].

Evaluation of injury

A normal growth response following freezing may be the most reliable means of testing for most types of cold injury. This can be achieved by forcing detached shoots or by growing whole plants if they are small enough. Problems may be encountered with woody plants in deep rest where forcing may not be possible. This may be overcome by providing an appropriate rest period before or after the freezing stress if time permits.

A combination of using a recovery method as outlined in the previous paragraph with evaluation of discoloration due to browning from polyphenol oxidation has been used with success. In many instances the browning reaction is used alone wherein it is possible to detect injury to woody stems and floral primordia if the samples are given a sufficient incubation following low temperature stress. The samples can be incubated in a moist environment to facilitate rapid browning prior to evaluation. A subjective scale is used with a careful comparison to an unfrozen control to determine survival or lethal temperature ranges. Subjective scoring may not always be appropriate for research purposes, consequently, a number of objective viability assays have also been developed.

Conductivity of electrolytes from plant extracts has been used for over forty years to provide objective data for evaluation of injury[49] but the results have not always been in total agreement. Harris[35] found conductivity to be more reliable than browning as a method of evaluating injury in strawberries but Stergios and Howell[78] preferred recovery for evaluating injury in strawberries, cherries, raspberries, and grapes. Details for the method and a discussion of the theoretical basis are provided by Levitt[49] as is information on other objective screening methods such as vital staining with neutral red and reduction of colorless triphenyltetrazolium chloride. The application of these methods as tests for viability in fruit breeding are also available in detail in a review by Quamme and Stushnoff.[68]

One artificial freezing method which may directly provide information on viability at the moment freeze injury occurs is the use of exotherm analysis. Thermal analysis produces an exotherm resulting from nucleation of a supercooled cell fraction which can be detected as an exotherm on a temperature recording of the sample during the freezing process.[8] The same principle is used in differential thermal analysis except that a dried reference sample is compared to add greater sensitivity.

This system is a powerful research technique for studying basic mechanisms but it has definite limitations as a tool for screening seedlings in a breeding program. There are limitations as to the number of samples which can be handled and very sensitive and costly recorders are necessary. In addition, because a physical process is involved one should be aware of the fact that although an exotherm in a live sample depicts death very precisely an exotherm can also be generated from a sample which was previously killed by freezing. Therefore caution must be exercised when exotherm analysis is used as a viability assay. Details in implementation for *Prunus* are available in[7,66] and for *Vaccinium*.[5]

Development of cold hardy cultivars and breeding programs

Undoubtedly, the greatest number of woody plant cultivars used by mankind in the past arose from selection of superior individuals from natural populations. This was possible because of both the heterozygous nature of most woody plants which provided a great diversity for selection, and because asexual propagation could be used to great advantage in preserving genotypes which were best suited for commercial purposes. A number of such selections are still being made in the development of woody ornamentals. The earliest breeding programs, particularly with fruit trees, were based on growing out large number of open pollinated seedlings from the best maternal parents available. This approach was logical and successful in developing a rather large number of new cultivars, especially pioneer cultivars in areas where cold completely prevented production from non-resistant cultivars. Controlled crosses soon followed as breeding programs grew and were designed to provide information on breeding behaviour as well as to produce new cold resistant cultivars.[68,79]

Only limited progress has been obtained in determining the inheritance of cold resistance in woody plants. This is not surprising because heterozygosity, long life cycles (up to 10 years) and complex expression of the character with strong environmental interactions do not make such studies simple or appealing to most researchers. In spite of these restraints Watkins and Spangelo[85] developed a well controlled artificial freezing study of very hardy seedling apples and concluded that with the possible exception for root damage, epistasis or dominance were not major factors. Shoot tip injury, stem damage, time and percentage of leaf emergence and composite scores for shoot damage exhibited 90—100% additive variance. Extreme hardiness was obtained from *Malus baccata* and they postulated that this north European species may have contributed a gene complex for plant survival. Rapp and Stushnoff[73] also found evidence for additive variance in flower bud survival of *Rubus chamaemorus*. This study was based on differential thermal analysis of flower bud vascular connective tissue. Time induced acclimation was also estimated in this study. Quantitative inheritance of cold hardiness is also supported by several other studies including apple,[6,46,59] peach,[56] raspberry[1] and strawberry.[65]

One study by Dorsey and Bushnell[12] strongly supports maternal inheritance for hardiness with interspecific crosses of *Prunus americana* (hardy) X *Prunus salicina* (non-hardy). Within species crosses in *Malus* by Harris[34], Wilner[88] and Thiele[83] show less striking maternal tendencies for inheritance of winter hardiness.

Interspecific hybridization and back crossing have been broadly relied on to incorporate cold hardiness as in apricots and raspberries,[67] blueberries,[80] grapes[58,63] and plums.[12]

With most woody plants breeders can deal with maximum fitness in the heterozygous condition, therefore, it is particularly important to design breeding programs to maintain variability with these crops and to develop accurate identification procedures to select desired individuals for propagation. Transgressive segregants have been identified for hardiness in blueberries.[80]

Perhaps the most successful ultimate approach might be to develop plants which avoid cold stress completely, such as: flowering and fruiting from current season's primordia rather than from the more common types which initiate flower primordia in the year previous to fruiting; low growing types which are protected from low temperature by snow cover; late blooming types to avoid spring frost; and modified heat threshold requirements to avoid stressful situations at critical phases of development.

It is evident from the foregoing that contemporary plant breeders have a number of alternatives available when considering implementation of a program for cold hardiness. Each crop, geographic location and breeder will have special opportunities and limitations so it is not easy or prudent to define one standard approach for all situations. General guidelines, however, may be useful for consideration in a new venture and a discussion along these lines is provided in Quamme and Stushnoff.[68] A summary of these guidelines follows:

1.   Identify and characterize primary and secondary physiological problems from cold stress. Give priority to, and determine emphasis of the program depending on losses and probability of success.

2.   Determine the nature and extent of environmental control on acclimation and deacclimation.

3.   Develop and evaluate data on seasonal temperature fluctuations as this might impact on a selection procedure.

4.   Develop an appropriate pre-conditioning or handling routine for standardized testing.

5.   Incorporate an artificial freezing procedure.

6.   Standardize freezing and thawing rates for testing purposes.

7.   Select and develop appropriate viability assays. Use standard cultivars and appropriate statistical analyses.[6]

8.   Develop a data bank to characterize germplasm.

9.   Develop crossing plans based on known breeding behaviour.

10. Utilize screening at juvenile seedling stages to the extent possible, to facilitate rapid progress in the breeding program.

REFERENCES

1.   Aalders, L.E. and Craig, D.L. Progeny performance of seven red raspberry varieties in Nova Scotia. *Can. J. Plant Sci.* **41**, 406–408, 1961.

2.   Andrews, C.J., Pomeroy, M.K. and de la Roche, I.A. Changes in cold hardiness of overwintering winter wheat. *Can. J. Plant Sci.* **54**, 9–15, 1974.

3.   Andrews, J.E. Controlled low temperature tests of sprouted seeds as a measure of cold hardiness of winter wheat varieties. *Can. J. Plant Sci.* **38**, 1–7, 1958.

4.   Arny, D.C., Lindow, S.E. and Upper, C.D. Frost sensitivity of *Zea mays* increased by application of *Pseudomonas syringae*. *Nature* **262**, 282–284, 1976.

5.   Biermann, J., Stushnoff, C. and Burke, M.J. Differential thermal analysis and freezing injury in cold hardy blueberry flower buds. *J. Amer. Soc. Hort. Sci.* **104**, 444–449, 1979.

6.   Bittenbender, H.C. and Howell Jr., G.S. Adaption of the Spearman-Karber method for estimating the T_{50} of cold stressed flower buds. *J. Amer. Soc. Hort. Sci.* **99**, 187–190, 1974.

7. Burke, M.J. and Stushnoff, C. Frost hardiness: A discussion of possible molecular causes of injury with particular reference to deep supercooling of water. *Stress Physiology in Crop Plants.* H. Mussell and R.C. Staple (eds.), Wiley-Interscience, New York, 510, 1979.
8. Burke, M.J., Gusta, L.V., Quamme, H.A., Weiser, C.J. and Li, P.H. Freezing and injury in plants. *Ann. Rev. Plant Physiol.* 27, 507–528, 1976.
9. Cahalan, C. and Law, C.N. The genetic control of cold resistance and vernalization requirement in wheat. *Heredity* 42, 125–132, 1979.
10. Clarke, J.A., Martin, J.H. and Parker, J.H. Comparative hardiness of winter wheat varieties. *USDA Dept. Circ.* No. 378, 19, 1926.
11. Dix, P.J. Cell culture manipulations as a potential breeding tool. *Low Temperature Stress in Crop Plants — The Role of the Membrane.* J.M. Lyons, D. Graham and J.K. Raison (eds.), Academic Press, New York, 565, 1979.
12. Dorsey, M.J. and Bushnell, J. Plum investigations. II. The inheritance of hardiness. *Univ. Minn. Tech. Bul.* 32, 34, 1925.
13. Dvorak, J. and Fowler, D.B. Cold hardiness potential for triticals and tetraploid rye. *Crop Sci.* 17, 477–478, 1978.
14. Erickson, J.R. Inheritance of winterhardiness. *Proc. 15th Hard Red Winter Wheat Workers Conference. Ft. Collins, Colorado.* 24–29, 1980.
15. Eunus, A.M., Johnson, L.P. and Aksel, R. Inheritance of winter hardiness in an eighteen-parent diallel cross in barley. *Can. J. Genet. Cytol.* 4, 356–376, 1962.
16. Fejer, S.O. Combining ability and correlation of winter survival, electrical impedance and morphology in juvenile apple trees. *Can. J. Plant. Sci.* 56, 303–309, 1976.
17. Fowler, D.B. Selection for winterhardiness in wheat. II. Variation within field trials. *Crop Sci.* 19, 773–776, 1979.
18. Fowler, D.B. and Gusta, L.V. The influence of fall growth and development on cold tolerance of rye and wheat. *Can. J. Plant Sci.* 57, 751–755, 1977a.
19. Fowler, D.B. and Gusta, L.V. Dehardening of winter wheat and rye under spring field conditions. *Can. J. Plant Sci.* 57, 1049–1054, 1977b.
20. Fowler, D.B. and Gusta, L.V. Selections for winterhardiness in wheat. I. Identification of genotypic variablity. *Crop Sci.* 19, 769–772, 1979.
21. Fowler, D.B., Gusta, L.V., Bowren, K.E., Crowle, W.L., Mallough, E.D., McBean, D.S. and McIver, R.N. Potential for winter wheat production in Saskatchewan. *Can. J. Plant Sci.* 56, 45–50, 1976.
22. Fowler, D.B., Gusta, L.V. and Tyler, N.J. Selection for winterhardiness in wheat. III. Screening methods. *Crop. Sci.* 21, 896–901, 1981.
23. George, M.F., Burke, M.J., Pellett, H.M. and Johnson, A.G. Low temperature exotherms and woody plant distribution. *HortScience,* 9, 519–522, 1974a.
24. George, M.F., Burke, M.J. and Weiser, C.J. Supercooling in overwintering azalea flower buds. *Plant Physiol.* 54, 29–35, 1974b.
25. Grafius, J.E. Winter hardiness in barley: Breeding for resistance. *Mich. State Univ. Research Rep.* 247, 16–20, 1974.
26. Graham, P.R. and Mullin, R. The determination of lethal freezing temperatures in buds and stems of deciduous azalea by a freezing curve method. *J. Amer. Soc. Hort. Sci.* 101, 3–7, 1976.
27. Gullord, M. Genetics of freezing hardiness in winter wheat (*Triticum aestivum* L.). *Ph.D. Thesis.* Michigan State University, 1974.
28. Gullord, M., Olien, C.R. and Everson, E.H. Evaluation of freezing hardiness in winter wheat. *Crop Sci.* 15, 153–157, 1975.
29. Gusta, L.V. *Personal communication,* 1982.
30. Gusta, L.V. and Fowler, D.B. The effect of temperature on dehardening and rehardening of winter cereals. *Can. J. Plant Sci.* 56, 673–678, 1976a.
31. Gusta, L.V. and Fowler, D.B. Dehardening and rehardening of spring collected winter wheats and a winter rye. *Can. J. Plant Sci.* 56, 775–779, 1976b.
32. Gusta, L.V. and Fowler, D.B. Factors affecting the cold survival of winter cereals. *Can. J. Plant Sci.* 57, 213–219, 1977.
33. Gusta, L.V., Boyachek, M. and Fowler, D.B. A system for freezing biological materials. *HortSci.* 13, 171–172, 1978.

34.   Harris, R.E.   The hardiness of progenies from reciprocal *Malus* crosses. *Can. J. Plant Sci.* **45**, 159–161, 1965.

35.   Harris, R.E.   Laboratory technique for assessing winter hardiness in strawberry (*Fragaria* X *ananassa* Duch.). *Can. J. Plant Sci.* **50**, 249–255, 1970.

36.   Hayes, H.K. and Aamodt, O.S.   Inheritance of winter hardiness and growth habit in crosses of Marquis and Minturki wheats. *J. Agric. Res.* **35**, 223–236, 1927.

37.   Ishikawa, M. and Sakai, A.   Freezing avoidance mechanisms by supercooling in some Rhododendron flower buds with reference to water relations. *Plant and Cell Physiol.* **22**(6), 953–967, 1981.

38.   Janick, J. and Moore, J.N.   *Advances in Fruit Breeding.* Purdue University Press, Lafayette, Indiana, 1975.

39.   Jenkins, G.   Transgressive segregation for frost resistance in hexaploid oats (*Avena spp.*) *J. Agric. Sci. Camb.* **73**, 477–482, 1969.

40.   Junttila, O. and Stushnoff, C.   Dehardening in flower buds of saskatoon berry *Amelanchier alnifolia* in relation to temperature, moisture control and spring bud dormancy (*in preparation*), 1982.

41.   Kaku, S. and Iwaya, M.   Deep supercooling in xylems and ecological distribution in the genera *Ilex, Viburnum* and *Quercus* in Japan. *Oikos,* **33**, 402–411, 1979.

42.   Kaku, S., Iwaya, M. and Kinishige, M.   Supercooling ability of Rhododendron flower buds in relation to cooling rate and cold hardiness. *Plant and Cell Physiol.* **21**(7), 1205–1216, 1980.

43.   Kinbacher, E.J. and Jensen, N.F.   Weather records and winter hardiness. *Agron. J.* **51**, 185–186, 1959.

44.   Kir'yan, M.V. and Barashkova, E.A.   Evaluation of winter hardiness and frost resistance in the first and second generations. *Pl. Br. Abst.* **51**, 9481, 1981.

45.   Lafever, H.N.   A progress report. . . . Hybrid Wheat. Ohio Rep. Res. and Devell. Center. *Wooster,* **53**(1), 8–10, 1968.

46.   Lantz, H.L. and Pickett, B.S.   Apple breeding: Variation within and between progenies of 'Delicious' { with respect to freezing injury due to the November freeze of 1940. *Proc. Amer. Soc. Hort. Sci.* **40**, 237–240, 1940.

47.   Law, C.N. and Jenkins, G.   A genetic study of cold resistance in wheat. *Gent. Res. Camb.* **15**, 197–208, 1970.

48.   Levitt, J.   *Responses of plants to environmental stresses.* Academic Press. New York, 697, 1972.

49.   Levitt, J.   *Responses of plants to environmental stresses* Vol. 1 *Chilling, freezing and high temperature stresses.* Academic Press. New York. 497, 1980.

50.   Li, P.H. and Sakai, A.   *Plant Cold Hardiness and Freezing Stress — Mechanisms and Crop Implications.* Academic Press. New York. 416, 1978.

51.   Limin, A.E. and Fowler, D.B.   The expression of cold hardiness in *Triticum* species amphiploids. *Can. J. Genet. Cytol.* **24**, 51–56, 1982.

52.   Lyfenko, S.F.   Some patterns of the inheritance of frost resistance in hybrids of winter bread wheat. *Pl. Br. Abst.* **19**, 6832, 1979.

53.   Lyons, J.M., Graham, D. and Raison, J.K.   *Low Temperature Stress in Crop Plants — The Role of the Membrane.* Academic Press. New York, 565, 2979.

54.   Martin, J.H.   Comparative studies of winterhardiness in wheat. *J. Am. Soc. Agron.* **35**, 493–535, 1927.

55.   Muehlbauer, F.J., Marshall, H.G. and Hill, Jr., R.R.   Winter hardiness in oat populations derived from reciprocal crosses. *Agron. J.* **10**, 646–649, 1970.

56.   Mowry, J.B.   Inheritance of cold hardiness of dormant peach flower buds. *Proc. Amer. Soc. Hort. Sci.* **85**, 128–133, 1964.

57.   Mussell, H. and Staples, R.C.   *Stress Physiology in Crop Plants.* Wiley-Interscience. New York. 510, 1979.

58.   Ning, H., Baozhong, Z., Yufeng, F., Ion, F.Y. and Shurong, L.   The inheritance of some characters in interspecific hybrids of grapes, with special reference to cold resistance. *Acta Horticulturae Sinica* **8**(1), 1–7, 1981.

59.   Nybom, N., Bergendal, P.O., Olden, E.L. and Tamas, P.   On the cold resistance of apples. *Eucarpia* **182**, 66–73, 1962.

60.   Olien, C.R.   Freezing stresses and survival. *Ann. Rev. Plant Physiol.* **18**, 387–408, 1967.

61.   Olien, C.R. and Smith, M.N.   *Analysis and improvement of plant cold hardiness.* CRC Press Inc. Bocca Raton Fla. 215, 1981.

62. Paulsen, G.M. Effect of photoperiod and temperature on cold hardening in winter wheat. *Crop Sci.* **8**, 29–32, 1968.

63. Pierquet, P. and Stushnoff, C. Variation and breeding potential of some northern clones of *Vitis riparia*, Michx. *Fruit Var. Jour.* **32**, 74–84, 1978.

64. Pierquet, P., Stushnoff, C. and Burke, M.J. Low temperature exotherms in stem and bud tissues of *Vitis riparia* Michs. *J. Amer. Soc. Hort. Sci.* **102**, 54–55, 1977.

65. Powers, L. Strawberry breeding studies involving crosses between cultivated varieties (*Gragaria* X *ananassa*) and the native Rocky Mountain strawberry (*F. ovalis*). *J. Agr. Res.* **70**, 95–122, 1945.

66. Quamme, H.A. An exothermic process involved in the freezing injury to flower buds of several *Prunus* species. *J. Amer. Soc. Hort. Sci.* **99**, 315–318, 1974.

67. Quamme, H.A. Breeding and selecting temperate fruit crops for cold hardiness. 313–332. *Plant Cold Hardiness and Freezing Stress — Mechanisms and Crop Implication.* P.H. Li and S. Sakai (eds.), Academic Press, New York, 1978.

68. Quamme, H.A. and Stushnoff, C. Breeding for resistance to environmental stress. *Methods in Fruit Breeding.* J. Janick and J.N. Moore (eds.). (In press), 1982.

69. Quamme, H.A., Stushnoff, C. and Weiser, C.J. The relationship of exotherms to cold injury in apple stem tissues. *J. Amer. Hort. Sci.* **97**, 608–613, 1972a.

70. Quamme, H.A., Evert, D.R., Stushnoff, C. and Weiser, C.J. A versatile temperature control system for cooling and freezing biological materials. *Hort. Sci.* **7**, 24–26, 1972b.

71. Quisenberry, K.S. Inheritance of winter hardiness, growth habit and stem-rust reaction in crosses between Minhardi winter and H-44 spring wheats. *USDA Tech. Bull.* **218**, 1–45, 1931.

72. Rajashekar, C., Gusta, L.V. and Burke, M.J. Frost damage in hardy herbaceous species. 565. *Low Temperature Stress in Crop Plants — The Role of the Membrane.* Lyons *et al.* (eds.), Academic Press, New York, 1979.

73. Rapp, K. and Stushnoff, C. Artificial freezing of *Rubus chamaemorus* L. for estimation of genetic components of cold hardiness. *Meldinger fra Norges Landbrukshøgskole* (*Scientific Reports of the Agricultural University of Norway*) **58**, 1–14, 1979.

74. Reitz, L.P. and Salmon, S.C. Hard red winter wheat improvement in the plains. A 20-year summary *USDA Tech. Bull.* **1192**, 1–117, 1959.

75. Rosenquist, C.E. Winter hardiness in the first generation of several wheat crosses. *J. Amer. Soc. Agron.* **25**, 528–533, 1933.

76. Sakai, A. Freezing tolerance of evergreen and deciduous broadleaved trees in Japan with reference to tree regions. *Low Temp. Sci. Ser-B.* **36**, 1–19, 1978.

77. Sprague, V.G., Neuberger, H., Orgell, W.H. and Dodd, A.V. Air temperature distribution in the microclimatic layer. *Agron. J.* **46**, 104–108, 1954.

78. Stergios, B.G. and Howell Jr., G.S. Evaluation of viability tests for cold stressed plants. *J. Amer. Soc. Hort. Sci.* **98**, 325–330, 1973.

79. Stushnoff, C. Breeding and selection methods for cold hardiness in deciduous fruit crops. *HortSci.* **7**, 10–13, 1972.

80. Stushnoff, C. Development of cold hardy blueberry hybrids. *Fruit Vars. Journal* **30**, 28–29, 1976.

81. Sutka, J. Genetic studies of frost resistance in wheat. *Theor. Appl. Genet.* **59**, 145–152, 1981.

82. Svec, L.V. and Hodges, H.F. Cold hardening and morphology of barley seedlings in controlled and natural environments. *Can. J. Plant Sci.* **52**, 955–963, 1972.

83. Thiele, I. Progressive frostversuche mit kiemender und eing jukrigen kernobstamlingen. *Zuchter* **27**, 161–172, 1957.

84. Vallejos, C.E. Genetic diversity of plants for respónse to low temperature and its potential use in crop plants. 565. *Low Temperature Stress in Crop Plants — The Role of the Membrane.* Lyons *et al.* (eds.), Academic Press, New York, 1979.

85. Watkins, R. and Spangelo, L.P.S. Components of genetic variance for plant survival and vigor of apple trees. *Theor. App. Gen.* **40**, 195–203, 1970.

86. Weaver, G.M., Jackson, H.O. and Stroud, F.D. Assessment of winter hardiness in peach cultivars by electric impedance, scion diameter and artificial freezing studies. *Can. J. Plant Sci.* **48**, 37–47, 1968.

87. Weiser, C.J. Cold resistance and injury in woody plants. *Sci.* **169**, 1269–1278, 1970.

88.   Wilner, J. The influence of maternal parent on frost-hardiness of apple progenies. *Can. J. Plant Sci.* **45**, 67–71, 1964.

89.   Worzella, W.W. Inheritance of cold resistance in winter wheat, with preliminary studies on the technique of artificial freezing test. *J. Agr. Res.* **50**, 625– 635, 1935.

90.   Worzella, W.W. and Cutler, G.H. Factors affecting cold resistance in winter wheat. *J. Am. Soc. Agron.* **33**, 221–230, 1941.

91.   Yelonosky, C. Freeze survival of citrus trees in Florida. 416. *Plant, Cold Hardiness and Freezing Stress — Mechanisms and Crop Implications.* P.H. Li and A. Sakai (eds.), Academic Press, New York, 1977.

CHAPTER 6

SCREENING AND SELECTION TECHNIQUES FOR IMPROVING DROUGHT RESISTANCE

J.M. Clarke and T.F. Townley-Smith
Research Station,
Research Branch, Agriculture Canada,
Swift Current, Saskatchewan, S9H 2X2 Canada

INTRODUCTION

In developing a breeding program to improve the drought resistance of a crop, it is first necessary to gain an understanding of how the crop reacts to drought. This is best done under field conditions in the area where the crop is grown, since the seasonal timing of drought stress varies from one location to another. Once the overall drought reaction has been established, the next step is to identify mechanisms of drought escape, avoidance and tolerance which contribute to performance under drought conditions. In some instances these can be inferred from our general knowledge of the drought responses of plants. Once the mechanism has been identified, different genotypes must be studied in order to determine genetic variability, heritability and relationship to performance under drought stress. After this has been done, screening procedures can be developed to select for the desired characteristics in either parental or segregating material.

Since the literature abounds with results of empirical studies of drought stress on crops, it is unnecessary to repeat any of it here. Not that such studies are unimportant, indeed, they

137

provide a foundation from which screening procedures can be developed. Instead, we will concentrate on providing examples of drought resistance characteristics that have been identified in particular crops and how these can be screened for in breeding programs.

SCREENING METHODS

(i) Definition of terms

One of the simplest mechanisms by which plants deal with drought is drought escape. This, as the name implies, is accomplished by rapid phenological development or by developmental plasticity. Drought resistance, by its simplest definition, is the sum of drought avoidance and drought tolerance.[82] Avoidance, called drought tolerance with high tissue water potentials by May and Milthorpe[86], consists of mechanisms to reduce water loss from the plant and of mechanisms to maintain water uptake. Drought tolerance refers to the ability of the plant to withstand low tissue water potentials.

There seems to be some controversy as to which component of drought resistance, avoidance or tolerance, is of most importance in a crop. Boyer and McPherson[21] concluded that although drought avoidance might permit a longer growth period in the crop through reduced water use or increased water uptake, drought avoidance mechanisms often operate at the expense of photosynthesis and reduce top growth by increasing root development. They suggest that tolerance would be more desirable since the crop could produce more yield at lower water potentials. Levitt[82], however, stated that in general, drought avoidance is more important than drought tolerance in higher plants. Fischer and Turner[43] suggested that mechanisms favoring drought survival and those favoring productivity are mutually opposed. In reality, a mixture of both avoidance and tolerance mechanisms is required. Even the best drought avoiding species requires tolerance, since some reduction in plant water potential is unavoidable during severe stress. Indeed, there is evidence to suggest that drought avoidance is operative during the vegetative phase, while tolerance comes into play during the reproductive phase in crops such as cereals. In winter wheat, Keim and Kronstad[75] found both avoidance and tolerance traits to be important in drought resistant genotypes.

(ii) Drought escape

Drought escape is perhaps most dramatic in the 'ephemeral' plants of desert regions which complete life cycles in as few as 4 to 6 weeks. However, drought escape also plays a significant role in crop species, both through earliness of maturity and developmental plasticity. In wheat (*Triticum aestivum* L.), Derera *et al.*[30] found that there was a negative correlation between grain yield and days to ear emergence. They concluded that 40 to 90% of the variation in the yield of the genotypes studied could be ascribed to earliness. Fischer and Maurer[42] also observed superior yields in early-flowering genotypes of wheat under stress. Early maturity was shown to be associated with reduced water use in sorghum (*Sorghum bicolor* L.),[12] resulting from lower leaf area index and root density.[18] Alessi and Power[2] concluded that grain yield of an early-maturing maize hybrid (*Zea mays* L.) would be less affected by severe drought than a later hybrid. In rice (*Oryza* spp.) breeding for drought escape has taken the form of developing short-season cultivars for regions where rainfall is marginal for production.[97] Under adequate moisture conditions, however, yield is correlated with time to maturity in crops such as sorghum.[27] Late-maturing cultivars may outyield early cultivars under certain drought conditions as well, such as when the drought occurs early in the season but is relieved before anthesis.[42]

There is considerable developmental plasticity exhibited by determinate annual crops such as wheat. Wheat commonly produces more tillers than required; the number which survive to produce spikes is reduced by water deficits between floral initiation and spike emergence.[9] Hurd[57] suggested that excessive tillering would be wasteful of soil moisture in moisture-restricted environments. Indeed, artificial de-tillering of a spring wheat genotype increased yield by 14 to 22% in one study.[65] Keim and Kronstad[75], however, found that tiller number was important to the yielding ability of drought-resistant winter wheats. Water stress can also hasten anthesis and maturity of wheat.[3]

In other crops, developmental plasticity is exhibited by varying degrees of determinancy of growth habit. Quisenberry and Roark[104] found that indeterminate cotton (*Gossypium hirsutum* L.) genotypes yielded more than determinate genotypes in a semi-arid environment. Under more favorable moisture regimes, there was less difference between the two types. Indeterminancy allows the crop to adapt to variable moisture regimes. In rice, selection for genotypes with their reproductive stage photoperiodically controlled to coincide with peak rainfall times has been used in order to ensure adequate available moisture during grain filling.[97]

Screening and selection for drought escape through maturity or developmental plasticity is relatively simple compared to screening and selection for some other drought resistance characteristics. However, in areas where the length of the growing season is clearly defined by cold temperatures, there is little opportunity for manipulation of maturity.

(iii) Drought avoidance

(a) *MAINTENANCE OF WATER UPTAKE*

The fundamental importance of root systems in the drought avoidance of plants has been well established. Species and genotypic differences in root system size and efficiency, in particular, have been widely researched and reported.[49,54,106,119] The high labor requirement for root system studies has limited the amount of work done in this area, although there is a continuing search for simple and effective screening techniques.

Root studies using glass-faced boxes in artificial environments revealed genotypic differences in rooting pattern and extent in wheat.[49] The *T. turgidum* variety Pelissier had a more extensive root system than the *T. aestivum* varieties Thatcher and Cypress, particularly at lower depths. This characteristic of Pelissier appeared to be responsible for its superior performance under drought conditions. In further studies of wheat root systems, Hurd[55] found the pattern of rooting to be related to yield performance. Hurd concluded that an extensive root system which rapidly penetrated the soil profile was essential for wheat cultivars grown in semi-arid areas. Crosses between Pelissier and Lakota, a variety with a smaller root system (Fig. 6.1), produced lines with root systems more like that of Pelissier than of Lakota,[61] indicating that root system size is heritable.

There has been some controversy as to whether an extensive root system, or a more restricted system which conserves water for late-season use, is preferable. Hurd's results support the former, while Passioura[101] has suggested the latter type may be superior. Passioura pruned the root systems of wheat plants to leave a single seminal root, and grew them entirely on stored soil moisture in a controlled environment. Seed yield of pruned plants was double that of intact plants, which was attributed to the superior water conservation of the pruned plants. More recently, Richards and Passioura[109] have proposed to reduce the xylem vessel size in wheat seminal root axes as a water conservation measure. The character is being introduced into adapted cultivars from land-race varieties. The superiority of either extensive or restricted root systems will depend upon the amount and location of stored soil moisture. In the North American Great Plains region, stored soil moisture exists at depths of up to 2 or more meters. Available water is still present in the 60- to 120-cm depth range after the wheat harvest in southern Saskatchewan.[58] Cultivars with deep root systems should be able to utilize this water and have a yield advantage over more shallowly-rooted cultivars.

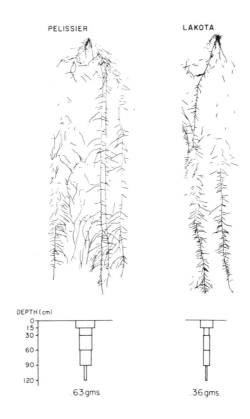

Fig. 6.1. Root systems of two *Triticum turgidum* cultivars: tracings from
glass-faced boxes (top) and spatial distribution of root mass (bottom).

Investigations have been made of the relationship between plant height type and extent of root systems in wheat. Observations indicated that semi-dwarf wheats were better adapted to moist growing conditions than to the dry plains area.[22] This led to speculation that perhaps the semi-dwarf wheats had inadequate root systems. However, subsequent comparisons of root system size and dwarfing in wheat have shown no relationship.[84,102] Similarly, Irvine *et al.*[64] found no relationship between rooting capability and height in normal and semi-dwarf barley (*Hordeum vulgare* L.).

Having established that genotypic differences in wheat root systems exist, research was begun on development of screening techniques to more easily determine these differences. The major requirement of a screening technique is to reduce the labor requirement associated with field excavation methods or in the use of glass-faced boxes in controlled environments. One approach has been to correlate seedling root mass with that of mature plants. Hurd[54] observed a relationship between seedling and mature plant root mass in Thatcher wheat. Townley-Smith and McBean (cited in Hurd[58]) found that root length of wheat seedlings grown for 5 to 7 days in sand was related to root mass at maturity. Recent results from our laboratory (Table 6.1)

TABLE 6.1. Root mass of *Triticum turgidum* genotypes sampled at 15, 30, 45
and 60 days from seeding

Genotype	Root mass (g)			
	15 days	30 days	45 days	60 days
Cando	0.042	0.450	1.163	3.315
DT363	0.070	0.513	1.570	3.390
DT367	0.054	0.618	1.295	4.255
Lakota	0.057	0.413	1.710	3.808
Pelissier	0.054	0.668	2.145	4.873
Wascana	0.052	0.610	2.055	3.533
HB5D	0.053	0.455	1.840	2.890
67B5	0.573	0.495	1.565	3.628
CF3C	0.048	0.410	1.510	4.613
FK2C	0.041	0.588	1.835	2.558

indicated that root mass at 30 days from seeding provides a good estimate of mature root mass. Kirichenko[78] reported successful selection for yield by selecting for vigorous root systems after 16 to 20 days growth in nutrient solution. If such screening procedures are indeed successful, many segregating lines could be screened during the inter months.

Recently, a method for screening of sorghum genotypes for root system characteristics has been reported.[119] The technique involves the hydroponic culture of plants in 10×100 cm tubes, using Carbowax 600 to induce stress. Although results are still preliminary, correlations with field performance under drought conditions have been observed. In rice, O'Toole and Soemartono[99] found that the force required to pull seedlings from paddy soil was correlated with root weight, branching and number. This technique shows good potential as a mass-screening technique for root-system size in rice.

(b) *REDUCTION OF WATER LOSS*

By far the majority of drought research investigations have dealt with the aerial portions of the plant because of their ease of study. Among the avoidance mechanisms which have been studied are stomatal control of transpiration, and the shape, morphology and orientation of leaves in relation to water loss. A broad range of direct and indirect screening techniques have been investigated.

The control of stomatal aperature is one of the major methods by which plants regulate water loss. There is a considerable range among species in sensitivity of stomata to water stress, while within species factors such a leaf position and age, and prior stress history of the plant affect stomatal sensitivity.[126] In sorghum, genotypic differences in stomatal response to leaf water stress have been reported.[14,53] Kazemi et al.[74] found genotypic differences in stomatal number in wheat, but it was suggested that genotype X environment interactions would make stomatal number difficult to select for.

Direct observation of stomata to determine response to water deficit is tedious, which limits the number of observations which can be made. The methods for direct observation of stomata have been reviewed by Meidner and Mansfield.[88] Indirect measurement of stomatal behaviour by means of porometers has gained wide acceptance because of the ease and rapidity with which determinations can be made. Porometers are of two basic types: viscous flow porometers which determine stomatal conductance by measuring the rate at which air can be drawn through the leaf, and diffusive flow porometers which measure the rate of diffusion of water vapour out of the leaf.[88] The former type gives information on stomatal aperature, while the latter type gives information on stomatal control of transpiration and assimilation. More recently, indirect measurement of leaf water status and stomatal activity by means of infrared thermometers has been attempted.[34] The use of this method is based on the premise that leaves with closed stomata will have higher temperatures than those with open stomata because of the evaporative cooling effect associated with transpirational water loss.

Studies of field-grown wheat have demonstrated varietal differences in stomatal behaviour. Shimshi and Ephrat[113] found differences in leaf permeability as measured by a viscous-flow porometer. Leaf permeability was positively rank-correlated with short-term transpiration and photosynthesis, and with yield. Similarly, Jones[69] demonstrated genotypic differences in conductance of wheat leaves measured with a diffusive flow porometer. He reported a negative correlation between yield and conductance for the period 7 to 14 days before anthesis. Genotypic differences in leaf diffusive resistance have also been reported in sorghum.[14] Ackerson et al.[1], however, could not find genotypic differences at high leaf water potentials or after anthesis.

Leaf temperature differences have been detected in studies of limited numbers of genotypes of soybean[24], wheat[79], maize[91], alfalfa[63] and cotton[33]. Blum[15] has proposed a method using infrared photographs of breeding nurseries taken from an elevation of 200 to 300 m. Image analysis for color density is then used to separate entries with high or low water potentials.

The usefulness of porometers or infrared thermometers as screening tools in breeding programs depends upon their accuracy and speed. Rapid measurement is necessary in order to take as many readings as possible during a short time period to avoid errors caused by diurnal changes in stomatal aperatures and weather conditions. Fischer et al.[41] claimed it was possible to make 200 measurements per hour with an airflow porometer in wheat. They suggested that a minimum of 10 observations per entry were required in order to reduce the standard error sufficiently to detect meaningful differences between treatments. At a rate of 200 observations per hour, some 20 cultivars should be screened each hour. Our experience with a portable diffusive resistance porometer suggests a rate of 50 to 60 observations per body. Jones[69] questioned the usefulness of screening for leaf conductance on the growth that although cultivar differences can be detected at specific stages of development, the differences are not strongly related to yield. Similarly, Fischer and Wood[44] found few significant correlations between wheat yield and diffusive resistance. Blum et al.[20] found no relationship between diffusive resistance and number of grains per spike in wheat.

Another approach to the study of stomatal behavior has been to measure plant abscisic acid (ABA) levels, as this hormone is thought to have an effect on stomatal aperature. Increases in ABA content of leaves undergoing water stress have been found in barley[50] and wheat[103,132]. Exogenously applied ABA will close wheat stomata[132] and there is much evidence to suggest that ABA performs this function endogenously as well. Quarrie and Jones[103] found differences in ABA content among water-stressed wheat genotypes which were associated with genotypic differences in water potential. Of 26 genotypes studied, the Canadian *T. turgidum* cultivar Wascana accumulated the least ABA, while the German *T. aestivum* cultivar Sirius accumulated the most. Maturity differences between cultivars could interact with the water stress effect on ABA level and lead to erroneous conclusions since ABA levels in the plant change seasonally.[77] Quarrie and Jones[103] noted that it is technically easier to measure leaf water potential than ABA level, so it is easier to screen for drought avoidance by measuring leaf water potential rather than ABA.

Increases in ethylene biosynthesis have been reported in stressed cotton,[87] in Valencia orange (*Citrus sinensis*, Osbeck)[10] and wheat[132]. Wright[132] found that ethylene evolution by excised leaves increased with time of wilting. T.N. McCaig of our laboratory (personal communication), however, found little measurable difference in ethylene emanation by stressed and non-stressed wheat leaves. The seasonal variation in ethylene evolution seemed to be much greater. Further research is required if ethylene production is to be properly evaluated as a screening procedure.

Water loss rates of excised leaves or plants is another method which has been used to assess genotypic differences in water retention capability. Salim *et al.*[112] found that drying wheat leaves over calcium chloride was so rapid that there was poor differentiation between cultivars. Dedio[28] demonstrated water retention differences between genotypes of common and durum wheats by weighing leaves drying in the laboratory at 24 and 48 hours after excision. The differentiation between genotypes was great enough that no special drying conditions were required. Dedio[28] also evaluated the water retention of the progeny of a cross between a cultivar with high water retention capacity, Pitic 62, and one with low retention capacity, ACEF-125. The results suggested that water retention is simply inherited and that the ability to retain water is controlled by dominant gene action.

More recent work with water loss rates of excised leaves at our laboratory indicates that the response of field-grown plants differs from that of controlled-environment grown plants. Water loss rates are greater in the field-grown plants, necessitating weighing of leaves at shorter time intervals after excision as well as controlled drying conditions. Good separation of cultivars has been achieved by weighing 6 to 10 hours after excision (Fig. 6.2). A system utilizing tared weighing dishes and drying in a room with controlled temperature (20C) and humidity (50% RH) has been developed. Leaf age and the water regime under which the plants are grown influence water retention capability. Differentiation between cultivars tends to be best when the plants are subjected to some water stress during growth.

Progeny of a Hercules X Pelissier cross, these cultivars having low and high water retention capability, respectively, showed a range of leaf water contents between the parental values (Fig. 6.3). Yield was positively correlated with water retention capability. These lines had undergone some yield selection and were tested at the F_9 generation. Evaluation of F_2 plants from a series of Pelissier crosses showed that the leaf water contents of progeny 24 hours after excision tended to fall between the parental values (Table 6.2).

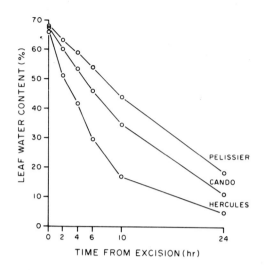

Fig. 6.2. Leaf water contents of *Triticum turgidum* genotypes at various times
after excision.

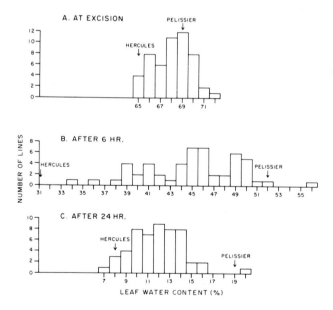

Fig. 6.3. Frequency distributions of excised-leaf water contents of 52 *Triticum turgidum* lines
from a cross of high (Pelissier) and low (Hercules) water retention capability parents.

TABLE 6.2. Leaf water content 24 hours after excision of nine bulk F_2
populations of *Triticum turgidum* from crosses with Pelissier

Parents	24 hr leaf water content (%)	
	Parent	F_2 bulk
Pelissier	25.8	
Cando	16.4	20.9
Wascana	10.9	17.1
DT363	8.2	13.4
DT367	17.5	21.6
AQ5C	12.5	16.0
EZ2E	16.4	24.8
Q4A	14.9	15.9
HB5D	15.2	17.8
67B5	25.4	23.6

Measurements of leaf water status, principally leaf water potential and osmotic potential, have been popular in many drought investigations. Reviews of the methods of making these determinations can be found elsewhere (e.g. Barrs.[8] Numerous studies have shown cultivar differences in leaf water status in wheat[69,71,115], sorghum[1,13,118], and rice[96], among other species.

Although cultivar differences in leaf water status have been reported in many species, the exploitation of these differences as a screening technique is still somewhat controversial. In wheat, Jones[69] detected cultivar differences in leaf water potential at certain growth stages, but seasonal differences were not significant and did not relate to yield. Blum et al.[20], however, found that leaf water potential was correlated with final number of grains per spike. Similarly, Sojka et al.[115] concluded that osmotic potential gave a good indication of relative water stress differences between wheat cultivars. These differences of opinion may arise from differing environmental conditions. Leaf water status is closely linked to soil water status, so uniform distribution of soil moisture and uniformity of soil type is required in order for genotypic differences to be detected. Direct measurement of water or osmotic potential tends to be too slow to screen many cultivars at once.

In addition to measurements of leaf water potential and osmotic potential, leaf water status can be evaluated by indices calculated from the difference between turgid and fresh weights of leaves. Leaves are removed from the plant, weighed, floated on water until fully turgid or saturated, and reweighed. From these weights, indices such as relative water content (RWC), water saturation deficit (WSD) and relative saturation deficit (RSD) can be calculated.[8] The use of RSD is often preferred, since RWC and WSD require determination of leaf dry weight as well. Todd and Webster[121] proposed a method for screening cereal seedlings for drought resistance based on a relative turgidity index. This method was somewhat cumbersome in that several cycles of drought were employed, and large populations were required in order to detect cultivar differences. Dedio[28] found differences in RSD between wheat cultivars, but the differences were not well correlated with known drought resistance. Leaf water content was more related to known drought resistance. Blum[14] found differences in WSD between sorghum geno-types which were consistent over two years.

Visual signs of leaf water status such as wilting and leaf rolling have been investigated in some crops. Leaf rolling can indicate stress, but it can also serve as a drought avoidance mecha-nism. Reductions in transpiration rate of 46 to 83% resulting from leaf rolling have been re-ported in grasses of the Mediterranean area.[95] Hurd[59] reported leaf rolling in durum wheat cultivars; rolling seemed to be greatest in drought resistant cultivars, and did not appear to impair the photosynthetic capability of the leaves. Leaf rolling is an obvious symptom of drought in rice, and has been evaluated as a screening technique for drought resistance.[96] Visual scoring for leaf rolling and leaf tip drying was done 39 days after the last irrigation. The visual score was correlated with leaf water potential; the genotypes with low osmotic potential could be identified by greater leaf rolling and tip drying. The visual scoring technique is now used for screening some 2000 rice lines per year.[97]

Plant water status can also be influenced by morphological features such as leaf size and shape, leaf angle, cuticular waxiness and leaf reflectance. Leaves of drought resistant species are generally smaller and have a lower surface area volume ratio than species adapted to higher moisture environments. Reduced leaf canopy area also reduces water loss. Artificial defoliation of barley has been shown to reduce the amount of water required to produce a crop.[45] Both native plant and crop species accelerate natural leaf drop under drought stress, but evidence of genotypic differences in leaf drop is lacking. Although leaf drop may reduce transpirational water loss, it also reduces yield in crops such as wheat, wherein leaf area duration tends to be positively correlated with yield.[40,116] Leaf movement to orient the leaf parallel to the incident radiation reduces effective leaf area and energy load on the plant.[9] Severe wilting of leaves results in limpness which also causes a change in the interception of radiation.

The leaf cuticle plays an important role in limiting water loss from leaves. In native species, cuticular transpiration has been shown to range from 2 to 50% of total leaf transpiration.[95] Cuticular thickness has often been cited as a drought avoidance characteristic, but cuticular structure and chemical composition are probably of more significance than thickness alone.[100] Deposition of lipids in the cuticle or of waxes on the leaf surface help to reduce cuticular trans-piration. O'Toole et al.[98] demonstrated the effectiveness of epicuticular waxes in increasing cuticular resistance in two rice cultivars. Removal of the waxes decreased the cuticular resist-ance by 50 to 60%.

Changes in the reflectance of leaves can be effected by pubescence or reflective coatings on the leaf surface. In the California shrub genus Encelia, Ehleringer et al.[32] found a 56% reduc-tion in light absorbance of leaves of a pubescent species as compared to a nonpubescent species. In sorghum, genotypes with a reflective waxy 'bloom' on the leaf surface had higher water use efficiency than bloomless genotypes.[25] Although increased reflectance lowers transpiration rate, it also reduces photosynthetic rate.[25,32] Genotypes with greater reflectance would, there-fore, only be adapted to arid conditions. If genotypic differences in pubescence or leaf color exist, screening and selection would be relatively straightforward. Gravimetric methods for measuring epicuticular waxes tend to be too slow for selection purposes,[15] but a colorimetric method has proven to be faster.[31]

Once a suitable screening procedure for plant or leaf water status is decided upon, the breeder must determine which type of plant response, in terms of stomatal sensitivity, he is looking for. For example, is a cultivar with stomata sensitive to water stress, i.e. stomata close at high leaf water potential, or one with relatively insensitive stomata preferable? The answer probably will vary with the environment and stage of crop growth. Where the crop is growing on stored moisture alone, water conservation during the early part of the season can leave more water available for grain filling. Sensitive stomata could be detrimental during grain filling if photosynthesis is reduced enough to limit yield. Recently, Blum et al.[20] have reported that some wheat genotypes seem to be able to maintain high leaf water potential without completely closed stomata. Genotypes such as these should be able to sustain higher rates of productivity than genotypes which maintain high water potential by closing stomata.

The growth stage at which measurements are made affects the observed stomatal sensitivity, since ontogenic changes in response have been reported.[170,90] Jones[70] raised the point that reliance upon leaf conductance alone as a screening technique does not necessarily identify genotypes with good control of leaf water potential. After all, transpiration rate of the crop depends upon other factors such as total canopy area as well as leaf conductance.

(iv) Drought tolerance

(a) GERMINATION AND EARLY SEEDLING DEVELOPMENT

The germination of the seed is the first stage of crop development at which drought stress can be encountered. This has led to investigations of the moisture requirements for germination in regions where dry seedbeds are a common problem. Germination percentage and rates of seedling growth differed between winter wheat lines when planted in dry soil or when stressed artificially by the use of osmoticums.[52] Similarly, another study[51] revealed differences in germination rate of 93 breeding lines tested in soils at water potentials of -0.22 to -1.44 MPa. Genotypic variation in germination has been reported in other crops, including spring wheat[11] and maize[131].

Once the genetic variability for germination under stress has been established, it is relatively easy to develop screening techniques. The simplest method is to germinate seeds in osmoticums using D-mannitol, polyethylene gycol (PEG) or carbowax. Initially, the appropriate osmotic stress level must be established. In winter wheat, a level of -1.0 MPa was found to be appropriate,[52] while in maize[131] the optimum level was -1.5 MPa. The rapidity at which germination tests can be performed makes this a screening technique which can be used in segregating populations. However, confirmation that the laboratory screening procedure parallels field performance is necessary.

Various authors have proposed that screening for germination and early seedling development under water stress conditions can be used to predict drought response of the mature plant. In maize, Williams et al.[131] found that germination in -1.5 MPa mannitol solutions was positively correlated (P = 0.01) with yield under field conditions. In wheat, Kaul (cited in Hurd[58]) found that germination at -2.0 MPa in mannitol solutions was related to field performance in a limited number of genotypes.

Other investigators have been unable to demonstrate a relationship between laboratory germination tests and field performance. Bhatt[11], found no relationship between yield stability of wheat cultivars and laboratory germination rate at -1.0 or -1.2 MPa, using D-mannitol and PEG as osmoticums. Similarly, Blum et al.[19] found germination of wheat under stress was not related to seedling growth under stress. Further, they reported differences in cultivar germination response from one stress level to another, as did Johnson and Asay[66] in crested wheatgrass. Our results with wheat also show no consistent relationship between germination rate on mannitol at -1.5 MPa and field emergence (Table 6.3) or yield. Germinating wheat seeds are

TABLE 6.3. Comparison of laboratory germination rate in mannitol with field emergence
rates in 16 *Triticum aestivum* lines

Line	Laboratory		Field	
	Time to 50% germination (hr)	Rank	Time to 50% emergence (days)	Rank
Neepawa	54.0	9	12.5	5
RL4137	80.6	15	12.3	1
NB112	47.6	7	12.4	3
Kenya 321	66.6	14	12.6	6
Fielder	57.4	12	12.7	7
Pitic 62	25.1	1	12.3	1
ACEF-125	56.1	11	12.9	10
Pitic X ACEF −1	54.3	10	13.1	14
−2	83.8	16	13.0	12
−3	46.2	5	12.8	8
−4	45.8	3	13.2	15
−5	60.0	13	13.2	15
−6	48.4	8	12.4	3
−7	47.5	6	12.9	10
−8	43.4	2	13.0	12
−9	46.1	4	12.8	8

very tolerant to dessication, but this tolerance is lost by emergence.[19,89] Blum *et al.*[19] concluded that seedling stress responses of germinating wheat seeds and of seedlings are not related.

Attempts have been made to relate the drought or heat tolerance of young seedlings to that
of adult plants. In maize, Williams *et al.*[131] subjected seedlings of several genotypes to heat and
moisture stresses. Recovery from a heat stress at $52°C$ for 6 hr or from a drought stress of 14
days duration at permanent wilting was positively correlated with field performance under
drought. Similarly, the degree of injury of maize seedlings by a $54.5°C$ artificial heat stress was
found to be correlated with degree of leaf firing in field-grown droughted plants.[76] Drought
tolerance of seedlings of sorghum genotypes has also been shown to differ.[119]

Screening procedures such as these are relatively simple and are less labor-intensive than
growing plants to maturity in the field. Heat stress procedures are straightforward, and can be
performed on plants grown on whatever medium is most convenient. Applying drought stress to
plants in pots is somewhat more difficult, since it is difficult to develop and maintain a realistic
stress level. Development of stress is somewhat easier in hydroponically-grown plants; water
stress can be induced by adding an osmotic agent to the nutrient solution. Sorghum seedlings
have been screened in this manner, using quantitative measurements of factors such as plant
height and dry weight or by qualitative comparisons of plant appearance when large numbers of
genotypes are involved.[119] Blum *et al.*[19] found that adding PEG to hydroponic media provided a good method for screening wheat seedlings for drought resistance.

(b) *PHOTOSYNTHESIS AND OTHER METABOLIC PROCESSES*

In crop plants, most studies of drought tolerance have focused on the response of the photosynthetic apparatus to drought stress. The process of photosynthesis is, of course, dependent upon gas exchange between the photosynthetic tissue and the atmosphere. Hence, drought avoidance mechanisms such as stomatal closure can interfere with photosynthesis. On the other hand, maintenance of favorable tissue water potentials can help to maintain photosynthesis during droughts.

Measurements of photosynthetic parameters have been evaluated as screening techniques in wheat[29,72,73]. Kaul and Crowle[73] measured leaf water potential, stomatal opening and potential net photosynthesis of six spring wheat cultivars (*T. aestivum* and *T. turgidum*). Seasonal change in leaf water potential was similar for the six cultivars. Net photosynthesis also declined as drought developed, but the cultivars differed significantly. Integration of the net photosynthesis rate measurements taken from the third leaf to the flag leaf over the growing season for each variety produced an index which was related to yield. Subsequent investigations of the same cultivars under more severe water stress[72] produced similar results, but only integrated net photosynthesis measurements of the flag leaf were correlated with yield. The correlation coefficients were highly significant and ranged from 0.80 to 0.91. Dedio et al.[29] assessed the use of photosynthetic rate measurements by means of a differential respirometer, an infrared gas analyzer and by $^{14}CO_2$ incorporation in four *T. aestivum* cultivars. They concluded that the infrared gas analyzer method was too slow and did not adequately differentiate between cultivars. The respirometer was somewhat faster, but sample-to-sample variation tended to be high. The use of $^{14}CO_2$ feeding was found to be satisfactory, particularly if all cultivars were treated at the same time. Our results have shown differences in $^{14}CO_2$ assimilation in *Triticum aestivem* genotypes (Table 6.4). Comparison of a broad range of genotypes is difficult where there are variations in time to maturity, since leaf assimilative capacity changes with leaf age and plant growth stage.

TABLE 6.4. Incorporation of $^{14}CO_2$ by flag leaves of stressed *Triticum aestivum* genotypes at anthesis

Genotype	$^{14}CO_2$ incorporated (% of Neepawa)
Neepawa	100
Amy	123
Canuck	125
Columbus	101
Echo	117
Glenlea	97
Manitou	88
Marquis	84
Napayo	110
NB320	103
Sinton	110
Thatcher	96

Although numerous studies of the effects of water stress on photosynthesis have been made in other crops, very few investigations of genotypic differences have been made. Genotypic differences in photosynthetic rate have been reported in sorghum[16] and in tall fescue (*Festuca arundinacea*).[4] In sorghum, a drought resistant line was found to have a higher photosynthetic rate at low leaf water potential than a less drought resistant line.[16] Asay *et al.*[4] found differences in net carbon exchange rate between tall fescue lines. Expression of the genetic variability tended to be greater under favorable moisture conditions than under water stress. In maize, Crosbie *et al.*[26] found that photosynthetic rate could be improved by recurrent selection. Rates were improved in both the vegetative and reproductive growth phases.

Sullivan and Eastin[118] noted that some caution must be used in interpreting data on photosynthetic rates in relation to water stress, since previous environmental conditions can affect plant response. Indeed, Asay *et al.*[4] expressed the same concern because they found variations in relative differences between genotypes grown in high and low moisture regimes. Most of the common methods for measuring photosynthetic rates are slow and/or expensive, limiting the amount of screening of genetic material that can be done. The approach of Kaul,[72] to integrate several measurements of net photosynthesis during the season, is too cumbersome for practical use. Radioactive tracer techniques are somewhat faster and may prove to be of some value in screening potential parental material.

The study of osmotic potential in crop plants has undergone renewed interest recently because of its importance in osmotic adjustment or osmoregulation. Part of the reduction in osmotic potential as drought stress develops is due to a net increase in cell solute concentration and not just due to the loss of water from the cell.[9] This adjustment of osmotic potential helps to maintain pressure potential (turgor), which is necessary for normal cell function and growth. Osmotic adjustment can also result in lowering of the leaf water potential level, which triggers stomatal closure in response to water deficit.[126]

Osmotic adjustment has been observed in sorghum grown in the field[1,127] and in a controlled environment,[117] and in field-grown sunflower.[127] Ackerson *et al.*[1] found differences in osmotic adjustment of sorghum genotypes after anthesis but not before anthesis. They note that making the measurements of leaf water potential and osmotic potential required to detect osmotic adjustment limits the number of comparisons that can be made. Osmotic adjustment during seed-filling could help to maintain photosynthetic production under water stress, resulting in improved drought resistance and yield.[1]

Another indirect approach to screening for drought resistance has been to search for readily-measured plant constituents such as proline and nonstructural carbohydrates which are responsive to drought stress level in the plant. These substances may also play a role in osmoregulation. Proline levels have been shown to increase with water stress in crops such as Bermuda grass (*Cynodon dactylon*)[7], barley[114], sorghum[129], and wheat[105]. Dedio (personal communication 1974) found that proline levels in 16 wheat lines increased with water stress, but there was little difference between cultivars. In sorghum, Blum and Ebercon[17] found significant cultivar differences in proline accumulation, but there was no relationship between proline and drought tolerance.

Investigations of nonstructural carbohydrate (NSC) levels in wheat stems at our laboratory indicate genotypic differences. Of the cultivars studied, Pitic 62 has shown the greatest NSC levels. Differences in NSC levels between stressed and nonstressed plants have not been found. Ontogenic changes in NSC levels may complicate the use of NSC content as a screening technique if indeed NSC level is found to be related to drought tolerance. Some carbyhydrate compounds may be involved in osmoregulation in wheat.[92]

(v) Field vs. controlled environments for drought studies

Concern has been stressed about discrepancies between results of drought studies done in controlled environments and those done under natural field conditions.[9] These differences can arise from the differences in rate of development of stress: a stress level which can take 4 to 5 weeks to develop in the field may only take 10 days in a controlled environment.[83] Different conclusions were drawn as to the tolerance of photosynthesis to water stress in wheat plants grown in the field[68] or in a controlled environment[67].

Excised leaf water retention of *Triticum* cultivars differed when measured on field-grown or growth chamber-grown plants (Table 6.5). These differences call into the question the results of Dedio[28] regarding inheritance of the water retention trait, since his work was conducted in the greenhouse and based on a study of the cross Pitic 62 X ACEF-125. In our field work, Pitic 62, ACEF-125 and four advanced lines from the cross were not significantly different.

TABLE 6.5. Initial and 24-hour leaf water contents of excised flag leaves of growth chamber and field-grown *Triticum* cultivars

	Water content (%)			
	Growth chamber		Field	
Cultivar	0 hr	24 hr	0 hr	24 hr
Pitic 62	64	38	65	5
Canuck	53	7	69	11
Hercules	63	25	69	5
Pelissier	70	38	71	17

Such discrepancies emphasize the need to corroborate controlled-environment results in the field before using these results to develop screening programs. These differences notwithstanding, the use of controlled environments in screening is a valuable tool, particularly in seedling studies and in cold climates where field studies are limited to one crop cycle per year. Some control over soil moisture levels in the field may be possible by using excavation and drainage techniques,[108] rain-out shelters, or stubble-cropping.

(vi) Practical applications of screening procedures

The speed at which particular screening procedures can be carried out has a great bearing on how the techniques can be incorporated into drought breeding programs. Slower techniques are limited to empirical drought studies or to the screening of potential parents. With a knowledge of the drought-resistance characteristics of crop genotypes, the breeder can plan his crosses with a better change of recovering progeny with improved drought resistance. Mass-screening techniques, on the other hand, will allow the breeder to screen segregating breeding populations for desired traits. This reduces the amount of material which must be carried forward each year. Mass-screening techniques are also valuable in screening large numbers of potential parents, such as those from world collections.

Present indications in the published literature reveal only a few mass-screening techniques in use in breeding programs, and a further few which show potential. This former category includes studies of sorghum seedling and root growth in osmotica (Sullivan and Ross[119]), comparative drought avoidance of sorghum genotypes by aerial infrared photography (Blum[15]) and visual drought response ratings of rice genotypes (O'Toole and Chang[97]). The seedling-pulling technique of O'Toole and Soemartono[99] may prove to be useful in mass-screening of root systems in rice. Screening for excised leaf water retention capability in wheat can be done on large numbers of lines, but further work on heritability and relationship of the trait to drought resistance is required.

BREEDING AND SELECTION

(i) Environmental effects

The environment under which plants are grown will affect the expression of certain morphological and physiological characters, and thus affect the ease with which they can be selected. The question is often raised as to whether it is possible to select for high yield under optimal moisture conditions and obtain genotypes which will also do well in dry environments.

Turner[126] suggested that yield potential of a cultivar under favorable moisture conditions is important in determining yielding ability under water-limited conditions. This suggestion is supported by the finding that semi-dwarf wheat lines selected under adequate moisture conditions generally yield well under limited moisture conditions as well.[81] Other research[46,110] indicates that selection for yield and its components is more efficient under optimal than under sub-optimal growing conditions.

Other evidence, however, points to the opposite conclusion: non-drought yields and drought yields are not necessarily connected. In other words, both yield potential and stability must be considered. Hurd[56] reported that a cultivar that was high yielding under adequate moisture was lowest yielding under drought stress. Similarly, Fischer and Wood[44] found that drought susceptibility increased with increased non-drought yield. Our own work with *T. turgidum* and *T. aestivum* cultivars reveals that some cultivars have a similar yield rank under both dry and moist environments, while others have a substantially different ranking under optimal and stressed environments (Fig. 6.4). The cultivar Pelissier provides a good example of the latter response: yield tends to rank considerably higher in dry than in moist environments.

The differing results regarding selection environment are perhaps in part due to generalizations being drawn from observations of too few genotypes. For example, had we tested fewer genotypes, such as DT363, DT367, Sinton, Neepawa and a few others (Fig. 6.4), our results may have been suggestive of selection in optimal environments. This stresses the point that a significant range of genotypes must be tested prior to making generalizations about crop response.

Given that non-drought and drought yields are not universally related, the breeding approach must be to either select under stress environments only, or consider both yield and stability by yield-testing over a broad range of environments. The former approach may be limiting if yield heritability is universally lower in low yielding environments,[46,110] while the latter approach greatly increases the cost of breeding programs.

The solutions to the problem tend to be somewhat environment-specific. Selection for a stable environment, such as one where the time of drought stress varies little from year to year is simpler than where the time and degree of drought varies widely. Blum[15] notes that rainfall instability tends to be a problem in semi-arid areas. This necessitates selection for both yield and stability, which increases the size of breeding populations which must be handled. Roy and Murty[110] suggested early generation testing under stress and non-stess environments, followed by selection of the families which show widest adaptation under the non-stress environment only.

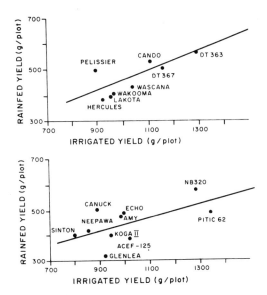

Fig. 6.4. Linear regression of rainfed yield on irrigated yield for *Triticum turgidum* genotypes (top) and *T. aestivum* genotypes (bottom); genotypes above the regression lines are better adapted to rainfed environments, while those below the line are better adapted to moist environments.

There seems to be a strong consensus that breeding for drought resistance should include both yield and stability improvement. Broadly adapted cultivars which have a high yield under all or most environmental conditions are preferable to specifically adapted cultivars, particularly where times and degree of stress vary from year to year.

(ii) Development of the plant ideotype

There are two general approaches to determining the effects of various physiological processes to crop performance under drought. The first of these, which Fischer[39] called the 'Black Box' approach, is to proceed from observed yield differences to possible underlying physiological processes. This is a necessary first approach to understanding the nature of drought resistance in a particular species. The second approach, which Fischer[39] called the 'Ideotype' approach, is to develop predictions of yield differences from an understanding of physiological processes and their interactions with the type of drought stress likely to be encountered.

We believe that it is essential for the breeder to have as clearly defined an ideotype as possible before embarking on a hybridization and selection program.

Development of a useful ideotype demands a clear understanding of the type of drought likely to be encountered. Drought varies in timing, duration, and magnitude, so the ideotype will depend on the environment. For example, screening for the ability of seed to germinate against an osmoticum is not essential for a cultivar normally seeded into soil with an adequate moisture supply. However, this characteristic would be required where drought is frequently encountered at seeding time. The selection of appropriate characteristics to include in the ideotype also depends on the year-to-year variation present in the target environment. However, the important first step in the breeding procedure is to define and characterize the target environment and then develop an appropriate ideotype for that environment.

(iii) Selection of parental material

A clearly defined ideotype forms the basis for selection of parental materials. It is necessary to ensure that each of the characteristics that we desire in a new cultivar are present in at least one of the parental cultivars. Hurd[56,58,59] stressed the importance of careful selection of parental materials in order to make most effective use of limited resources.

Crosses selected on the basis of parental physiological complementation will give, on average, more high yielding progenies than crosses between parents for which little is known about their physiological attributes.[130]

Yield testing in the target environment will not be effective in identifying lines which possess useful characteristics unless one or more of the parents are well adapted to the target environment. Thus, when screening genotypes from diverse origins as potential parents, it is essential to screen for the particular characteristics required. In most breeding programs, other than those employing repeated backcrosses, the selection of superior genotypes from a segregating population consumes most of the resources and time utilized by the program. A reduction in the number of crosses by careful selection of parents will greatly improve the effectiveness of the utilization of these resources. Thus, any screening technique, regardless of how cumbersome it may be, can be usefully applied to select parental material. Those techniques which are rapid and relatively simple can be used to screen large numbers of potential parents while those which are more cumbersome must be restricted to screening fewer potential parents.

Several workers[56,58,85] have pointed out that yield level in a particular environment depends on a balance between many diverse processes. In order to ensure a higher probability of success it is essential that a major portion of the parental material entering a cross be well adapted to the target environment. Since desired physiological traits may be identified in rather unadapted lines, careful consideration must be given to methods of incorporation of desirable traits into an adapted background. Two basic strategies are apparent: (a) incorporate the desired characteristic into an adapted background via backcrossing, and (b) utilize three-way double crosses or convergent crosses. The choice between these strategies depends on the presence of a suitable recurrent parent and on the ease of identifying the presence of plants with the desired trait among segregating populations. Hurd[56] suggested backcrossing simply inherited characteristics into proven varieties and then using these backcross derivations as parents. MacKey[85] stated that the balance of well-adapted cultivars was upset when crossed to unadapted types, making it unlikely that superior types could be produced from such a cross. Using one unadapted parent with several adapted parents in a multiple cross tends to overcome this difficulty.

Perhaps the most important step in the breeding procedure is the development of the plant ideotype and selection of suitable parental material. Therefore, considerable effort should be expended in the study of the environment, the physiology of yield under dry conditions, and the selection of appropriate parents.

(iv) Selection of superior genotypes from segregating populations

Blum[15] noted that there are two general approaches to selection for drought resistance via yield in segregating populations: improvement of yield potential under non-stress conditions, with the hope that yield will also be greater under stress conditions, or to select for yield in the stress environment. He suggested a third approach which involved incorporation of specific drought resistance characteristics into genotypes with high yield potential to buffer them against drought stress.

Selection for yield potential in a non-stress environment, as noted previously, does not always produce genotypes that yield well under drought stress. Selection for drought-buffered genotypes in this manner would require an approach such as that of Roy and Murty,[110] which involved selection for both yield potential and yield stability over environments. The expression of certain drought-resistance mechanisms may be best when tested under stress than under non-stress conditions.

Selection for drought resistance in the stress environment can take two forms, either direct selection for the drought-resistance mechanism or indirect selection for drought resistance through selection for yield. Direct selection in segregating populations is only possible where simple, rapid techniques are available. A few such procedures have been used by breeders[15,97,119]. Early generation selection for heritable drought resistance traits reduces the number of lines to be tested in later generations. Direct selection for drought resistance could produce acceptable cultivars directly, or produce lines which would be useful in back-cross or conventional crossing programs.

Indirect selection for drought resistance through yield selection in stress environments is advocated on the basis of the premise that superior yield under such conditions must be a result of superior drought resistance[60,61,62,106,107]. Hurd[56,57,58,59] and Townley-Smith and Hurd[125] have outlined a testing procedure for selecting high-yielding genotypes from segregating populations. Briefly, this procedure is based on early generation testing of large numbers of lines at more than one location.

Where crosses of wheat lines with complementary mechanisms for drought resistance have been used in either simple, double or three-way crosses the F_1 population must be large to provide large F_2 populations. Hurd[59] suggested F_2 populations of 20,000 to 50,000 plants from each cross. Strong selection for highly heritable characteristics such as height, awnedness, maturity, disease and insect resistance and straw strength will cut this population size to perhaps 1000 plants acceptable for advancement. Where rapid direct selection techniques exist for those physiological or morphological traits conferring drought resistance, they may be applied to help reduce the numbers of F_2 progenies to be subsequently yield tested.

Shebeski[111] examined the population sizes required to combine varying numbers of genes in a self-pollinating crop. If two parents differ by 25 independent genes, only one F_2 plant in 1330 would be expected to carry all 25 desirable alleles, and that plant is most likely to have 17 of those 25 alleles in the heterozygous state. Clearly, large populations are required if the traits to be combined are controlled by a number of genes or if several traits are to be combined.

Direct selection for yield in F_2 has been found ineffective by many workers[80,111]. Recently, however, there has been renewed interest in selecting for yield in F_2. Two different approaches have been made to improve the effectiveness of selection at this stage. The first of these attempts to reduce the effects of environmental variability and thus improve the heritability of F_2 plant yield. Fasoulos[35,36,37,38] has proposed the use of a 'honeycomb' design in which plants are carefully planted in a 'honeycomb' pattern such that they are equidistant. Only those plants which exceed the yield of their closest neighbors are selected. Other workers[48,128] have suggested stratifying the nursery into small segments and selecting an equal number of superior genotypes from each segment. The second approach has been to select indirectly for yield by selecting for high harvest index.[93,120] Although the honeycomb and stratification techniques do reduce environmental variability, and harvest index does tend to be more closely correlated with F_3 yields than does F_2 yield, it is questionable whether the gains are sufficient to justify the extra work necessary to carry out these procedures. Fischer[39] argues that harvest index is related to drought tolerance as it reflects the distribution of dry matter production (and hence water use) between the pre- and postanthesis periods. This could be a second justification for selecting for high harvest index.

Shebeski[111] suggested that yield testing should begin in F_3, however, F_2 plants grown in a semi-arid climate may not produce sufficient seed for replicated yield tests. Frey[47] and other workers suggested the use of hill plots in early generations. However, much more replication is required to obtain the same precision.[6] Labor requirements to seed and harvest the additional replications required have prevented many breeders from utilizing hill plots. If yield testing is to be delayed until F_4, the progeny of a single F_2 plant could be multiplied in an optimum environment out of season so that adequate F_4 seed is available to conduct replicated yield tests at several locations.

Similarly, single plant selections in F_4 and F_6 generations can be increased in an out-of-season nursery in F_5 and F_7 to supply adequate seed for F_6 and F_8 yield trials.

The breeding procedure outlined by Hurd[56,58,59] and Townley-Smith and Hurd[125] utilized replicated yield tests of many varieties at two or more locations within the target area. The use of several locations allowed for the selection of stable lines as suggested by Roy and Murty[110] and also protected against the possibility of the test environment being abnormally wet, which would provide limited opportunities to select for drought resistance, or too dry, which could lead to complete loss of the crop. It is important that the test be grown under conditions as close as possible to the target environment. Fischer[39] issued a strong warning against using data from small plots in determining the physiological basis for drought resistance. To provide uniform stress in small plots, Hurd[59] suggested the growing of an alternate crop, such as spring-sown winter wheat in a spring wheat nursery, between plots to compete for moisture and to facilitate mechanical harvest. In fact, at our station winter wheat is sown completely around each small spring cereal plot at the time of sowing of the plots.

Where large numbers of lines are to be tested in replicated trials control of environmental error is very important. Shebeski[111] and Briggs and Shebeski[23] concluded that frequent systematically distributed control plots were effective in controlling environmental differences. Baker and McKenzie[5] have questioned the value of repeated controls on both experimental and practical grounds. They suggested the use of co-variance to adjust the yields of experimental lines in order to avoid over-corrections resulting from use of repeated controls. Townley-Smith and Hurd[122] found that the moving mean of adjacent experimental plots gave a greater reduction in experimental error than the use of repeated controls, even when the controls were grown in every third plot. Knott[80] found no differences between these two methods of error control. The use of repeat controls requires the growing of a larger number of plots to test the same number of experimental lines. Other workers have suggested stratifying the population into small tests and selecting an equal number of superior lines from each test. Regardless of the methods used to reduce the experimental errors, they cannot replace replication. Townley-Smith et al.[123] presented a table which illustrates the necessity of replication in yield plots, unless very low selection intensities are used. Shebeski[111] suggested that at least 100 plants should be selected from each selected F_3 line in order to sample adequately the variability present. Unless the test of the succeeding generation is to greatly exceed the size of the F_3 yield test, very intense selection must be practised. Because of the relatively low heritability of yield measurements even with replication,[94] it is probably better to select fewer individuals from more lines than to select many plants from one or two percent of the lines.

Modern computerized and mechanized handling techniques allow the breeder to grow large numbers of yield plots with relatively few support staff. As many as 30,000 plots in yield trials were handled by as few as 7 man-years of support.[59]

This system of early generation yield testing using replication at several locations in the target environment has been successful in selecting cultivars with higher yield under semi-arid conditions[60,62,124]. This system combined with selection for heritable drought resistance traits, would perhaps be even more successful. At the present time there are only a few physiological screening techniques for drought resistance which are more rapid and less consuming of resources than yield testing. Such rapid tests that can be run in conjunction with yield testing include seedling and root growth in osmotica[119], aerial infrared photography[15], visual drought response ratings[97], seedling pulling techniques[99] and perhaps screening for excised leaf water retention.

CONCLUSIONS

It is impossible to make a firm recommendation of a method for breeding for drought resistance that will apply to all crops and environmental conditions. The best approach in a particular crop and situation will probably be a combination of some of the approaches discussed here. Most of the approaches of screening and selecting for drought resistance are not mutually exclusive, rather, they overlap quite considerably.

This review cited only a very small proportion of the published works dealing with drought stress, specifically, only those which reached the stage of meaningful evaluation in a range of genetic material. The practicality of drought stress research is perhaps greatest where there is a term approach to the problem, with plant physiologists and breeders inter-acting directly. A team which carried the problem from identification of drought resistance mechanisms through selection for these characteristics is more likely to make progress than where physiologists and breeders work in isolation from each other.

Such an approach, however, is not without its cost in manpower and facilities. Although research support is a continuing problem at most institutions, this is being partially offset by technological advances in instrumentation and electronic data capture and processing.

REFERENCES

1.   Ackerson, R.C., Krieg, D.R. and Sung, F.J.M. Leaf conductance and osmo-regulation of field-grown sorghum genotypes. *Crop Sci.* **20**, 10–14, 1980.

2.   Alessi, J. and Power, J.F. Water use by dryland corn as affected by maturity class and plant spacing. *Agron. J.* **68**, 547–550, 1976.

3.   Angus, J.F. and Moncur, M.W. Water stress and phenology in wheat. *Aust. J. Agric. Res.* **28**, 177–181, 1977.

4.   Asay, K.H., Nelson, C.J. and Horst, G.L. Genetic variability for net photosynthesis in tall fescue. *Crop Sci.* **14**, 571–574, 1974.

5.   Baker, R.J. and McKenzie, R.I.H. Use of control plots in yield trials. *Crop Sci.* **7**, 335–337, 1967.

6.   Baker, R.J. and Leisle, D. Comparison of hill and rod row plots in common and durum wheats. *Crop Sci.* **10**, 581–583, 1970.

7.   Barnett, N.M. and Naylor, A.W. Amino acid and protein metabolism in Bermuda grass during water stress. *Plant Physiol.* **41**, 1222–1230, 1966.

8.   Barrs, H.D. Determination of water deficits in plant tissues. *Water Deficits and Plant Growth*, T.T. Kozlowski (ed.), Vol. I, pp. 235–368. Academic Press, New York, 1968.

9.   Begg, J.E. and Turner, N.C. Crop water deficits. *Advan. Agron.* **28**, 161–217, 1976.

10.   Ben-Yehoshua, S. and Aloni, B. Effect of water stress on ethylene production by detached leaves of Valencia orange (*Citrus senensis* Osbeck). *Plant Physiol.* **53**, 863–865, 1974.

11.   Bhatt, G.M. Effect of simulated drought on germination of wheat cultivars. *Cereal Res. Comm.* **7**, 123–133, 1979.

12.   Blum. A. Effect of plant density and growth duration on grain sorghum yield under limited water supply. *Agron. J.* **62**, 333–336, 1970.

13.   Blum, A. Genotypic responses in sorghum to drought stress. I. Response to soil moisture stress. *Crop Sci.* **14**, 361–364, 1974a.

14.   Blum, A. Genotypic responses in sorghum to drought stress. II. Leaf tissue water relations. *Crop Sci.* **14**, 691–692, 1974b.

15.   Blum, A. Genetic improvement of drought resistance in crop plants: A case for sorghum. *Stress Physiology in Crop Plants*, H. Mussell and R.C. Staples (eds.), John Wiley and Sons, New York, 1979.

16.   Blum, A. and Sullivan, C.Y. A laboratory method for monitoring net photosynthesis in leaf segments under controlled water stress. Experiments with sorghum. *Photosynthesis* **6**, 18–23, 1972.

17.   Blum, A. and Ebercon, A. Genotypic responses in sorghum to drought stress. III. Free proline accumulation and drought resistance. *Crop Sci.* **16**, 428–431, 1976.

18.   Blum, A., Arkin, G.F. and Jordan, W.R. Sorghum root morphogenesis and growth. I. Effect of maturity genes. *Crop Sci.* **17**, 149–153, 1977.

19.   Blum, A., Sinmena, B. and Ziv, O. An evaluation of seed and seedling drought tolerance screening tests in wheat. *Euphytica* **29**, 727–736, 1980.

20.   Blum, A., Gozlan, G. and Mayer, J. The manifestation of dehydration avoidance in wheat breeding germplasm. *Crop Sci.* **21**, 459–499, 1981.

21.   Boyer, J.S. and McPherson, H.G. Physiology of water deficits in cereal crops. *Adv. Agron.* **27**, 1–23, 1975.

22.   Briggle, L.W. and Vogel, O.A. Breeding short-stature, disease resistant wheats in the United States. *Euphytica Suppl.* **1**, 107–130, 1968.

23.   Briggs, K.C. and Shebeski, L.H. Implications concerning the frequency of control plots in wheat breeding nurseries. *Can. J. Plant Sci.* **48**, 149–153, 1968.

24.   Carlson, R.E., Yarger, D.N. and Shaw, R.H. Environmental influences on the leaf temperatures of two soybean varieties grown under controlled irrigation. *Agron. J.* **64**, 224–229, 1972.

25.   Chatterton, N.J., Hanna, W.W., Powell, J.B. and Lee, D.R. Photosynthesis and transpiration of bloom and bloomless sorghum. *Can. J. Plant Sci.* **55**, 641–643, 1975.

26.   Crosbie, T.M., Pearce, R.B. and Mock, J.J. Recurrent phenotypic selection for high and low photosynthesis in two maize populations. *Crop Sci.* **21**, 736–740, 1981.

27.   Dalton, L.G. A positive regression of yield on maturity in sorghum. *Crop Sci.* **7**, 271, 1967.

28.   Dedio, W. Water relations in wheat leaves as screening tests for drought resistance. *Can. J. Plant Sci.* **55**, 369–378, 1975.

29.   Dedio, W., Stewart, D.W. and Green, D.G. Evaluation of photosynthesis measuring methods as possible screening techniques for drought resistance in wheat. *Can. J. Plant Sci.* **56**, 243–247, 1976.

30. Derera, N.F., Marshall, D.R. and Balaam, L.N. Genetic variability in root development in relation to drought tolerance in spring wheats. *Expl. Agric.* **5**, 327–337, 1969.

31. Ebercon, A., Blum, A. and Jordon, W.R. A rapid colorimetric method for epicuticular wax content of sorghum leaves. *Crop Sci.* **17**, 179–180, 1977.

32. Ehleringer, J., Bjorkman, O. and Mooney, H.A. Leaf pubescence: Effect on absorptance and photosynthesis in a desert shrub. *Science* **192**, 376–377, 1976.

33. Ehrler, W.L. Cotton leaf temperatures as related to soil water depletion and meteorological factors. *Agron. J.* **65**, 404–409, 1973.

34. Ehrler, W.L., Idso, S.B., Jackson, R.D. and Reginato, R.J. Wheat canopy temperature: Relation to plant water potential. *Agron. J.* **70**, 251–255, 1978.

35. Fasoulos, A. A new approach to breeding superior yielding varieties. *Publ.* 3, pp. 42, Arist. Univ., Thessaloniki, 1973.

36. Fasoulos, A. Principals and methods of plant breeding. *Publ.* 6, pp. 55, Arist. Univ., Thessaloniki, 1976.

37. Fasoulos, A. Field designs for genotypic evaluation and selection. *Publ.* 7, pp. 61, Arist. Univ., Thessaloniki, 1977.

38. Fasoulos, A. The honeycomb field designs. *Publ.* 9, pp. 69, Arist. Univ., Thessaloniki, 1979.

39. Fischer, R.A. Optimizing the use of water and nitrogen through breeding of crops. *Plant and Soil* **58**, 249–278, 1981.

40. Fischer, R.A. and Kohn, G.D. The relationship of grain yield to vegetative growth and post-flowering leaf area in the wheat crop under conditions of limited soil moisture. *Aust. J. Agric. Res.* **17**, 281–295, 1966.

41. Fischer, R.A., Sanchez, M. and Syme, J.R. Pressure chamber and air flow porometer for rapid field indication of water status and stomatal condition in wheat. *Expl. Agric.* **13**, 341–351, 1977.

42. Fischer, R.A. and Maurer, R. Drought resistance in spring wheat cultivars. I. Grain yield responses. *Aust. J. Agric. Res.* **29**, 897–912, 1978.

43. Fischer, R.A. and Turner, N.C. Plant productivity in the arid and semi-arid zones. *Ann. Rev. Plant Physiol.* **29**, 277–317, 1978.

44. Fischer, R.A. and Wood, J.T. Drought resistance in spring wheat cultivars. III. Yield associations with morpho-physiological traits. *Aust. J. Agric. Res.* **30**, 1001–1020, 1979.

45. Foutz, A.L., Dobrenz, A.K. and Massengale, M.A. Effect of leaf area reduction on the water requirement of barley (*Hordeum vulgare* L.). *J. Ariz. Acad. Sci.* **9**, 51–54, 1974.

46. Frey, K.J. Adaptation reaction of oat strains selected under stress and non-stress environmental conditions. *Crop Sci.* **14**, 55–58, 1964.

47. Frey, K.J. The utility of hill plots in oat research. *Euphytica* **14**, 196–208, 1965.

48. Gardiner, C.O. An evaluation of effects of mass selection and seed irradiation with thermal neutrons on yield of corn. *Crop Sci.* **1**, 241–245, 1961.

49. Garkavy, P.F., Danilchuk, P.V. and Linchevsky, A.A. The selection of malting varieties of spring barley according to the vigor of root development. *Vop., Genet.* Sel'ksii, Semenovodstva, pp. 53–65, 1970.

50. Goldbach, H. and Goldbach, E. Abscisic acid translocation and influence of water stress on grain abscisic acid content. *J. Exp. Bot.* **28**, 1342–1350, 1977.

51. Gul, A. and Allan, R.E. Stand establishment of wheat lines under different levels of water potential. *Crop Sci.* **16**, 611–615, 1976.

52. Helmerick, R.H. and Pfeifer, R.P. Differential varietal responses of winter wheat germination and early growth to controlled limited moisture conditions. *Agron. J.* **46**, 560–562, 1954.

53. Henzell, R.G., McCree, K.J., Van Bavel, C.H.M. and Shertz, K.F. Sorghum genotype variation in stomatal sensitivity to leaf water deficit. *Crop Sci.* **16**, 660–662, 1976.

54. Hurd, E.A. Root study of three wheat varieties and their resistance to drought and damage by soil cracking. *Can. J. Plant Sci.* **44**, 240–248, 1964.

55. Hurd, E.A. Growth of roots of seven varieties of spring wheat at high and low moisture levels. *Agron. J.* **60**, 201–205, 1968.

56. Hurd, E.A. A method of breeding for yield of wheat in semi-arid climates. *Euphytica* **18**, 217–226, 1969.

57.   Hurd, E.A. Can we breed for drought resistance? *Drought Injury and Resistance in Crops*. K.L. Larson and J.D. Eastin (eds.), CSSA Spec. Publ. No. 2, Crop Sci. Soc. Amer., Madison, Wis., 1971.

58.   Hurd, E.A. Phenotype and drought tolerance in wheat. *Agric. Meteor.* **14**, 39–55, 1974.

59.   Hurd, E.A. Plant breeding for drought resistance. *Water Deficits and Plant Growth*. T.T. Kozlowski (ed.), Vol. IV, pp. 317–353, Academic Press, New York, 1976.

60.   Hurd, E.A., Townley-Smith, T.F., Patterson, L.A. and Owen, C.H. Wascana, a new durum wheat. *Can. J. Plant Sci.* **52**, 687–688, 1972.

61.   Hurd, E.A., Townley-Smith, T.F., Patterson, L.A. and Owen, C.H. Techniques used in producing Wascana wheat. *Can. J. Plant Sci.* **52**, 689–691, 1972.

62.   Hurd, E.A., Townley-Smith, T.F., Mallough, D. and Patterson, L.A. Wakooma durum wheat. *Can. J. Plant Sci.* **53**, 261–262, 1973.

63.   Idso, S.B., Reginato, R.J., Reicosky, D.C. and Hatfield, J.L. Determining soil-induced plant water potential depressions in alfalfa by means of infrared thermometry. *Agron. J.* **73**, 826–830, 1981.

64.   Irvine, R.B., Harvey, B.L. and Rossnagel, B.G. Rooting capability as it relates to soil moisture extraction and osmotic potential of semi-dwarf and normal-statured genotypes of six-row barley (*Hordium vulgare* L.). *Can. J. Plant Sci.* **60**, 241–248, 1980.

65.   Islam, T.M.T. and Sedgley, R.H. Evidence for a 'uniculm effect' in spring wheat (*Triticum aestivum* L.) in a Mediterranean environment. *Euphytica* **30**, 277–282, 1981.

66.   Johnson, D.A. and Asay, K.H. A technique for assessing seedling emergence under drought stress. *Crop Sci.* **18**, 520–522, 1978.

67.   Johnson, R.R., Frey, N.M. and Moss, D.N. Effect of water stress on photosynthesis and transpiration of flag leaves and spikes of barley and wheat. *Crop Sci.* **14**, 728–731, 1974.

68.   Johnson, R.R. and Moss, D.N. Effect of water stress on $^{14}CO_2$ fixation and translocation in wheat during grain filling. *Crop Sci.* **16**, 697–701, 1976.

69.   Jones, H.G. Aspects of water relations of spring wheat (*Triticum aestivum* L.) in response to induced drought. *J. Agric. Sci.* **88**, 267–282, 1977.

70.   Jones, H.G. Stomatal behavior and breeding for drought resistance. *Stress Physiology in Crops*. H. Mussell and R.C. Staples (eds.), John Wiley and Sons, New York, 1979.

71.   Kaul, R. A survey of water suction forces in some prairie wheat varieties. *Can. J. Plant Sci.* **47**, 323–326, 1967.

72.   Kaul, R. Potential net photosynthesis in flag leaves of severely drought-stressed wheat cultivars and its relationship to grain yield. *Can. J. Plant Sci.* **54**, 811–815, 1974.

73.   Kaul, R. and Crowle, W.L. An index derived from photosynthetic parameters for predicting grain yields of drought-stressed wheat cultivars. *Z. Pflanzenzuecht* **71**, 42–51, 1974.

74.   Kazemi, H., Chapman, S.R. and McNeal, F.H. Variation in stomatal number in spring wheat cultivars. *Cer. Res. Comm.* **6**, 359–365, 1978.

75.   Keim, D.L. and Kronstad, W.E. Drought response of winter wheat cultivars grown under field stress conditions. *Crop Sci.* **21**, 11–15, 1981.

76.   Kilen, T.C. and Andrew, R.H. Measurement of drought resistance in corn. *Agron. J.* **61** 669–672, 1969.

77.   King, R.W. Abscisic acid in developing wheat grains and its relationship to grain growth and maturation, *Planta* **132**, 43–51, 1976.

78.   Kirichenko, F.G. The effect of plant selection according to root strength on increasing the yield and improving its quality in the progeny. *J. Vestvik Sel Skokboyaistvennoi Nauki* No. 4 (USSR), 3–20, 1963.

79.   Kirkham, M.B. and Ahring, R.M. Leaf temperature and internal water status of wheat grown at different root temperatures. *Agron. J.* **70**, 657–662, 1978.

80.   Knott, D.R. Effects of selection for F_2 plant yield on subsequent generations in wheat. *Can. J. Plant Sci.* **52**, 721–726, 1972.

81.   Laing, D.R. and Fischer, R.A. Adaptation of semi-dwarf wheat cultivars to rainfed conditions. *Euphytica* **26**, 129–139, 1977.

82.   Levitt, J. *Responses of Plants to Environmental Stresses*. Academic Press, New York, 1972.

83.   Ludlow, M.M. and Ng, T.T. Effect of water deficit on carbon dioxide exchange and leaf elongation rate of *Panicum maximum* var. *trichoglume*. *Aust. J. Plant Physiol.* **3**, 401–413, 1976.

84. Lupton, F.G.H., Oliver, R.H., Ellis, F.B., Barnes, B.T., Howse, K.R., Wellbank, P.J. and Taylor, P.J. Root and shoot growth of semi-dwarf and taller winter wheat. *Ann. Appl. Biol.* **77**, 129–144, 1974.

85. MacKey, J. Autogamous plant breeding based on already hybrid material. *Recent Plant Breeding Research.* A. Akerberg and A. Hagberg (eds.), Wiley and Sons, New York, 1953.

86. May, L.H. and Milthorpe, F.L. Drought resistance of crop plants. *Field Crop Abstr.* **15**, 171–179, 1962.

87. McMichael, B.L., Jordan, W.R. and Powell, R.D. An effect of water stress on ethylene production by intact cotton petioles. *Plant Physiol.* **49**, 658–660, 1972.

88. Meidner, H. and Mansfield, T.A. *Physiology of stomata.* McGraw-Hill, London, 1968.

89. Milthorpe, F.L. Changes in the drought resistance of wheat seedlings during germination. *Ann. Bot.* **14**, 79–89, 1950.

90. Morgan, J.M. Changes in diffusive conductance and water potential of wheat plants before and after anthesis. *Aust. J. Plant Physiol.* **4**, 75–86, 1977.

91. Mtui, T.A., Kanemasu, E.T. and Wasson, C. Canopy temperatures, water use, and water use efficiency of corn genotypes. *Agron. J.* **73**, 639–643, 1981.

92. Munns, R. and Weir, R. Contribution of sugars to osmotic adjustment in elongating and expanded zones of wheat leaves during moderate water deficits at two light levels. *Aust. J. Plant Physiol.* **8**, 93–105, 1981.

93. Nass, H.G. Harvest index as a selection criterion for grain yield in two spring wheat crosses grown at two population densities. *Can. J. Plant Sci.* **60**, 1141–1146, 1980.

94. O'Brien, L., Baker, R.J. and Evans, L.E. Comparison of hill and row plots for F_3 yield testing. *Can. J. Plant Sci.* **59**, 1013–1017, 1979.

95. Oppenheimer, H.R. Adaptation to drought: Xerophytism. *Plant Water Relationships in Arid and Semi-Arid Conditions. Reviews of Research.* UNESCO, Paris, 1960.

96. O'Toole, J.C. and Moya, T.B. Genotypic variation in maintenance of leaf water potential in rice. *Crop Sci.* **18**, 873–876, 1978.

97. O'Toole, J.C. and Chang, T.T. Drought resistance in cereals – rice: A case study. *Stress Physiology in Crop Plants,* H. Mussell and R.C. Staples (eds.), John Wiley and Sons, New York, 1979.

98. O'Toole, J.C., Cruz, R.T. and Seiber, J.N. Epicuticular wax and cuticular resistance in rice. *Physiol. Plant.* **47**, 239–244, 1979.

99. O'Toole, J.C. and Soemartono. Evaluation of a simple technique for characterizing rice root systems in relation to drought resistance. *Euphytica* **30**, 283–290, 1981.

100. Parker, J. Drought resistance mechanisms. *Water Deficits and Plant Growth.* T.T. Kozlowski (ed.), Vol. I, pp. 195–234, Academic Press, New York, 1968.

101. Passioura, J.B. The effect of root geometry on the yield of wheat growing on stored water. *Aust. J. Agric. Res.* **23**, 745–752, 1972.

102. Pepe, J.F. and Welsh, J.R. Soil water depletion patterns under dryland field conditions of closely related height lines of winter wheat. *Crop Sci.* **19**, 677–680, 1979.

103. Quarrie, S.A. and Jones, H.G. Genotypic variation in leaf water potential, stomatal conductance and abscisic acid concentration in spring wheat subjected to artificial drought stress. *Ann. Bot.* **44**, 323–332, 1979.

104. Quisenberry, J.E. and Roark, B. Influence of indeterminate growth habit on yield and irrigation water-use efficiency in upland cotton. *Crop Sci.* **16**, 762–764, 1976.

105. Rajagopal, V., Balusubramanian, V. and Sinha, S.K. Diurnal fluctuations in relative water content, nitrate reductase and proline content in water-stressed and non-stressed wheat. *Physiol. Plant.* **40**, 69–71, 1977.

106. Raper, C.D. and Barber, S.A. Rooting systems of soybeans. I. Differences in root morphology among varieties. *Agron. J.* **62**, 581–584, 1970.

107. Raper, C.D. and Barber, S.A. Rooting systems of soybeans. II. Physiological effectiveness of nutrient absorption surfaces. *Agron. J.* **62**, 585–588, 1970.

108. Reetz, H.F., Hodges, H.F. and Dale, R.F. Managed soil moisture system for studying plant water relations under field conditions. *Agron. J.* **71**, 861–865, 1979.

109. Richards, R.A. and Passioura, J.B. Seminal root morphology and water use of wheat. II. Genetic variation. *Crop Sci.* **21**, 253–255, 1981.

110. Roy, N.N. and Murty, B.R.   A selection procedure in wheat for stress environment. *Euphytica* **19**, 509−521, 1970.

111. Shebeski, L.H.   Wheat and breeding. *Can. Centennial Wheat Sym.*, K.F. Nielsen (ed.), pp. 253−277, Modern Press, Saskatoon, 1967.

112. Salim, M.H., Todd, G.W. and Stutte, C.A.   Evaluation of techniques for measuring drought avoidance in cereal seedlings. *Agron. J.* **61**, 182−185, 1969.

113. Shimshi, D. and Ephrat, J.   Stomatal behavior of wheat cultivars in relation to their transpiration, photosynthesis and yield. *Agron. J.* **67**, 326−331, 1975.

114. Singh, T.N., Aspinall, D. and Paleg, L.G.   Proline accumulation and varietal adaptability to drought in barley: A potential metabolic measure of drought resistance. *Nature, New Biol.* **236**, 188−190, 1972.

115. Sojka, R.E., Stolzy, L.H. and Fischer, R.A.   Comparison of diurnal drought response of selected wheat cultivars. *Agron. J.* **71**, 329−335, 1979.

116. Spiertz, J.H.J., ten Hag, B.A. and Kupers, L.J.P.   Relationship between green area duration and grain yield in some varieties of spring wheat. *Neth. J. Agric. Sci.* **19**, 211−222, 1971.

117. Stout, D.G. and Simpson, G.M.   Drought resistance of *Sorghum bicolor*. Drought avoidance mechanisms related to leaf water status. *Can. J. Plant Sci.* **58**, 213−224, 1978.

118. Sullivan, C.Y. and Eastin, J.D.   Plant physiological response to water stress. *Agric. Meteor.* **14**, 113−117, 1974.

119. Sullivan, C.Y. and Ross, W.M.   Selecting for drought and heat resistance in grain sorghum. *Stress Physiology in Crop Plants*. H. Mussell and R.C. Staples (eds.), John Wiley and Sons, New York, 1979.

120. Syme, J.R.   Single plant characters as a measure of field plot performance of wheat cultivars. *Aust. J. Agric. Res.* **23**, 753−760, 1972.

121. Todd, G.W. and Webster, D.L.   Effects of drought periods on photosynthesis and survival of cereal seedlings. *Agron. J.* **57**, 399−404, 1965.

122. Townley-Smith, T.F. and Hurd, E.A.   Use of moving mean in wheat field trials. *Can. J. Plant Sci.* **53**, 447−450, 1973.

123. Townley-Smith, T.F., Hurd, E.A. and McBean, D.S.   Techniques of selection for yield in wheat. *Fourth Int. Wheat Genet. Sym.*, E.R. Sears and L.M.S. Sears (eds.), 605−609, Univ. of Missouri, 1974.

124. Townley-Smith, T.F., Hurd, E.A. and Leisle, D.   Macoun durum wheat. *Can. J. Plant Sci.* **55**, 317−318, 1975.

125. Townley-Smith, T.F. and Hurd, E.A.   Testing and selecting for drought resistance in wheat. *Stress Physiology in Crop Plants*, H. Mussell and R.C. Staples (eds.), John Wiley and Sons, New York, 1979.

126. Turner, N.C.   Drought resistance and adaptation to water deficits in crop plants. *Stress Physiology in Crop Plants*, H. Mussell and R.C. Staples (eds.), John Wiley and Sons, New York, 1979.

127. Turner, N.C., Begg, J.E. and Tonnet, M.L.   Osmotic adjustment of sorghum and sunflower crops in response to water deficits and its influence on the water potential at which stomata close. *Aust. J. Plant Physiol.* **5**, 597−608, 1978.

128. Verhalen, L.M., Baker, J.L' and McNew, R.W.   Gardner's grid system and plant selection efficiency in cotton. *Crop Sci.* **15**, 588−591, 1975.

129. Waldren, R.P., Teare, I.D. and Ehler, S.W.   Changes in free proline concentration in sorghum and soybean plants under field conditions. *Crop Sci.* **16**, 447−450, 1974.

130. Wallace, D.H., Ozbun, J.L. and Munger, H.H.   Physiological genetics of crop yield. *Adv. Agron.* **24**, 97−146, 1972.

131. Williams, T.V., Snell, R.S. and Ellis, J.F.   Methods of measuring drought tolerance in corn. *Crop Sci.* **7**, 179−182, 1967.

132. Wright, S.T.C.   The relationship between leaf water potential (leaf) and the levels of abscisic acid and ethylene in excised wheat leaves. *Planta* **134**, 183−189, 1977.

CHAPTER 7

PHOTOSYNTHESIS AND ASSIMILATE PARTITIONING
IN RELATION TO PLANT BREEDING

P. Apel

Akademie der Wissenschaften der DDR,
Zentralinstitut für Genetik und Kulturpflanzenforschung,
4325 Gatersleben, DDR

INTRODUCTION

The worldwide increasing activities to apply basic knowledge in plant physiology, and especially photosynthesis, to practical agriculture and plant breeding have been documented in a number of symposia, monographs and reviews[9,16,22,28,35,39,45,55,60,76,79,85,87,91,114,116,123,124,140,145]

At first it may seem an obvious suggestion that an increase in photosynthesis should be a desirable breeding aim *per se*, and some proposals in the literature, e.g. inhibition of photorespiration, manipulation of ribulose, 1,5-diphosphate carboxylase etc. seem to tacitly presuppose this. It will be shown that this is by no means the case. In most species only a distinct part of the plant, often a storage organ, is the economic yield. Economic yield (Y_e) is the function of total dry matter production, the biological yield (Y_b) and the harvest index (h), so that $Y_e = Y_b \times h$. Therefore problems of partitioning of assimilates and the interactions between photosynthesis and the use of photosynthates for growth and storage (source − sink relationships) must be considered. Furthermore, the biomass of a canopy of a given crop at

163

TABLE 7.1. Some processes involved in yield formation in a hierarchical order

System level	Processes	Products	Response time
molecules	absorption of quanta energy transfer between molecules, primary charge separation	excited states of molecules, reduced primary acceptor	$10^{-12} - 10^{-9}$ s
thylacoids chloroplasts	electron and ion transport, water decomposition, photo-phosphorylation	O_2, ATP, NADP·H biopotentials	10^{-3} s
organelles cells	reduction of CO_2, NO_3^-, SO_4^{--}	carbohydrates, amino acids	$10 - 10^2$ s
leaves and other green organs	biosyntheses, transport both connected with respiration	organic material of different composition, mainly transportable	minutes, days, weeks
whole plants	transport, differentiation, growth, loss of material by senescing parts	organic material of different composition, structural and non-structural	hours, days, weeks
biocoenoses	interactions between plants and between plants and environment	harvestable crop	month, years

harvest is the integral of the apparent net assimilation during the ontogenetic life time of that crop. So the duration of the photosynthetic apparatus and its changes in size and specific activity are of conclusive importance. Size as well as activity of the photosynthetic apparatus and the pattern of assimilate distribution are genetically determined and also depend on changing environmental conditions and the adaptation to the given environment. As in other scientific fields, where complicated systems have to be analysed, mathematical models for the productivity of canopies have been developed.[33,128,138] For plant breeding these models are of particular interest insofar as they allow us more or less to assess the relative importance of subsystems or single processes. This may be helpful in characterising the 'ideotype' of a given species, including physiological as well as morphological parameters.

How complex the system of crop production is (in terms of photosynthesis and assimilate distribution) can be demonstrated by the following scheme, in which the processes are placed in a hierarchical order (Table 7.1).

The yield capacity of a given variety or breeding line is usually given as a value which relates economic yield to unit ground area (e.g. tons per ha), and represents the mean of several experiments and years. Yield is, therefore, always determined by a canopy, i.e. a population of plants during their ontogenetic life cycle. Physiological processes such as photosynthesis, transloca-tion, storage of assimilates were very often investigated with single plants, growing isolated or in wide stands, often in growth chambers under near optimum conditions. This makes the experiments easier and without doubt improves the reproducibility of the measured values, but it makes it difficult to transfer the results to a closed canopy under field conditions. In the practice of agriculture the smallest effective unit is a canopy and each parameter, measured at a lower level of organization such as single plant, organ, or enzyme, should be assessed with a view to its role in the hierarchy of processes. This is true for physiological processes (rates of photosynthesis and respiration, light curves of photosynthesis, senescence, translocation etc.) as well as for morphological parameters (leaf area, leaf inclination, plant height, etc.).

It follows that a major improvement in yield can hardly be expected by genetic improve-ment of single processes, especially at the lower levels of organization, e.g. enzyme activities, rates of photosynthesis, photorespiration or respiration,[99] because compensation mechanisms, such as duration of the activity of photosynthesis, or different response to environmental changes, are always possible.[131] Nevertheless, the processes at each level of the system should be considered, as to whether they could be improved by breeding, bearing in mind that the law of limiting factors means that such improvement will only be effective in relation to all the other factors involved. In accordance with the author's experience most of the following examples are related to the physiology of cereals.

Some arguments in favour of improved photosynthesis are: in cereals, the grain yield per area is closely correlated with the number of kernels per m^2. The last is the product of the number of kernels per ear and the number of ear bearing culms per m^2. An increase in the kernel number per ear will presumably be a permanent breeding aim. In this context results are of interest that demonstrate a close correspondence between assimilate production during the differentiation and preanthesis growth of ear, and number of fertile florets at anthesis. Short-age of assimilates, induced by preanthesis shading, reduced kernel number per spike in wheat,[48,108] whereas enhanced photosynthesis by CO_2 treatment at preanthesis increased the kernel number per spike in rice and wheat.[2,72,141] Selection for a high photosynthetic poten-tial in the preanthesis stage should therefore indirectly enhance the sink capacity of the ear.

The number of ears which can grow per m^2 without severe competitive inhibition is greatly dependent on morphology, mainly leaf area and leaf angle. Increased density seems possible only with a reduced leaf area per culm. This corresponds with the recent trend to select for small, erect, leaves. However, to provide the assimilate production for optimal grain filling this loss of leaf area must be compensated for by enhanced photosynthesis per unit leaf area. A valuable side effect of smaller but photosynthetically very efficient leaves should be improved water use efficiency, if the higher photosynthetic activity is achieved mainly by diminishing the mesophyll resistance.

Arguments in favour of improved photosynthetic efficiency come additionally from exam-ining nitrogen metabolism. In non-legumes the main source of nitrogen is nitrate, taken up by the roots. It is either reduced in the roots or translocated to the shoot and reduced in the leaves.[13,63,84] In the roots the energy source for nitrate reduction is carbohydrates. In the leaves the majority of reactions can occur in the chloroplasts. In this case NO_3^- competes with CO_2 for ATP and NADPH, the products of the light driven electron transport of photosyn-thesis. Nitrate reduction is a very energy consuming process[84] and therefore a high efficiency of photosynthetic electron transport may positively influence nitrogen metabolism and, in consequence, protein biosynthesis.[20]

Some results of experiments with legumes are very impressive. In legumes the symbiosis with *Rhizobium* enables the uptake of atmospheric nitrogen by the root nodules. The energy costs for binding of one mol N_2 was assessed as 12—15 mol ATP,[98] whereas other authors[1] estimated 40 mol consumed per mol N_2 reduced. Cultivation of legumes at increased CO_2 concentrations not only produced significant increases in dry matter production but also in N_2 fixation.[56,57,83]

POSSIBILITIES FOR IMPROVEMENT OF PHOTOSYNTHESIS

(i) Primary processes and electron transport

If, with the aim of improving the efficiency of energy transformation, we look at a canopy as a light harvesting system, we have to analyse the losses inevitably connected with transformations at different levels of the system. A theoretical analysis[90,91] yielded the following values:

Energy absorbed = 100%

Energy stored after losses			Energy stored in form of
— in electron transport	32%		ATP, NADPH
— in the Calvin cycle	28%		products of the Calvin cycle
— by respiration of leaves	23%		leaf biomass
— by respiration of the whole plant	17	— 18%	plant biomass
— by reflection, transmission and diffusion	13	— 14%	canopy biomass

These values were calculated for apparent photosynthesis of a well adapted crop under optimum conditions. As a theoretical mean value for a whole growing season under optimum conditions an energy conversion efficiency of 4—5% for recent varieties was calculated, and improvement up to 6—8% for varieties in the future was predicted. Tables with energy conversion coefficients for different crops under different climatic conditions are given by Cooper.[36]

Such overall calculations however are of limited value for practical breeding especially in relation to the primary processes. Most investigations in this field were carried out with standard objects such as spinach chloroplasts or algae and nearly nothing is known about the genetically determined intraspecific variability of the efficiency of electron transport and the energy costs of ATP and NADPH production. Hieke,[64] in an extensive work, investigated electron transport in relation to biomass production in *Triticum* species of different ploidy level and in three *Aegilops* species. Positive correlations were found between electron transport rates of isolated chloroplasts and biomass production per shoot among these species. The photochemical activity, calculated per unit of chlorophyll, was higher in the polyploid in comparison to diploid species. Because of differences in chlorophyll concentration per unit biomass a 'potential biomass productivity' was defined. It is the potential dry matter increase of a given organ assuming a 100% use of reduction equivalents from the electron transport chain in subsequent dark reactions. Under CO_2 saturating conditions the empirical values were close to the theoretical, indicating a high efficiency in the use of NADPH for biosynthesis. It should be noticed that such measurements are not only time consuming but can also be a source of serious experimental errors, because the optimum conditions for isolation of chloroplasts, composition of the reaction medium etc. differ from species to species. Nevertheless the concept of using the efficiency of primary reactions of photosynthesis, with ATP and NADPH production as a basis of comparison between genotypes seems sound unless many difficulties in methodology have to be surmounted.

(ii) Biochemistry of CO_2 assimilation

Among the around 300,000 species of higher plants, about 1,000 in 18 families and 196 genera are known to represent the C_4-type of photosynthesis[43,104] and about 600 in 26 families and 159 genera the Crassulacean acid metabolism (CAM) type of carbon dioxide assimilation.[125] Because among the CAM-species only a small number is cultivated, and up till now of limited economic importance, the following considerations will be restricted to C_3 and C_4 plants.

Of the economically most important species, wheat, rice, barley, potato, sugar beet and all the cultivated legumes are C_3, while maize, sugar cane, sorghum and millet are C_4-species. It now seems well established that the C_3-type of photosynthesis is the evolutionary ancestor of C_4 as well as CAM photosynthesis.[148,149,150,151,152,153,154]

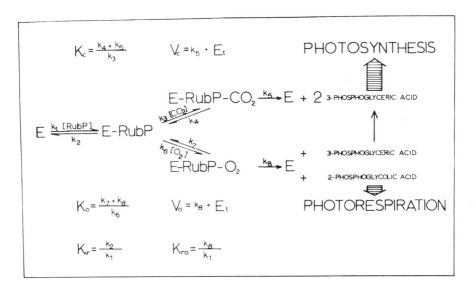

Fig. 7.1. Scheme of the reaction sequences, catalyzed by ribulose 1,5-diphosphate carboxylase/oxygenase (Bauwe, *Photosynthetica* 14 (4) 1980).

Much of the properties of C_3 photosynthesis are causally related to the features of the ribulose 1,5-diphosphate carboxylase/oxygenase (RubPCO), the most abundant enzyme on earth.[94,135] This enzyme catalyses the carboxylations as well as the oxygenation of ribulose-diphosphate (Fig. 7.1). From the double function of the RubPCO and from its kinetic properties as well as from the normal atmospheric concentrations of O_2 and CO_2 it follows that photo-respiration is an inevitable attribute of C_3 photosynthesis. One consequence among others at the physiological level is a CO_2 compensation concentration (Γ) of about $40 - 60\ \mu l.\ 1^{-1}CO_2$.

In C_4 species phosphoenolpyruvate (PEP) is the primary acceptor of CO_2 and PEP carbo-xylase the primarily carboxylating enzyme. A special compartmentalization, accomplished by the so-called 'Kranz anatomy' is responsible for a sequence of reactions by which the primary fixation products (oxalacetate, malate, aspartate) are translocated to the bundle sheath cells. This and the subsequent decarboxylation work as a CO_2 concentration mechanism, enabling the refixation of CO_2 by RubPCO in the bundle sheath with high efficiency. This sequence of reactions, the high *in vivo* affinity of PEP carboxylase for CO_2 and its oxygen independence are responsible for the absence of an apparent photorespiration and consequently Γ-values below $10\ \mu l.\ 1^{-1}CO_2$.

The division into C_3 and C_4 types by means of Kranz anatomy is not always clearcut, as intermediate forms exist. For examples, Crookston and Moss[157] described a variation in the leaf anatomy of the C_4 plant *Arundinella hirta*, Shomer-Ilan *et al.*[158] found that *Suaeda monoica* lacks the typical bundle sheath, and recently the seagrass *Cymodocea nodosa* has been reported[156] to have C_4 photosynthesis but to lack completely the Kranz type anatomy. Both C_3 and C_4 photosynthesis have been reviewed in recent years.[58,67,111]

Most speculations about an improvement of the photosynthetic efficiency at the bio-chemical level were concentrated on photorespiration, properties of the RubPCO and the C_4-pathway of photosynthesis. Photorespiration was thought to be a wasteful process diminishing the rate of photosynthesis and dissipating energy. Inhibition of photorespiration is possible in experiments with enhanced CO_2 or diminished O_2 concentrations and in short term experi-ments an enhanced rate of CO_2 uptake was observed. Long term exposure under such condi-tions results in remarkable dry matter increase.[24,56,106] A recent review on photorespiration has been given by Rathnam and Chollet.[110]

Genetic control of photorespiration in C_3 plants in principle should be possible: (a) by change of the kinetic properties of RubPCO in favour of carboxylation, or (b) by a deficiency in an enzyme of the glycolate pathway. It seems rather uncertain, whether this latter type of 'control' could in any way improve photosynthesis. Deleterious side effects induced by accumu-lation of metabolites before the block seems more likely. Inhibition of enzymes of the glycolate pathway, chemically or by mutagenesis, has been demonstrated,[119,120,145,146] but only in short term experiments was a stimulation of photosynthetic CO_2 uptake observed.[147] It has yet to be shown, that alterations of the glycolate pathway, not induced by inhibition of the primary formation of glycolate, are not deleterious in C_3 plants.

The catalytic sites at the large subunit of the enzyme for both carboxylation and oxygena-tion of RubPCO are perhaps identical.[74,97] Therefore a mutagenic change of the one without the other seems very unlikely. The only statement of the separation of the two activities *in vitro*[25] has so far not been confirmed by other laboratories. But there are at least two reports, that the carboxylase/oxygenase ratio can be modified by factors not yet identified. Garrett[52] reported an influence of ploidy level on $K_m (CO_2)$ of RubPCO in ryegrass cultivars. In tetra-ploid lines the $K_m (CO_2)$ was about half the value in diploids, whereas the $K_m (O_2)$ was not changed. The lower $K_m (CO_2)$ was connected with a diminished post-illumination CO_2 burst and a lower value of CO_2 compensation concentration, indicating a lower rate of photorespira-tion. This change in the kinetic properties of RubPCO was fully supported in experiments with mesophyll protoplasts and isolated chloroplasts from diploid and tetraploid ryegrass cultivars.[112]

A twofold increase in specific activity of RubPCO was also found in decaploid in comparison to hexaploid genotypes of tall fescue.[109] The higher specific activity was correlated with an enhanced rate of net CO_2 uptake but there was no significant difference in the carboxylase/ oxygenase ratio *in vitro*. Kung and Marsho[73] estimated a significant change in the carboxylase/ oxygenase ratio in favour of carboxylase activity in mutants of tobacco. However in these experiments both activities were significantly lower than in the wild type. In seven spring barley varieties the specific carboxylase activity of the fully developed 3rd leaf varied between 0.91 and 1.53 μmol $CO_2 \cdot mg^{-1}$ protein. min^{-1}. No correlation was found to the rate of photo-synthesis of those leaves *in vivo*.[10] An unusually low value was estimated for *Panicum mili-oides*, a $C_3 - C_4$ intermediate species (0.26 μmol$\cdot mg^{-1} \cdot min^{-1}$, Bauwe, unpublished).

The utilization of differences in the specific activity of RubPCO in plant breeding should be assessed with some scepticism. The *in vitro* measurement of the amount and the activity of the enzyme does not mean too much about the activity *in vivo*, which in part can be determined by the microenvironment and regulative mechanisms in the chloroplasts of intact leaves. How-ever, genotypes with an extreme high specific activity could be of interest for pilot experiments studying the heritability, the consequences for photosynthesis and productivity etc. The problem of modification of RubPCO is discussed in more detail in [92,93].

The discovery of the C_4 pathway of photosynthesis in the Sixties raised the hope of detect-ing a single C_4 plant among populations of C_3 plants.[29,82,86] The stimulation to perform screenings with this aim came from the significantly higher rates of photosynthesis and the better water use efficiency as well as the high productivity of C_4 plants. Somewhat euphoric-ally C_4 plants were called "Kranz-Typ Wirtschaftswunder".[23] However, it should be noted that the superiority of C_4 species is expressed mainly under tropical and subtropical conditions with at least temporary water stress. Today it is well established that the C_4 condition develop-ed during evolution by the mutation of some, perhaps most regulatory genes. Therefore it seems extremely unlikely to detect C_4 mutants among populations of mutagen treated C_3 plants.

In recent years species have been described and analysed with intermediate C_3 and C_4 anatomical and biochemical features. They belong to the genera *Panicum, Mollugo, Moricandia* and *Flaveria*.[6,7,11,27,69] These species have CO_2 compensation concentrations lower than 40 μl. l^{-1}, a lower rate of apparent photorespiration and a diminished sensitivity of photo-synthesis against oxygen in relation to C_3 plants. It cannot be excluded that the mutation of only one or two genes is responsible for the expression of the intermediate character. This would give us a chance to select plants with intermediate character in populations of C_3 plants. However, screening of 12,000 seedlings of spring barley (M_2) was without succes.[155] Another way to select photorespiratory mutants was proposed by Zelitch and co-workers.[21,146] They proposed the use of photoautotrophically grown haploid single cells, the selection of mutated cells and the regeneration of whole plants. The efficiency of such a system would of course be much higher than in a system with whole plants. Nevertheless, because the influence of the leaf anatomy, i.e. the compartmentalization in the form of a Kranz or Kranz-like structure on photorespiration is neglected in single cell systems, and up till now the regeneration of whole plants from single cells is not possible in all species, a general use of this method seems questionable.

(iii) Photosynthetic gas exchange

The CO_2 exchange rate (CER) of a leaf in the light is the result of biochemical CO_2 fixation, by ribulose 1,5-diphosphate or phosphoenolpyruvate, and the evolution of CO_2 by photorespiration and dark respiration. Besides the biochemical processes the value of CER is also determined by diffusion resistances. The situation may be illustrated by a model, first developed by Gaastra[49] and later modified by many authors, as in Fig. 7.2. In C_3 plants 30—50% of the CO_2 fixed is subsequently evolved by photorespiration. Dark or mitochondrial respiration is in the light inhibited to about 30% of the rate in the dark.[102] In C_4 plants the apparent photorespiration is near zero.

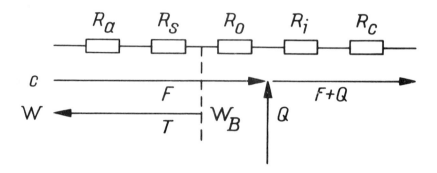

Fig. 7.2. Model of CO_2 exchange and transpiration of a leaf (Waggoner[132]).
R_a — boundary layer resistance; R_s — stomatal resistance; R_0, R_i — diffusion
resistances in the liquid phase of mesophyll cells; R_c — carboxylation resistance; Q — rate of photorespiration; F — net CO_2 exchange rate; c — CO_2
concentration around the leaf; T — rate of transpiration; W — water
vapour concentration around the leaf; W_b — water vapour concentration at the surface of mesophyll cells.

Of especial ecological significance is the relationship between gas phase resistance (boundary layer resistance + stomatal resistance) and the so-called residual resistance (mesophyll resistance + carboxylation resistance). It determines the water use efficiency, by which is meant the relationship between ratio of water loss by transpiration and CO_2 uptake. The 2—3 times better water use efficiency of C_4 in relation to C_3 plants is caused by their extremely low residual resistance[42,55,77], shown in Table 7.2.

It is a consequence of this situation that genetic improvements of CER would be most valuable if they were brought about by a diminished residual resistance, i.e. by means of more efficient carboxylation and/or facilitated transport of CO_2 from the mesophyll cell walls to the chloroplasts. Carboxylation and its possible enhancement were discussed under biochemistry. The transport across the mesophyll cells is dependent on factors such as permeability of the membranes, plasma streaming, activated transport, equilibrium between CO_2, HCO_3^- and CO_3^-, pH and activity of carbonic anhydrase, and all of these are not yet well understood. The relationship between transport and carboxylation resistance within the mesophyll is discussed more in detail in[8,32,129].

CER as well as gas phase and residual resistances can easily be estimated by infrared gas analysis in combination with measurements of transpiration and leaf temperature as carried out in many laboratories. Nevertheless, in the near future such measurements will mainly be of

TABLE 7.2. Gas phase resistances and residual resistances in *Gramineae* species with C_4 and C_3 pathway. The values were estimated on fully developed attached leaves by means of infrared gas analysis in combination with measurement of transpiration and leaf temperature (8).

	resistances (s · m^{-1})	
	gas phase	residual
C_4-species		
Cenchrus echinatus	260	140
Eragrostis curvula	450	120
Panicum esculentum	220	70
Sorghum bantuorum	290	100
Sorghum japonicum	210	80
Panicum miliaceum	320	170
C_3-species		
Bromus catharticus	390	590
Festuca pratensis	430	560
Poa chaixii	570	400
Hordeum vulgare var. Lami	310	350
Hordeum vulgare from Ethiopia	740	930

significance in basic research and only indirectly in plant breeding. Each of such values always provides only an instantaneous picture from the ontogenetic life cycle of the leaf. It depends on the ontogenetic age, the environmental conditions during development and the conditions during measurement. In inter- or intraspecific comparisons of CER values it is also important as to whether the value is calculated on leaf area or dry weight basis. In most cases leaf area is used. This is in part justified, because leaf area is the basis for calculations within the framework of growth analysis, and the data could be compared directly. Ludwig[78] and Charles-Edwards[30,31] pointed out with good reasons that much of the apparent genetic variablity in CER is due to simple differences in the site of the assimilatory apparatus beneath unit leaf area, rather than to differences in the intrinsic photosynthetic activities of contrasting genotypes. Nevertheless, some data about the intraspecific variability of CER, measured at light saturation and atmospheric O_2 and CO_2 content and comparable leaf age may be of interest (Table 7.3).

An interesting case of interspecific variability of CER should be mentioned in this context. It was shown[44,46,70,71,104] that during the evolution of *Triticum aestivum* L. CER per unit leaf area decreased with increasing ploidy level and the values in *Triticum aestivum* are lower than in its primitive ancestors. From this it was concluded that CER is not a limiting factor in yield formation. As a description of what happened during the evolution of *Triticum aestivum* this is obviously true. The loss of activity per unit leaf area was compensated for by an

TABLE 7.3. Intraspecific variability of CO_2 exchange rates in
different crop species

species	range of CER* $(mg\ CO_2.dm^{-2}.h^{-1})$	source
Medicago sativa	20—50	(101)
Glycine max	29—43	(41)
Phaseolus vulgaris	13.5—18.5	(65)
Triticum aestivum	26.7—37	(123)
Triticum aestivum	32.5—41.4	(38)
Triticum aestivum	38—43	(14)
Triticum aestivum	18.5—25.8	(104)
Hordeum vulgare	14—22	(5)
Hordeum vulgare	28.4—36.6	(38)
Zea mays	25—89	(61)
Zea mays	35—59	(86)

* CER = CO_2 exchange rate.

increased area per culm, and perhaps a longer leaf area duration. But it seems to be a mistake to conclude that CER is of no significance in the future of wheat breeding. The desirable high density of the canopies in modern agriculture makes it inevitable that a small but very efficient photosynthetic apparatus is developed, instead of the large but less efficient as in recent cultivars. Planchon[104] emphasized the cross of T. aestivum with T. spelta. Gaudillere[53] concluded from his results that the S genome of Aegilops speltoides is responsible for the expression of high rates of CER per unit leaf area.

During recent years the method of measuring CER in the field by means of short time (15 sec) $^{14}CO_2$ application has been developed considerably.[159,160,161,162] It is assumed that during 15 sec none of the activity taken up by the leaf will be evolved by photorespiration, and so the true or gross photosynthesis rate is measured. Because of the ease of handling, a large number of leaves can be measured in a relatively short time, and perhaps this method could be used to compare breeding lines under field conditions.

Another methodical approach for selection of genotypes with high rates of CER is the search for correlations between photosynthetic gas exchange and parameters of the leaves, which can be measured more easily. Positive correlations between CER and specific leaf weight (dry weight per unit leaf area) were reported for alfalfa[101], Aegilops and Triticum species[70], rice[89] and soybean[95]. Dunstone et al.[44] found a positive correlation between these two parameters among diploid and tetraploid species of Triticum but not in Aegilops and hexaploid Triticum species. The N-content per unit leaf area was positively correlated in experiments with rice[126] and in Aegilos and Triticum species[70]. Mesophyll cell size was shown to be negatively correlated with CER in Lolium perenne genotypes.[137,144]

Stomatal frequency in relation to CER has been investigated often, and in some cases positive correlations were found. Yoshida[142] estimated a correlation coefficient of $r = +0.94$ between CER and stomatal frequency in seven spring barley varieties, and because of the high heritability of the stomatal frequency he emphasized the use of this trait for screening high photosynthetic capacity in barley. Almost the same close correlation $r = +0.82$ was found in seven other spring barley varieties.[10] A close correlation between CER and the number of stomata per mm^2 was also found in a comparison of *Triticum* species of different ploidy level.[104] However, Dunstone *et al.*[44] also investigating *Triticum* species of different ploidy level plus four *Aegilops* lines, failed to detect a significant correlation between stomatal frequency and gas phase resistance. As long as we do not have additional and detailed data available about stomatal frequency and transpiration rates, especially under conditions of water stress, it could be a Trojan horse to emphasize stomatal frequency as a selection criterium having regard to the deteriorated water use efficiency. At times it seems, indeed, that the type of reaction of the stomata against water stress is much more important for water use efficiency than their conductance of CO_2 and water vapour at water saturating conditions.

A positive correlation was found between leaf vein density and net assimilation rate, as determined by growth analysis among 17 varieties of spring barley.[3] CER was not determined in these experiments. It should be noticed in this context that the photosynthetically more efficient C_4 grasses have twice as great a leaf vein density as C_3 grasses[4,37,68] and a 1.5 times higher stomatal frequency.[4]

INTERNAL REGULATING MECHANISMS

One cannot discuss photosynthetic gas exchange of leaves without mentioning internal regulating mechanisms. This problem is of conclusive significance, insofar as it would be unwise to recommend photosynthetic activity *per se* as a desirable breeding aim, if this activity is mainly regulated by the demand on assimilates in the whole plant. In the literature the problem is discussed in terms of sink–source relationships. The two terms have been used in various senses. Here the definitions given by Warren Wilson[134] will be applied whereupon sources and sinks are characterized in terms of losses and gains of a particular substance in a particular plant part. Recent reviews of sink–source interactions, including possible regulation mechanisms were given by Herold[62] and Pinto[103].

Summarizing the different and often contrasting results it can be stated first, that during the ontogenesis of a given species the limitation of crop growth by source or sink may change, depending on the development of the plant and the environmental conditions. For instance, the early developmental stages of cereals are perhaps sink limited because of the different Q_{10} values for growth and photosynthesis. Another stage of sink limitation seems to be at the early phase of grain filling in cereals.

Secondly, manipulation of the normal sink–source relationship (by defoliation, shading, removal of sink organs, etc.) in some experiments influenced the apparent rate of photosynthesis but not in others. The results of such experiments are therefore only of particular interest for the assessment of the equilibrium between assimilate production and the demand on assimilates by intact plants in a closed canopy, because of many possible side effects.[66] Development of isogenic lines, differing in sink or source at a given developmental stage would be helpful for assessing sink or source limitation of crop growth in single developmental stages. In some cases the direct connection between source and sink seems to be uncoupled or buffered by the possibility of temporary storage of assimilates in organs which normally don't act as competing sinks.[9]

Finally, in a discussion of sink—source relationships we need to distinguish between short and long term effects. The latter seem to be more important for characterizing physiological trait of 'future ideotypes'. As a long term effect in cereals we may see the influence of a large source strength (source size X source activity) on the development of spikelets and florets, the potential sinks during grain filling. Other examples are known, in which the number of flowers and developing fruits (sink strength X sink activity) stimulates the activity of the photosynthetic apparatus developed previously.[75]

Considering photosynthesis in relation to crop production it is of interest that one of the environmental factors limiting photosynthesis, the CO_2 content of the atmosphere, increases steadily. Sources of this increase are worldwide deforestation, mainly in the tropics, and the increase in burning fossil fuels.[117] Inevitably this will enhance photosynthetic productivity and, as a desirable side effect, improve water use efficiency especially in C_3 plants.[54,139] Among others, this should be a strong argument that partitioning of assimilates requires special attention of plant physiologists and breeders.

GROWTH, DEVELOPMENT AND PARTITIONING OF ASSIMILATES

A basic principle of life in higher plants is the interaction between autotrophic and heterotrophic organs, involving flows of materials and information between them. Normally plants are programmed to grow and develop with the 'aim' of reproduction. In species in which the reproductive organs, such as grains or tubers, are the economically useful parts this natural programme has been changed by breeding only in a quantitative manner, to obtain a relatively higher economic yield. It can be assumed that in other species, in which the whole vegetative shoot or the root are the harvest products, the changes by breeding in the evolutionary programme assimilate partitioning were more severe. Nevertheless, the success in breeding both types of species demonstrate that genetic manipulation of assimilate partitioning and storage in a harvest organ is possible within relatively broad limits.

In cereals the changes in the harvest index between old and new varieties can be seen as the result of breeding for improved partitioning in favour of enhanced economic yield. In the earliest days of breeding, simple selection for high grain yield may perhaps involuntarily have affected the grain/straw ratio as can be seen from a comparison between *Triticum aestivum* and its primitive ancestors.[46] Ruckenbauer[113] comparing a wheat variety from 1877 with two modern varieties estimated a harvest index of 0.37 in the old but 0.44 in the modern varieties. Aufhammer and Fischbeck[15] in extensive field and pot experiments compared genotypes of oat and barley from 1831 with modern varieties. From the results they concluded that about 25% of the higher yield in modern varieties was due to changed assimilate partitioning. In barley the range in harvest index between old and new varieties was from 0.38 to 0.53.[124]

Harvest index is strongly influenced by breeding for short strawed varieties. Bingham[22] found a close negative correlation between plant height and grain/straw ratio in 4 winter wheat varieties. In an extensive study at the Plant Breeding Institute in Cambridge, including 12 varieties of winter wheat (introduced between 1908 and 1978) it was convincingly demonstrated that the total dry matter production of the varieties was similar, "the increase in grain yield due to variety improvement being associated mainly with greater harvest index".[17] It should be noticed that the harvest index is not only genetically determined but also depends on environmental conditions and the adaptation of the given genotype. Passioura[100] found a close positive correlation between water use after anthesis and harvest index in wheat. This relates the problem not only to water use efficiency but also to problems of root geometry and its consideration in breeding.

Leaving out of consideration such improvements as disease resistance, lodging resistance, grain quality, etc. there seems to be agreement that most if not all, variety improvements in the past were due to increased harvest index. Therefore some questions are raised in respect of the strategy for future breeding: is a further increase in harvest index possible and desirable? What is the limit and how close are the modern varieties to this limit? Austin et al.[17] on the basis of the above mentioned experiment concluded that, with no change in above-ground biomass, breeders may be able to increase harvest index in wheat to about 0.6, achieving a gain in yield of 25%.

The final distribution of dry matter of crops at harvest between grain, straw and roots in cereals is the result of the developmental pattern during ontogenesis. This pattern for its part depends on the distribution and use of currently produced or temporarily stored products of photosynthesis and protein biosynthesis, for growth of new organs and respiration. Unfortunately, in the past a lot of empirical work was done in analysing the yield components at harvest time, and these are only of limited value for the understanding of internal mechanisms determining the final distribution of dry matter. There is one exception to this statement. The physiological processes of grain filling in cereals were extensively studied. However, during this stage only grain weight is determined. The other and seemingly more important yield components (grains per ear and ears per area) are differentiated at earlier developmental stages. The understanding of factors, genetic or environmental, influencing ear development, seems to be only just beginning.

The development of an adequate leaf area index by spring sown species, such as spring wheat, spring barley or sugar beet is a main requirement for efficient interception and utilization of solar radiation. Normally after the first leaf has reached maturity and acts as an assimilate exporting organ, roots and the newly developing leaves are the main sinks. The use of currently produced assimilates for formation of new photosynthetic active tissue consequently results in exponential crop growth, supposing other factors are not limiting. The shape of the growth curve depends on the partitioning coefficient.[55] Wareing[133] emphasized that in temperate regions early crop growth in spring is in most cases sink-limited because of the higher optimum for growth compared with photosynthesis. Light in the open canopies at that time is unlikely to be limiting. Consequently, selection of genotypes with low temperature optima for early growth should be a desirable aim. Studies of temperature/growth relations in different grass species[34,96] indicated a sufficiently broad intraspecfic variability as a basis for selection. In a screening of 900 spring barley varieties large differences in seedling growth rate (14 days at $5°C$) were estimated. Interestingly, not only the whole dry weight of the seedlings differed but also shoot/root ratio (Apel, unpublished).

Concerning the further development till anthesis only some ideas can be given, which in most cases still have to be proven experimentally. Tillering, or more precisely the genetic capacity to develop tillers, does not seem to be a serious problem in cereal breeding. Donald's[40] uniculm ideotype for wheat may be well founded from theoretical reasons but was hardly ever accepted by breeders. The reasons for this seem clear. The number of developing tillers is one of the traits which mainly depends on competition in a canopy, and at normal seed rates only 1–3 tillers in wheat or 3–5 in spring barley develop to maturity, independent of the capacity for tillering of singly grown plants of a given genotype. Also a minimum tillering capacity seems necessary for compensation of losses by frost injury, diseases, etc. In addition, nutrients stored in tillers without bearing ears are normally translocated to ear bearing culms.

Much more important seems the synchrony in tillering and tiller development.[16] Emergence of tillers is closely linked to that of leaves, thus tillering is synchronized with the development of leaves. It was recently shown that phytohormones and nutrients supplied by the roots may influence tiller development.[115] The danger of too rapid development of first leaves and development of tillers at low temperatures is damage by late frost in spring. But if low temperature optimum for growth could be combined with frost hardiness some advantage for the later stages of development could result.

In cereals, tillers become autotrophic when they have three leaves and three to five roots. Translocation of nutrients between tillers can be induced by some type of manipulation (defoliation of the main culm or other tillers, removal of the ear, etc.) but does not play an important role under normal conditions.

Besides the development of an optimal density of the canopy, the number of ear bearing culms per m^2, ear initiation, differentiation of spikelets and flowers, and grain filling, are the main processes determining grain yield. Ear formation begins immediately after the differentiation of the last leaf primordium at the shoot apex. Primordia destined to become spikelets and florets are initiated at an average rate 3—4 times faster than leaf primordia.[51] During the development of the ear, between initiation and anthesis, three distinct phases can be distinguished: 1. initiation of spikelets and florets; 2. growth and dry matter increase of the ear; 3. reduction of spikelets and florets to a definite number of fertile florets. During the first phase in which the maximum number of florets is determined, the dry matter increase of the shoot apex is so small (0.5 mg dry matter in barley if the terminal spikelet primordium is initiated) that it can hardly be imagined that this process is influenced by competition for assimilates. If nevertheless it was shown that a restricted source activity during this early stage of development inhibits the growth of the shoot apex,[47] it must be assumed that either it is a weak competitor or that some other mechanisms (perhaps so-called feed forward regulation) is involved. In a comparison of 10 spring barley genotypes including old and new high yielding varieties no differences in the rate of floret primordia initiation could be detected (Frank and Apel, unpublished).

Ear growth and dry matter increase coincide under normal conditions with the development of photosynthetically active leaves, and a shortage of assimilates seems likely only under stress conditions. Quite another situation governs the following stage between the time of maximum floret number and anthesis. Depending on genotype and environmental conditions only a given part of initiated florets become fertile. The causes of floret death are uncertain, but Gallagher[51] summarized arguments in favour of the hypothesis that shortage of assimilates is probably the main reason for the reduction in the number of florets. Strong arguments for this hypothesis are results in which experimentally changed photosynthesis during this phase significantly influenced the number of kernels per ear.[2,48,136] Babenko and Lyong[18] found a higher content of soluble carbohydrates in the shoot apex of high yielding than in low yielding varieties at preanthesis. An intensified study especially of this stage of ontogenesis seems justified, considering the fact that in two-rowed barley, 30—40%, and in wheat, 40—70%, of potential kernels per ear are lost by this reduction.

At anthesis the number of kernels per ear is determined. The final yield depends fundamentally on processes of translocation of carbohydrates and amino acids to the growing kernels, and their storage in the form of starch and proteins. It now seems clear that the main source of carbohydrates stored in the grain is current photosynthesis, mainly of the flag leaf. In awned varieties a considerable amount is produced by the awns. It is a crucial question as to whether the final grain weight is limited by supply of assimilates or internal mechanisms of grain growth. Severe stress apart, the early stages of grain growth are surely not limited by availability of assimilates, but are influenced by the pattern of ontogenetic development of cereals. The photosynthetic apparatus before and during anthesis is fully developed and the demand of the growing ear is still low. Therefore, large amounts of soluble carbohydrates are temporarily stored in the culm. A part of these are later translocated to the grains while another is lost by respiration. The interpretation of the significance of these temporarily-stored assimilates is not the only one possible. Some authors assume that they are reserves ensuring an optimum grain growth under conditions of temporary stress, others argue that a large part is lost by wasteful respiration and large stem reserves may be an undesirable character.[127] The problem of stem reserves and their use has recently been reviewed by Stoy.[122]

It seems nearly impossible to give a single answer to the question as to whether the later stages of grain filling are source-limited. Leaf area duration between anthesis and maturity seems to play the most important role. Austin et al.[17] concluded that a further increase in post-anthesis photosynthesis may be obtained by selecting genotypes in which anthesis is closer to the time of maximum leaf area index. However, results of experiments in which final grain size was not influenced by enhanced photosynthesis (induced by increased CO_2 concentration)[2,127] indicate that, at least in some genotypes, internal control mechanisms are involved, determining maturation and final grain weight independently of assimilate availability. Phytohormones[50,88] and connected to these the number of endosperm cells[26], and perhaps restrictions in transport[118] may be involved.

CONCLUSIONS

The position of plant physiology in relation to breeding is in principle not different from other scientific fields such as genetics, phytopathology, etc. This position can be formulated by the question: how can the results of analytical research be integrated into the synthetic task of the breeder? A general and not exclusive answer would be that the breeder should know which processes at the different stages of yield formation may, under given circumstances, be rate limiting. Furthermore, for nearly all the factors involved an often large genetic variability is available in the world collections. This could be used for selection and combination. New developments in methods (e.g. hybridization of somatic cells and genetic engineering) will surely open up possibilities for unconventional combinations. But whichever traits must be selected and then combined, and whatever the technique, it will in the end depend on our knowledge of yield formation processes. It must be an important task for physiologists to develop selection methods for single processes which can easily be used in practical breeding work.

Specifically in relation to photosynthesis and assimilate partitioning, the following points could be of interest to plant breeders:

(1)   Sink and source should be near equilibrium during all developmental stages. A small surplus in source strength seems advisable in spite of formation of reserves for stress periods. Source or sink limitation can be tested in experiments with enhanced CO_2

concentration.[57,80] A positive response of the genotype investigated would indicate source-limitation under natural conditions. No response indicates sink-limitation. Results of defoliation or shading experiments should be interpreted with care because many side-effects are possible.

(2)   A high source strength is especially important during initiation and differentiation of sink organs.

(3)   Improved source strength should be brought about by source activity rather than source size. The extremely high CO_2 uptake rates of $60-70$ mg $CO_2.dm^{-2}.h^{-1}$ in primitive ancestors of *Triticum aestivum*[46] should be a challenge to wheat breeders to study how this trait could be used in modern wheats.[131]

(4)   Synchrony in the development of storage organs is important. Each developing sink is a potential inhibitor for the development of other sinks, and this not only by competition for assimilates.

(5)   In cereals the borderline for a further increase in the harvest index may be reached in the near future. Therefore enhanced biomass production (per unit ground area) should be a breeding aim. This may be reached by optimal geometry of the canopy (leaf area, leaf angle, plant height), high rates of photosynthesis per unit leaf area, and a duration of the photosynthetic apparatus in accordance with the mean climatic conditions.

(6)   Water use efficiency could be improved by decreasing mesophyll resistance and by achieving a better understanding of control mechanisms of stomata.

(7)   Highly speculative, but perhaps not quite absurd, are ideas in connection with the C_4-pathway of photosynthesis. First, it seems possible to screen for C_4 genotypes with lower temperature optima (a programme for the selection of cold tolerance in maize has been started in New Zealand). Secondly, the genetic manipulation of temperate C_3-species should be tried, with the aim of getting C_3/C_4 intermediates with somewhat diminished photorespiration.

(8)   Close cooperation between good breeders and good physiologists is necessary if physiological knowledge is to greatly influence further breeding.

REFERENCES

1.   Anderson, K., King, W. and Valentino, R.C. Catalytic mutants of ribulose diphosphate carboxylase/ oxygenase. In: (116), 379—390, 1978.

2.   Apel, P. Untersuchungen über die Beeinfluessbarkeit des Rinzelährenertrags von Sommerweizen — Experiments in Pflanzenwachstumskarmmern. *Kulturpflanze* 24, 143—157, 1976.

3.   Apel, P. Leitbündeldichte und Nettoassimilationsrate bei Sommersgerste. *Arch. Züchtungsforsch.* 9, 179—184, 1979.

4.   Apel, P. Leitbündeldichte und Stomatafrequenz von Gramineen-Arten mit C_3 beziehungsweise C_4-pathway der Photosynthese. *Kulturpflanze* 27, 91—95, 1979.

5.   Apel, P. und Lehmann, Chr. O. Variabilität und Sortenspezifität der Photosyntheserate bei Sommergerste. *Photosynthetica* 3, 255—262, 1969.

6.   Apel, P. and Maass, I. CO_2 compensation concentration, O_2 influence on photosynthetic gas exchange and ^{13}C values in species of *Flaveria* (*Asteraceae*). *Biochem. Physiol. Pflanzen* 176, 398—401, 1981.

7.   Apel, P. und Peisker, M. Pflanzenarten mit intermidiärer Merkmalsausprägung in bezug auf den C_3- und C_4-pathway der Photosynthese. *Kulturpflanze* 27, 49—66, 1979.

8. Apel, P. und Peisker, M. CO_2-Versorgung der photosynthetischen Carboxylierungszentren. *Colloquia Pflanzenphys. der Humbold Universität zu Berlin*, 3, 181–204, 1980.
9. Apel, P., Peisker, M. und Tschäpe, M. Beiträge der Photosyntheseforschung zur Ertragsphysiologie, speziell bei Getreide. *Kulturpflanze* 20, 43–62, 1972.
10. Apel, P. und Schalldach, I. Sortenunterschiede in der Photosynthese bei Gerste (*Hordeum vulgare* L.): CO_2-Abhängigkeit, Aktivität und Menge der Ribulose 1,5-diphosphate Carboxylase. *Biochem. Physiol. Pflanzen*. In press, 1981.
11. Apel, P., Ticha, I. and Peisker, M. CO_2 compensation concentration in leaves of *Moricandia arvensis* (L.) DC. at different insertion levels and O_2 concentrations. *Biochem. Physiol.* Pfl. 172, 547–552, 1978.
12. Apel, P., Tschäpe, M., Schalldach, I. und Aurich, O. Die Bedeutung der Karyopsen für die Photosynthese und Trockensubstanzproduktion bei Weizen. *Photosynthetica* 7, 132–139, 1975.
13. Aslam, M., Huffaker, R.C., Rains, D.W. and Rao, K.P. Influence of light and ambient carbon dioxide concentration on nitrate assimilation by intact barley seedlings. *Plant Physiol*. 63, 1205–1209, 1979.
14. Aslam, M. and Hunt, L.A. Photosynthesis and transpiration of the flag leaf in four spring-wheat cultivars. *Planta* 141, 23–28, 1978.
15. Aufhammer, G. und Fischbeck, G. Ergebnisse von Gefäss- und Foldversuchen mit dem Nachbau keinfähiger Gersten- und Haferkörner aus dem Grundstein des 1932 errichteten Nürnberger Stadttheaters. *Z. Pflanzenzüchtung* 51, 354–373, 1964.
16. Austin, R.B. Physiological limitations to cereal yield and ways of reducing them by breeding. *Opportunities for increasing crop yields*. R.G. Hurd *et al.* (ed.), Pitman, Boston-London-Melbourne, 3–19, 1980.
17. Austin, R.B., Bingham, J., Blackwell, R.D., Evans, L.T., Ford, M.A., Morgan, C.L. and Taylor, M. Genetic improvements in winter wheat yields since 1900 and associated physiological changes. *J. Agric. Sci.* 94, 675–689, Cambridge, 1980.
18. Babenko, V.I. and Lyong, May Suan. On carbohydrate content of growing points and upper leaves in ontogenesis of winter wheat varieties of different productivity. *Sjelsk. biologija* 1, 70–73 (in Russian), 1981.
19. Bauwe, H., Apel, P. and Peisker, M. Ribulose 1,5-diphosphate carboxylase/oxygenase and CO_2 exchange characteristics in C_3 and $C_3–C_4$ intermediate species checking mathematical models of carbon metabolism. *Photosynthetica* 14, 550–556, 1980.
20. Bhatia, C.R. and Rabson, R. Bioenergetic considerations in cereal breeding for protein improvement. *Science* 104, 1418–1421, 1976.
21. Berlyn, M.B. A mutational approach to the study of photorespiration. In: (116), 153–164, 1978.
22. Bingham, J. Physiological objectives in breeding for grain yield in wheat. *Proceed. of 6th Eucarpia Congr.* 15–29, Cambridge, 1971.
23. Bishop, D.G. and Reed, M.L. The C_4-pathway of photosynthesis: Ein Kranz-Typ Wirtschaftswunder? *Photochemical and Photobiological Reviews*, 1, 1–69. K.C. Smith (ed.), 1976.
24. Björkman, O., Hiesey, W.M., Nobs, M.A., Nicholson, F. and Hart, R.W. *Carnegie Inst. Yearbook*, 228–232, 1968.
25. Bränden, R. Ribulose 1,5-diphosphate carboxylase and oxygenase from green plants are two different enzymes. *Biochem. Biophys. Res. Com*. 81, 539–546, 1978.
26. Brocklehurst, P.A. Control of grain morphogenesis in wheat and its relation to grain yield. (121), 41–44, 1979.
27. Brown, R.H. and Brown, W.V. Photosynthetic characteristic of *Panicum milioides*, a species with reduced photorespiration. *Crop Sci*. 15, 681–685, 1975.
28. Burris, R.H. and Black, C.C. (eds.). *CO_2 metabolism and plant productivity*. Univ. Park Press, Baltimore-London-Tokyo, 1976.
29. Cannell, R.Q., Brun, W.A. and Moss, D.N. A search for high net photosynthetic rate among soybean genotypes. *Crop Sci*. 9, 840–841, 1969.
30. Charles-Edwards, D.A. An analysis of the photosynthesis and productivity of vegetative crops in the United Kingdom. *Ann. Bot*. 42, 717–731, 1978.
31. Charles-Edwards. D.A. Photosynthesis and crop growth. In: (79), 111–124, 1979.

32. Chartier, Ph. and Catsky, J. Photosynthetic systems — conclusions. In: (35), 425—433, 1975.

33. *Computer simulation of a cotton production system — users manual.* Agric. Res. Service, U.S. Dept. Agriculture, 1975.

34. Cooper, J.P. Climatic variation in forage grasses. I. Leaf development in climatic races of *Dactylis glomerata* in controlled environments. *J. Appl. Ecol.* 1, 45—61, 1964.

35. Cooper, J.P. (ed.) *Photosynthesis and productivity in different environments.* Cambridge Univ. Press, 1975.

36. Cooper, J.P. Control of photosynthetic production in terrestrial systems. (35), 593—621, 1975.

37. Crookston, R.K. and Moss, D.N. Interveinal distance for carbohydrate transport in leaves of C_3 and C_4 grasses. *Crop Sci.* 14, 123—125, 1974.

38. Dantuman, G. Rates of photosynthesis in leaves of wheat and barley varieties. *Neth. J. Agric. Sci.* 21, 188—198, 1973.

39. Day, P.R. Plant Genetics: Increasing crop yield. *Science* 197, 1334—1339, 1977.

40. Donald, C.M. The breeding of crop ideotypes. *Euphytica* 17, 385—403, 1968.

41. Dornhoff, G.M. and Shibles, R.M. Varietal differences in net photosynthesis of soybean leaves. *Crop Sci.* 10, 42—45, 1970.

42. Downes, R.W. Effect of light intensity and leaf temperature on photosynthesis and transpiration in wheat and sorghum. *Aust. J. Biol. Sci.* 23, 775—782, 1970.

43. Downton, W.J.S. The occurrence of C_4 photosynthesis among plants. *Photosynthetica* 9, 96—105, 1975.

44. Dunstone, R.L., Gifford, R.M. and Evans, L.T. Photosynthetic characteristics of modern and primitive wheat species in relation to ontogeny and adaptation to light. *Aust. J. Biol. Sci.* 26, 295—307, 1973.

45. Evans, L.T. (ed.) *Crop physiology — some case histories.* Cambridge Univ. Press, 1975.

46. Evans, L.T. and Dunstone, R.L. Some physiological aspects of evolution in wheat. *Aust. J. Biol. Sci.* 23, 725—741, 1970.

47. Felippe, J.M. and Dale, J.E. Effects of shading the first leaf of barley plants on growth and carbon nutrition of the stem apex. *Ann. Bot.* 37, 45—56, 1973.

48. Fischer, R.A. and Stockman, Y.M. Kernel number per spike in wheat (*Triticum aestivum* L.): Responses to preanthesis shading. *Aust. J. Plant Physiol.* 7, 169—180, 1980.

49. Gaastra, P. Photosynthesis of crop plants as influenced by light, carbon dioxide, temperature, and stomatal diffusion resistance. Mededel. Landbouwhogesch. Wageningen 59, 1—68, 1959.

50. Gale, M.D. Genetic variation for hormonal activity and yield. In: (121) 29—34, 1979.

51. Gallagher, J.N. Ear development: Processes and prospects. In: (121) 3—9, 1979.

52. Garrett, M.K. Control of photorespiration and RuDP carboxylase/oxygenase level in ryegrass cultivars. *Nature* 274, 913—915, 1978.

53. Gaudillere, J.P. Caracteristiques photosynthetiques d'especes appartenant aux genres *Aegilops* et *Triticum. Ann. Ameriol. Plantes* 29, 523—533, 1979.

54. Gifford, R.M. Carbon dioxide and plant growth under water and light stress: Implications for balancing the global carbon budget. *Search* 10, 316—318, 1979.

55. Good, N.E. and Bell, D.H. Photosynthesis, plant productivity and crop yield. *The biology of crop productivity.* 3—51, P.S. Carlson (ed.), Academic Press, 1980.

56. Hardy, R.W.F. Chemical plant growth regulation in world agriculture. In: (114) 165—206, 1979.

57. Hardy, R.W.F., Havelka, U.D. and Quebedeaux, B. The opportunity for and significance of alteration of ribulose 1,5-diphosphate carboxylase activities in crop production. In: (116), 165—178, 1978. 1978.

58. Hatch, M.D. The C_4 pathway of photosynthesis: Mechanism and function. In: (28), 59—81, 1976. 81, 1976.

59. Hayashi, K., Yamamoto, T. and Nakagahra, M. Genetic control for leaf photosynthesis in rice, *Oryza sativa* L. *Japan. J. Breeding* 27, 49—56, 1977.

60. Heichel, G.H. Agricultural production and energy resources. *Amer. Scientist* 64, 64—72, 1976.

61. Heichel, G.H. and Musgrave, R.B. Varietal differences in net photosynthesis of *Zea mays* L. *Crop Sci.* **9**, 483–486, 1969.

62. Herold, A. Regulation of photosynthesis by sink activity – the missing link. *New Phytologist* **86**, 131–144, 1980.

63. Hewitt, E.J. Primary nitrogen assimilation from nitrate with special reference to cereals. In: (121), 139–155, 1979.

64. Hieke, B. Photosynthetischer Elektronentransport und Biomassebildung bei ausgewählten Evolutionsformen des Weizens. *Dissertation*, Humboldt-Univesität, Berlin, 1980.

65. Izhar, S. and Wallace, D.H. Studies on the physiological basis for yield differences. III. Genetic variation in photosynthetic efficiency of *Phaseolus vulgaris* L. *Crop Sci.* **7**, 457–460, 1967.

66. Jenner, C.F. Grain filling in wheat plants shaded for brief periods after anthesis. *Aust. J. Plant Physiol.* **6**, 629–641, 1979.

67. Jensen, R.G. and Bahr, J.T. Regulation of CO_2 incorporation via the pentose phosphate pathway. In: (28), 3–18, 1976.

68. Kanai, R. and Kashwagi, M. *Panicus milioides*, a *Gramineae* plant having Kranz leaf anatomy without C_4 photosynthesis. *Plant and Cell Physiol.* **16**, 669–679, 1975.

69. Kennedy, R.A. and Laetsch, W.M. Plant species intermediate for C_3, C_4 photosynthesis. **Science 184**, 1087–1089, 1974.

70. Khan, M.A. and Tsunoda, Sh. Evolutionary trends in leaf photosynthesis and related characters among cultivated wheat species and its wild relatives. *Japan. J. Breeding* **20**, 133–140, 1970.

71. Khan, M.A. and Tsunoda, Sh. Leaf photosynthesis and transpiration under different levels of air flow rate and light intensity in cultivated wheat species and its wild relatives. *Japan. J. Breeding* **20**, 305–314, 1970.

72. Krenzer, E.C. and Moss, D.N. Carbon dioxide enrichment effects upon yield and yield components in wheat. *Crop Sci.* **15**, 71–74, 1975.

73. Kung, S.D. and Marsho, R.V. Regulation of RuDP carboxylase/oxygenase activity and its relationship to plant photorespiration. *Nature* **259**, 325–326, 1976.

74. Kung, S.D. and Rhodes, P.R. Interaction of chloroplast and nuclear genomes in regulating RuDP carboxylase activity. In: (116), 307–324, 1978.

75. Lenz, F. Fruit effects on photosynthesis, light and dark respiration. In: (79), 271–281, 1979.

76. Loomis, R.S., Williams, W.A. and Hall, E.A. Agricultural productivity. *Ann. Rev. Plant Physiol.* **22**, 431–468, 1971.

77. Ludlow, M.M. Ecophysiology of C_4 grasses. *Water and Plant Life.* O.L. Lange, L. Kappen, E.D. Schulze (eds.), 364–386. Springer Verl. Berlin-Heidelberg-New York, 1976.

78. Ludwig, L.J., Charles-Edwards, D.A. and Withers, A.C. Tomato leaf photosynthesis and respiration in various light and carbon dioxide environments. *Environmental and biological control of photosynthesis.* R. Marcelle (ed.), The Hague, 29–36, 1975.

79. Marcelle, R., Clijsters, H. and van Pouke, M. *Photosynthesis and plant development.* Dr. W. Junk bv. Publ. The Hague-Boston-London, 1979.

80. Mauncy, J.R., Fry, K.E. and Guinn, G. Relationship of photosynthetic rate to growth and fruiting of cotton, soybean, sorghum and sunflower. *Crop Sci.* **18**, 259–263, 1978.

81. McCashin, B.G. and Canvin, D.T. Photosynthetic and photorespiratory characteristics of mutants of *Hardeum vulgare* L. *Plant PHysiol.* **64**, 354–360, 1978.

82. Menz, K.M., Moss, D.N., Cannell, R.Q. and Brun, W.A. Screening for photosynthetic efficiency. *Crop Sci.* **9**, 692–694, 1969.

83. Merbach, W. and Schilling, G. Bezichungen zwischen C and N-Stoffwechsel bei höheren Pflanzen. *Lolloquia Pflanzenphysiologie der Humboldt-Universität zu Berlin*, **3**, 134–150, 1980.

84. Miflin, B.J. Nitrogen metabolism and amino acid biosynthesis in crop plants. *The biology of crop productivity.* P.S. Carlson (ed.), 255–296, Academic Press, New York, 1980.

85. Mitsui, A., Miyachi, Sh., San Pietro, A. and Tamura, S. (eds.) Biological solar energy conversion. Academic Press, New York-San Francisco-London, 1977.

86. Moss, D.N. Laboratory measurement and breeding for photosynthetic efficiency. *Prediction and measurement of photosynthetic productivity. Proceedings of the IBP/PP Technical Meeting, Trebon*, 14–21, 9, 1069; 323–330, 1970.

87. Moss, D.N. and Musgrave, R.B. Photosynthesis and crop production. *Adv. in Agron.* **23**, 317—336, 1971.

88. Mounla, M.A.Kh. Phytohormones and grain growth in cereals. In: (121), 20—28, 1979.

89. Murata, J. Studies on the photosynthesis of rice plants and its culture significance. *Bull. Nat. Inst. Agr. Sci.* (Japan), D 9, 1—70, 1961.

90. Nichiporovich, A.A. Potential plant productivity and the principles of its optimal use. *Sjelschos. biologija* **14**, 683—694 (in Russian), 1979.

91. Nichiporovich, A.A. Energetic efficiency of photosynthesis and productivity of plants. Pushcino (37 p., in Russian), 1979.

92. Ogren, W.L. Search for higher plants with modifications of the reductive pentose phosphate pathway of CO_2 assimilation. In: (28), 19—29, 1976.

93. Ogren, W.L. and Hunt, L.D. Comparative biochemistry of ribulose diphosphate carboxylase in higher plants. In: (116), 127—138, 1978.

94. Øhmann, E. Ribulose diphosphate carboxylase. *Colloquia Pflanzenphysiologie der Humboldt-Universität zu Berlin*, 3, 99—117, 1980.

95. Ojima, M. and Kawashima, R. Studies on the seed production of soybean. 5. Varietal differences in photosynthetic rate of soybean. *Proc. Crop Sci. Soc. Japan,* 37, 667—675, 1968.

96. Ollerenshaw, J.H., Stewart, W.S., Gallimore, J. and Baker, R.H. Low temperature growth in grasses from Northern latitudes. *J. Agric. Sci.* **87**, 237—239, Cambridge, 1976.

97. Paech, Ch., McCurry, S.D., Pierce, J. and Tolbert, N.E. Active site of ribulose 1,5-diphosphate carboxylase/oxygenase. In: (116), 227—243, 1978.

98. Parthier, B. Die biologische Fixierung des atmosphärischen Stickstoffs. *Biol. Rundschau* **16**, 345—364, 1978.

99. Passioura, J.B. Accountability, philosophy and plant physiology. *Search* **10**, 347—350, 1979.

100. Passioura, J.B. Grain yield, harvest index and water use of wheat. *J. Aust. Inst. Agric. Sci.* **43**, 117—120, 1977.

101. Pearce, R.B., Carlson, C.E., Barnes, D.K., Hart, R.H. and Hanson, C.H. Specific leaf weight and photosynthesis in alfalfa. *Crop Sci.* **9**, 423—426, 1969.

102. Peisker, M. and Apel, P. Dark respiration and the effect of oxygen on CO_2 compensation concentration in wheat leaves. *Z. Pflanzenphysiol.* **100**, 389—395, 1980.

103. Pinto, M.C. Regulation de la photosynthèse para la demande d'assimilats: mécanisms possibles. *Photosynthetica* **14**, 611—637, 1980.

104. Planchon, C. Assimilation nette par unité de surface chez diverses espèces du genre *Triticum. Ann. Amélior. Plantes* **24**, 201—207, 1974.

105. Planchon, C. Essai de détermination de critères physiologiques en vue de l'amélioration du blé tendre: Les facteurs de la photosynthèse de la derniere feuille. *Ann. Amélior. Plantes* **26**, 717—744, 1976.

106. Quebedeaux, B. and Hardy, R.W.F. Reproductive growth and dry matter production of (*Glycine max* L.) Merr. in response to oxygen concentration. *Plant Physiol.* **55**, 102—107, 1975.

107. Raghavendra, A.S. and Das, V.S.R. The occurrence of C_4-photosynthesis: A supplementary list of C_4 plants reported during late 1974 mid 1977. *Photosynthetica* **12**, 200—208, 1978.

108. Rahman, M.S., Wilson, J.H. andAitken, Y. Determination of spikelet number in wheat. II. Effect of varying light level on ear development. *Aust. J. Agric. Res.* **28**, 575—581, 1977.

109. Randall, D.D., Nelson, C.J. and Asay, K.H. Ribulose diphosphate carboxylase. Altered genetic expression in tall fescue. *Plant Physiol.* **59**, 38—41, 1977.

110. Rathnam-Chaguturu and Chollet, R. Regulation of photorespiration. *Current Advances in Plant Science* 12, 38.1—38.19, 1980.

111. Rathnam, C.K.M. and Chollet, R. Photosynthetic carbon metabolism in C_4 plants and C_3—C_4 intermediate species. *Progr. in Phytochem.* **6**, 1—48, 1980.

112. Rathnam, C.K.M. and Chollet, R. Photosynthesis and photorespiratory carbon metabolism in mesophyll protoplasts and chloroplasts isolated from isogenic diploid and tetraploid cultivars of ryegrass (*Lolium perenne* L.). *Plant Physiol.* **65**, 489—494, 1980.

113. Ruckenbauer, P. *Keimfähiger Winterweizen aus dem Jahre 1877. Beobachtungen und Versuche.* K.u.K. Hochschule für Bodenkultur, Wien, 372—386, 1971.

114. Scott, T. (ed.) *Plant regulation and world agriculture.* Plenum Press, New York-London, 1979.

115. Sharif, R. and Dale, J.E. Growth regulating substances and the growth of tiller buds in barley: Effects of cytokinina. *J. Exp. Bot.* **31**, 921−930, 1980.

116. Siegelman, H.W. and Hind, G. (eds.). *Photosynthetic carbon assimilation.* Plenum Press, New York-London, 1978.

117. Siegenthaler, U. and Oeschger, H. Predicting future atmospheric carbon dioxide levels. *Science* **199**, 388−395, 1978.

118. Sofield, J., Wardlaw, J.F., Evans, L.T. and Zee, S.Y. Nitrogen, phosphorus and water contents during grain development and maturation in wheat. *Aust. J. Plant Physiol.* **4**, 799−810, 1979.

119. Somerville, C.R. and Ogren, W.L. A phosphoglycolate phosphatase deficient mutant of *Arabidopsis. Nature* **280**, 833−836, 1979.

120. Somerville, C.R. and Ogren, W.L. Photorespiration mutants of *Arabidopsis thaliana* deficient in serine-glyoxylate amino-transferase activity. *Proc. Nat. Acad. Sci.* USA **77**, 2684−2687, 1980.

121. Spiertz, J.H.J. and Kramer, Th. (eds.) *Crop physiology and cereal breeding.* Centre for Agricultural Publishing and Documentation, Wageningen, 1979.

122. Stoy, V. The storage and re-mobilization of carbohydrates in cereals. In: (121), 55−59, 1979.

123. Stoy, V. Photosynthesis, respiration and carbohydrate accumulation in spring wheat in relation to yield. *Physiol. Plantarum, Suppl.* IV, 5−125, 1965.

124. Stoy, V. *Pflanzenphysiologische Merkmale als Selektionskriterium in der Züchtung auf Ertrag bei Getreide. − Bericht über die Arbeitstagung 1976 der Arbeitsgemeinschaft der Saatzuchtleiter in Gumpstein.* 11−29, 1976.

125. Szarek, S.R. The occurrence of crassulacean acid metabolism: A supplementary list during 1976−1979. *Photosynthetica* **13**, 467−473, 1979.

126. Takano, Y. and Tsunoda, Sh. Curvilinear regression of the leaf photosynthetic rate on leaf nitrogen content among strains of *Oryza* species. *Japan. J. Breeding* **21**, 69−76, 1971.

127. Thomas, S.M., Thorne, G.N., Kendall, A.C. and Pearman, I. Stem and ear respiration, and leaf photo-respiration during grain growth: their significance to yield. In: (121), 96−101 1979.

128. Tooming, H. and Kallis, A. Productivity and growth calculations of plant stands. Tartu 1972 (in Russian), 1972.

129. Troughton, J.H. Photosynthetic mechanisms in higher plants. In: (35), 357−391, 1975.

130. UK−ISED. *Biomass for energy.* Conference (C 20) at the Royal Soc., July 1979, London, 1979.

131. Vos, de N.M. Cultivar differences in plant and crop photosynthesis. In: (121), 71−74, 1979.

132. Waggoner, P.E. Predicting the effect upon photosynthesis of changes in leaf metabolism and physics. *Crop Sci.* **9**, 315−321, 1969.

133. Wareing, P.F. Temperature response and yield in temperate crops. In: (114), 129−139 1979.

134. Warren Wilson, J. Control of crop processes. *Crop processes in controlled environments.* A.R. Rees, K.M. Cockshull, D.W. Hand and R.C. Hurd (eds.), 7−30, Academic Press, London-New York, 1972.

135. Wildman, S.G. Aspects of fraction 1 protein evolution. *Arch. Biochem. Biophys.* **196**, 598−610, 1979.

136. Willey, R.W. and Holliday, R. Plant population and shading studies in barley. *J. Agric. Sci.* **77**, Camb. 445−452, 1971.

137. Wilson, D. and Cooper, J.P. Assimilation of *Lolium* in relation to mesophyll. *Nature* **214**, 989−992, 1967.

138. Wit de, C.T. *et al. Simulation of assimilation, respiration and transpiration of crops.* Centre for Agric. Publ. and Documentation, Wageningen, 1978.

139. Wittwer, S.H. The shape of things to come. *The biology of crop production.* P.S. Carlson (ed.), 413−459, Academic Press, 1980.

140. Yoshida, Sh. Physiological aspects of grain yield. *Ann. Rev. Plant Physiol.* **23**, 437−464, 1972.

141. Yoshida, Sh. Effects of CO_2 enrichment at different stages of panicle development on yield components and yield of rice (*Oryza sativa* L.). *Soil Sci. Plant Nutr.* **19**, 311−316, 1973.

142. Yoshida, T. Effect of stomatal frequency on photosynthesis and its use for breeding in barley. *Bull. Kyushu Nat. Agr. Exp. Sta.* **20**, 129—193, 1978.

143. Wilson, D. and Cooper, J.P. Assimilation of *Lolium* in relation to leaf mesophyll. *Nature* **214**, 989—992, 1973.

144. Wilson, D. and Cooper, J.P. Effect of selection for mesophyll cell size on growth and assimilation in *Lolium perenne* L. *New Phytologist* **69**, 235—245, 1978.

145. Zelitch, I. Improving the efficiency of photosynthesis. *Science* **188**, 626—633, 1975.

146. Zelitch, I. Biochemical and genetic control of photorespiration. In: (28), 343—358, 1976.

147. Zelitch, I. Measurement of photorespiratory activity and the effect of inhibitors. *Methods in Enzymology* **69**, 453—464, 1980.

148. Zindler-Frank, E. Oxalate biosynthesis in relation to photosynthetic pathway and plant productivity — a survey. *Z. Pflanzenphysiol.* **80**, 1—13, 1976.

149. Björkman, O. Adaptive and genetic aspects of C_4 photosynthesis. *CO_2 Metabolism and Plant Productivity.* Burris, R.H. and Black, C.C. (eds.), 287—309, University Park Press, Baltimore, 1976.

150. Osmong, C.B., Björkman, O. and Anderson, D.J. Physiological processes in plant ecology. Towards a synthesis with *Atriplex. Ecological Studies* **36**, Springer Verlag, 1980.

151. Powell, A.M. Systematics of *Flaveria* (*Flaveriinae-Asteraceae*). *Ann. Missouri Bot. Garden* **65**, 590—636, 1978.

152. Smith, B.N. Natural abundance of the stable isotopes of carbon in biological systems. *BioScience* **22**, 226—231, 1972.

153. Smith, B.N. Evolution of C_4 photosynthesis in response to changes in carbon and oxygen in the atmosphere through time. *BioSystems* **8**, 24—32, 1976.

154. Smith, B.N. and Robbins, M.J. Evolution of C_4 photosynthesis: An assessment based on $^{13}C/^{12}C$ ratios and Kranz anatomy. *Proc. 3rd Int. Congr. Photosynthesis.* Avron, M. (ed.), **2**, 1579—1587, Elsevier, Amsterdam, 1975.

155. Apel, P. Dark and photorespiration. *Crop physiology and cereal breeding. Proc. Eucarpia Workshop, Wageningen 1978.* Spiertz, J.H.J. and Kramer, Th. (eds.), 102—105, Pudoc, 1979.

156. Beer, S., Shomer-Ilan, A. and Waisel, Y. Carbon metabolism in seagrasses. II. Patterns of photosynthetic CO_2 incorporation. *J. Expt. Bot.* **31**, 1019—1026, 1980.

157. Crookston, R.K. and Moss, D.N. A variation of C_4 leaf anatomy in *Arundinella hirta* (*Gramineae*). *Plant Physiol.* **52**, 397—402, 1973.

158. Shomer-Ilan, A., Beer, S. and Waisel, Y. *Suaeda monoica*, a C_4 plant without typical bundle sheath. *Plant Physiol.* **56**, 676—679, 1975.

159. Strebeyko, P. Rapid method for measuring photosynthetic rate using $^{14}CO_2$. *Photosynthetica* **1**, 45, 1967.

160. Shimski, D. A rapid method for measuring photosynthesis with labelled carbon dioxide. *J. Expt. Bot.* **20**, 381, 1969.

161. Incoll, L.D. and Wright, W.H. A field technique for measuring photosynthesis using $^{14}CO_2$. *Spec. Bull.* Soils Conn. Agri. Expt. Sta. XXX/100, 1969.

162. Sondahl, M.R., Crocomo, O.J. and Sodek, L. Measurements of ^{14}C incorporation by illuminated intact leaves of coffee plants from gas mixtures containing $^{14}CO_2$. *J. Expt. Bot.* **27**, 1189—1195, 1976.

CHAPTER 8

POTENTIAL FOR ENHANCING BIOLOGICAL NITROGEN FIXATION

J.M. Vincent
Department of Microbiology,
University of Sydney,
Sydney, Australia

INTRODUCTION

The ecologically and productively important capacity to utilise atmospheric nitrogen (N_2) appears to be restricted to procaryotic microorganisms (diverse bacteria and blue-green algae). The most important of these make their greatest contribution in some form of association with a higher plant. These associations may take the form of the intimate specialised root-nodule, or be relatively unstructured on, or to a limited extent, within the root. The most studied close symbiosis is that between members of Family *Leguminosae* and *rhizobia* which, besides being economically pre-eminent, has the facility of easy isolation and pure culture study of the microsymbiont. Root nodulation outside the *Leguminosae* is almost entirely due to actino-mycete-like organisms, a few of which only now appear to be yielding to isolation and cultivation procedures. The loose N_2-fixing associations with plant roots ('diazotrophic rhizoco-enoses')[142] have recently come into prominence because they extend the potential of N_2-fixation to be widespread and economically important Family *Gramineae*. The detection of these associations and a good deal of their study was made possible by the introduction of the extremely sensitive gas chromatographic technique to follow the reduction of acetylene to ethylene by the same enzyme as is responsible for N_2-fixation.

ASSESSMENT OF N_2-FIXATION CAPACITY

Fixation by the legume-*Rhizobium* system is quantitatively so significant that it has never offered difficulty in its demonstration with even the simplest facilities. Increment in yield or total N of specifically nodulated plants against an appropriate non-nodulated control, under conditions of limiting N, suffices for quantitation of symbiotic efficiency, and lends itself to cheap convenient screening of many hosts, bacterial strains and host X strain interactions. Non-destructive indicators, such as leaf dimension and colour, can be useful at the stage when selected plants need to be kept for crossing or seed increase. It may be associated with the increase of tested plants by cloning. Growth of single plants on agar within an enclosed tube, or with only the roots enclosed, may be useful for preliminary screening of smaller seeded legumes. Sand culture in Leonard jars, or in pots if rhizobial control is not necessary, permits greater differentiation and is necessary for larger seeded species.

Soil in cores, pots or in the field will have to be resorted to eventually to the stage when field trials in diverse localities become the ultimate test. It will be advantageous if some at least of the sites used in these trials are sufficiently short of available N, and for some purposes effective specific rhizobia, to permit assessment of N_2-fixation benefit. For special purposes there could be a case for obtaining controls in which effective nodulation has been prevented by swamping with heavy inoculum of an ineffective nodulating strain.[103] Such controls can give a reasonably good measure of fixed-N in the variously treated cultivars. Modifications and refinements of field testing might include provision for determining the nature of the strains which are responsible for nodulation. Such strain identification may be achieved serologically, by phage typing or by the experimental use of an antibiotic-resistance label.

Considerable differences can be expected on some occasions between results for performance of plant lines in the greenhouse and in the field. Indeed it will be no surprise to find significant differences between field sites and some inconsistency between years. Some part of these discrepancies reflect the influence of diverse environmental factors and some, differences in the nodulating rhizobia. It is not unexpected therefore to find in the literature some reports of success following greenhouse selection, and others where this is not so. A similar low level technology is applicable to root nodule systems of non-legumes although, in this case, control of the endosymbiont has been restricted by difficulties in its isolation and cultivation.

The acetylene reduction technique for demonstrating nitrogenase activity has been able, because of its extreme sensitivity, to demonstrate N_2-fixing potential in many other systems. It has certain limitations in application and interpretation but is of considerable research value for all N_2-fixation systems, particularly for the short interval study of factors influencing fixation. It is generally difficult to maintain long term experiments under sufficiently natural conditions for a large number of treatments, but improved continuous *in situ* C_2H_2 — reduction has been recently described by working with low concentrations C_2H_2 (0.2%),[126] sufficient to achieve 90% of the potential rate, and open bottom chambers of a design which permit non-nodulated parts of the root to explore deeper for water and minerals. Intra-varietal variability in C_2H_2-reduction remained a problem to be understood and overcome.

Uptake of nitrogen itself is of course the final definitive evidence of N_2-fixation. Three methods are available but all require access to a mass or emission spectrophotometer and present certain difficulties in their widespread application. $^{15}N_2$ techniques have their use as definitive proof for fixation but are much less sensitive than acetylene reduction, are short-termed, of limited availability and expensive. For field purposes ^{15}N isotope dilution techniques (dependent on fixation of unlabelled N_2) is capable of giving a quantitatively integrated estimate. In assessing the suitability of the several methods, it has been pointed out[113] that the greatest practical difficulty with this technique is the provision of a non-fixing plant as control. Genotypes definitely unable to support N_2-fixation will be valuable, the more so if they are otherwise genetically very like the test material. This has been achieved with some legume genotypes but not otherwise. Change in the ^{15}N: ^{14}N ratio permits the estimate of N_2-fixation at levels of ^{15}N natural abundance, and depends on preferential uptake of ^{14}N. Disturbance of the system by the use of fertilizer containing the expensive ^{15}N is avoided, but it depends on having a biologically stabilised isotope balance in the substrate and this imposes a limit on the soils which can be used.

Related publications

Several recent publications dealing with biological N_2-fixation will provide background to this account.[57,91,93,102,111,128,130]

THE LEGUME-*RHIZOBIUM* SYMBIOSIS

Nature of the symbiosis

The symbiosis involving rhizobia is highly structured, involves an intimate relationship between host and invading bacterium, and requires many steps in its achievement. The resultant nodules constitute prominent outgrowths of the root which develop as a result of the establishment of extra meristematic regions within the cortex and the production of a large volume of specialised tissue. Within this tissue the rhizobia, modified morphologically and physiologically as bacteroids, occupy most of intracellular space but remain enclosed, singly or in groups, in a plant-produced 'peribacteroid' membrane.

The prime reductive step, N_2 to NH_3, rests with the bacteroid; the energy needed for the process, as well as carbon skeletons required for amino-acid synthesis, derives from products of photosynthesis. At some stage in the incorporation of fixed-N, the host takes on the major role as it does in the partitioning and utilisation of the early products.

The effective N_2-fixing nodule of the legume always contains leghaemoglobin. The exact cytological location and function of this unusual plant component have yet to be determined[42] but it is certainly closely associated with the functioning bacteroid and seems to have a controlling role in the dispensing of O_2 essential to the energy yielding process whilst avoiding such a level as would divert an excessive number of electrons away from N_2 reduction.[144] The detailed chemical structure of leghaemoglobin reflects the intimacy of this symbiosis in that its globin moiety is determined by host genome; the haem portion seemingly, by that of the microsymbiont.

There are several recent accounts which describe the establishment, structure and function of the leguminous root nodule.[26,27,76,90] An expanded codified phenotypic sequence has been proposed to give more precision to the influence of host, rhizobial or environmental factors.[138] In this account the emphasis will be on the influence of host genotype.

STEPWISE ANALYSIS

The process which begins at the root surface and results in a N_2-fixing nodule is a sequence of many successively dependent steps, each of which is subject to control by host and rhizobial genotypes, separately or interacting. It follows that the establishment of an effective symbiosis can be expected to involve many genes so that the common use of expressions as broad as 'non-invasibility' or 'ineffectiveness' fail to define that part of the process at which failure has occurred. Direct evidence is needed as to affected stages and this will require closer study than has generally been brought to bear.

The scheme presented in this account is based on plants with root-hair infections; the earlier stages for legumes with different portals of entry and method of invasion within the developing nodule[26,27] will have to be appropriately modified; some of the presently designated stages will undoubtedly have to be further divided as more information is secured.

The eleven detailed stages in the present scheme have been grouped to conform with three major steps: preinfection, infection and nodule formation, and nodule function.

(i) PREINFECTION

Colonisation of the root surface by the rhizobia (Roc) is likely to be the first step, whether one starts with a minority indigenous population or an applied inoculum. It seems to be important for the rhizobia to secure a place for themselves in close proximity to the possible sites of infection (Roa). The extent to which host-produced lectins, or related molecular forms, are necessary to achieve this is still a matter for decision and has been discussed in various accounts.[8,17,33,89,138]

Root hair response to the presence of invasive rhizobia takes two major forms: branching (Hab), which is relatively non-specific, and 'marked' curling (Hac), which is much more specific to both host and *Rhizobium*,[55,145] and appears to be a prerequisite or common condition associated with invasion.

(ii) INFECTION AND NODULE FORMATION

The ability of a particular rhizobium to form a morphologically recognisable nodule is the most common evidence of host invasibility but earlier, closer observation, reveals the infection thread (Inf) laid down and developed by the plant in response to invading rhizobia.

Nodule initiation (Noi) is achieved generally by only a few of the infections; many, sometimes most, of the infection threads abort. Success depends on the establishment of one or more meristems in the region of the host cortex. The possible cause or effect relationship of polyploidy to this initiation remains an area of uncertainty, but most of the nodule tissue is at least tetraploid.

Two more steps are needed for the development of a potentially functional nodule. Rhizobia need to be released into the host cell (Bar) and to develop into the bacteroid form (Bad), in which the 'released' bacteria are still found to have a degree of isolation from host cytoplasm by virtue of a surrounding envelope of plant origin ('peribacteroid' membrane).

Recently developed mutants of *R. trifolii* permit the separate recognition of 'nodule initiation' (Noi; development of meristem in root cortex), and the additional steps of 'infection-thread branching' (Inb) and 'nodule development and differentiation' (Nod) (B.G. Rolfe, pers. comm.).

(iii) *NODULE FUNCTION*

The Nif coding, although often used loosely for the total function, is best restricted to the capacity to produce the enzyme complex directly responsible for the reduction of N_2 to NH_3. Various complementary functions (Cof) are needed to provide an apt environment for fixation. These will include the synthesis of leghaemoglobin, supply of photosynthates and enzymes involved in passing fixed-N to other molecular forms (amines, allantoin, various amino-acids). Partition of acquired-N between vegetative parts and grain will often be an important aspect of practical significance. The total fixation success will depend on nodule persistence (Nop), which in turn is likely to be affected by time to seed set and maturation.

Influence of the host genotype

(i) *INCIDENCE OF NODULATION DUE TO RHIZOBIA*

In the broadest terms one sees the influence of host in the general but not absolute failure of non-legumes to nodulate with *Rhizobium*, and the common nodulating success of legumes. Although only a small proportion of the known species in the Family Leguminosae have been examined for capacity to form nodules, most of these have yielded a positive result. The notable exception is in the Caesalpinioideae, particularly in the genus *Cassia*, for which there have been more negative cases reported than positive.[2,24] Although root-nodulation occurs widely outside the family there are few cases in which this has been attributed to *Rhizobium*, and only one for which the evidence is fully compelling.[133]

Some leguminous hosts, such as species of *Trifolium*, *Pisum*, *Vicia* and *Medicago*, are nodulated only by faster-growing rhizobia; others such as soybean, only, or more consistently, by slow-growers. Other hosts (such as *Lotus*) nodulate freely with both of the two distinctive rhizobial forms; in that case a considerable degree of interaction specificity is still to be seen when one assesses the N_2-fixing effectiveness of the respective associations.

(ii) *TAXONOMIC RELATEDNESS AMONG HOSTS NODULATED BY THE SAME RHIZOBIA*

Within the nodulating members of the family it is possible to see some taxonomic relatedness among hosts which share susceptibility to the same rhizobia, as for example the wide cross-reactivity amongst species of *Trifolium* (though with sub-divisions that accord with development in distinct geographical regions). Generic barriers are not uncommonly breached (e.g. some strains able to nodulate both *Pisum* and *Trifolium*, others *Medicago sativa* and *Leucaena latisiliqua*). In other cases cross invasion may be more readily achieved between than within genera, e.g. restriction between *Medicago polymorpha* and *Medicago laciniata* with strains all able to nodulate *Melilotus alba*. The reaction of clover hosts to *R. leguminosarum* RCR 1020 is also of interest. Some lines of *Trifolium pratense* and *T. subterraneum* cv. Mt. Barker were freely nodulated by this 'pea rhizobium' but several other species of *Trifolium* were resistant to invasion across this usual inoculation group boundary.[60]

(iii) *SELECTIVE AND PROMISCUOUS HOSTS*

There are indications that less selective (more promiscuously nodulating) hosts are likely to be those which are naturally out-crossed, and hence have had the opportunity to collect a wider gene spectrum. Nodulation of clovers by *Rhizobium leguminosarum* has not conformed to this trend in that the markedly selfed *Trifolium subterraneum* had all its test plants nodulated with the pea rhizobium, whereas only a small proportion of out-crossing *T. pratense* nodulated with the pea rhizobium.[103]

Typical examples of nodulating promiscuity are *Macroptilium atropurpureum*, *Macroptilium lathyroides*, *Vigna unguiculata*; of restriction: *Stylosanthes guyaneasis*, *Centrosema pubescens*, *Glycine max*, *Leucaena latisiliqua* and *Lotononis bainessi*. Specialised nodulating requirements of cv1R1—1002 of *Stylosanthes gracilis* have been hard to match with isolated rhizobial cultures, although effective nodulation could be obtained with rhizobia already in some soils.[32,127] Specialised rhizobial requirements of hexaploid lines of Caucasian clover (*Trifolium ambiguun*) also demonstrate difficulties likely to be imposed on host lines developed for otherwise desirable agronomic features.[148]

From a practical point of view the promiscuously nodulating host has maximum opportunity to form an association with one or several strains already to be found in an area. On the other hand if the local rhizobial population is sub-optimal, or frankly ineffective, there may be difficulty in suppressing less effective nodulation by the introduction of a fully effective inoculum. The difficulty is rather less for the grain legumes, for which each seeding gives the opportunity to reinforce the benefit with effective inoculum, than it is for the perennial or re-seeding forage and pasture legume. Conversely the selective host species or variety permits establishment of an effective inoculum relatively free of competition. But then an inoculum could be more serious. It has been suggested that soybean lines with the 'non-nodulating' recessive gene could be used to secure selective nodulation by a more aggressive (rhizobitoxine-producing) *R. japonicum*.[39] However the fully invasive effective strain has yet to be found or developed for this non-nodulating host.

(iv) *THE IMPORTANCE OF SPECIFICITY BETWEEN HOST AND RHIZOBIUM*

Interpretation of data obtained on the legume-*Rhizobium* symbiosis cannot go far without constant recognition of the common over-ruling effect of specific interaction between the symbionts' genotypes. In the case of ordinary nodulating cultivars it is not possible to point to a host line which is completely unable to form nodules with other rhizobial strains. Even 'non-nodulating' isolines of soybean can be nodulated under more favourable conditions by rhizobitoxine-producing strains of *R. japonicum*[39] which also permit entry by otherwise non invading strains supplied as a mixture with the invasive form.[37] Cv Afghanistan field pea (sym2 sym2) has been nodulated by a few of the tested European strains.[77] Examination of commercial seed lots of *Centrosema pubescens* tested against certain rhizobial strains showed extreme heritable variation in nodule abundance but other rhizobial strains were able to nodulate the sparsely nodulating lines satisfactorily.[14]

The importance of interaction between the two partners is even more apparent when we consider the effectivity of such nodules as are formed. The common experience is that the establishment of a fully effective N_2-fixing symbiosis is markedly dependent on mutual compatibility of the symbionts. Nodulation which crosses usual host barriers is likely to be ineffective; on the other hand a host gene change which affects some stage of the symbiosis with one strain is likely to leave its capacity to function with another strain unimpaired. This suggests either an array of genes providing separately for specific compatibility requirements set by different strains, or degrees of demand between strains that would leave the less fastidious able to function with a host which fails to provide for the more exacting.

Cultivars within *Trifolium subterraneum* have shown various degrees of specific strain requirements for effectiveness, but it is not possible to relate cultivar response to a collection of strains to the taxonomically accepted sub-species.[51,53] Specificity has also been found with lines of *Trifolium repens*[80,84,125], *Medicago sativa*[50], *Pisum sativum*[78], and *Desmodium aparines*[109].

Several dominant genes have been recognised which cause incompatibility with some strains of *R. japonicum* which results in ineffective nodule formation. These are: R_{j2} in cultivars CNS and Hardee *vs.* USA strains c1 and 122;[19] R_{j3} in Hardee *vs.* USA strain 33;[135] R_{j4} in Dunfield and Hill *vs.* USA strain 61.[136] Devine and Breithaupt[36] have reported on the occurrence of R_{j2} and R_{j4} in a collection of 851 soybean genotypes from many regions. R_{j4} was relatively common (in 31% of the lines), particularly those from southeast Asia. R_{j2} response was found in only 2%, which were almost all from China and Japan. The evidence was taken to suggest coevolution of host and rhizobial genotypes which may have been disrupted by separate introduction of host and rhizobia to a new environment such as the USA. The authors put forward the thought that fixation by these 'difficult' hosts could be improved by reassembling host/ strain combinations that had evolved in Asia. The converse of this points up the undesirability of taking 'westernised' strains into areas which have a high incidence of incompatible host genotypes. This is a real risk when a technologically advanced country sets out to help the less developed by importing its inocula as part of an aid programme.

Host genotype may also exercise an effect on the relative nodulating success of competing separately invasive strains. This can occur under field or laboratory conditions and need bear little relationship to the strains' ability to establish in the vicinity of the plant root[21,48,73,82,117,139]. Heritable differences have also been found with white clover.[56]

Host control may also be reflected in the distribution and form of nodules, even when the symbiosis is fully effective. The species of lupin determined whether nodules were evenly distributed over the root system, or were restricted to, or dominant on, crown, tap or lateral roots.[74]

Although it is sometimes difficult to make the distinction, the genetic constitution of the host legume will affect the N_2-fixing symbiosis in two ways: first, by its general growth habit and vigour; secondly by its direct influence on the symbiosis.

(v) HOST EFFECTS NOT DIRECTLY INVOLVED IN THE SYMBIOSIS

The growth habit and vigour of the host plant, affected by its ability to express its inherent capacities in a particular environment, will be reflected in such factors as the amount of photosynthate available for translocation, the extent of the root system and time to flowering and seed set. Tolerance of limiting deficiencies or excesses (such as temperature, H^+ and other ions)[66] is likely to be important in a general way, but can be more critical for the symbiotically dependent plant than for one fed on soil or fertilizer N.

Large seededness and juvenile leaf vigour can be a highly inheritable character able markedly to affect the yield of fixed N_2[110] as well as increasing the opportunity for successful establishment of an over-sown legume.[81]

Difference in ability amongst soybean cultivars to mobilise N from leaflets, petioles and stems (expressed as 'harvest N' index) was found with both N-fed and symbiotically provided plants.[67] While N-fertilizer increased all yield criteria it did not affect the harvest N index. However it cannot be assumed that this close parallelism in the handling of two sources of N will always be maintained. For example in some hosts the preferred form of translocated nodule-fixed-N may be allantoin.[62]

TABLE 8.1. Within species control by host genotype

Host	Observed lesion	Reference
Soybean (*Glycine max*)	Nodulation failure in cv Haberlandt	22
	"Non-nodulating" rj_1rj_1 isolines[a]	143
	Ineffective lines R_{j2} (cv Hardee etc.)[b]	19
	″　　　″　R_{j3} (cv Hardee)[b]	135
	″　　　″　R_{j4} (cv Hill etc.)[b]	136
	″　　　″　cv. Peking[b]	22
Lucerne (*Medicago sativa*)	Degrees of effectiveness-ineffectiveness with range of cultivars[b]	18,4,47,50
Red clover (*Trifolium pratense*)	Nodulation failure in rr (*0*) lines	95
	Ineffective nodulation due to i_1i_1[b]	96
	″　　　″　　　″　ie[b]	97,10
	″　　　″　　　″　nn[b]	100
	″　　　″　　　″　dd[b]	100
Subterranean clover (*Trifolium subterraneum*)	Nodulation difficulty in cv. Woogenellup[b]	52
	Reduced root-hair infections in cv. Yarloop[b]	120
	Degree of effectiveness in cv. Northam First Early[b]	51
Field pea (*Pisum* sp.)	Nodulation restricted by low temperature in cv. Iran[b]	77,63
	Larger number of nodules dependent on 2 genes: no, nod	
	Non-nodulating due to sym2, sym2 in cv. Afghanistan[a]	77,63
	Ineffective nodulation due to sym3, sym3	63

Notes:　a: nodulated sparsely with some strains

　　　　b: strain-specific response

　　　　c: Independent alleles:

　　　　　　associated with inadequate bacteroid development (Table 2); n with sparse
　　　　　　infection of central zone of small nodules and absence of bacteroid; d producing
　　　　　　fewer and smaller nodules, cells mostly uninfected but some tumourized.

(vi) THE DIRECT EFFECT OF HOST GENOTYPE ON THE SYMBIOSIS

　　Earlier work on the genetics of host influence on the symbiosis between legume and *Rhizobium* has been comprehensively reviewed by Nutman.[101] The better studied cases will be included in the present account.

　　Table 8.1 lists better studied cases of host genotype control for five host species. They are generally described in broad terms such as nodulation failure or ineffectiveness in the association. Table 8.2 endeavours to relate described genes to more closely analysed steps in the total symbiotic process. Unfortunately cases for which the presently available information permits

TABLE 8.2. Influence of host genotype on particular stages of symbiosis

Phenotypic stage	Recorded effect	Reference
I Preinfection		
1. Root colonisation (Roc)	Inhibition by seed coat toxin	131,13,25
	Stimulation by homoserine in *Pisum*	134
2. Root adhesion (Roa)	Lectin-facilitated attachment in *Trifolium* and *Medicago*	34
3. Root-hair branching (Hab)	Abundant with homologous, less (or absent) with heterologous, hosts	55,145
4. Root-hair curling (Hac)	"Marked" curling with homologous and near-homologous hosts	55,145
II Infection and nodule formation		
5. Infection (Inf)	Abundant infection threads with homologous, some with heterologous	
	Number of infection threads	27,99
	Mode of entry	
	Resistant lines of *T. pratense* (rr (O))	95
	" " *Glycine max* (rj$_1$)	143
	Strain specific resistance in *T. subterraneum* cv. Woogenellup	52
6. Nodule initiation (Noi)	Proportion of successful infection threads	98,99
	Degree of polypoidy	27
	Location of meristem	
7. Bacterial release (Bar)	Failure of *R. leguminosarum* to release from infection thread in *Trifolium pratense*	61
8. Bacteroid development (Bad)	i$_1$ and ie lines of *Trifolium pratense*	27,95,10
	Ineffectivity with *Trifolium ambiguum*	9
III Nodule function		
9. Nitrogen fixation (Nif)	Non-functional nitrogenase with *Pisum* cv. Afghanistan (sym3)	77,63
10. Complementary functions (Cof)	Globin moiety of leghaemoglobin	41
	Carbon sources (energy and post-fixation syntheses)	106
	NH$_3$-utilising enzymes	106
	N-compounds, nature, distribution and partition	106,67
11. Nodule persistence (Nop)	Growth habit and onset of maturation	106

this to be done are exceptional and it is to be hoped that future work will make good present gaps in such genetic information and avoid confusion that must arise when invalid expectations are aroused by uninformative broad categorisation.

PREINFECTION

The plant may influence colonisation of the root surface (Roc) adversely if its seed coat contains a toxic factor[13,25,131] or favourably if it produces a stimulatory substance.[134] From what few data are available it appears that legumes are more stimulatory of rhizobia (amongst other Gram-negative bacteria) than are non-legumes,[121] but evidence as to specificity between legumes and rhizobia at the Roc stage is slight. *Rhizobium meliloti* seems to be more stimulated by its homologous host (*Medicago*) than by *Trifolium*; on the other hand *R. trifolii* is stimulated by both.[116] Selective stimulation of *R. leguminosarum* by homoserine produced in the vicinity of lateral root emergence may provide a degree of specificity in the colonisation of pea roots.[134] The failure of 'non-nodulating' lines of soybean cannot be attributed to their inability to support rhizobial growth,[23] nor does relative colonisation of the root necessarily explain host-dependent nodulating success of competing strains. Root adhesion (Roa), if facilitated by host-specific lectins, could be the second host-controlled step to determine or modify the invasive success of appropriately matched rhizobia.

Root-hair branching (Hab) is generally more abundant in the homologous (nodulating) situation but is less specific than marked hair curling (Hac) which is almost restricted to the homologous partnership.[55,145]

INFECTION AND NODULE FORMATION

Mode of entry and number of threads are host dependent. The nodulation failure with red clover due to rr(σ) was apparently due to failure to establish infection threads in the root hair, even though the hairs did curl.[95] This is not unlike the failure of *Pisum sativum* to nodulate with strains of *R. trifolii*,[55] or of *Trifolium glomeratum* to nodulate with *R. leguminosarum*,[145] when in both cases marked root-hair curling had been observed. In such cases the genetic deficiency appears to relate to the actual infection (Inf) stage. The number of infection threads formed, and mode of entry and the proportion persisting into the cortex are aspects of the early infection process demonstrably affected by host genotype.[99,120]

A good deal of the incompatibility between host and *Rhizobium* will be expressed in the subsequent stages leading up to the recognisable nodule. Determination of the precise stage at which the symbiosis fails demands carefully matched use of light and electron microscopy and can be fairly difficult. Familiarity with the normal (fully effective) course of development for a particular host will be the essential basis for decision as to meaningful abnormality. Nodule initiation (Noi) involves a host-controlled reaction to the advancing infection thread which, if positive, results in the establishment of one or several meristems (number and location according to host) which produce polyploid nodule tissue. The proportions of infections which result in recognisable nodules differ with host genotype. In fact all infections may abort without the formation of recognisable nodules (as with the non-nodulating Afghanistan cultivar of *Pisum sativum*[35]).

There appear to be few reports of host control of the 'release' of bacteria from the infection thread into the host cells (Bar) but failure at this stage has been reported with *Rhizobium leguminosarum* invading red clover;[61] further development into bacteroids (Bad) can be host controlled as with cases of ineffectiveness reported with *Trifolium ambiguum*,[9] and i_1 and ie lines of red clover (*T. pratense*). These independently inherited genes are responsible for distinctive breakdown in the normal bacteroid development. With ie there is failure of the bacteria to multiply after release; i_1 prevents development to the normal bacteroid form.[10] Host genotype also determines the number of bacteroids within the host-produced peribacteroid membrane.[28]

The time taken to produce nodules is an important parameter which is largely under heritable host control, determined multifactorially or influenced by a major gene. It can be affected by the time to first infection, delays caused by abortion of earlier infection threads, as well as growth rates of infection thread and developing nodules. Selection for early nodulating habit could be advantageous, particularly for quick establishment on N-poor soils.

Nodule number[101] is a complex response to rhizobial strain and environment as well as host genotype. To some extent a deficiency in nodule number can be offset by nodule size. A large number of nodules in an effective association may be balanced out by smaller individual size and so maintain a similar nodule mass; ineffective associations are often marked by an excessive number of very small nodules.

NODULE FUNCTION

The capacity to produce the Nif enzyme complex with the potential of reducing N_2 to NH_3 appears to reside completely in the rhizobial genotype but its effective operation, requiring diverse 'complementary functions' (Cof), seems to be largely host determined and to range from the detailed constitution of the globin moiety of leghaemoglobin to the way the plant distributes and partitions the early N-compounds arising from fixation. The sym_3 condition in *Pisum, cv.* Afghanistan was responsible for nodules which, although they contained nitrogenase, were only able to show N_2-fixation capacity (as determined by acetylene reduction) when supplied with appropriate C-source.[63]

(vii) *SITE OF CONTROL OF NODULATION RESPONSE*

Grafting studies to determine whether the host control of nodulation depends on root or shoot have given different results according to the host system and symbiotic criteria concerned. The resistance of *Trifolium ambiguum* to nodulation by rhizobia adapted to Mediterranean type clovers has been overcome by grafts on *T. ambiguum* roots, using *T. repens* or *T. hybridum* as scion.[60] On the other hand resistance to nodulation in red clover, in *Pisum* and in soybean, as well as R_{j2} and R_{j3} ineffectiveness in the latter, were controlled from the roots.[20,95] Lawn & Brun[75] also found that total nodule activity (measured as acetylene reduction) was independent of shoot genotype but reflected that of the root on which the constant scion had been grafted. Nodule mass was influenced by the scion.

Selection and manipulation of host genotype

(i) *GRAIN LEGUMES*

Increased productivity with grain legumes has lagged behind that of cereals. Whilst improvement of soybean yield, particularly in the technologically developed countries, has been more satisfactory, grain legumes such as pigeonpeas and cowpeas, typical of low technology regions, have remained almost stationary.[71] World average yields in developing countries, quoted from FAO, 1974 even including soybeans, was only a little more than half that of the developed; in the cases of peanut (groundnut) and cowpea, it was in the order of a third. In developed countries also the World average fell far short of reported high yields, so that there, too, there is plenty of scope for varietal, along with technological, improvement.

Although plant breeding is inherently time consuming, the reliable introduction of an improved host line is likely to be more readily achieved than that of an improved micro-symbiont, and is likely to give more scope for marked improvement over existing productivity.

Khan[71] sees the twin aims of a breeding programme for grain legumes as improvement of grain yield with sufficient species and varietal diversity to provide long term stability through adaptability to variable seasons and diverse environments. The twin objectives have to be balanced to a degree that can be expected to differ between technologically more sophisticated practices when there can be more emphasis on yield, and simpler systems for which adaptability may take on greater significance.

Knowledge of and ability to handle the rhizobial component needs to be included in these considerations. Restrictive symbiotic capacity can be better handled when the supply and use of specialised inocula is efficiently provided for. Low technology farming may on the other hand be better served by the use of legumes well adapted to indigenous or naturalized rhizobia. In this connection a comprehensive investigation of soybean specificity in relation to place of origin has been particularly illuminating.[36]

(ii) PASTURE AND FORAGE LEGUMES

Exploitation of new environments in pasture development has generally been a matter of introduction of new genera and species. This has often been accidental, as with subterranean and other European clovers in the post European settlement in Australia. There has been con-siderable genetic improvement of pasture and forage species by the selection, and some breed-ing, and the establishment of well adapted cultivars of *Trifolium* and *Medicago*. It seems that the exploitation by tropical pasture legumes has so far been largely by the introduction of new genera and species rather than by selection or breeding.[70] Jones cites some examples of genetic improvement by selection from a wide range of material from different geographical areas. Selection of cultivars of *Stylosanthes* and *Centrosema* and improvement due to breeding and selection of *Macroptilium atropurpureum* and *Leucaena latisiliqua* are noted.

(iii) NEED FOR DIVERSE GERMPLASM

There is a widely recognised need for well recorded and maintained collections of germplasm for all forms of utilisable and related legumes. International agencies are increasingly recognis-ing and taking steps to provide the needed source of genetic variability from which selections can be made and new genotypes constructed through hybridization, mutation breeding and other genetic manipulation. Extension of this facility from international to national centres is necessary in the interests of testing in a wider environmental range as well as a sensible sharing of the work load.[141]

(iv) OBJECTIVES AND PROCEDURES

The plant breeder concerned with leguminous species will need to satisfy the same agro-nomic criteria as those dealing with other crop, forage or pasture plants. These will require that the selected or bred lines be suited to climate and soil conditions of the region, have appro-priate capacity to escape or resist disease and pests and so provide a satisfactory yield of good quality grain, forage or pasture. But he will also have a prime responsibility to provide a legume able to fill its N-independent role as it supplies protein and improves soil fertility. The agronomic objectives need to be determined, and parameters of assessment chosen accordingly. It has to be decided to what extent to aim to secure more and better plant tops for forage and grazing, a higher yield of quality grain, or improve soil fertility.

There could be some scope in selecting for earlier and greater nodule mass, for earlier initiation and later shut-down of fixation, for greater availability of net photosynthate to the nodule (including less competition between pod and nodule at the time of seed-fill) for greater capacity to nodulate and fix N_2 in the presence of appreciable levels of soil N. Delayed lodging will help to maintain higher levels of photosynthesis. Lines without, or with less, toxic seed coat factors could be advantageous in assisting nodulation.

Expertise needs to be coordinated so that an agronomically superior host line retains sufficient compatibility with rhizobia already in the soil or used as inoculant. Adequate communication between the plant breeder, the rhizobiologist and the agronomist, is essential to define objectives and solve problems of research logistics that will otherwise develop.

The possibility that advantage might be taken of selective resistance to nodulation to secure better results when the rhizobia already in the soil are ineffective or clearly sub-optimal in their fixation capacity with the species concerned, presupposes the availability, concomitant introduction and persistence of a superior strain and places a premium on the success of the inoculation step. Less drastically, lines may be selected for preferential nodulation with competing strains. In these and like attempts to control nodulation, full account has to be taken of longer term survival of introduced rhizobia. This is still an under-studied ecological problem that merits full and urgent investigation. Techniques are now to hand for such investigations to be undertaken and hopefully we can look forward to a good deal of useful, much needed information.

Clearer, more analytical recognition of the separate and distinctive symbiotic 'lesions' due to host genotype will provide better understanding of the phenomena associated with legume-*Rhizobium* N_2-fixation. Hidden genes will be brought to light when no longer masked by the plant's ready access to combined-N. It would be idle to claim that the presently proposed 11 step analysis is sufficiently detailed. That described as 'complementary functions' (Cof) seems most likely one for which several further divisions will need to be made. It also seems to be a stage at which improvement in the host genotype can be expected to have marked beneficial effects. Persistence of the functioning nodule (Nop) also seems a likely host-modifiable property that could be a major contributor to total fixed-N.

(v) *SELECTION FROM A HETEROZYGOUS POPULATION*

This remains a major means of genetic improvement and, as with other techniques, requires careful consideration of procedures that will in combination permit the recognition of the more general agronomic criteria as well as providing in due course for assessment in relation to the rhizobial requirements. An approach to distinguish host and rhizobial genotype effects and the interaction between them is illustrated with white clover[86] and *Vicia faba*.[87]

The capacity of a selected line to nodulate early, possibly selectively, with the more beneficial rhizobia, and produce sufficient mass of long-lasting fully effective nodule tissue will of course be of great importance. The inherent vigour of the host plant may well be reflected, in interacting fashion, in its nodule status. Large seededness and juvenile leaf vigour can be expected to influence the total plant performance, if not the efficiency of a given amount of nodule tissue. Some of the seed isoenzyme patterns in *Stylosanthes guyanensis* seem to have predictive value for effectivity with a collection of *Rhizobium* strains.[110] Workers from the same laboratory who have also assessed affinities between species of *Stylosanthes* on the basis of soil pH and geographical relationships, as well as rhizobial affinities, look to this approach as a possible help in planning the use of genetic resource material.[31]

To be successful selection needs to start with a large number of genetically diverse lines and their systematic examination on a sufficiently large scale. Two cycles of phenotypic recurrent selections of lucerne (alfalfa), tested against a 'mixed' inoculum increased N_2-fixation by about two thirds but such an exercise is likely to require N_2-fixation to be evaluated against 5,000 or more plants.[7] With the same host under North American conditions the condition of dormancy affected both growth and fixation at the onset and termination of effective symbiotic reaction.[58] A lucerne 'line' used in South Africa was found to be very variable in yield and to have about a quarter of the plants failing to nodulate with either of three strains. Selection reduced variability and removed non-nodulation. Higher nitrogen content in the non-nodulating plants indicated a possible correlation between tissue N levels and nodulation in those cases.[129]

Phaseolus vulgaris has shown considerable variation in N_2-fixing ability with the indeterminate climbing cultivars at the top and the bushy types at the bottom and mostly dependent on supplied-N for their yield.[54] It is possible that the superior fixation by climbing varieties reflects a greater supply of soluble carbohydrates to the nodule as distinct from starch. Early flowering of bush beans could also shorten the N_2-fixing period (Nop). The benefit of delayed senescence has also been demonstrated with soybean plants, found in a segregating nursery, on which the leaves remained green and the plant maintained both photosynthesis and nitrogenase activity after pod maturity.[108]

The screening of a collection of lines of pigeonpea, chickpea and lentils showed promise for improvement in nodule number, weight, and ability to retain nodule function and form fresh nodules as the surface soil dries out and roots penetrate more deeply. Large differences in lentil varieties would also justify a selection and breeding programme to obtain more uniform nodulation and maintenance of fixation throughout the season.[112] Eight varieties of *Vicia faba* showed considerable variation in their yield with either supplied N or nodulated with a 'standard' strain. Similarly there was up to a tenfold difference amongst a large collection of breeding lines of cowpea tested against a 'standard' mixed rhizobial inoculant. A collection of 11 accessions of *Glycine wightii* showed about eightfold range in effectiveness, twofold range in nodule number and a difference up to 7 days in the time to the first nodule.

Selection for effectiveness with a particular strain of *R. trifolii* out of a heterozygous population of red clover and crosses within selections from the highest class, displaced the response towards effectiveness, but not strikingly. Carried into the second generation the results were similar but some families showed the opposite to expected responses although there were a few families (high X high) which were exceptionally effective. The effect of selection was less, though in the same direction with two other strains.[105] Selections from a line of white clover based on total nodule volume resulted in a 20%, barely significant, improvement in yield but this was not maintained in subsequent generations.[88] Vegetatively propagated plants of the same host showed generally closer adaptation to rhizobia which had come from nodules of the parent or closely related plants.[84] Such adaptation is reminiscent of better assessment of local rhizobial strains when tested with indigenous rather than with an introduced line of white clover.[80,125] Phenotypic selection combined with diallele crossing from a population of *Trifolium ambiguum*, which showed great variation in time to nodulation, resulted in the development of a cultivar suitable for higher altitudes under Australian conditions.[59] On the other hand experience with selection of lines of subterranean clover for desirably low oestro-genic content indicates the kind of difficulties that can, without necessary causal relationship, be introduced at the same time.[16]

Host-determined genetic variation was found in a large collection of pea lines from the USDA Plant Introduction Collection.[64] About 7% of 211 samples had fixation unimpaired when grown with 200 ppm nitrate-N but this resistance did not carry over to the field.[65]

(vi) *BREEDING*

Hybridisation and stabilisation of new lines by self pollination during selection is of course a regular means whereby the plant breeder constructs new genotypes with desired combination of properties. It seems to have been applied to legumes with varying success. Precautions need to be taken in the selection of parent lines and in the testing of progeny, lest overt (dominant) or concealed (recessive) symbiotic lesions be introduced. Non-nodulating (rj_1) and strain-specific ineffectiveness genes in soybean are well known examples. A wild relative such as *Vicia anarbonensis* if used to improve disease or insect resistance in *Vicia faba* may cause difficulty with nodule persistence (Nop) during early pod fill in contrast to the situation with *V. faba*. Crosses between a rust resistant parent and agronomically suitable lines of groundnut, both well nodulated, segregated to both non-nodulating and nodulating progeny (P.J. Dart, pers. comm.).

Other crossing experience has given more positive results. Crosses among lucerne clones, whose nodules showed high acetylene reduction, were more than twice the crosses from low activity clones.[124] Crossing within groups of *Medicago sativa* selected for higher C_2H_2-reduction activity resulted in superior symbiotic performance over those crossed from the low C_2H_2-reduction group.[45] The F_2 population of a cross between lines of *Glycine wightii* showed greater mean nodulation and growth, as well as a greater range of variation than did either parent.[94] However with white clover the significant increase shown by the F_1 was too small to be significant in the F_2 generation.[68] As noted above a bred line of diploid *Trifolium ambiguum* has been obtained with good nodulation prospects as well as high productivity and cold and drought hardiness.[147]

A breeding procedure for pigeonpeas which provides conditions for natural outcrossing has been put forward and is discussed by Khan.[71] Composites obtained in this way were used as reservoirs of genetic variability which after several generations of random mating served as a basis for population improvement. Gamblin and Morton are quoted by the same author as pre-ferring simple crosses in a programme with *Phaseolus vulgaris*. Cases of male sterility, cited by Khan for cowpea and soybean, open up the possibility of obligatory outcrossing and a hybrid breeding line to exploit heterosis found, for example, with *Desmodium*.[110]

A recurrent selection system with bush cultivars of *Phaseolus vulgaris* has been described and used at CIAT.[54] This method uses plants already selected for ability to fix nitrogen and permits lines with desirable agronomic properties and disease resistance to be introduced in each cycle of crossing. Hybrid seed incorporating genes from the freshly introduced material is then available for subsequent cycles at the same time as those flowers which self-fertilize produce F_2 (half sib) lines which can be progressively tested in the field.

PROVISION FOR THE RHIZOBIAL FACTOR

The interaction between plant and rhizobial genotypes complicates selection and breeding for N_2-fixation; plant breeders and agronomists need to consider rhizobial compatibility in their breeding and testing programmes. Testing against a single strain of *Rhizobium* seems inappropriate for a crop sown into an area where competition with other strains is a major problem. There is however the possibility of developing cultivars sufficiently selective against strains already in the soil to favour a superior inoculum.[56,69,83] Decision as to the stage in the breeding or selection programme at which the rhizobial factor should be investigated is particularly important. It can be argued that the greatest improvement in N_2-fixation will be achieved by the simultaneous selection of both symbionts[87] but the logistics of such a programme have to be kept in mind. This will require that some of the work at least will have to be done under greenhouse conditions with protection against contamination with 'wild' rhizobia. Out of the greenhouse, screening can be against natural soil populations in pots and in the field. Such screening would need to be against more than one such population and under conditions which permit appropriate parameters of symbiotic performance to be assessed. The amount of unfamiliar work involved in keeping the rhizobial factor under adequate control has been enough to discourage most plant breeders from the attempt, and the rhizobiologist will not be without hesitation as he remembers how host *Rhizobium* specificity can limit the general applicability of selection against a restricted rhizobial background.

Agronomically excellent material could be lost early in a programme against too narrow a rhizobial background. There may therefore be a good case for earlier large scale screening to be directed to a more general agronomic evaluation under conditions of adequate supplied-N. This the plant breeder has commonly done, with some justification. Difficulties arise when the programme remains 'hooked' on soil/fertiliser-N, so that selection may continue without giving prominence to the achievement of symbiotic competence. There has to be a stage, as soon as this step becomes practicable, when the lines under test have the opportunity of demonstrating a high level of compatibility with relevant rhizobia.

In deciding the stage at which the host genotype is to be challenged by the demands of an effective symbiosis, it needs to be kept in mind that it is not likely to be sufficient merely to substitute a rhizobial inoculum for supplied-N at the end of a selection or breeding programme. The symbiotic system can be more sensitive than its N-fed host to the operation of several factors such as temperature, soil pH, some features of mineral nutrition (e.g. molybdenum, cobalt and calcium) and the capacity of the host to maintain a supply of photosynthate to the aging nodule. Repeated experience has not inured the rhizobiologist to the feeling of frustration at being given the task of locating suitable rhizobia for use with an unexpectedly recalcitrant host. Nor will the plant breeder appreciate being directed to an impossibly large screening programme under the conditions needed for precise determination of rhizobial compatibility.

A large tray device has been developed which permits a large number of plants to be tested simultaneously[38] under conditions which give good differentiation of plant response.

INTERACTION WITH THE ENVIRONMENT

Because various environmental conditions can modify the ability of a host-*Rhizobium* combination to establish an effective N_2-fixing symbiosis, it becomes necessary to test selected and bred material in a good many regions. At the same time it will be realised that steps need to be taken to restrict the genotype combinations so tested and to arrive at the preliminary screening by simpler greenhouse or light room procedures. This will rule out those obviously not worth the cost of more extensive testing. There is little to suggest that a combination in which the symbiosis is clearly ineffective under the reasonably favourable conditions which can be provided near a research centre will achieve a favourable reversal of form under diverse field conditions. However it is true that better performers, not split by greenhouse tests, may well be spaced out in the less growth-restricted and diverse environmental situation of the field.

Even with such preliminary screening, adequate field testing will be expensive and time-consuming. Cooperation within and beyond the national level will be needed and the trials they conduct have to be sufficiently standardised to permit valid combination and comparison of results. 'Network' trials of the kind sponsored by the NifTAL project of the University of Hawaii, College of Tropical Agriculture, could provide such an opportunity to expand information on host genotype environment interaction.

In his emphasis on adaptability as an important breeding objective, particularly for diverse tropical environments and having regard for the need of subsistence farmers to avoid fluctuations in yield, Khan[71] discusses a regression technique for defining this character in a genotype its modification and use for cereals. The regression is however often of limited use so that an alternative parameter has been introduced and successfully applied to cowpeas. The techniques of numerical taxonomy can also be used to group soybean genotypes which have a similar response to the environment as a means of recognising parental material for a crossing programme in which the aim is to breed for adaptability. There seems to be no reason why these approaches should not be directed to those aspects of the host genotype which, one way or another, affect the N_2-fixing function of the nodulated legume, including interaction with local rhizobial strains as an element in the plant's total environment.

Tolerance of the N_2-fixing system to soil or applied N is a most important objective in selection and breeding of host genotype, but progress in this direction is very limited. There were some early data showing different tolerance between rhizobial strains,[107] but results like this are not generally to be found — presumably because others have failed to find cases of like tolerance. The host genotype could be expected to exercise a more direct influence at least to the extent that it might be able to match the production of photosynthate and mop up additional combined-N. Promising results with a test of more than 200 cultivars of the World Pea Collection showed that 13% showed less than 30% reduction of nitrogenase activity (measured as acetylene reduction) by 200 ppm nitrate-N, when tested in the greenhouse.[65] However these differences did not stand up in field tests. Differential sensitivity to combined-N has been found with different species of host; nodulation of cowpea was found to be relatively tolerant to 24 mgN/plant whereas that of *Vicia atropurpurea* was depressed.[30]

EVALUATION

Where evaluation of a host line's ability to utilise N_2 relates to measurements more concerned with general agronomic qualities is a matter for decision by the breeder. Too early or complete dependence on fixed-N might mean that otherwise promising material is prematurely discarded from the programme. Leaving N_2-fixation assessment too late may at the worst favour selection against that quality, or at least waste resources. The use of isotopic N or acetylene reduction has been briefly touched on already (p. 187). These techniques could have a justified place in a breeding programme but not instead of simpler methods which permit more extensive testing of a large amount of genetic material under diverse conditions, and which have to be applied at some stage in the evaluation of a new host line.

Vegetative propagation[1,79,85] when possible, provides good material for evaluation of symbiotic characteristics against a constant genetic background.

Indirect indicators which might help predict N_2-fixing usefulness are seedling vigour (p. 186) and, in the case of *Stylosanthes guyanensis*, isozyme patterns (p. 197).

The way ahead: needs and possibilities

Needs and possibilities for the selection and breeding of legumes for enhanced N_2-fixation were extensively considered by a workshop attended by workers from several disciplines.[15] These points were among, or are related to, the recommendations which emerged:

(i)   A multidisciplinary approach is necessary. The team needs to include a plant breeder, crop physiologist and rhizobiologist and be able to co-opt such other specialists as are needed to complement the programme. These could be expected to include those concerned with crop management, plant biochemistry, mineral nutrition, plant hormones, morphology and cytology, pathology, entomology and virology.

(ii) N_2-fixation capacity needs to be incorporated with other desirable agronomic traits related to the appropriate environment in which production can be expected to take place. To this end programmes aimed at improving N_2-fixation should be located at centres with established breeding programmes in the species concerned.

(iii) Because agronomically important legumes differ in their genetic structure, breeding behaviour, growth characteristics, metabolic behaviour, root pattern and the plant part of economic value, programmes for the collection and development of wider germplasm source material should be established in several major legume species. Naturally the nature, scope and location of such collections will be largely determined by importance to the region concerned and suitability for its testing, maintenance and increase.

The same group put forward the following priorities:

— Investigation of the allocation of C and N compounds to various plant tissues in order to identify useful traits for a N_2-fixation selection programme.
— Plant breeding and genetic research based on several germplasm sources and using alternative selection strategies (based on yield, N_2-fixation, nodule mass and specific activity).
— Development of genetic control systems which can selectively favour *Rhizobium* strains.
— Improvement of the legume's input to soil fertility by a better rotational contribution, excretion of N within intercropping, grazing and forest systems, and capacity to provide an effective symbiosis in the presence of a reasonable level of soil N.
— Improvement of techniques for determination of N_2-fixation in large scale programmes which are: non-destructive, rapid, inexpensive in labour and materials, and applicable in the field.

SYMBIOSIS BETWEEN ACTINOMYCETE-LIKE ORGANISMS AND NON-LEGUMINOUS PLANTS

There is no doubt as to the ecological and, in relation to forested areas, the economic significance of N_2-fixation by nodulated non-leguminous plants.[11]

Almost half of 342 species listed as 13 or 14 genera of 8 families of non-leguminous plants have been reported as having root nodules.[12,132] Those that have been examined closely are all occupied by an actinomycete-like organism and form the nodules by a process following root-hair infection which has some points of similarity with the legume-*Rhizobium* symbiosis. There are however distinctive differences between the two processes.[3,49]

Difficulties in obtaining pure cultures of the endophyte have for a long time constituted a barrier to precise allocation of respective roles to the symbionts, although some degree of specificity appears to have been established between them.[11,132] Recent reports on the cultivation of what appears to be the microsymbiont of a *Comptomia* (syn. *Myrica*) appear to be well based and, if more generally applicable, could open up useful lines of information.

Already certain limits of cross-invasiveness have been demonstrated with the cultivated *Comptomia* isolate. Beyond this however there seems as yet no proper basis for defining the respective contribution of the two genotypes or attempting its improvement by genetic manipulation. Even when information is available the nature of the host plants and their extensive, rather than intensive, utilisation will continue to offer barriers to rapid progress.

ASSOCIATIVE NITROGEN-FIXING SYSTEMS ('DIAZOTROPHIC BIOCOENOSES')

Considerable interest has developed in systems where there is a looser form of association between the host plant and microbiological partners than that which characterises the readily recognisable, fully symbiotic, root-nodule. Although fixation by these looser systems is low compared with the nodulated legume, the extent of grasslands could make it as great or greater on the global scale.[113] A recent international workshop covered many aspects of associative N_2-fixation and addressed itself to the introduction of better descriptive terms. These deliberations resulted in the recommendation of 'diazotrophic biocenosis' as the general term for N_2-fixing associations (of the looser kind) and such a term as 'diazotrophic rhizocoenosis' when the association is in, on or close to the root. Workers have been asked to adopt this more precise terminology.[142]

Nature of the associations

Rhizocoenoses able to reduce N_2 have been clearly demonstrated in many ecological systems in many parts of the world. Members of the Gramineae so involved include wheat, maize, sorghum, millet, sugar cane and many grasses. Those with the more energy-efficient 4 carbon photosynthetic systems are in a clear majority, but there are also some 3 carbon representatives. In some cases the bacteria have been shown to invade the root but it seems that the exact site of nitrogenase activity remains undetermined.[113]

The causal microbial agents are known in relatively few cases, but these are widely representative of aerobic, facultatively anaerobic and anaerobic bacteria (Table 8.3). Species of *Azospirillum* are widely distributed and are taken to be responsible for major N_2-fixing and hormonal effects. There is definite plant genotype specificity. The role of other associated (not necessarily N_2-fixing) microorganisms needs to be determined. Although clear evidence of infection of roots of grass and maize has been obtained[6,43,44] there are problems that still need to be investigated relative to such aspects as cause of initial attraction of N_2-fixing bacteria to the root, their build up, the efficiency of N_2-fixation, operation of control by the plant in relation to the supply and partition of substrates and products of fixation, the role of non-fixing microbial associates and mode of protection from too high a level of oxygen.

TABLE 8.3. Major reported diazotrophic rhizocoenoses

| Host | Aerobic | | | | | | Facultative | | Anaerobic |
	Azotobacter	Beijerinckia	Derxia	Azospirillum	Bacillus	Enterobacter	Klebsiella	Rhodopseudomonas	Clostridium
Paspalum notatum	A								
Cynodon dactylon	A	A							
Digitaria				C					A
Other grasses		A	A	A	A	A	A		A
Wheat	A			A	A	A			A
Maize		A		A	A	A			A
Rice		A		A		A		A	A
Sugar cane	B	A		B	B	B			B
Sorghum				C					

A: Balandreau et al.[5] ; B: Ruschel[122] ; C: Baldani & Döbereiner[6] .

The fixation system associated with sugar cane, as found in Brazil, seems to include several bacteria notably *Klebsiella*, *Enterobacter*, *Erwinia*, *Azotobacter*, *Beijerinckia*, *Bacillus* and *Clostridium*. *Azospirillum* in this case appears of less significance. Fixation could well be affected by cane variety, locality, soil type and climate as well as by the kinds of bacteria that happen to be in a particular area. Much more detailed work is required along lines concerned with the bacteria: identification, location, site of fixation, possibility and procedure of inoculation. The N_2-fixing capacity of varieties and genetic evidence relating to this capacity is needed, as well as information on plant physiology and biochemistry and field studies.[122]

Genetic specificity in the association

The case which attracted attention particularly to the potential value of a grass association (that between *Paspalum notatum* and *Azotobacter paspali*) is markedly specific: for host at the cultivar level and for the bacterium, at species level.[43] Other specificities have since come to light such as *Bacillus polymyxa* and a strain of *Azospirillum brasilense*, favoured by wheat and *Azospirillum lipoferum* favoured by maize.[141] A pot experiment with rice, wheat and maize inoculated with a mixture of *Azospirillum lipoferum* and two substrains of *Azosp. brasilense* showed that, whilst all three could be recovered from soil and/or the unsterilized roots of all hosts, the surface-sterilized roots of wheat and rice carried predominantly the nir⁻ strain of *Azosp. brasilense* whereas maize had about equal representation of *Azosp. lipoferum* and the other (nir⁺) strain of *Azosp. brasilense*. The selective effect of host was substantially the same whether the experiment was done in soil in pots, inoculated with a mixed inoculum or left uninoculated, or in field trials, again whether inoculated or uninoculated. The failure of *Azosp. brasilense* nir⁻ to replace *Azosp. lipoferum* on maize, and the converse situation on wheat were clearly demonstrated.[6]

Another 7 Gramineae and one member of family *Cyperaceae*, preferred *Azosp. lipoferum* to *Azosp. brasilense* on their roots, although the latter was well represented in the soil.[119] These were all C_4 plants as is sugar cane which preferred *Azosp. brasilense*.

Contribution of host genotype

Nitrogenase activity (measured as acetylene reduction by root species) shows considerable host cultivar effects in grasses such as *Paspalum notatum*, *Pennisetum purpurem*[44] and many others.[141]

Crop plants have also shown clear cultivar effects. These include wheat, maize, rice, sugar cane, and sorghum and millet[5,6,29,113,115,123,140,141]. Disomic substitutions lines in Spring wheat have shown close control exercised by the host genome on N_2-fixation with diazotrophic rhizocoenoses.[114] In phytotron experiments with *Bacillus* culture (C–11–25) (Table 8.4), cv. Cadet was better than cv. Rescue, but was improved by disomic substitutions R2D and R5D (from Rescue). Other Rescue substitutions had little (R2A) or an adverse (R5B) effect. The inferiority of associative capacity of Rescue was strikingly reversed by C5B and C5D substitutions from Cadet, but not by substitutions C2D and C2A. In a field experiment, benefit due to inoculation with *Bacillus* was small but the response generally agreed in the trend with the phytotron result. Two lines (Rescue–C5D and Cadet) showed clear reversal in that both failed to respond to inoculation in the field. No explanation is apparent for the first, the second has been attributed to the use of a different Cadet line in the phytotron and field experiments. A further pot experiment supported this explanation for the behaviour of Cadet (Rennie, pers. comm.).

TABLE 8.4. Influence of disomic chromosome substitution on response of wheat to inoculation[a]

| Parent line | Substitution | Percent plant nitrogen attributable to inoculation with: | |
		Bacillus C−11−25[b]	Azospirillum ATCC[c] 29145
Cadet	nil	21	34
	R2A	28	54
	R2D	41	42
	R5B	0	22
	R5D	47	25
Rescue	nil	_[d]	0
	C2A	−	0
	C2D	0	33
	C5B	60	32
	C5D	59	31

(a)   Data from phytron experiments[114]

(b)   Isolated from rhizosphere of Cadet−R5D

(c)   *Azosp. brasilense* from J. Döbereiner

(d)   − : significantly inferior to uninoculated control.

Results of the phytotron experiment with *Azospirillum* 29145 again showed effects due to disomic substitution but not necessarily related to those found with *Bacillus* (Table 8.4). In this case the performance of Cadet was improved by both R2 substitutions but not by either R5; Rescue was improved by C2D and both C5 substitutions, but not C2A. There was no clear pattern of agreement or disagreement with the two inoculum species.

Nitrogenase activity, measured by acetylene reduction with incubated root segments taken from nonsterile soil, has been compared with group-5 disomic substitutions in Chinese Spring wheat.[72] Some of the substitutions increased nitrogenase activity others, even replicates of the same group from a different source, were quite without effect. Although the results show that a relatively restricted change in host genome can affect this activity, there appears to be no easy interpretation, the more so because so little seems to be known about the precise nature of the associated bacteria obtained from such a heterogeneous source.

Two maize cultivars differed in their respective ability to maintain nitrogenase activity in the presence of as much as 80 kg/ha fertilizer N.[141] However it needs to be noted that grain production by the two cultivars reversed their relative performance in this respect, so that selection for better nitrogenase activity without taking account of grain yields, could lead to a negative result.

Two rice mutants grown in one soil showed marked differences in nitrogenase activity, both between themselves and in comparison with the parent cultivar. These differences were however abolished with plants grown in a different soil, thus emphasising the lack of definition which handicaps definitive studies of these loose associations.[115]

Six cultivars of sugar cane fell into high and low nitrogenase activity groups when tested as intact whole plant in soil.[123] The relative activity of cultivars ranged from 8 fold (best to worst) in normal air as atmosphere, to 26 fold when O_2 was restricted. The relative performance of cultivars was generally consistent under both sets of conditions but in one case nitrogenase activity was disproportionately improved with reduced aeration.

SELECTION AND MANIPULATION OF THE HOST GENOTYPE

SELECTION

Whilst it is not difficult on a particular occasion in a particular location to detect marked differences in nitrogenase activity amongst different host lines, the loose and often poorly defined nature of the association often results in marked variation and even reversals of performance. There can be poor agreement between greenhouse and field results and in the case of grain crops, there is often poor correlation between nitrogenase activity measured in the vegetative phase and grain yield.[141] Certainly selections from comprehensive germplasm collections and progeny derived from a breeding programme will have to be tested in diverse locations.

Selection for host lines able to establish superior N_2-fixing associations would be facilitated if more were known of their exact nature and mode of operation. Then it might be possible to select for significant predictor characteristics such as, perhaps, the availability of excess carbohydrates to the roots.

TABLE 8.5. Nitrogenase activity[a] in maize lines and crosses (data from von Bulow[141])

	Piranão	UR-1	Comp. Amplo	Cateto Nort
1. Piranão X	260	218 (191)[b]	238 (208)	321 (258)
2. UR-1 X		122	140 (139)	201 (190)
6. Comp. Amplo			156	270 (206)
8. Cateto Nort X				257

(a) Average n moles C_2H_4 (h^{-1}, g^{-1}, dry roots) for 5 plantings to cover hot and cool seasons.

(b) Bracketed values: intermediate values 'expected' from parents.

BREEDING

Crosses amongst four maize lines showed nitrogenase activity in the progeny as generally intermediate between the parents, but with some effect of heterosis (Table 8.5). Grain yield was entirely unrelated however to the average determined nitrogenase activity.

Two crosses have been reported for sugar cane,[123] one (Co331 X Co290) between parents having much in common in their background; in the other (CP 36–105 X 38–34) they are very different. Germinated cuttings of F_1 and F_2 and clones from F_1 seedlings were analysed for nitrogenase activity in an intact system (plant + soil medium). Cuttings of the same material were assayed with a year between and on the later occasion they were in vermiculite and sand instead of soil. The results were consistent between the two occasions and generally gave an F_1 which lay between the parents (although tending towards the lower). On one occasion the F_1 of the close cross marginally exceeded both parents. The data are taken to indicate that ordinary breeding methods with this host can be applied to enhance N_2-fixation.

Future prospects

It is apparent that there is genetic diversity available to provide the basis for a selection and breeding programme. Less clear is how such selection is to be made against a background of environmental and bacterial diversity. Important to this decision will be determination of the roles of the respective genotypes and, hopefully, the capacity of that of the host to prove broadly compatible. Selections and progenies will need to be tested in soil, ultimately *in situ* to provide answers to these questions; in the process useful application might emerge.

Some points made in considering such a programme[141] emphasise that selections should be made against a background of low soil-N (natural or applied) under conditions (e.g. with adequate molybdenum and phosphorus) as will permit the N_2-fixation to be expressed. A long history of N-fertilizer application in the technologically more advanced areas can be expected to have prevented natural selection for greater fixation efficiency. It follows then that not only should stored germplasm be screened for outstandingly efficient genotypes but plant populations, not yet selected out against high levels of soil or applied N, should be looked for as a good source of genes for better N_2-fixing ability. Given limited resources, selection between cultivars seems more likely to succeed than between species.

The introduction of the very sensitive acetylene reduction assay method has permitted low levels of nitrogenase to be detected in plant-bacterial associations that had not been convincingly demonstrated before. Most of the work has been done with short term determination, often under highly artificial conditions, which do not readily permit precise estimates of N_2-fixation over a total growing period in the soil. Various *in situ* methods have been devised and these include provision for extended continuous exposure to a low, non-toxic, level of acetylene. However it will be apparent that the available number of such units is more suited to specific physiological studies than to a comprehensive selection and breeding programme. Various methods are available for using the ^{15}N isotope to provide definitive evidence of fixation. These involve the direct use of $^{15}N_2$, dilution of ^{15}N isotope supplied as fertilizer or the change of ^{15}N: ^{14}N ratio at levels of natural abundance (p. 187). All of these have some limitations in their application to a large screening programme and are more suited to research programmes as such.

In looking towards a means for large scale screening it should be possible to provide a sufficiently low level of soil N for the plants themselves to reflect in the amount, colour and, if necessary, N content, their capacity to fix significant amounts of N_2. If the effect is too small to be demonstrated under those conditions it will be hardly worth the trouble of incorporating the genotype into a breeding programme. A bacterial component might also be ensured by the provision of a 'shot-gun' application of a crude inoculum as suspension prepared from a suitable mixture of soil.

Comparison within an array of tested genotypes against the same set given a suitable level of N fertilizer will provide a guide to a contribution being made by the association; this might be determined at several N levels for a better approximation. The adequate-N plots will also provide for the evaluation of genotypes under conditions of N non-limiting conditions.

GENERAL CONCLUSIONS

There is enough already known about the nodulated legume to permit the selection and breeding of better crop, forage and pasture plants able to obtain their nitrogen from the atmosphere and ultimately enhance the fertility of the soil. However, prospects for improvement of nodulated non-legumes by these means appear remote at the present stage of our knowledge and ability to handle these symbioses. The looser nitrogen-fixing associations require a lot more information about the genetic inter-actions between the participants before significant plant breeding progress towards better acquisition of atmospheric nitrogen can be expected.

Even with the legumes, more analytical basic information about the many steps involved in securing a fully effective symbiosis is urgently required. Its acquisition will require the co-operative efforts of workers in several disciplines and, most importantly, better communication amongst them and between them and the plant introduction officer, plant breeder and agronomist. Mutual awareness of what can and should be done, as well as the limitations and logistic demands made by a selection and breeding programme can be expected to provide a sound basis for profitable cooperation. Such an end will not be achieved so long as the experimentalist and the plant breeder go their own way unheeding, leaving it to the occasional attempt to plug gaps or 'mend fences' towards a programme's end.

REFERENCES

1. Agamutu, P., Tan, C.S., Ramadasan, K. and Broughton, W.J. Vegetative propagation of legumes. *Proc. Natl. Plant Propagation Symp.* Rubber Res. Inst., Univ. Malaya, Kuala Lumpur, Malaysia, 1976.

2. Allen, E.K. and Allen, O.N. The nodulation profile of the genus *Cassia*. *Symbiotic Nitrogen Fixation in Plants, International Biological Programme* 7. P.S. Nutman (ed.), 113–121. Cambridge Univ. Press, Cambridge, 1976.

3. Angulo, A.F., van Dijk, C. and Quispel, A. Symbiotic interactions in non-leguminous root nodules. *Symbiotic Nitrogen Fixation in Plants. International Biological Programme* 7. P.S. Nutman (ed.), 475–483. Cambridge Univ. Press, Cambridge, 1976.

4. Aughtry, J.D. Effect of genetic factors in *Meidcago* in symbiosis with *Rhizobium*. *Cornell Univ. Agr. Exp. Sta. Mem.* **280**, 19, 1948.

5. Balandreau, J., Ducerf, P., Hamadfares, I., Weinhard, P., Rinaudo, G., Millier, C. and Dommergues, Y. Limiting factors in grass nitrogen fixation. *Limitations and Potentials for Biological Nitrogen Fixation in the Tropics*. J. Dobereiner, R.H. Burris, A. Hollander (eds.), Basic Life sciences **10**, Plenum Press: New York, 1978.

6.   Baldani, V.L.D. and Dobereiner, J.   Host plant specificity in the infection of maize, wheat and rice with *Azospirillum* spp. *Associative Dinitrogen Fixation*. P.B. Vose and A.P. Ruschel (eds.), CRC Press, Boca Raton, Fla., 1981.

7.   Barnes, D.   *Selecting and Breeding Legumes for Enhanced Nitrogen Fixation*. Workshop: Boyce Thompson Institute, Cornell, 1978.

8.   Bauer, W.D.   Role of soybean lectin in the soybean-*Rhizobium japonicum* symbiosis. *Nitrogen Fixation* Vol. II. W.E. Newton and W.H. Orme-Johnson (eds.), 205–214, Univ. Park Press, Baltimore, 1980.

9.   Bergersen, F.J.   The occurrence of a previously unobserved polysaccharide in infected cells of root nodules of *Trifolium ambiguum* M. Bieb. and other members of the *Trifolieae*. *Austral. J. Biol. Sci.* 10, 17–24, 1957.

10. Bergersen, F.J. and Nutman, P.S.   Symbiotic effectiveness in nodulated red clover IV. The influence of the host factors i_1 and ie upon nodule structure and cytology. *Heredity* 11, 175–184, 1957.

11. Bond, G.   Root-nodule symbiosis with actinomycete-like organisms. *The Biology of Nitrogen Fixation*. North Holland Research Monographs, Frontiers of Biology, 33. A. Quispel (ed.), 342–378, North Holland: Amsterdam, 1974.

12. Bond, G.   The results of the IBP survey of root-nodule formation in non-leguminous angiosperms. *Symbiotic Nitrogen Fixation in Plants. International Biological Programme 7*. P.S. Nutman (ed.), 443–474, Cambridge Univ. Press, Cambridge, 1976.

13. Bowen, G.D.   The toxicity of legume seed diffusates toward rhizobia and other bacteria. *Plant and Soil* 15, 155–165, 1961.

14. Bowen, G.D. and Kennedy, M.   Heritable variation in nodulation of *Centrosema pubescens* Benth. *Queensland J. Agric. Sci.* 18, 161–170, 1961.

15. Boyce Thompson Institute.   Selecting and breeding legumes for enhanced nitrogen fixation. Workshop, Cornell, October 23, 24, 1978.

16. Brockwell, J. and Robinson, A.C.   Symbiotic competence of *Trifolium subterraneum* lines of low oestrogenic isoflavone content. *Aust. J. Exp. Agric. An. Husb.* 10, 555–561, 1970.

17. Broughton, W.J.   A Review. Control of Specificity in Legume-*Rhizobium* Associations. *J. Appl. Bact.* 45, 165–194, 1978.

18. Burton, J.C. and Wilson, P.W.   Host plant specificity among the *Medicago* in association with root-nodule bacteria. *Soil Sci.* 47, 293–303, 1939.

19. Caldwell, B.E.   Inheritance of a strain-specific ineffective nodulation in soybeans. *Crop Sci.* 6, 427–428, 1966.

20. Caldwell, B.E., Hinson, K. and Johnson, H.W.   A strain-specific ineffective nodulation reaction in the soybean *Glycine max* L. Merrill. *Crop Sci.* 6, 495–496, 1966.

21. Caldwell, B.E. and Vest, G.   Nodulation interactions between soybean genotypes and serogroups of *Rhizobium japonicum*. *Crop Sci.* 8, 680–682, 1968.

22. Caldwell, B.E. and Vest, H.G.   Genetic aspects of nodulation and dinitrogen fixation by legumes: The macrosymbiont. *A Treatise on Dinitrogen Fixation, Section III Biology*. R.F.W. Hardy and W.S. Silver (eds.), 557–576. John Wiley and Sons, New York, London, Sydney, Toronto, 1977.

23. Clark, F.E.   Nodulation responses of two near isogenic lines of the soybean. *Can. J. Microb.* 3, 113–123, 1957.

24. Corby, H.D.L.   Systematic implications of nodulation among Rhodesian legumes. *Kirkia* 9, 301–329, 1974.

25. Dadarwal, K.R. and Sen, A.N.   Inhibitory effect of seed diffusates of some legumes on rhizobia and other bacteria. *Indian J. Agric. Sci.* 43, 82–87, 1973.

26. Dart, P.J.   Development of root-nodule symbioses: The infection process. *The Biology of Nitrogen Fixation*. A. Quispel (ed.), *Frontiers of Biology 33*, 381–429, North Holland, Amsterdam, 1974.

27. Dart, P.J.   Infection and development of leguminous nodules. *A Treatise on Dinitrogen Fixation, Section III Biology*. R.W.F. Hardy and W.S. Silver (eds.), 367–472, John Wiley and Sons, New York, 1977.

28. Dart, P.J. and Mercer, F.V.   Fine structure of bacteroids in root nodules of *Vigna sinensis*, *Acacia longifolia*, *Viminaria juncea* and *Lupinus angustifolius*. *J. Bact.* 91, 1314–1319, 1966.

29. Dart, P.J. and Subba Rao, R.V.   Nitrogen fixation associated with soybean and millet. *Associative Dinitrogen Fixation*. P.B. Vose and A.P. Ruschel (eds.), CRC Press, Boca Raton, Fla., 1981.

30. Dart, P.J. and Wildon, D.C. Nodulation and nitrogen fixation by *Vigna sinensis* and *Vigna atropurpurea*. The influence of concentration, form and site application of combined N. *Austral. J. Agric. Res.* **21**, 45—56, 1970.

31. Date, R.A., Burt, R.L. and Williams, W.T. Affinities between various *Stylosanthes* species as shown by rhizobial, soil pH and geographic relationships. *Agro-Ecosystems* **5**, 57—67, 1979.

32. Date, R.A. and Norris, D.O. Rhizobium screening of *Stylosanthes* species for effectiveness in nitrogen fixation. *Austral. J. Agric. Res.* **30**, 1—19, 1979.

33. Dazzo, F.B. Determinants of host specificity in the *Rhizobium*-clover symbiosis. *Nitrogen Fixation*. Vol. II. W.E. Newton and W.H. Orme-Johnson (eds.), 165—187, Univ. Park Press, Baltimore, 1980.

34. Dazzo, F.B., Napoli, C.A. and Hubbell, D.H. Adsorption of bacteria to roots as related to host specificity in the *Rhizobium*-clover symbiosis. *Appl. and Environ. Microbiol.* **32**, 166—171, 1976.

35. Degenhardt, T.L., LaRue, T.A. and Paul, E.A. Investigation of a non-nodulating cultivar of *Pisum sativum*. *Can. J. Bot.* **54**, 1633—1636, 1976.

36. Devine, T.E. and Breithaupt, B.H. Significance of incompatibility reactions of *Rhizobium japonicum* strains with soybean host genotypes. *Crop Sci.* **20**, 269—271, 1980.

37. Devine, T.E., Kuykendall, L.D. and Breithaupt, B.H. Nodulation of soybeans carrying the nodulation-restrictive gene, rj_1, by an incompatible *Rhizobium japonicum* strain upon mixed inoculation with a compatible strain. *Can. J. Microbiol.* **26**, 179—182, 1980.

38. Devine, T.E. and Reisinger, W.W. A technique for evaluating nodulation response of soybean genotypes. *Agron. J.* **70**, 510—511, 1978.

39. Devine, T.E. and Weber, D.F. Genetic specificity of nodulation. *Euphytica* **26**, 527—535, 1977.

40. Diatloff, A. and Ferguson, J.E. Nodule number, time to nodulation and its effectiveness in eleven accessions of *Glycine wightii*. *Tropical Grasslands* **4**, 223—228, 1970.

41. Dilworth, M.J. The plant as a genetic determinant of leghaemoglobin production in the legume root nodule. *Biochem. Biophys. Acta* **184**, 432—441, 1969.

42. Dilworth, M.J. Host and *Rhizobium* contributions to the physiology of legume nodules. *Nitrogen Fixation* Vol. II, W.E. Newton and W.H. Orme-Johnson (eds.), 3—31, Univ. Park Press, Baltimore, 1980.

43. Döbereiner, J. *Azotobacter paspali* n. sp. uma bacteria fixadora de nitrogenio na rhizofera de Paspalum. *Pesquisa Agropecuario Brasileira* **1**, 357—365, 1966.

44. Döbereiner, J. and Day, J.M. Associative symbioses in tropical grasses: Characterisation of microorganisms and Dinitrogen-fixing sites. *Proc. 1st Intern. Symp. on Nitrogen Fixation* **2**. W.E. Newton and C.J. Nyman (eds.), 518—538, Washington State Univ. Press, 1976.

45. Duhigg, P., Melton, B. and Baltensperger, A. Selection for acetylene reduction rates in 'Mesilla' alfalfa. *Crop Sci.* **18**, 813, 1978.

46. El-Sherbeeny, M.H., Lawes, D.A. and Mytton, L.R. Symbiotic variability in *Vicia faba*. *Euphytica* **26**, 377—386, 1977.

47. Erdman, L.W. and Means, U.M. Strain variation of *Rhizobium meliloti* on three varieties of *Medicago sativa*. *Agron. J.* **45**, 625—629, 1953.

48. Franco, A.A. and Vincent, J.M. Competition amongst rhizobial strains for the colonization and nodulation of two tropical legumes. *Plant and Soil* **45**, 27—48, 1976.

49. Gardner, I.C. Ultrastructural studies of non-leguminous root nodules. *Symbiotic Nitrogen Fixation in Plants. International Biological Programme* 7. P.S. Nutman (ed.), 485—495, Cambridge Univ. Press, Cambridge, 1976.

50. Gibson, A.H. Genetic variation in the effectiveness of nodulation of lucerne varieties. *Austral. J. Agric. Res.* **13**, 388—399, 1962.

51. Gibson, A.H. Genetic control of strain specific ineffective nodulation in *Trifolium subterraneum* L. *Austral. J. Agric. Res.* **15**, 37—49, 1964.

52. Gibson, A.H. Nodulation failure in *Trifolium subterraneum* L. cv Woogenellup (syn. Marrar). *Austral. J. Agric. Res.* **19**, 907—918, 1968.

53. Gibson, A.H. and Brockwell, J. Symbiotic characteristics of subspecies of *Trifolium subterraneum* L. *Aust. J. Agric. Res.* **19**, 891—905, 1968.

54. Graham, P. Some problems on nodulation and symbiotic nitrogen fixation in *Phaseolus vulgaris* L. *Field Crops Res.* **4**, 93—112, 1981.

55.   Haack, A.   Uber den Einfluss der Knöllchenbakterien auf die Wurzelhaare von Leguminosen und Nichtleguminosen. *Zentbl. Bakt. Parasitkde* (Abt II) **117**, 343–366, 1964.

56.   Hardarson, G. and Jones, D.G.   The inheritance of preference for strains of *Rhizobium trifolii* by white clover (*Trifolium repens*). *Ann. Appl. Biol.* **92**, 329–333, 1979.

57.   Hardy, R.W.F. and Silver, W.A. (eds.).   *A Treatise on Dinitrogen Fixation: III Biology.* 675. John Wiley and Sons, New York, 1977.

58.   Heichel, G.H.   *Selecting and Breeding Legumes for Enhanced Nitrogen Fixation.* Workshop, Boyce Thompson Institute, Cornell, 1978.

59.   Hely, F.W.   Genetic studies with wild diploid *Trifolium ambiguum* M. Bieb. with respect to time of nodulation. *Austral. J. Agric. Res.* **23**, 437–446, 1972.

60.   Hely, F.W., Bonnier, C. and Manil, P.   Effect of grafting on nodulation of *Trifolium ambiguum*. *Nature* London **171**, 884–885, 1953.

61.   Hepper, C.M.   Physiological studies on nodule formation, the characteristics and inheritance of abnormal nodulation of *Trifolium pratense* by *Rhizobium leguminosarum*. *Ann. Bot.* London **42**, 109–116, 1977.

62.   Herridge, D.F., Atkins, C.A., Pate, J.S. and Rainbird, R.M.   Allantoin and allantoic acid in the N economy of the cowpea (*Vigna unguiculata* (L) Walp.) *Plant Physiology* **62**, 495–498, 1978.

63.   Holl, F.B.   Host-determined genetic control of nitrogen fixation in the *Pisum-Rhizobium* symbiosis. *Can. J. Genet. Cytol.* **15**, 659, 1973.

64.   Holl, F.B.   Host plant control of the inheritance of dinitrogen fixation in the *Pisum-Rhizobium* symbiosis. *Euphytica* **24**, 767–770, 1975.

65.   Holl, F.B. and La Rue, T.A.   Genetics of legume plant hosts. *Proceeding of the First International Symposium on Nitrogen Fixation.* W.E. Newman and C.H. Nyman (eds.), 391–399. Washington State Univ. Press, 1976.

66.   Jefren, L.D., Ellis, R. and Paulsen, G.M.   Nodulation and nitrogen fixation by two soybean varieties as affected by phosphorus and zinc nutrition. *Agron. J.* **64**, 566–568, 1972.

67.   Jeppson, R.G., Johnson, R.E. and Handley, H.H.   Variation in mobilization of plant nitrogen to the grain in nodulating and non-nodulating soybean genotypes. *Crop Sci.* **18**, 1058–1062, 1978.

68.   Jones, D.G. and Burrows, A.C.   Breeding for increased nodule tissue in white clover. *J. Agric. Sci.* **71**, 73–79, 1968.

69.   Jones, D.G. and Hardarson, G.   Variation within and between white clover varieties in their preference for strains of *Rhizobium trifolii*. *Ann. Appl. Biol.* **92**, 221–228, 1979.

70.   Jones, R.J.   Yield potential for tropical pasture legumes. *Exploiting the Legume-Rhizobium Symbiosis in Tropical Agriculture.* J.M. Vincent, A.S. Whitney and J. Bose (eds.), 39–65. Coll. Trop. Agric., Misc. Publ. **145**, 1977.

71.   Khan, T.N.   Yield potential for tropical legumes from a geneticist's point of view. *Exploiting the Legume-Rhizobium symbiosis in Tropical Agriculture.* J.M. Vincent, A.S. Whitney and J. Bose (eds.), 21–37. Coll. Trop. Agric. Misc. Publ. **145**, 1977.

72.   Klucas, R.V., Pedersen, W., Shearman, R.C. and Wood, L.V.   Nitrogen fixation associated with winter wheat, sorghum and Kentucky blue grass. *Associative Dinitrogen Fixation.* P.B. Vose and A.P. Ruschel (eds.), CRC Press, Boca Raton, Fla., 1981.

73.   Labandera, C.A. and Vincent, J.M.   Competition between an introduced strain and native Uruguayan strains of *Rhizobium trifolii*. *Plant and Soil* **42**, 327–347, 1975.

74.   Lance, R.T. and Parker, C.A.   Nodulation patterns on legumes. *Nature* London **186**, 178–179, 1960.

75.   Lawn, R.J. and Brun, W.A.   Symbiotic nitrogen fixation in soybeans. I. Effect of photosynthetic source-sink manipulations. *Crop Science* **14**, 11–16, 1974.

76.   Libbenga, K.R. and Bogers, R.J.   Root-nodule Morphogenesis. *The Biology of Nitrogen Fixation.* A. Quispel (ed.), North Holland Research Monographs: Frontiers of Biology, Vol. 33, 439–472. North Holland, Amsterdam, 1974.

77,   Lie, T.A.   Symbiotic fixation under stress conditions. *Plant and Soil, Spec. Vol.* 117–127, 1971.

78.   Lie, T.A.   Symbiotic specialization in pea plants. The requirements of special *Rhizobium* strains for peas from Afghanistan. *Ann. Appl. Biol.* **88**, 462–465, 1978.

79.   Lindsay, C.R. and Jordan, D.C.   Use of stem cuttings to reduce plant variation in *Rhizobium*-leguminous plant investigations. *Can. J. Soil Sci.* **56**, 495–497, 1976.

80. Lowe, J.F. and Holding, A.J. Influence of clover source and of nutrient manganese concentrations on the *Rhizobium*-white clover association. *White Clover Res. Occas. Symp.* 6, 79–89, 1970.

81. Ludlow, M.M. and Wilson, G.L. Relationship between seed and seedling dry weight of tropical pasture grasses and legumes. *J. Austral. Inst. Agric. Sci.* 38, 65–67, 1972.

82. Marques-Pinto, C., Yao, P.Y. and Vincent, J.M. Nodulating competitiveness amongst strains of *Rhizobium meliloti* and *R. trifolii*. *Austral. J. Agric. Res.* 25, 317–329, 1974.

83. Materon, L.A. and Vincent, J.M. Host specificity and interstrain competition with soybean rhizobia. *Field Crops Res.* (in press).

84. Mytton, L.R. Plant genotype rhizobium strain interactions in white clover. *Ann. Appl. Biol.* 80, 103–107, 1975.

85. Mytton, L.R. The relative performance of white clover genotypes with rhizobial and mineral nitrogen in agar culture and in soil. *Ann. Appl. Biol.* 82, 577–587, 1976.

86. Mytton, L.R. Some problems and potentialities in breeding for improved white clover-rhizobium (*Trifolium repens* – *Rhizobium trifolii*) symbiosis. *Ann. Appl. Biol.* 88, 445–448, 1978.

87. Mytton, L.R., El-Sherbeeny, M.H. and Lawes, D.A. Symbiotic variability in *Vicia faba*. Part 3. Genetic effects of host plant, *Rhizobium* strain and host X strain interaction. *Euphytica* 26, 785–792, 1977.

88. Mytton, L.R. and Jones, D.G. The response to selection for increased nodule tissue in white clover (*Trifolium repens* L). *Plant and Soil, Spec. Vol.* 17–25, 1971.

89. Napoli, C., Sanders, R., Carlson, R. and Albersheim, P. Host-symbiont interactions: recognizing *Rhizobium*. *Nitrogen Fixation* Vol. II. W.E. Newton and W.H. Orme-Johnson (eds.), 189–203, Univ. Park Press, Baltimore, 1980.

90. Newcombe, W. Control of morphogenesis and differentiation of pea root nodules. *Nitrogen Fixation*. Vol. II. W.E. Newton and W.H. Orme-Johnson (eds.), 87–102, Univ. Park Press, Baltimore, 1980.

91. Newton, W.E. and Nyman, C.J. Proceedings of the 1st International Symposium on Nitrogen Fixation. Vol. 2, 717. Washington State Univ. Press, 1976.

92. Newton, W.E. and Orme-Johnson, W.H. *Nitrogen Fixation*. Vol. II. Symbiotic Associations and Cyanobacteria, 325. Univ. Park Press, Baltimore, 1980.

93. Newton, W.E., Postgate, J.R. and Rodriguez-Barrueco, E. (eds.). *Recent Developments in Nitrogen Fixation*. Academic Press, London, 1977.

94. Nicholas, D.B. Genotypic variation in growth and nodulation in *Glycine wightii*. *J. Austral. Inst. Agric. Sci.* 37, 69–70, 1971.

95. Nutman, P.S. Nuclear and cytoplasmic inheritance of resistance to infection by nodule bacteria in red clover. *Heredity* 3, 263–271, 1949.

96. Nutman, P.S. Symbiotic effectiveness in nodulated red clover. II. A major gene for ineffectiveness in the host. *Heredity* 8, 47–60, 1954.

97. Nutman, P.S. Symbiotic effectiveness in nodulated red clover. III. Further studies on inheritance of ineffectiveness in the host. *Heredity* 11, 157–173, 1957.

98. Nutman, P.S. Some observations on root-hair infection by nodule bacteria. *J. Exp. Bot.* 10, 250–263, 1959.

99. Nutman, P.S. The relation between root-hair infection by *Rhizobium* and nodulation in *Trifolium* and *Vicia*. *Proc. Roy. Soc. Ser. B.* 156, 122–137, 1962.

100. Nutman, P.S. Symbiotic effectiveness in nodulated red clover. V. The n and a factors for ineffectiveness. *Heredity* 23, 537–551, 1968.

101. Nutman, P.S. Genetics of symbiosis and nitrogen fixation in legumes. *Proc. Roy. Soc. Ser. B.* 172, 417–437, 1969.

102. Nutman, P.S. (ed.) *Symbiotic Nitrogen Fixation in Plants*. *International Biological Programme* 7, 584. Cambridge Univ. Press, Cambridge, 1976.

103. Nutman, P.S. (ed.). IBP Field experiments on nitrogen fixation by legumes. *Symbiotic Nitrogen Fixation in Plants, International Biological Programme* 7, 211–237. Cambridge Univ. Press, Cambridge, 1976.

104. Nutman, P.S. *Rothamsted Rep.*, Soil Microbiology Dept., 1977.

105. Nutman, P.S., Mareckova, H. and Raicheva, L. Selection for increased nitrogen fixation in red clover. *Plant and Soil Spec. Vol.* 27–31, 1971.

106. Pate, J.A. Functional biology of dinitrogen fixation by legumes. *A Treatise on Dinitrogen Fixation*. Section III. R.W.F. Hardy and W.S. Silver (eds.), 473–517. John Wiley and Sons, New York, 1977.

107. Pate, J.S. and Dart, P.J. Nodulation studies in legumes. IV. The influence of inoculum strain and time of application of ammonium nitrate on symbiotic response. *Plant and Soil* 15, 329–346, 1961.

108. Phillips, D. *Selecting and Breeding Legumes for Enhanced Nitrogen Fixation*. Workshop, October 23, 24. Boyce Thompson Institute, Cornell, 1978.

109. Pinchbeck, B.R. Strain-specific ineffective nodulation of the tropical legume *Desmodium* by *Rhizobium. Crop Sci.* 17, 513–514, 1978.

110. Pinchbeck, B.R. A genetic study of symbiotic nitrogen fixation in the tropical legume *Desmodium. Ph.D. Thesis.* Univ. Alberta, 1979.

111. Quispel, A. (ed.) The Biology of Nitrogen Fixation. *North Holland Research Monographs, Frontiers of Biology.* Vol. 33, 769. North Holland, Amsterdam, 1974.

112. Rai, R. and Singh, S.N. Interaction between chickpeas (*Cicer arietinum* Linn) genotypes and strains of *Rhizobium* sp. *J. Agric. Sci.* 92, 437–441, 1979.

113. Rennie, R.J. Diazotrophic biocoenosis – The Workshop Consensus Paper. *Associative Dinitrogen Fixation.* P.B. Vose and A.P. Ruschel (eds.), CRC Press, Boca Raton, Fla., 1981.

114. Rennie, R.J. and Larson, R.I. Dinitrogen fixation associated with disomic chromosome substitution lines of Spring wheat in the phytotron and in the field. *Associative Dinitrogen Fixation.* P.B. Vose and A.P. Ruschel (eds.), CRC Press, Boca Raton, Fla., 1981.

115. Rinaudo, G., Gauthier, D. and Dommergues, Y. Enhancement of associative N_2-fixation through manipulation of the rhizosphere microflora. *Associative Dinitrogen Fixation.* P.B. Vose and A.P. Ruschel (eds.), CRC Press, Boca Raton, Fla., 1981.

116. Robinson, A.C. The influence of host on soil and rhizosphere populations of clover and lucerne root-nodule bacteria in the field. *J. Austral. Inst. Agric. Sci.* 33, 207–209, 1967.

117. Robinson, A.C. Competition between effective and ineffective strains of *Rhizobium trifolii* in the nodulation of *Trifolium subterraneum. Austral. J. Agric. Res.* 20, 827–841, 1969.

118. Robinson, P.J., Date, R.A. and Megarrity, R.G. Prediction of effective *Rhizobium* strains for *Stylosanthes guyanensis* using iso-enzyme patterns. *Austral. J. Agric. Res.* 27, 381–389, 1976.

119. Rocha, R.E.M., Baldani, J.I. and Dobereiner, J. Specificity of infection by *Azospirillum* spp. in plants with C_4 photosynthetic pathway. *Associative Dinitrogen Fixation.* P.B. Vose and A.P. Ruschel (eds.), CRC Press, Boca Raton, Fla., 1981.

120. Roughley, R.J., Dart, P.J., Nutman, P.S. and Clarke, P.A. The infection of *Trifolium subterraneum* root hairs by *Rhizobium trifolii. J. Exp. Bot.* 21, 186–194, 1970.

121. Rovira, A.D. Rhizobium numbers in the rhizospheres of red clover and paspalum in relation to soil treatment and the numbers of bacteria and fungi. *Austral. J. Agric. Res.* 12, 77–83, 1961.

122. Ruschel, A.P. Sugar cane: a position paper. *Associative Dinitrogen Fixation.* P.B. Vose and A.P. Ruschel (eds.), CRC Press, Boca Raton, Fla., 1981.

123. Ruschel, R. and Ruschel, A.P. Inheritance of N_2-fixing ability in sugar cane. *Associative Dinitrogen Fixation.* P.B. Vose and A.P. Ruschel (eds.), CRC Press, Boca Raton, Fla., 1981.

124. Seetin, M.W. and Barnes, D.K. Variation among alfalfa genotypes for rate of acetylene reduction. *Crop Sci.* 17, 783–787, 1977.

125. Sherwood, M.T. and Masterson, C.L. Importance of using the correct test host in assessing the effectiveness of indigenous populations of *Rhizobium trifolii. Irish J. Agric. Res.* 13, 101–109, 1974.

126. Sinclair, I.N., Hannagan, R.B., Johnstone, P. and Hardacre, A.K. Evaluation of a non-destructive acetylene reduction assay of nitrogen fixation for pasture legumes grown in pots. *N.Z. J. Exptl. Agric.* 6, 65–68, 1978.

127. Souto, S.M., Coser, A.C. and Döbereiner, J. Host plant specificity of a native variety of *Stylosanthes gracilis. Pesquisa Agropecuario Brasileira Ser. Zootec.* 7, 1–5, 1972.

128. Sprent, J.I. *The Biology of Nitrogen-fixing Organisms*. European Plant Biology Series. 196, McGraw Hill, London, 1979.

129. Staphorst, J.L. and Strijdom, B.W. Nodulating ability and growth of South African Standard lucerne. (*Medicago sativa* L.). *Agroplantae* 2, 29–32, 1970.

130. Stewart, W.D.P. (ed.). Nitrogen fixation by free-living micro-organisms. *International Biological Programme* 6, 471. Cambridge Univ. Press, Cambridge, 1975.

131. Thompson, J.A. Inhibition of nodule bacteria by an antibiotic from legume seed coats. *Nature* Lond. **187**, 619—620, 1960.

132. Torrey, J.G., Baker, D., Callaham, D., Del Tredici, P., Newcombe, W., Peterson, R.L. and Tjepkema, J.D. On the nature of the endophyte causing root nodulation in *Comptonia*. *Nitrogen Fixation*. Vol. II. W.E. Newton and W.H. Orme-Johnson (eds.), 217—227. Univ. Park Press, Baltimore, 1980.

133. Trinick, M.J. Symbiosis between *Rhizobium* and the non-legume, *Trema aspera*. *Nature* London, **244**, 459—460, 1973.

134. Van Egeraat, A.W.S. The possible role of homoserine in the development of *Rhizobium leguminosarum* in the rhizosphere of pea seedlings. *Plant and Soil* **42**, 381—386, 1975.

135. Vest, G. Rj$_3$ — a gene conditioning ineffective nodulation in soybean. *Crop Sci.* **10**, 34—35, 1970.

136. Vest, G. and Galdwell, B.E. Rj$_4$ — a gene conditioning ineffective nodulation in soybean. *Crop Sci.* **12**, 692—693, 1972.

137. Vincent, J.M. *A Manual for the Practical Study of the Root-nodule Bacteria*. IBP Handbook No. 15. Blackwell Scientific Publ., Oxford, 1970.

138. Vincent, J.M. Factors controlling the legume-*Rhizobium* symbiosis. *Nitrogen Fixation*. Vol. II. W.E. Newton and W.H. Orme-Johnson (eds.), 103—129. Univ. Park Press, Baltimore, 1980.

139. Vincent, J.M. and Waters, L.M. The influence of host on competition amongst clover root-nodule bacteria. *J. Gen. Microbiol.* **9**, 357—370, 1953.

140. Vlassak, K. and Reynders, L. Agronomic aspects of biological dinitrogen fixation by *Azospirillum* spp in temperate region. *Associative Dinitrogen Fixation*. P.B. Vose and A.P. Ruschel (eds.), CRC Press, Boca Raton, Fla., 1981.

141. Von Bulow. Plant influence in symbiotic nitrogen fixation. *Limitations and Potentials for Biological Nitrogen Fixation in the Tropics*. J. Döbereiner, R.H. Burris, A. Hollaender *et al.* (eds.), Basic Life Sciences, Vol. 10, 398. Plenum Press, New York, 1978.

142. Vose, P.B. and Ruschel, A.P. (eds.). Associative Dinitrogen Fixation, *Proc. Int. Workshop Associative N$_2$-Fixation, 2—6 July 1979 Piracicaba, Brasil*. CRC Press, Boca Raton, Fla., 1981.

143. Williams, L.F. and Lynch, D.L. Inheritance of a non-nodulating character in the soybean. *Agron. J.* **46**, 28—29, 1954.

144. Wittenberg, J.B. Utilization of leghaemoglobin-bound oxygen by *Rhizobium* bacteroids. *Nitrogen Fixation*. Vol. II. W.E. Newton and W.H. Orme-Johnson (eds.), 53—67. Univ. Park Press, Baltimore, 1980.

145. Yao, P.Y. and Vincent, J.M. Host specificity in the root hair "curling factor" of *Rhizobium* spp. *Austral. J. Biol. Sci.* **22**, 413—423, 1969.

146. Zary, K.W. and Miller, J.C. Intraspecific variability for dinitrogen fixation in cowpea (*Vigna unguiculata* (L) Walp). *Hortscience* **12**, 402, 1977.

147. Zorin, M., Dear, B.S. and Hely, F.W. Young plant vigour and nodulation studies in diploid forms of *Trifolium ambiguum* Bieb. *Austral. C.S.I.R.O. Div. Plant Indus. Fld. Stn. Rec.* **15**, 35—49, 1976.

148. Zorin, M., Hely, F.W. and Dear, B.S. Host-strain relationships in symbiosis between hexaploid *Trifolium ambiguum* Bieb. (Caucasian clover) and strains of *Rhizobium trifolii*. *Austral. C.S.I.R.O. Div. Plant Indus. Fld. Stn. Rec.* **15**, 63—71, 1976.

CHAPTER 9

ROLE OF INDUCED MUTATIONS

C.F. Konzak

*Department of Agronomy and Soils, and Program in Genetics
and Cell Biology, Washington State University,
Pullman, Washington 99164, U.S.A.*

INTRODUCTION

Mutations provide the fundamental variability required for plant improvement by breeding. Most genetic variability available today in plant collections is the result of prior evolution involving some genetic recombination and exposure to forces of natural selection. Artificial selection pressures relating to cultivation practice and uses have been imposed by man toward the development of our crop plants. Controlled or artificial cross breeding is now the principal means to reassort and recombine genetic variability from the different germplasm sources usually within a crop. This is followed by selection procedures aimed at identifying genotypes with improved characteristics, such as yield, disease resistance, flower or fruit shape color or quality.

Using such modern technologies as embryo cultures, germplasm combinations are now possible from related species, e.g. cherry: sour cherry X Japanese bird cherry[447]; *Citrus*: mandarin X grapefruit, sweet orange X grapefruit, etc.[389]; *Lilium*: *L. candidum* X *L. chalcedonicum*, *L. candidum* X *L. Parryi*, *L. tigrinum* X *L. croceum*, *L. croceum* X *L. Willmottiae*, *L. philadelphicum* X *L. dauricum*, *L. paradalium* X *L. Humboldtii*, *L. paradelinum* X *L. Parryi*,

L. Sargentiae X *L. regale, L. myriophyllum* X *L. regale,* etc.[385]; Rosa spp. (common) *Chrysanthemum* spp. (common); Orchid spp. (common), *Saccharum* spp. *Sorghum* (common), and entergeneric crosses such as *Triticum* X *Agropyron, Triticum* X *Hordeum*[203], *Triticum* X *Secale* (now common, including the synthetic species hexaploid and octoploid *Triticale*), *Citrus* X *Poncirus*[389] etc. However, recombinants with improved characteristics from such wide crosses are seldom recovered even though the F_1 hybrid often may be useful.

Modern methods for inducing mutations in plants, has led to the development of two related applications for improving plants. The most important of these involves the artificial induction of new mutant genetic variation, supplementary and complementary to that already available for use in cross-breeding. Important to this application is the fact that the mutant genetic variation can be induced in already adapted, modern genotypes, but it also has become apparent that even the allelic variations induced at specific gene loci may differ slightly to markedly from one another when analysed in terms of their genetic fine structure.[143,145] Therefore, it should no longer be assumed that any mutant, spontaneous or induced, can be reinduced at will. Rather, exact allelic variants may be extremely rare.

The second important application is termed mutation breeding. In this application, mutagens are employed to induce marketable genotypes in the sexual or vegetative progeny. Following the direct mutation breeding approach does not preclude exploitation of any useful mutants as germplasm. But, with the 'germplasm induction' approach, the much higher mutagen doses often employed may induce accessory mutations limiting the direct commercialization of selections carrying a useful mutant trait.

A variety of physical and chemical mutagens are widely available and methods for their safe handling, application, and disposal are now well understood. The IAEA Manual on Mutation Breeding, 2nd Edition is a standard work on practical mutation induction.[248] X-ray, gamma and neutron radiations have proved most useful for mutation induction in vegetatively propagated species, especially when whole propagules or organ segments, i.e. stem segments, leaves, scales, root segments, etc. are treated.[62] However, with tissue or protoplast culture also ultraviolet radiation and at least some chemicals may be effective.

With seed propagated plants, both radiation and chemical mutagens are effective, but chemicals are often the most efficient mutation induction agents. They include Azide, diethyl sulphate (DES), ethyl methanesulphonate (EMS), N-nitroso-N-methyl urea (MNH) and N-nitroso urea (MNU). Other chemicals such as isopropyl methanesulphonate (IPMS) and propane sultone show promise but have not been much used. All these chemicals are extremely hazardous and require great care in use. Often, chemicals may induce a somewhat different mutation spectrum than radiations.

Guidelines for practical mutation breeding in seed plants have been given by Yonezawa and Yamagata,[443] especially for cereals, and by Dellaert[93] particularly for dicotyledonous plants, while Brock[52] has considered constraints to the induction and selection of mutations in crop plants.

Applications of mutagens for mutation breeding and germ plasm induction may differ markedly for vegetatively propagated plants and for seed reproduced sexually propagated plants. For this reason the two groups of plants are considered separately.

VEGETATIVELY PROPAGATED PLANTS

Mutation breeding methods are already widely employed for improving vegetatively propagated plant crops, because the generally high heterozygosity and often polyploid nature of these plants complicate the inheritance of traits and may impose intolerable requirements for number of progeny necessary to recover desirable recombinants. However, these handicaps to conventional cross breeding often prove advantageous with mutation breeding, as they increase the genetic base for the release of induced variability. In obligate apomicts or seed sterile plants, mutation breeding is the only means to artificially increase genetic variation. Yet, the principal advantage of mutation breeding is that it offers the breeder a rapid method to change one or a few characters of an otherwise outstanding cultivar or selection without affecting the other special features of the genotype.

Thus, mutation breeding is uniquely complementary to conventional cross breeding, making it easier for the breeder to more fully exploit new improved genetic combinations which often have required many years to achieve. Mutants in new cultivars or breeding lines which make particular loci homozygous or otherwise improve the genotype may complement the products of conventional cross breeding also in other ways: (1) they may displace the original as the preferred genotype as germ plasm for further cross breeding work, and (2) they may be preferable for further exploitation by mutation breeding.[63]

While the interest in and use of mutation breeding for improving vegetatively propagated plants is rapidly increasing, much already has been accomplished (Table 9.1). The true impact of the method on improvement of vegetatively propagated crops is certainly underestimated, because few data are available from mutation breeding programs of private firms. In the Netherlands alone, hundreds of private breeders (nearly all) now use mutation breeding methods for ornamentals[63] (Harten, personal communication 1981). As a consequence, about 40% of the chrysanthemums, including 98% of the white and yellow cultivars currently in production are induced mutants while spontaneous mutants have little importance.[63] Considering that there were practically no commercial mutants in vegetatively propagated crops 20 years ago,[164] the 297 mutants commercialized, largely in the past 10 years, represent strong documentary evidence for the success and future potential of the method.

(i) Tissue and cellular origin of mutations

With vegetatively propagated plants, mutations may originate as chimeras in one or more of the 3 (L–I, L–II, L–III) ontogenic tissue layers of shoot apices (Fig. 9.1a, b), or from adventitious buds which develop from one or a small number of initial cells. Mutations only in L–III or corpus tissue will not be expressed even if they affect the whole of the L–III tissue layer. Those only in L–II (sub epidermal cell layer) may be expressed only if they pass through a sexual stage, and those only in L–I (epidermal – outermost cell layer) will be expressed in the phenotype but not in the genotype or germ line (sexual) tissues. Chimeral mutants affecting part of one or more tissue layers are common when multicellular meristematic buds are irradiated.

A mutation occurring only in one of the tissue layers can be periclinal (affecting the whole of either the L–I, L–II, or L–III tissue layer, or mericlinal) affecting only part of a tissue layer (Fig. 9.1a). A sectorial chimera usually affects a segment of all tissue layers, while the solid mutant or homohistant affects the whole of all tissue layers.[62] Convincing evidence for the existence of three or more histogenic tissue layers involved in apical meristem construction has recently been presented by Kukimura et al.[225] They obtained cytochimeras in mulberry with identifiable tissue layers differing in ploidy. The frequency of 2–4–4 ploidy levels in periclinal cytochimeras of histogenic shoot apex tissues was very high, followed by those with 4–2–2 and 2–4–2 ploidy construction, while homohistonts of 2–2–2 (diploid) and 4–4–4

TABLE 9.1. Commercial mutants of various vegetatively propagated crops (October 1981)[2]

Plant group	Number of commercial mutants in the period					Total # of mutants (in cooperation with the ITAL foundation)	Crops of which (radiation-induced) mutants were commercialized (number)
	Before 1950	1950-1960	1960-1970	1970-1980	1980-		
ROOT AND TUBER CROPS				1		1 ( )	Potato (1)
ORNAMENTALS							
Tuber and bulb crops	1	1	13	48		63 (45)	Dahlia (23), Lilium (2), Polyantha (2), Tulip (25)
Pot plant			7	48		55 (26)	Achimenes (8), Azalea (10), Begonia (21), Guzmania (1), Streptocarpus (15)
Cut flowers			14	119	2	135 (68)	Alstroemeria (22), Carnation (2), Chrysanth (107), Euphorbia fulgens (1), Roses (7)
Other ornamentals				17		17 (2)	Abelia (1), Azalea (1), Bougainvillea (2), Hibiscus (4), Malus (1), Populus (1), Portulaca (7)
FRUIT CROPS			4	14		18 ( )	Apple (4), Apricot (1), Cherry (7), Peach (1), Black Currant (1), Grapefruit (1) (Fig. 9.1), Pomegranate (2)
OTHER CROPS				10	1	11 ( )	Mint (3), Sugarcane (8)
TOTALS	1	1	38	257	3	300 (141)	

(1) Additions: 1. Ficus benjamin cv. Golden Princess (Belgium) 1980; 2. Chrysanthemum morifolium – numerous (Netherlands) 1980; 3. Peach cv. Plovdiv 6 (Bulgaria) 1981; 4. Bermudagrass cv. Tifway 2 (USA) 1981; 5. Olive cv. Briscola (Italy, 1982)

(2) Sources: 1. C. Broertjes, currently Institute of Plant Breeding, The Netherlands. 2. Mutation Breeding Newsletters 1–18, 1972–1981.

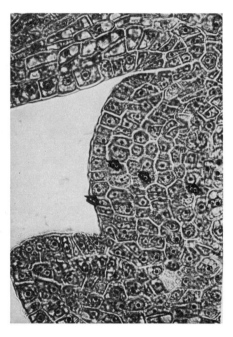

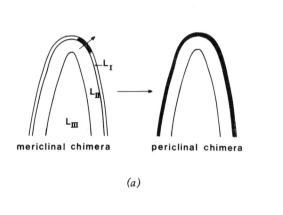

merichinal chimera   periclinal chimera

(a)

(b)

Fig. 9.1. (a) Schematic drawing of plant apical meristem, showing tissue layers
and structural basis for mericlinal and periclinal chimeras affecting the L–I
layer. Courtesy C. Broertjes. (b) Photomicrograph of a median section
through an apical bud of potato, *Solanum tuberosum*, showing the
layered character of the L–I (1) and L–II (2) tissues, and the
larger irregular, L–III (3) zone. Courtesy A.M. van Harten.

(tetraploid) also occurred. However, cytochimeras with 2–2–4 also were observed and some
others had 2–4–4–2, 2–2–2–4 or 2–2–2–2–4 chimeral structure indicating the possibility
that even more than 3 histogenic tissue layers may be identifiable. Leaf traits were notably
influenced by cytochimeral construction such that variants with 2–4–4, 2–4–2, or 4–2–2
construction were distinguishable from the original diploid 2–2–2 homohistont. In principle,
the 3 histogenic layer concept is valid for all practical purposes since the more complex corpus

tissues observed by Kukimura *et al.*[225] will not be expressed without tissue reorganization which typically does not occur. However, the number of initial cells in the shoot apex may differ with the species and possibly to some extent with varieties, perhaps depending on the degree of apical dominance expressed. Moh[267] for example, concluded that the shoot apex of the *Coffea* embryo is controlled by a single initial cell.

Reorganization of tissues within apical or axillary buds to occur, as a consequence of the disruption of meristem initials following irradiation treatments severe enough to kill some but not all of the cells of a shoot apex. This type of response to irradiation is not uncommon. Hence, periclinal chimeras can be made complete by irradiation and it is possible to use irradiation as a tool to analyse the histogenic structure of vegetatively propagated plants[62,366].

Mutation is, nevertheless, a single cell event (Fig. 9.2a, b). The mutagen may "hit" or alter a specific site in a chromosome or induce a chromosome segmental selection, resulting in a change expressable in the phenotype as a mutation. More than one mutational event can occur within a single cell, but generally each occurrence is at random. The probability for expression of a mutation is very low, but is increased by increasing the radiation dose applied, but only to the point that tissue destruction, including diplontic selection (competition from non- or less-injured cells) results in a reduced mutation yield[62,131,197]. For that reason, there is an optimum mutagen dose and treatment condition for each plant species.

The differences in response, termed **radiosensitivity** for radiation treatments, have a complex basis, including such genetic factors as interphase chromosome volume (ICV) in which ploidy is a factor[442] and heterozygosity of gene loci[62] as well as such biological factors as metabolic and meristematic cell activity, tissue hydration, dormancy and physiological stage and exposure to such environmental factors as oxygen and temperature during radiation treatment.[219,307] Conditions provided during a 'recovery' period following radiation treatment also can be important, with better results generally achieved if conditions most favorable for plant growth are provided.

The fate of a mutated cell during organogenesis of a shoot apex is largely determined by its position within the apex.[101] Mutations occurring in cells of leaf initials, for example, usually will be lost. Mutations induced in apical initial cells usually will produce mutated tissue sectors. If the mutated initial cell is stable, retaining its position in the apex, long mutated sectors or chimeras will be produced. With long sectors the chance is increased that later-formed axillary buds will occur partially or completely within the mutated tissue. As noted earlier, if the mutated cell is in the L—I layer of cells in the shoot apex, it is more likely that the resulting mutant tissue can be detected and 'captured' for evaluation and potential exploitation.

Many apparently solid mutants isolated by vegetative propagation prove to be periclinal chimeras, affecting only the L—I or sometimes the L—II tissue layer, depending on the trait and its mode of expression. *In vitro* tissue culture methods for isolating complete mutants from L—I periclinal or mericlinal chimeras now seem feasible. Mutations in the L—II tissue layer may be expressed in the germ line, and tissue reorganization to include the mutation in L—I tissue might be induced by a second cycle of mutation breeding. Periclinal chimeras in some cases can be directly useful, and may have advantages over solid or complete mutants. Negative mutations for example, as lower vigor or yield associated with an improved flower color, may be buffered by the unchanged underlying tissues, or the mutation can affect only a subepidermal L—II layer, and produce effects on leaves, flowers, or fruits that do not occur when the tissue genotypes are separated.

The recovery of relatively stable periclinal chimeras generally can be achieved by cutting

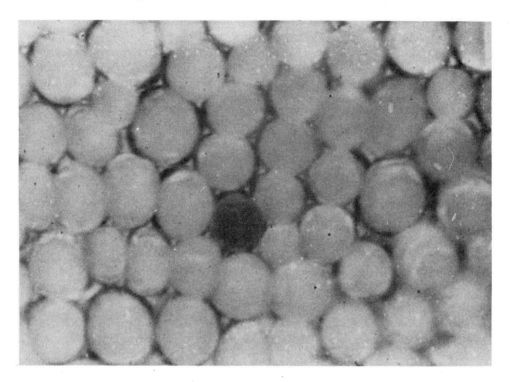

Fig. 9.2. Mutant chimeras affecting: (a) a single cell in a petal; (b) segments of a flower,
in *Dahlia variabilis*. After Broertjes, 1981.

back a shoot with a mericlinal chimera as is frequently done with *Chrysanthemum*.[62] However, periclinal chimeras tend not to be fully stable, and often back-sport. True-to-type *in vitro* propagation of periclinal chimeras can be obtained only from existing buds, not via adventitious plantlet formation. Moreover, periclinal chimeras will not show transmission of the improved trait in cross breeding.[164] Therefore, it is more efficient from a breeding point of view when plants to be screened for mutations can be grown from a small number of cells, or even better, from a single cell derived from the mutated cell layer.[62]

(ii) Propagation techniques for isolating mutations

Because mutation is a single cell event, mutation breeding technique development has been focused on ways to facilitate the development of tissues derived from only one or a few cells. This is necessary so that mutant chimeras (if they occur at all) will be large enough to allow the application of methods capable of isolating complete mutants. Successful techniques now widely employed for isolating vegetative mutants include various adventitious bud techniques, disbudding, cutting back, grafting or budding of tissues taken from first year shoots of irradiated explants, and development of cuttings from shoots of irradiated explants, etc. *In vitro* propagation systems are developing rapidly.[86]

(b)

(1) ADVENTITIOUS BUD TECHNIQUES

Adventitious bud techniques used for propagation of tissues shortly after they are mutagenized have the advantage that nearly all genetic variations recovered will be complete mutants. Methods and tissues for adventitious bud induction vary greatly for different plant species. For some species, adventitious bud techniques have not been developed; while for others, this method of propagation is too slow and time-consuming compared with other *in vivo* propagation methods with which similar results can be achieved.[62,63,164]

Perhaps the most extensive, yet detailed investigations on methods for propagating plants from adventitious buds have been conducted by Broertjes and colleagues.[53—55,57—62]

In *Achimenes, Saintpaulia, Streptocarpus, Nicotiana* and *Kalenchoe*, it has been shown that adventitious buds develop from a single epidermal cell (the L—I layer) of detached leaves, petioles or in the case of *Lilum* from bulb scales. In all cases, the neoformation of meristems involves several histogenic tissue layers. An L—I epidermal cell becomes dedifferentiated and meristematic to develop into a shoot primordium after the formation of a meristem. In the L—III tissue layer below the developing L—I shoot primordium another meristem is formed which develops into a root primordium.[58]

With some species, as *Lilium* and presumably other true bulb species — onion, garlic, hyacinth, tulip, daffodil, muscari, iris, etc., adventitious buds can be made to form on artificial bulb scales made by cutting the bulbs into pieces following their exposure to radiation. Damage from fungi can be controlled with fungicides, such as benomyl, (trade name, Benlate).[59]

Even many root crops, such as sweet potato *Ipomea batatas* will produce adventitious plantlets from detached leaves.[62] Methods for producing adventitious buds have not yet been developed for some plant species, even after considerable research. This was formerly the case with potato and tomato, but with persistent effort, Roest and Bokelman[359] and Harten *et al.*[165] have succeeded using an *in vitro* adventitious bud propagation technique. Similar methods are now being applied with other solanaceae (S.S. Chase, personal communication 1981). Broertjes *et al.*[61] listed more than 350 species, representing several plant families, in which adventitious plantlets can be produced on detached leaves.

(2) *OTHER* IN VIVO *AND* IN VITRO *TECHNIQUES*

With some species, adventitious bud formation occurs naturally, while with some others, alternative techniques may be faster and more efficient. With *Alstroemeria*, for example, rhizomes are normally used for propagation. Unexpectedly, virtually all mutants propagated from irradiated rhizomes have proved to be complete, even though multicellular apices are irradiated.[164] Presumably, apical cell dominance is sufficiently strong in this species for the method to be successful, and since the method is highly efficient, it should continue in use.[65,164] With mint, *Mentha spp.* and a number of vegetatively propagated or apomictic grasses, mass scale propagation from irradiated rhizomes, stolons or apomictic seeds appears to be relatively efficient, so long as the post-irradiation conditions are made favorable for growth and attention is given to the identification and isolation of sectorial chimeras as they become distinguishable. Murray[275], Burton[75] and Burton *et al.*[78], Powell and Murray[329], and Powell and Toler[330] could screen large numbers of propagules in field trials. More recently, Green *et al.* (personal communication, 1981) found it more efficient to grow plants from irradiated mint rhizomes under better controlled greenhouse conditions before transplanting them, than to start treated rhizomes directly in the field. The use of new highly effective fungicides for control of soil rot fungi, however, may improve the efficiency of field screening, although limitations of irradiation equipment for efficient treatment of large quantities of vegetative material might be the determining factor with regard to field mass screening methods.

On the other hand, efficient adventitious bud techniques have not yet been developed for propagating many of the woody plant species. Consequently, methods such as disbudding, cutting back, rooting of subdivisions from shoots allowed to grow from irradiated buds even if inadequate, appear to be practical. Donini (1975) has illustrated procedures found effective for isolating mutations of grape and of fruit trees, as cherry (Fig. 9.3), apple, pear. Similar methods are applicable to apricot, peach, etc.[229,416]

With *Chrysanthemum*, both *in vitro* and *in vivo* adventitious bud techniques have been developed, and both are successful.[64] However, Broertjes[56] concluded that because the technique of cutting back has proved both practical and highly efficient, the commercial production of mutants from irradiated cuttings is preferable. The production of adventitious plantlets in *Chrysanthemum* proved to take more time, thus was not as efficient as the easier, faster method of propagating from cuttings, from which complete mutants can be isolated by repeatedly cutting back shoots carrying chimeras. With a number of other species, however, *in vitro* methods, including tissue cultures of sugar cane[167] and mint[310], and protoplast culture of potato[381,430] appear to have exceptional promise for mutation breeding. Because of instabilities that occur, especially with protoplast culture, this method is unsuitable for plant propagation, whereas tissue or callus culture methods appear to be useful and effective for

(b)

(a)

Fig. 9.3. (a) Compact mutant of the sweet cherry, *Prunus avium*, cv. Bigarreau Napoleon induced in Italy. (b) Normal cv. Bigarreau Napoleon tree of same age. Courtesy B. Donini.

maintaining virus-free stocks.[164,359,360]

(iii) Role of genetics

As indicated earlier, gene heterozygosity and genome polyploidy, both of which complicate genetic segregations and the recovery of improved progeny in cross breeding programs, contribute favorably to success in mutation breeding programs. Because most mutations involve changes from the dominant to the recessive state of gene action, including changes from the dominant to the null condition as the result of genetic deletions, mutations of those loci in heterozygous genotypes will be most frequently observed. To a large extent, this is because mutational changes in each one of a pair of alleles occurs at the same frequency. Thus, changes from dominant to recessive, e.g. A — a, in a genotype that is already Aa will result in genetic homozygosity aa and the aa condition will be expressed. In contrast, a change of one allele A — a in a genotype that is homozygous AA will not be expressed if a single A locus is sufficiently dominant in its action to control or suppress the effect of the mutated a allele. For many gene loci, this simple concept surely pertains, but all gene loci do not interact in such a simple manner. Other genetic phenomena such as epistasis in which a gene at one locus controls the expression of another gene at a different locus is also common, though in principle the overall effects of mutations at the controlling locus sometimes may simulate the more simple situation. However, incomplete dominance in gene expression and polygenic inheritance of traits increases the number of possibilities for mutant gene expression.

Because greater gene heterozygosity increases the number of possible mutations readily expressed by a genotype mutation breeding provides the perfect complement to cross breeding when that means for developing new gene combinations is available. Thus, with mutation breeding a breeder now has the tools available to exploit fully each hard won advance made in cross breeding. An efficient use of mutation breeding is to induce at least those "sports" or mutations most likely to occur in a new genotype before it is placed on the market.[62,164] Ordinarily, with ornamental plants, especially, this can be accomplished in 1—2 years or less, or while the new improved genotype is being multiplied for the initial marketing.[164] Broertjes et al.[63] have demonstrated the success of a like scheme, using only mutation breeding with a recent cross bred *Chrysanthemum* cultivar.

The influence of polyploidy on the yield of mutations in vegetatively propagated plants varies considerably with the species, but in general, it can be said that increased ploidy increases the number of gene loci that potentially can be heterozygous, translating into a greater number of possible mutant traits recoverable via mutation breeding. This appears to be the case with *Chrysanthemum*, *Dahlia* and *Saccharum* (sugar cane) for example, as these species commonly carry considerable stored (heterozygous gene) variability. Cultivars may differ greatly with regard to the range of inducible variability, because they differ in genetic heterozygosity vs. homozygosity, and because many gene loci have an additive rather than dominance action. Additivity in gene expression, amplified in polyploids results in a range of expressions, as might be illustrated by flower color intensity or pigment concentration. Little is yet known about conditions for the induction of mutations in the rhizomatous iris,[220] but efficient practical methods should now be feasible using modern technology (see also[62]).

While polyploidy increases the possible number of traits that can be heterozygous and thus releaseable by mutagenesis, the more complicated inheritance of traits in polyploids may also make it extremely difficult to mutate characteristics controlled by completely dominant genes present in the duplex to quadruplex conditions. For that reason, if genetic variability for a trait exists, mutation breeding is best used to complement cross breeding. On the other hand there are some surprises, which illustrate that conclusions about the possibilities to recover mutations even from autoploids should not be made without actual experimental results. Thus, Broertjes[53] discovered that a colchicine-induced autotetraploid *Achimenes* cv. 'Tango' yielded 20 to 40 times more mutant variation than its diploid form cv. 'Tarantella'. Similar results were obtained with two other induced autotetraploid cultivars 'Repelsteeltje' and 'Little Beauty', but their corresponding diploids were not studied. The results of this study appear difficult to explain, at least on the basis of dominant genes. However, it is more likely that the mutant variation isolated is largely that controlled by genes with additive effects, though the colchicine treatment for induction of autoploidy was suspected as possibly causing some mutations. Autoploidy appears easily induced by colchicine using the adventitious bud technique.[57]

(iv) Types of mutations induced

As mutagenesis is a means for creating new forms of genetic variability, the potentially inducible variation obtainable may encompass far more than can be discussed in this overview. However, relationships among plant species, genera, and families can be found in their similar or parallel range of potential genetic variability.[143,145] This is the modern-day interpretation of Vavilov's[413] Law of Homologous Variation among species, discussed on p. 263. With our greater base of genetic knowledge, we now know that many gene loci may control a specific character expression or phenotype. This is not only because of allelism, it also relates to the biosynthetic process basis for individual traits, such that genotypes with a genetic block at any stage in a sequence of several biosynthetic steps leading to a character expression will often result in the same or similar phenotype.

(1) *FOLIAGE AND FLOWER COLORS, PATTERNS, FORM, SIZE*

Grouped in this manner, the traits considered here are largely those of interest in breeding of ornamental plants for which mutation breeding is already commonplace.

Mutation breeding has led to a great burst of flower color, form, pattern and other variation in *Chrysanthemum*, *Dahlia*, *Begonia*, *Streptocarpus*, *Achimenes* and *Alstroemeria*, with smaller numbers of mutants reported from mutation breeding of *Rosa* spp., *Rhododendron* (Azaelea), *Dianthus* (carnation), *Gladiolus*, *Tulipa*, *Lilium* and *Portulaca grandiflora*. Examples of mutants induced are listed in Broertjes and Harten[62] and in issues of Mutation Breeding Newsletters, IAEA, Vienna. Broertjes *et al.*[63] also have demonstrated the efficacy of sequential mutation breeding in *Chrysanthemum morifolium*, first inducing four marketable mutants differing largely in flower color in the cv. 'Horim'. Then, all four of the mutants were subjected to further mutation breeding. One unusual mutant induced, marketed as cv. 'Mikrop' also had sturdier and longer stems, as well as larger inflorescences. Further mutation breeding with the cv. Mikrop led to the isolation of a desirable flower color mutant marketed as cv. 'Miros'. Still further mutation breeding with many of the Miros mutants led to the production of a whole 'family' of mutants which encompassed a wider range of flower colors than had been obtained from the original stock (Fig. 9.4). Broertjes *et al.*[63] noted that the production of a comparable family of color sports without mutation breeding could have required from 20—30 years instead of the 1—2 years to produce each of the mutant families. In a program started in 1975, 244 promising mutants had been selected for further testing by 1979. Over half of this number were of the Miros types first irradiated in 1977. The Miros cultivars were expected to dominate the market by 1981.[63]

Fig. 9.4. *Chrysanthemum morifolium* cv. Miros 'family' of flower color mutants, including primary, secondary and tertiary, etc., induced from the light pink flowered
mutant Miros cultivar (lower left) which originated as an induced mutant of Mikrop,
itself an induced mutant of Pink Horim (see text). After Broertjes *et al.* 1980.

Two variegated leaf mutants induced ornamental vine *Bougainvillia* cvs. 'Jayalakshmi varie-
gata' and 'Arjuna' in India; two gamma ray induced streaked leaf mutants induced in
Polyanthes tuberosa, cvs. 'Rajat Rekha' and 'Swarna Rekha' were introduced into cultivation in India and a variegated leaf mutant of apple (Malus) has been introduced in the Netherlands.[292,293] A striped leaf mutant of *Guzmania paecockii* cv. 'Edith' was induced by gamma
rays and introduced in Belgium.[295]

Mutation breeding with a number of other species has failed so far or have not proved as
successful. To a greater extent, this appears to be due to insufficient development of methods,
but in some cases adventitious bud development is too time-consuming. With these species,
Hyacinth, Iris, Narcissous, Polyanthes, Ranunculus, Fuchsia, etc. *in vitro* propagation and

mutation breeding methods may be more successful.[62]

Mutation breeding of some other species, *Euphorbia pulcherrima* (*Poinsettia*), for example, is efficient and the adventitious bud technique is highly effective[61,241], but because all breeding is done by private firms, and all cultivars are protected by Breeders Rights laws, the actual number of mutant varieties produced is yet unknown.

Changes in leaf, or foliage form and color also occur frequently in irradiated vegetative progeny of *Saintpaulia, Begonia masoniana, Begonia elatior*.[62] However, so far relatively little effort has been invested to produce such variations in foliage-type ornamentals, such as *Hedera, Ficus, Coleus, Begonia rex, Philodendron, Marantha, Papyrus, Sansevieria* etc. even though spontaneous mutations have played an important role in the origin of the cultivars of such species as *Hedera* and *Begonia rex*.[62]

(2) FRUIT, ROOT OR TUBER COLOR, FORM, SEEDLESSNESS, FRUIT SETTING AND FLESH CHARACTERISTICS

Mutation breeding of root and tuber crops, and of fruit trees, and berry crops has been more limited but due in large extent to the longer generation times of some species. Also, for most of these crop species mutation breeding techniques are still under development and do not have the level of efficiency desirable. However, potato breeding by *in vivo* or *in vitro* culture methods, now has reached a level of sophistication requiring serious consideration as a complement to cross breeding programs everywhere. Techniques developed for mass scale plantlet production by Roest and Bokelmann[359,360] and Harten *et al.*[165] permit the ready induction and isolation of high frequencies of mutations, with a wide mutation spectrum and with few chimeras. Protoplast culture techniques developed by Wenzel *et al.*[430] and by Shephard *et al.*[381], also achieve high yields of mutations, even without the use of mutagens. The protoplast culture techniques might permit use of a wider variety of mutagens as ultraviolet light and chemicals such as methylnitrosourea in addition to X or gamma radiations to amplify the spectrum and increase the yield of mutations induced.

Much less work has been done with sweet potato *Ipomea batatas* and the tropical root, crops, yams (*Dioscorea* spp.), taro etc. An efficient adventitious bud technique has been developed for sweet potato[439] and mutant variations affecting skin, flesh color and flesh sugar content have been obtained, none of which had commercial value. With continued effort, especially screening of large populations, successful mutation breeding results with many of these species are expected.[62] In *Manihot esculenta* (cassava or tapioca) Vasudevan *et al.*[414] obtained mutants, including one with a higher starch but lower hydrocyanic acid content after irradiation of stem cuttings. Abraham[2] and Nayar[299] also obtained promising mutants after irradiating cassava cultivars. Kartha *et al.*[199] have developed *in vitro* propagation methods which should aid in developing more efficient mutation breeding applications with these species.

In *Citrus*, most research has been concerned with technique development, hence relatively few mutants have been induced and evaluated. However, a nearly seedless mutant grapefruit named 'Texas Star Ruby' was induced by exposure of the apomictic seeds from a deep red fleshed sport to thermal neutrons.[170] This mutant cultivar already is in wide use. Considering enormous base of stored genetic variability now available, especially in the many *Citrus* species hybrids, there appears to be considerable future for mutation breeding in these crops. In the stone fruits, the only fruit character mutants described to date have been nectarine (fuzzless skin)[100], and larger fruit and a deeper red skin color mutants in peach, *Prunus persicae*. The latter mutant was isolated after gamma field irradiation of the cv. Magnif 43 and introduced as cv. Magnif 135 in Argentina.[279] In grape, Donini[100] obtained a self-thinning (lower fruit setting) mutant, which is still under evaluation. With apples, however, technique development work has been conducted more widely, and as a result more fruit character mutants have been induced. Some of these have been marketed. Lapins *et al.*[233] and Lapins[229] for example, described an induced mutant of McIntosh apple with a brighter red skin and white flesh color, apparently resulting from a reduction in chlorophyll content of the fruit. This mutant was introduced to cultivation in 1970.[279] A partially self-compatible mutant of cv. 'Cox Orange' apple was obtained by Scott and Campbell.[377] Decourtye and Latin[295] have marketed in France a russetting free mutant cv. 'Lysgolden' induced from cv. 'Golden Delicious'. They also induced deeply colored, larger fruited mutant, cv. 'Belrene', from cv. 'Reine des reinettes' and obtained a mutant, cv. 'Blackjoin BA2–520', with improved and more regularly red colored fruit after irradiation of cv. 'Johnathan Blackjoin'.

Many mutants from apple, pears and other tree fruits from these and other programs are still under evaluation. Current research with apples in the UK is aimed at improving skin color, fertility and fruit size in cv. Cox Orange Pippin; uniformly shaped fruit in cv. 'Bramley's Seedling', reducing fruit size in cvs. 'Mutsu' and 'Tydemans Worcester', improving skin finish in cv. 'Kidd's Orange Red', and improving skin color in cv. 'Spartan'.[227] A large proportion of mutant variation induced in these plants is of negative value, but these often can be discarded early in a mutation breeding program. No fruit character mutants have been reported from the limited research with pears and most of the research with stone fruits has been concerned with other traits (mainly dwarfing). However, an apomictic mutant was obtained in a cherry species hybrid of the cultivated sour cherry X the wild Japanese bird cherry. The apomictic trait of the mutant was derived from the wild bird cherry, but released by mutation breeding of the improved hybrid, which is heterozygous for the trait. This mutant has the advantage of higher fruit set during inclement weather, since fruit set is not dependent on pollination.[447]

The self-fruited sweet cherry cultivar Stella and its compact mutant cv. Compact Stella attained their self compatability from a radiation induced self fertility mutant isolated by Lewis and Crowe[238] in basic research studies on the nature of the self incompatability in this species. The induced self compatability allele is now extensively used in sweet cherry breeding.[229]

Ease in identification of useful fruit character mutants in these and other fruits appears to be a major problem. Most visible traits are easily screened for, but those concerned with internal fruit quality, for example, are still discovered largely by chance. Even so, the long time between treatment and fruiting of a vegetative clone is an obstacle, as it is with cross bred materials.

Inadequate attention has been given to mutation breeding of small fruits — brambles (*Rubus* spp.), gooseberries and currants (*Ribes* spp.) blueberries and cranberries (*Vaccinium* spp.) and to strawberries (*Fragaria* X *ananassa*). In each of these crops, particularly the blackberries, and the black, purple and red raspberries and strawberries, the stored genetic variability must be considerable, and in many cases improvements in fruit characteristics are needed.

(3) *PLANT STRUCTURE AND FORM*

Either because the trait is among the most mutable or because similar effects result from mutation at any of several loci, economically useful compact, dwarf or semidwarf, spur-type mutants have been readily induced in a wide range of vegetatively propagated crops. Compact growth habit cultivars have been marketed in sweet cherry, *Prunus avium* cvs. 'Compact Lambert' and 'Compact Stella'[229,230,231] and in sour cherry, *Prunus cerasus* cv. 'Polukarlik

Orlovskoi Rennei', 'Polukarlik Turgenevki', both semidwarfs and cv. 'Karlik Samorodka' a more compact dwarf have been released. Two compact pomegranate (*Punica granatum*) cultivars 'Karabakh' and 'Khyrda' have been introduced in the USSR.[296]

Numerous compact mutants in cherries were induced by Lapins[229] Zagaya and Przybla[445] and Donini[100]; a compact apricot (*Prunus armeniaca* L.) was induced by Lapins[229]. Similar mutants also have been induced in apples by several workers[229,416,445] but none so far have been marketed. However, Visser[416] obtained an exceptional semi-compact mutant with smooth skinned, bright red fruits in cv. Belle de Boskoop apple, and found that a number of compact mutants yielded only slightly less than normal taller trees. Considering that more of the compact trees can be grown in a given area than normal tall trees, compact mutants will likely supplant many currently grown forms which are made compact by their growth-reducing rootstock. Many spur-type compact sports of major apple cultivars are already widely grown. Compact mutants also have been induced in pears[416], but so far as known none have yet been marketed. Attempts are being made to induce compact forms of olive.[99,100] Compact forms also were induced in *Achimenes*, leading to the marketing of cv. 'Cupido' induced in the diploid cv. 'Paul Arnold', and cvs. 'Flamingo' and 'Lollipop', induced in the autotetraploid cv. 'Tango' and cv. 'Pink Attraction' induced in an autotetraploid of cv. 'Repelsteeltje'.[295]

In *Streptocarpus* many induced mutants have changed flower color, but one cv. 'Juwel' has relatively small leaves and short flower stalks which make it more showy, cv. 'Burgund' induced in the mutant cv. Juwel has a changed flower color and compact plant stature; cv. Weisse Glocke (white clock) retained the white flower color but gained its compact growth habit via mutation and induced mutant cv. 'Albatros' has a more sturdy flower stalk and a comparatively compact growth habit. The X-ray induced mutant cv. 'Nicky' has a darker blue flower color than the original cv. 'Neptun' but is also more free flowering, and has a compact growth and shorter leaves.[292,294,295] The compact mutant of *Begonia* cv. Mini-Mini Iron[279] has been introduced in Japan, while an upright growing, more compact slow growing Begonia mutant, cv. Fantasy was induced in the mutant cv. Aphrodite Rose and released in the USA.[295]

Compact mutants would be valuable additions in many more ornamentals, especially in *Lilium* spp. *Euphorbia pulcherrima* (Poinsettia) *Alstroemeria* and *Rhododendron* spp., among others. Compact forms of evergreens also have considerable economic value. Kukimura *et al.*[225] have in fact induced dwarf forms of *Cryptomeria* (Japanese cedar). Some mutants also had 'waxless' leaves. The *Chrysanthemum* mutant cv. 'Mikrop' its mutant cv. 'Miros' and a derived family of further mutants have a markedly improved stem strength and larger inflorescences which contribute in large measure to their success.[63] Sturdy stemmed mutants also were induced in *Chrysanthemum* by other workers.[62]

A spineless mutant has been induced in pineapple (*Ananas comosus* L.) by Singh and Iyer[384] using the chemical mutagen nitrosomethyl urethane (NMU), and a possible thornless mutant was induced recently in red raspberry cv. 'Heritage' by Merriman.[255] Compact and thornless forms are needed in all *Rubus* species. Some spontaneous variation is available for both traits, which should permit a combined cross breeding — mutation breeding effort to produce the desired forms in a minimum of time.

In European black currant, *Ribes Nigrum* L., Bauer[28] produced an upright-growing form in the cultivar 'Westwick choice'. The mutant has been introduced as cv. 'Westra'.

With strawberry, forms with strong upright fruit clusters to facilitate the development of mechanical harvesting. An *in vitro* propagation technique has been developed[46,428] which may help to reduce difficulties associated with chimera formation. Mutations affecting morphological traits as well as other properties appear to be readily induced.[1,365,373]

In grasses, semidwarf and dwarf forms have been readily induced in the apomictic Kentucky bluegrass, *Poa pratensis*[329], and a wide diversity including highly vigorous mutants have been induced in the seed sterile species hybrid turf bermudagrass[75,78] and in triploid turf St. Augustine grass, *Stenotaphrum secundatum.*[328,330] One of the St. Augustine grass mutants had a faster rate of establishment and shorter internodes than the original cultivar 'Floratam'. All but one of the mutants reproduced for evaluation were found to retain their resistance to southern chinch bug (*Blissus insularis*) and all had resistance to *Panicum* mosaic virus equal to that of Floratam.[349] Mutants also have been induced in Dallis grass *Paspalum dilatatum* and Bahia grass *Paspalum notatum.*[329] The results already reported show that mutation breeding with the seed sterile species hybrid, triploid and apomictic grasses can produce considerable useful genetic variability for improvement of these species otherwise dependent on spontaneous mutations.

(4) CHEMICAL COMPOSITION

Among the most significant of induced mutations affecting the chemical composition of an industrial crop is the recently introduced mutant cultivar, 'Rose Mint' induced in the Japanese mint, *Mentha arvensis* var *piperascens* by Ono.[310] The normal species is extensively cultivated for the production of Menthol. Ono[310] irradiated tissue cultured callus at a time just prior to until just after the callus began to produce intermediates to menthol. Mutation frequencies in as high as 96% of the calli were obtained if irradiation was done after 85 days growth, while at other times of treatment virtually no mutants were induced. One mutant with larger leaves proved to produce an essential oil that was greatly different from that of the original cultivar, and is completely devoid of menthol. Surprisingly, the aroma of the mint oil is similar to rose oil and also has a similar composition established by gas chromatography.[310] Because the mutant has larger leaves (the oil glands are located at the tips of leaf hairs) the mutant yields about twice the amount of oil produced by the original menthol-producing cultivar. Moreover, about 50 kg of Rose Mint herbage yields as much oil as a million fragrant rose flowers. The oil has a market value twenty times that of the original cultivar.[293] Another essential oil quality mutant has been induced in cv. RRL–59 of *Cymbopogon flexuosus* (lemon grass) which is grown commercially in India for the production of methyl-engenol. The new mutant induced by irradiating vegetative slips is devoid of methyl-eugenol, but produces an oil which closely resembles Java type citronella oil and can substitute for it.[83]

In sugarcane, mutants with higher sugar yield have been induced.[183,184,340,341,418] Many promising mutants also have been isolated using mutagenized sugarcane tissue cultures.[167]

(5) PHYSIOLOGICAL TRAITS, ONTOGENETIC PATTERNS

A number of mutant cultivars of vegetatively propagated plants have been introduced to cultivation because of improvements that affect some aspect of growth patterns best described in terms of physiological properties, even though a high proportion of other mutant forms could also be so classified. Many of the mutant cultivars flower or ripen earlier than the cultivar from which they were obtained (Table 9.2). In the fruit crops early ripening and annual bearing are important traits because earlier marketing often provides a higher economic return. Thus, early maturing apple, peach, cherry and apricot mutants have been marketed, and others still are under evaluation.[62,229,262] .

TABLE 9.2. Examples of mutant varieties of vegetatively propagated crops with altered
flowering and/or ripening times

Species and variety	Country and year of release	Characteristics in comparison with parent variety
Achimenes		
Cupido	Netherlands 1973	freely flowering, compact, sturdy
Orion	Netherlands 1973	early flowering, large flower, sturdy but taller plant
Alstromeria		
Red Sunset	Netherlands 1979	extended duration of flowering period, improved flower color
Malus pumila		
Belrine	France 1970	earlier maturity, deeper colored and larger fruit, yield reduced
Mentha piperita (peppermint)		
Todd's Mitcham	USA 1971	5–10 days earlier, resistant to *Verticillium*
Murray Mitcham	USA 1977	5–10 days earlier, resistant to *Verticillium*, plants branch from base more freely than Todd's Mitcham
Prunus armeniaca (apricot)		
Early Blenheim	Canada 1970	1 week earlier ripening, bears fruit every year
Prunus avium (sweet cherry)		
Compact Lambert	Canada 1972	very early, high yield, compact type
Prunus persicae (peach)		
Magnif 135	Argentina 1968	1 day earlier ripening, larger fruits
Streptocarpus		
Margaret	United Kingdom 1974	all year flowering
Nicky	Fed. Rep. Germany 1979	free flowering, darker blue flower color, compact growth habit and shorter leaves
Snow White	Netherlands 1973	compact, freely flowering
Mini Nymph	Netherlands 1969	free flowering
Achimenes		
Early Arnold	Netherlands 1971	1–2 weeks earlier flowering
Spring Time	Netherlands 1971	1–2 weeks earlier flowering
Populus Trichocarpa	Ukranian SSR 1977	more vigorous growth – 1.5 to 1.8 m annual growth under dry, over 3 m under wet conditions

Adapted from: A. Micke, *Use of mutation induction to alter the ontogenetic pattern of crop plants*, Gamma-Field Symposia No. 18, *Crop Improvement by Induced Mutation*, 1979 and *Mutation Breeding Newsletters* 13–18, 1979–1980.

With ornamentals, early flowering mutants have been introduced, and perhaps more important, mutants that flower more freely or all year round such as the *Streptocarpus* mutant 'Margaret' represent breakthroughs in ornamental crop improvement. In mulberry, mutants with a later time of initiation of spring growth have been induced in Japan.[225] The later 'sprouting' of the mutants may give them greater protection from frost injury in northern areas. Mutants with higher leaf yields due to changes in leaf shape, leaf thickness and number of leaves per branch have been induced in modern cultivars used for silkworm feeding in Japan.[297] Some of these mutants may carry tetraploid tissues.[225] Retreatment of mutant strains of mulberry also appears promising as a means to achieve additive improvements.[128]

Rapid growth is also an important trait in fiber crops, hence the new variegated leaf mutant of *Populus* introduced in the USSR is able to make more growth either in dry or wet conditions than the original cultivar.[293]

In wine grapes, *Vitis vinifera*, an induced mutant of cv. Perle with greater tolerance to low temperature than other cultivars is cultivated on a small scale in Germany.[51] Tifway 2 turf bermudagrass, recently released in the USA has greater frost tolerance than the original Tifway, has better quality and produces a more weed-free turf.[76]

In sugarcane, a great many mutants with altered vegetative growth features have been reported, including some that do not flower under growing conditions in India. The sugar yield is said to be significantly increased.[184,341,418] Another mutant reportedly has increased vigor and such accelerated development, that it can be harvested a month earlier and still produce a higher sugar yield than the original cultivar.[184] A major problem in past sugarcane mutation breeding has been the instability of mutant clones due to their chimeral structure. Difficulties encountered stabilizing mutants may have delayed their development as cultivars. However, highly efficient tissue culture techniques have been developed, and successfully applied in mutation breeding with this species.[168,169,244]

There is thus great promise that mutation breeding in sugarcane using the new culture techniques will make important contributions in the near future. In white potato (*Solanum tuberosum*) mutation breeding experiments have successfully produced photoperiod insensitive, earlier maturing clones permitting potato culture in regions of India, for example, in which the normal cultivars are not adapted.[207,411] Tissue and protoplast culture techniques, have also been applied successfully[381] to yield photoinsensitive clones of the commercially important Burbank potato. These clones are able to produce 100 fold more flowers and berries than the original clone and thus should prove exceptionally useful in cross breeding programs. Considerable variation in regenerated white potato protoclones has been obtained even without use of mutagens, though with the new culture systems, treatments even with chemical mutagens might prove efficient and easily employed. Maliga[246] notes that in tissue cultures of other plants mutagens typically induce 4 or more times the spontaneous rate of mutations.

A wide range of variation in the earliness of tuber setting also has been found among the Burbank protoclones. Considering the fact that other protoclones recovered have improved resistance either to the early or late blight diseases, and still other variants had more uniform tuber production, the basis now seems established for sequential improvement of potatoes, using protoplast culture with or without the application of artificial mutagens.

(6) DISEASE AND PEST RESISTANCE

While the mutagenizing techniques applied for the induction of disease resistance mutations do not differ notably from those applied for the induction of other traits, screening methods

applied may be similar for most species. Moreover, the economic importance of disease resistance, and the significant successes already obtained merit a special consideration in this review.

The classical example of disease resistance induction in a vegetatively propagated crop is that
of *Verticillium* wilt resistance in peppermint *Mentha piperita*. Two important cultivars Todd
Mitcham (Fig. 9.5) and Murray Mitcham[276,404] resulted from the work of Murray[274,275],

Fig. 9.5. Increase field of mutant peppermint cultivar Todd's Mitcham, compared to
the original Mitcham. Scattered surviving plants are not mutants, and will eventually
succumb to *Verticillium* disease. Courtesy M.J. Murray and W. Todd, A.M. Todd
Company.

initiated as a result of frustration over the failure of more than ten years of prior effort
attempting to recover essential oil quality in *Verticillium* wilt resistant lines, developed by
cross-breeding with the resistant native spearmint *M. spicata*. The commercial peppermint is
a pollen sterile tetraploid; the lack of seed formation may contribute to increased herbage and
thus oil yield, because the oil glands are located at the tips of hairs on the leaves. The *Verticillium* wilt disease had become so severe that production could be maintained only by shifting
new plantings to soils that had not grown mint before and it appeared impossible to avoid
transfer of the disease to new plantings. Consequently, Murray initiated a mass scale effort to
induce wilt resistance by irradiating millions of rhizome pieces.

The irradiated rhizomes were planted in wilt infected soil, and screened for wilt infection
over a period of years. Several lots of rhizomes were treated, resulting in the discovery that the
most promising selections came from mutagen treatments that were severe enough to kill all
exsiting buds and from which adventitious buds were able to regenerate after a long waiting
period. After evaluating over 50 promising selections, the two cultivars mentioned above have
been released. These cultivars are similar in that they not only have high disease resistance but
also the required properties of their essential oil. These new mutants have rescued mint production in the traditional areas (Murray & Todd 1975).

The success of this venture with peppermint stimulated efforts to induce resistance to *Verticillium* wilt in the similarly pollen sterile cultivated spearmint *M. cardiaca* var. Scotch.[176] In this new work, rhizome pieces were severely irradiated, and adventitious buds allowed to develop in part under greenhouse protected conditions and in part in the field prior to screening for wilt resistance. The greenhouse grown material produced 30 resistant selections while 13 selections were recovered from the field material after two years tests. All 43 selections are now undergoing replicated field tests and evaluation of essential oil quality (Green, personal communication 1981).

Attempts to induce rust *Puccinia mentha* resistance in *M. cardiaca* cv. Scotch also were made and 21 selections immune to rust were isolated from the field plantings. All selections have excellent vigor and winter hardiness. These 21 selections were tested in Washington State and in Oregon. Eleven of the 21 proved immune to rust in Washington, but only 6 were resistant in Oregon. Moreover, the rust fungus apparently has proved to have enough genetic variation so that at the present time all 21 formerly immune selections are now showing rust (Skotland and Green, personal communication 1981). This appears to be another classic example of the ephemeral nature of race specific disease resistance, for which selection is easy and attractive to breeders and plant pathologists. According to Green (personal communication, 1981) neither the *M. piperita* nor *M. spicata* cultivars in production are severely damaged by the mint rust disease. Regrettably, in the experiments completed so far, *M. cardiaca* variants showing reduced susceptibility or partial resistance rather than immunity were not saved from the mutagen treated material. Consequently, it would appear desirable to repeat the experiments using a more vigorous mutant as the initial material, aiming this time at isolating variants with lower rust readings rather than immunity in order to obtain a resistance that simulates that present in other related species.

Disease resistance appears to be readily induced also in apple cultivars. Lapins[229] induced variants of McIntosh apple with resistance to mildew (*Podosphaera leucotrica*) and to apple scab (*Venturia inaequalis*) one of which has been released.[62,260] Similarly, Campbell and Wilson[80] report variation in mildew infection of Cox Orange Pippin apple mutant clones. In European black currant, *Ribes nigrum*, Bauer[28] observed that about 2% of vegetative clones obtained following radiation treatment were less infected with rust (*Cronartium ribicola*) than the original cultivar 'Westwick Choice'. In mulberry, Nakajima[297] has induced resistance to Dogare disease (*Diaporthe nomurai*).

In sugar cane, Rao *et al.*[342] have induced resistance to red rot (*Physalospora tucumanensis*, spp. *Colletotrichum falcatum*) in the clone Co449 after gamma irradiation of stem cuttings. One of the mutants, was released as Co6602, red rot resistant mutants of another clone, Co997 are being tested widely in India, and new mutants have been induced in clones Co312 and Co527. The resistances of Co6602 and Co997 have proved stable and have not become susceptible after many generations of asexual propagation.[183] Haq *et al.*[161] in Burma have isolated red rot resistance in the sugarcane clone Co419. This same clone was treated without success in the experiments of Jagathesan *et al.*[183]. In Bangladesh, Haq *et al.*[160] obtained 9 mutants of Co633 and Co527 with various degrees of resistance to red rot, after gamma radiation. One of the mutants, Co527–85 was found resistant both in the field and laboratory. This mutant had the same sugar recovery percentage as the original clone. Darmodjo[89] in Java also using gamma rays, induced resistance to sugarcane mosaic (Kebonagung strain) in the cultivar POJ3016.

Gamma radiation of stem cuttings also was successfully used in India by Jagathesan[182] to induce resistance to smut (*Ustilago scitaminea*) in the high yielding clone C0740. The induced smut resistant mutants appear to be more like the original clone and appear to be better yielding than resistant selections developed by cross breeding.[182] Smut resistant mutants of sugar cane also have been induced in the clone C0528 from Pakistan and in varieties F146, F173 and F177 from Taiwan, whereas a mutant with improved resistance to downy mildew (*Sclerospora sacchari*) has been induced in the variety F160 in Taiwan.[182]

Heinz[167] using mutagenized tissue cultures has isolated an abundance of mutants resistant to *Helminthosporium sacchari* toxin in culture and to the fungus in the field. Resistant mutants lack the ability to bind the *H. sacchari* toxin protein.[391] Resistance to the stem borer *Procerus indicus* has been reportedly induced in an ethylmethanasulfonate (EMS) mutagenized sugarcane clone in India[12] and Darmodjo and Wirioatmodjo[89] using gamma radiation induced resistance in sugarcane to the top borer *Scirpophaga nivella*. Burton[76] has reported the release of Tifway 2 bermudagrass, a gamma ray-induced mutant with greater resistance to root knot, ring and sting nematodes than the original Tifway. In bermudagrass a sterile interspecific hybrid of *Cynodon dactylon* and *C. transvaalensis*, Burton and Hanna[78] and Burton[74] have reported the induction of mutants with resistance as well as mutants with tolerance to root-knot nematode, *Meloidogyne graminis*.

Shephard *et al.*[381] in a brilliant demonstration of new technology have recovered resistance to late blight (*Phytophthera infestans*) in protoclones from protoplast cultures of the Russet Burbank potato. Other protoclones appear to have increased resistance to early blight (*Alternaria solani*). The range of morphological and physiological trait variability in protoclones was unexpectedly high without using mutagens suggesting an influence of the culturing conditions on genetic variation.

Resistance to 'Dutch elm disease' of the American elm, *Ulmus Americana*, has been obtained by Canadian workers using an *in vitro* culture system.[109,110,111,334] The new resistant trees may permit the replanting of this exceptionally beautiful, tall shade tree, in many places where it was destroyed by disease. With this success, it would appear that a similar program to induce resistance to chestnut blight in the native American chestnut would now be feasible, especially since introduced species, such as the Chinese chestnut are resistant.

The examples presented above indicate there is considerable potential for the induction of disease resistance in vegetatively propagated crop species by mutagens, mainly radiation treatment. The use of cell and protoplast culture systems rapidly coming into practice should broaden the scope for mutagen applications and allows for better control over conditions for induction and selection of mutants.

(v) Use of induced asexually propagated mutants in further breeding

Experience using induced mutant clones as germ plasm for further breeding is yet limited, but for newly arisen variations, or useful combinations of traits, attempts to recover such traits in cross progeny appear to be worthwhile. Many of the mutant traits should be recoverable in progeny although special methods may be necessary when the mutant trait arises as a periclinal chimera. Results reported to date indicate that the 'upright growth' trait in *Ribes nigrum* L. cv. Westra is recoverable in progeny[28] as is the compact growth habit of some but not all such mutants in apples.[229,232,309] Many vegetative mutants also may be due to genetic deficiencies. These chromatin deficiencies may have reduced gametic transmission; on the other hand, if the initial material was heterozygous the genetic deficiency may favorably affect transmission of normal recessive alleles the mutant deficiency has exposed. Because vegetative mutants may occur as periclinal chimeras, and because generative tissues develop from the subepidermal layer of cells, it may be necessary to assure that mutants intended for use in cross breeding are nonchimeric by first propagating them using adventitious bud or cell culture methods.[99,294] However, vegetatively propagated mutants originating from adventitious buds, are generally complete and are more likely to transmit their improved trait to their progeny than mutants isolated from irradiation of multicellular meristem tissues.[62]

Induced mutants may have exceptional value for further mutation breeding. Broertjes et al.[63] obtained useful new mutant flower color and other variation in *Chrysanthemum morifolium* cv. Horim after irradiating each of a series of successive induced mutants. One of the commercialized secondary mutants cv. 'Miros' not only had a changed flower color but also sturdier stems and larger inflorescences. Further irradiation of this mutant has led to the commercialization of a whole new family of mutants expected to displace the original Horim types because of their overall improved characteristics. It is thus logical to exploit the plant improvements made by mutation breeding methods through cycles of mutation breeding, as well as in cross breeding.

SEED PROPAGATED SEXUALLY REPRODUCED SPECIES

The major portion of induced mutants now available have come from more theoretically oriented research on mechanisms of mutation induction, methods of mutagen application, range of induced variability, etc. More recently, the number of strictly practical applications has increased, even though refinements in methodology are still needed. In most situations, mutation breeding should be looked upon by the breeder as one of the viable options available for creating genetic variability, much as in crossbreeding. Each method has advantages and limitations. The range of variability released via each method is of course genotype-dependent; with the mutation method it is also mutagen and to some extent mutagen dose dependent; whereas with crossbreeding, the relatedness of parents is a determinant of the range of genetic variability released in progeny.

Whether to use cross breeding or mutation breeding or both is a choice now more available to the breeder than ever before, depending largely on the efficiency of screening methods and to some extent on the requirement for germplasm not otherwise available in plant genetic resource collections. However, having a mutation or trait in a suitable genetic background may also be important, so as to eliminate those of highly unadaptive types in breeding programs. In most cases, mutation breeding methods should be considered as complementary to cross-breeding methods, and mutants complementary to other germplasm sources. However, it has become increasingly clear that for most traits, the genetic variation available in current germplasm resource collections, including induced mutants represent only a portion of the potential genetic variation, which can include genes at other loci as well as alleles with similar phenotypic expression.[143,145] Thus, breeders often may find it advantageous to simultaneously exploit both crossbreeding and mutation breeding methods toward selected specific goals.

A significant number of crop cultivars have been developed via direct mutation breeding, demonstrating the capability of the technique. In the major cereals alone, Micke and Donini[263] listed 26 rice, 21 barley, 11 common wheat, 5 durum wheat, 4 oats, 1 rye and 1 pearl millet cultivar developed from the direct use of induced mutants (Table 9.3). Yet, only a few induced mutants with traits of potential value for crop improvement ever become cultivars. Thus, nearly

TABLE 9.3. Number of released varieties of seed propagated crops developed through
induced mutation

Crops	Total varieties No.	Direct mutant	Mutant used in cross No.	Varieties developed through crosses with mutant %
Cereal: barley, bread and durum wheat, rice, rye, oat, pearl millet, maize	137	69	68	49.6
Proteoleaginous: soybean, peanut, lupine, sunflower, castor bean	22	15	7	31.8
Fiber: flax, cotton, jute	8	8	–	–
Vegetable: onion, lettuce, spinach, pepper, bean, pea, tomato	26	20	6	23.0
Other crops: mustard, clover, tobacco, sesame, etc.	20	14	6	30.0
	213	126	87	40.8

After Micke and Donini 1981.

TABLE 9.4. Period of release of varieties in seed propagated crops developed through
induced mutation

Period of release	Method of breeding				
	Release of varieties No.	%	Direct mutant No.	Mutant used in cross No.	Varieties developed through cross with mutant %
Before 1954	5	2.4	5	–	0
1955–60	11	5.3	7	4	36.3
1961–65	20	9.6	15	5	25.0
1966–70	50	24.1	34	16	32.0
1971–75	53	25.6	27	26	49.0
1976–80	68	33.0	29	39	57.3
	207	100	117	90	43.5

After Micke and Donini 1981.

TABLE 9.5. Improved characters in mutant varieties

Improved characters	Cereal[1]		Other crops[2]	
	No.	%	No.	%
Plant morphology				
reduced plant height	88	64.3	7	9.2
stiff straw, lodging resistance	67	48.9	6	7.9
Increased earliness				
5—10 days	29	21.2	13	17.1
10 days	3	2.2	1	1.3
Increased yield				
3—10%	48	35.0	33	43.3
10%	16	11.7	8	10.5
Seed characteristics				
morphology (size, color)	11	8.0	2	2.6
quality (protein, oil, malting, baking)	47	34.3	8	10.5
Fruit, fiber and leaf				
morphology	—	—	1	1.3
quality	—	—	10	13.1
Resistance to disease				
fungi	34	24.8	7	9.2
bacteria	—	—	2	2.6
virus	—	—	2	2.6
others (nematode, insect)	—	—	3	3.9
Other characters				
adaptability	11	8.0	4	5.3
threshability	2	1.4	2	2.6
easy harvesting	—	—	7	9.2

1) total 137 varieties
2) total 76 varieties

After Micke and Donini 1981.

as many cultivars of these crops have been developed by crossbreeding using induced mutants. These results are much the same as for variability produced by crossbreeding. Any new trait combination, including induced changes from undesirable-desirable to desirable-desirable trait associations or gene complexes, may prove useful as germplasm. As with any other new methodology, successful applications leading directly to cultivar development have increased with time, as have the uses of mutants in crossbreeding to produce new cultivars (Table 9.4). The use of mutants as germplasm has become increasingly important in breeding, often effecting major changes in crop plant morphology or physiology.

(i) Mutant traits may have a major role in new cultivars

A wealth of new traits or combinations is readily obtainable via application of mutagens to already improved material, and mutation breeders often may complement their own working germplasm collections with unique stocks. Cultivars developed using induced mutants may carry a wide range of improved traits (Table 9.5). The wealth of new mutant genetic variation already available and inducible by mutagens has encouraged Davies[90] to develop new conceptual models for crop plants and suggest ways the concepts might prove useful in plant breeding. Adams[3] formulated a concept of plant architecture for field beans, and then sought genetic variation from which to construct model prototypes.[4] Blixt[32] has demonstrated use of a computer data bank in breeding peas. Induced mutant cultivars already have proved to be outstanding parents for further cultivar development (Table 9.4, Fig. a—f). In fact, an induced mutant trait often breaks through an important limitation of previous cultivars, hence may become a fixed component of many future cultivars even though the original mutant form may not be directly usable.

Such was the case for the early determinate bush habit trait incorporated in a series of drybean cultivars beginning with the *Phaseolus vulgaris* cv. Sanilac and ending with the cv. Seafarer. The induced mutant germplasm source of the early maturing bush habit trait was too abnormal for direct use, but provided the germplasm necessary for improving plant structure (Anderson, personal communication 1981). The success of this series of cultivars and especially the wide adaptability of Seafarer has more recently led Adams[3] to use the approach of factor analysis to develop new ideotype concepts of plant architecture necessary for maximizing physiological efficiency of field beans grown in monoculture. Because a critical genetic component of the ideotype, a relatively non-branching, or low basal branching but tall rather than bush plant form was unknown in *Phaseolus* germplasm,[4] mutation studies were initiated to induce the necessary mutation in locally adapted germplasm and several non-branching plants were in fact obtained.

A cross made between one of the mutants from Seafarer and a tropical black bean cv. Black Turtle Soup (BTS) led to the development of a tall but weak stemmed, indeterminate, but non-viny 6—8 podded plant about 80 cm tall identified as line # 61319. At the same time, an EMS-induced white seeded mutant from the tropical black bean cv. San Fernando also was crossed with BTS to combine its longer linear phase of seed filling and high yield potential traits with the many noded, strong basal stem traits of BTS. An improved, white seeded plant from this cross was selected in F_3 and used in crosses with line 61319. Selections from this final combination led to the isolation in 1979 of six recombinant lines currently undergoing yield tests. The new lines combine the features of a tall, erect, narrow plant profile, having a strong central axis with few erect basal branches, superior lodging resistance and excellent pod development, all apparently contributing to their exceptionally high yield performance.

The new plant type has pods with larger numbers of seeds than was obtained with cv. Seafarer. One of the 6 lines has proved to be the highest yielding entry in all Michigan tests.[4] Yet this new archetype may not incorporate all of the features considered desirable by the breeder and proved via field trials. Neither is it known whether the new plant type itself provides the major input to physiological efficiency or if these features were obtained from the BTS and NEP-2 germplasm. It is evident nevertheless that the improved mechanical harvesting ability of the new type is a breakthrough, and it is quite likely that this form will become the basis for future monoculture-grown field bean cultivars world wide.

In the tomato, *Lycopersicon esculentum*, induced mutants continue to play an important role alongside natural variation and germplasm from alien species in developing the crop for its many uses.[6,354] Many mutant germplasm stocks already have been induced in cultivated as well as the wild *L. pimpinellifolium* tomato.[394,395] A large number of wild forms including *L. peruvianum* and other derivatives of wild X cultivated tomato crosses also can now be exploited to complement the available cultivated tomato germplasm in cross breeding programs.[353] However, mutants have been induced in the cultivated tomato to provide unique germplasm previously unavailable in genetic resource collections. Alexander *et al.*[6] for example, were able to induce mutants for short style length, as needed to assure adequate pollination especially in greenhouse grown tomatoes. In addition they sought and obtained non-branching, single-stemmed mutants for exploitation in developing tomatoes better adapted for machine harvesting, as well as for developing staked tomatoes for culture out-of-doors or in greenhouses. They envisioned developing 30–40 cm tall tomato plants with a single uniformly maturing fruit cluster for mechanical harvesting. The plants could be grown at a higher density, and the terminal fruit cluster would hold the fruit away from soil to reduce losses from rot, and the uniform ripening would permit one-time harvesting. The phenomenal success recently of protoplast culture and regeneration techniques in the white potato by Wenzel *et al.*[430] and Shephard *et al.*[381] also suggests that protoplast culture methods, complemented by the use of mutagens, should greatly enhance opportunities to exploit more of the potential of this genetically plastic and economically very important food crop.

In the snapdragon, *Antirrhinum majus*, the mutant *eramosa* has a strongly dominant apical meristem, but the original mutant has abnormal flowers and is a relatively weak plant[393] of no direct use. However, crossbred descendants with normal flowers now provide home gardeners and commercial florists with lovely long flower spikes. Two modern cultivar derivatives are 'Vulkan' and 'Goldina' (F. Scholz, personal communication 1981). Also, the recently introduced family of snapdragon cultivars introduced as 'butterfly snaps' cvs. Bright Butterflies, Madam Butterfly and Little Darling, etc. owes its origin to a combination of three induced mutant genes from *A. divaricata* which together control the wide, open flower trait[285] (Goldsmith, personal communication 1981). Stubbe[392] has shown with crosses of mutants that the number of anthers per flower can be reduced to two, or increased to eight in response to selection. The wealth of *Antirrhinum* mutants induced and described by Stubbe[393,396] and by Harte[163] merit fuller evaluation by commercial plant breeders not only for their potential exploitation, but to serve as examples of mutants reinducible in modern successful cultivars.

A few additional examples will show further how the use of induced (or spontaneous) mutants as germplasm has changed or is currently changing in a major way the structure, properties, or cultivation practice of a plant species thus demonstrates the great potential of mutants as germplasm for crossbreeding. Among them:

(1) **Early ripening mutants.** In barley, the Svalof cultivar Mari was developed in 1960 by direct use of an induced early maturing mutant. Its use extended safe barley cultivation farther north than ever was possible before, and the mutant is a key parent in derivatives improved for mildew resistance and other traits, including combinations with other mutants[143,144,145] (Table 9.6). Even earlier maturing recombinants have been obtained by crossing with other early barleys[208] and more early maturing induced mutants have been obtained as well. Mari barley is the basis for an early maturing barley improvement program also in Iceland (Sigurbjornsson, cited by Gustafsson[143]).

In rice, induced earliness mutants already have been exploited by crossbreeding to develop new varieties with superior combinations of traits useful in further breeding. The early-maturing

TABLE 9.6. Use of induced mutants in the development of short stature and/or early maturing, high yielding barleys by crossbreeding. Courtesy A. Micke, T. Kawai, FAO/IAEA, Vienna

a)	Origin	b)	Origin
Bonus	Sweden	Bonus	Sweden
----- X-ray		----- X-ray	
Mari	Sweden 1962	Pallas	Sweden 1960
----- crossbreeding		----- crossbreeding	
— Kristina	Sweden 1969	— Hellas	Sweden 1967
— Mona	Sweden 1970	— Visir	Sweden 1970
— Eva	Sweden 1972	— Rupal	Sweden 1972
— Salve	Sweden 1974	— Senat	Sweden 1974
— Eero	Finland 1975	— Atlanta	Canada 1977
— Stange	Norway 1978		

c)	Origin	d)	Origin
Valticky	CSSR	Maythorpe	UK
----- X-ray		----- γ-radiation	
Diamant	CSSR 1965	Miln's Golden Promise	UK 1966
----- crossbreeding		----- crossbreeding	
— Ametyst	CSSR 1972	— Midas	UK 1970
— Trumpf	GDR 1973	— Goldspear	UK 1975
— Nadja	GDR 1973	— Minak	UK 1976
— Hana	CSSR 1973	— Jupiter	UK 1976
— Favorit	CSSR 1973	— Goldmaker	UK 1976
— Rapid	CSSR 1976		
— Atlas	CSSR 1976		
— Spartan	CSSR 1977		
— Diabas	CSSR 1977		
— Safir	CSSR 1978		

medium grain Japonica cultivar M—101 released in California has both a semidwarf mutant and a gamma ray induced early-flowering mutant D31 in its pedigree.[363,364] Early mutants of *Japonica* rice cv. Koshihikari 'R—151' and Mutant of Koshihikari have been intensively used for crossbreeding in Japan, resulting in the release in 1977 of the extremely early high grain quality cultivar Fujihikari.[372] Induced earliness mutants in locally adapted germplasm are now widely used for crossbreeding also in Hungary[264], in Indonesia[181], in the Peoples Republic of China[266], Taiwan[178], Pakistan[16], India[129,362], Bangladesh[159], or essentially wherever rice is grown, and in both *Japonica* and *Indica* types[261,262].

In soybeans, earliness is particularly important for successful culture in northern areas. Zacharias[444] using X-irradiation was able to induce mutants of soybeans with 5—7 days earlier maturity, thus improving the germplasm base for breeding higher yielding, more reliable soybean cultivars adapted to northern European conditions. Kwon *et al.*[226] were able to induce a mutant, now released as cv. 'KEX—2' which ripens 11 days earlier and produces higher yields than other comparable varieties in Korea. Two early maturing soybean varieties, cv. Raiko and Raiden (both have the same gamma irradiated material) have been released in Japan.[180] The mutants are respectively 15 and 25 days earlier than the original cv. Nemashirazu. Both outyield other varieties in their maturity class. These new varieties provide markedly improved germplasm for breeding, since their yield/maturity relationship has been markedly improved.

In chickpea, late maturity has limited the expansion of this valuable crop into more northern areas. An induced early mutant M669, ripens in 145 vs. 155 days and due to its altered plant type, the mutant has a higher pod and seed set and a higher yield.[379] Mutants of this type should prove of exceptional value as germplasm for crossbreeding. In lupins, Porsche[326] induced an abundance of earlier flowering mutants in *L. albus*, including one mutant that did not respond to increasing temperatures. In crosses, with the mutant, transgressive segregates were obtained that were as much as 20 days earlier ripening and higher yielding, but did not flower earlier than the parents. Early ripening mutants from *L. angustifolius* and *L. cosentinii* were selected for use in crossbreeding in Australia[136] and an early maturing mutant of *L. cosentinii* cv. Eregulla has been released.[11,261,262]

In cotton, early flowering mutants are expected to permit expansion of good quality locally adapted cotton varieties into more northern areas. Mutants of the cv. MCU5 were obtained that flower in 60—80 vs. 120 days in New Delhi. The mutants can be harvested in 160 days whereas capsules of the original cultivar did not even open due to the low temperatures during their later maturation period.[344] One of the mutants released as cv. Rasmi, is daylength tolerant, high yielding and of high quality. Another mutant cultivar Pusa Ageti was induced in the USA, cv. Stoneville 213 matures in 150 days.[346] These mutants provide exceptionally valuable new germplasm for use in further breeding, illustrating some of the potentials for use of earliness mutants in crossbreeding.

(2) **Erucic acid**. Induced mutants for low erucic acid as well as naturally occurring variation in rape-seed, *Brassica napus* L. and *B. campestris* L., have revitalized the cultivation of this species as an edible oil crop adapted to cool northern climates where other oil crops are not successful.[19] Total elimination of the toxic erucic and eicosenoic acids is being sought through the production and use as germplasm of additional mutants.[356,357] Jonsson[189] has already identified multiple alleles at 2 loci in rape and at locus in turnip rape. The alleles show additive and dominance effects for erucic acid content and dominance for high ecosinoic acid content. Once recessive allele results in zero erucic acid. Toxic glucosinolates present in the seed also limit the use of the rape-seed oil meal for animal feed. However, low glucoinolate variants have been identified[190] and can be induced.[357]

In contrast, other breeders are exploiting the available germplasm to achieve high erucic acid strains for production of an industrial oil.[17,79] It should also be possible to induce mutants to further improve the efficiency of rape-seed oil for use as a renewable vegetable oil substitute for diesel engine fuel. Peterson *et al.*[317,318] have shown that high erucic acid rape-seed oil could be substituted as much as 70% for diesel fuel in powering diesel engines with no evident short term or long term negative effects. While the cost of vegetable oil is still too high in comparison with fossil fuels, the cost differential could quickly disappear. The net energy return from a rape-seed crop already shows an advantage for this crop, but sunflower might also have potential since the crop is more widely adapted than rape. High erucic acid safflower oil already exists, but could be improved.

A mutant of flax, (linseed, *Linum usitatissimum*) with increased oil content and degree of unsaturation was induced by Srinivasachar *et al.*[390] using EMS and gamma rays. This mutant can be used to convert the linseed oil to an edible type clearly demonstrating that changes in oil content and quality are inducible and useful for modifying crop quality for almost any intended use.

(3) **Leaf and plant structure modifications in peas.** A small number of *Pisum* cultivars have been directly produced via mutation breeding,[263] and several almost grotesque mutant traits such as 'leafless', 'semileafless', and 'fasciated stem' from both spontaneous and induced sources are now in extensive use in *Pisum* breeding programs over the world. Use of these mutants is expected to result in a marked change in the appearance and agronomic performance capabilities of future pea cultivars.[33,91,388] Some of the more attractive features of the semileafless, highly tendrilled peas are their standing ability for easier harvest, reduced foliar disease sensitivity and yield capacity equivalent to standard leafy peas[91] (and F.J. Muehlbauer and J. Kraft, personal communication). An induced mutant of the breeding line 5/2 carrying the tendrilled trait has been released as a cultivar (named Wasata) in Poland.[293] A tendril type cross derivative, named Hamil was released in 1981. This mutant is more lodging resistant and suitable for combine harvest.[296]

A semileafless (af) cultivar processing (freezing) and garden pea named 'Novelle' derived from crossbreeding has been released recently by Rogers Brothers Seed Co., Idaho Falls, Idaho. Another semileafless pea named 'Lacy Lady' was released in New York State and numerous other cultivars can be expected soon from the numerous programs now exploiting mutants at the *af* and *st* loci. (J. Marx, personal communication 1981). Three tendrilled cultivars (Wensum, Filby and Barton) have been released recently in the UK.[91] A semileafless cultivar named Brasilia also has been released in Brazil by EMBRAPA/CNPH (Andeoli, personal communication 1981).

(4) **Proanthocyanidin-free mutants.** These have been induced in barley to prevent the formation of chill-haze in beer produced from the barley malt.[432] Use of the mutant barley renders superfluous the requirement of chemicals to stabilize beer, resulting in considerable economic saving to brewers. The 'anthocyaninless' mutants are readily induced, and easily selected in a first screen by looking for anthocyanin less (green) plants as the red pigment becomes evident on the normal barley spikes in the M_2 field population. Laboratory tests are then employed to distinguish those mutants with the appropriate genetic block in the biosynthesis of catechins and proanthocyanidins. A wide scale effort is now in progress to induce similar mutants in all major malting barley cultivars (Nilan *et al.*, personal communication 1981). It is likely that this trait will become fixed as a necessity in future malting barleys because of the significant benefit to brewers, the saving of chemical and energy costs as well as eliminating a likely pollutant.

(5) **Low alkaloid, non-shattering, early.** These and other spontaneous and induced mutants of *Lupinus angustifolius* L., *L. albus* L., and *L. luteus* L. have made it possible to resume the

cultivation of these species and the development of genetic improvement programs in Europe and Australia.[136] Similar mutants have been induced in *Lupinus mutabilis* as a base for breeding to bring about cultivation of this species as a high protein and oil seed plant.[312]

(6) **A long, high temperature insensitive vegetative phase.** This is important for achieving high yields of leafy vegetable species such as spinach. From EMS treatments, Handke[151] isolated a mutant with a long vegetative phase, and with ability to produce a good yield even during July—August when standard cultivars bolt quickly. Moreover, the mutant produces more than double the leaf yield of standard cultivars because it can be harvested 3 weeks later; the leaves have a high dry matter content, as desired by the canning industry; the leaves have about 4 times the iron content because of their later harvest, and have four times less oxalic acid. The mutant trait was shown due to a single incompletely recessive gene having an association with femaleness. However, by special conditions, with a 20 hour day some male flowers are formed permitting seed reproduction. This new mutant trait will likely become widely exploited in spinach breeding, and provides an example of variants inducible in other leafy vegetable, as well as in forage species.

Micke and Donini[263] listed 37 barley, 9 rice, 7 durum wheat, 7 common wheat, 5 oat and 3 maize cultivars developed by crossbreeding to exploit an induced mutant trait. The already extensive use of mutants as germplasm for further barley and rice improvement demonstrates the importance of newly created mutant germplasm in changing the character of these crops, and illustrates as well that breeders will quickly exploit valuable new genetic traits. In contrast to barley, the use of induced mutants as a gene resource for further bread wheat improvement has been comparatively limited. This situation is surely temporary, but does illustrate that breeders using the mutation breeding approach, or any other specific breeding method, must compete with advances made from the strong on-going research in the same or other programs using other more conventional methods. It is for this reason especially, that induced mutant alleles rather than the mutants themselves in the long run can be expected to make the greatest contributions to breeding of seed propagated crops.

In wheat, three major sources of semidwarfing genes (two of 'Daruma' origin and one from 'Akakomughi', a Japanese land race) originating from old Japanese land varieties (see review by Dalrymple[88] and Konzak[212,213]) have become widespread and have reduced the need for induced mutant sources. Also, the generally large and active bread wheat breeding programs, especially with germplasm inputs from the strong CIMMYT program centered in Mexico, have had less need for some other traits, as earliness and disease resistance, which are still readily found in the germplasm sources made available. However, there is an increasing concern over too intensive use of the two major semidwarfing genes, the somewhat limited sources of earliness genes currently available, and over the rapid surmounting of rust resistance genes in cereal cultivars by new fungus races. These concerns and the encounters by breeders with undesirable gene associations such as the short coleoptile association in the most widely used semidwarfing sources have stimulated increased interest in the use of induced mutations for cereal variety developement.

(ii) Specific mutant types frequently obtained or sought from mutation experiments

(1) SEMIDWARF STATURE MUTANTS IN CEREALS

In barley, the short, stiff straw trait of cv. Pallas, Mari, Diamant, and Miln's Golden Promise barleys has been incorporated into several successful European derivative cultivars (Table 9.6), while in Western USA the chemical mutagen induced semidwarfing trait of the mutant winter barley cv. Luther already has been incorporated into the new winter barley cultivars Boyer, Mal and Hesk as well as into many promising advanced breeding lines. Other new semidwarfing genes, as the dominant mutant of 'Jotun'[7,343] and a semidominant mutant in cv. Advance barley (Fig. 9.6), have been induced (Nilan and Ullrich, personal communication 1981) to broaden the genetic base of barley.

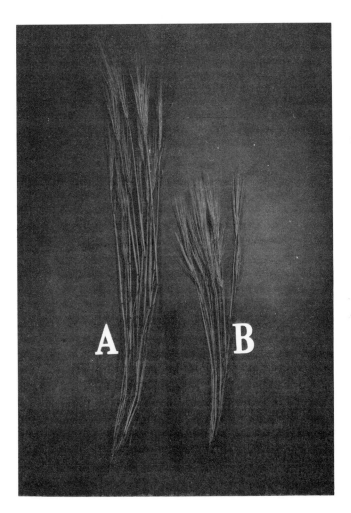

Fig. 9.6. A plant of the six-row malting barley cultivar 'Advance' (A) and a recently semidwarf height mutant (B) induced using azide. Courtesy R.A. Nilan and S.E. Ullrich.

The semidwarfing of the 'Jotun' mutant gene, now transferred into a better genetic background in the M_{21} and M_{22} semidwarf germplasm lines[343] has already become popular with

TABLE 9.7. Uses of induced mutants in the improvement of several crops by cross breeding

12a Dry bean (*P. vulgaris*)

Michelite   **Origin**

     X-irradiation
     Mutant (germplasm)
     cross breeding

Sanilac	USA—MI	1957
Seaway		1960
Gratiot		1962
Seafarer		1967

12b Rice *Oryza sativa*

        Origin

Calrose

    γ-irradiation

Calrose 76	USA—CA	1976

    crossbreeding

M7	USA—CA	1977
M101	USA—CA	1979
Calmochi 202	USA—CA	1980
	USA—CA	
M301	USA—CA	

12c Oats *Avena Sativa*

Floriland   **Origin**

    Nth

Florad	USA—FL	1959

    crossbreeding

Florida 500	USA—FL	1965
Florida 501	USA—FL	1967
Bates	USA—MO	1977
Bob	USA—MO	1977

12d Bread wheat *T. aestivum*

Bezostaya 1   **Origin**

    NMH

Karlik-1 (Germplasm release)

    crossbreeding

Odesskaia 75	USSR 1975
Odesskaia polukarlikovaja	USSR 1975

12e Durum wheat *T. turgidum*

     Origin

Capelli

    *N*th

Castelporziano (CPB 132)	Italy 1968

    crossbreeding

Tito	Italy 1975
Augusto	Italy 1976
Miradur	Austria 1978
Grandur	Austria 1980
Attila	Austria 1980

12f Rice* *Oryza sativa*

     Origin

Fijiminori

    γ-radiation

Reimei	Japan 1966

    crossbreeding

Mutsuhonami	Japan 1973
Hanahikari	Japan 1973
Hayahikari	Japan 1976
Akihikari	Japan 1976
Houhai	Japan 1976

* Source: T. Kawai, FAO/IAEA, Vienna
  H. Sato, MBNL, No. 15:2—4, 1980

plant breeders and is responsible for the semidwarfing in important new US cultivars such as 'Gus'. The original mutant from which the M_{21} and M_{22} germplasm was derived was isolated as a radiation induced short, but weak strawed mutant with normal spikes at the Norwegian College of Agriculture in Vollebekk. The characteristics of the background for the mutant semi-dwarfing gene were markedly improved after several breeding cycles.[343] The induction of new semidwarf mutants has become common in barley breeding programs.[408]

In rice, the semidwarf trait of the induced mutant 'Reimei' already has been exploited in six new Japanese cultivars (Table 9.7), whereas the induced semidwarf mutant of cv. 'Calrose 76' (Table 9.7) 'Calrose' M_3 has become a standard for California rice cultivars.[363] The tall, late maturing Basmati 370 rice is prized for its scented grain and excellent cooking qualities. The complexity of the quality traits is such that crossbreeding to improve the agronomic properties has produced a high yielding semidwarf strain with somewhat inferior organoleptic qualities. However, high yielding, agronomically improved semidwarf mutants of this tall, lodging susceptible, but high quality cultivar have been induced in India.[130] An early maturing photo-period tolerant semidwarf, of Basmati 370 was induced in Pakistan and has been released as cv. Kashmir Basmati. Its earliness has permitted culture of 2 rice crops in a single season, a seeded and a ratoon crop, thereby significantly increasing the amount of grain that can be pro-duced from a given area.[16]

Semidwarf mutants appear to be among the most frequently induced vital mutants in both tetraploid *T. turgidum durum* and hexaploid *T. aestivum* bread wheats[213,215], as in the diploid cereals barley and rice. A new dominant semidwarf mutant also has recently been induced in oats[69], and the mutant has proved useful in further breeding[73]. Both partially dominant and completely recessive mutants occur in the wheats, and there appears to be no obvious association of the induced mutant semidwarfing traits with short coleoptiles or GA insensitivity.[215] Some associations of mutant semidwarfing and GA insensitivity occur in barley[39,174,408] and may occur in rice (Kawai and Rutger, personal communication 1981). Epistatic relations especially among height-reducing and height-enhancing genes, as well as between height-reducing genes apparently may differ considerably, resulting in a rather con-fusing array of genetic results (Konzak, unpublished 1981). Several durum wheats carrying the induced partially dominant mutant semidwarfing trait from cv. Castelporziano (CPB 132) have already been released.

(2) DISEASE RESISTANCE

Mutation breeding practice for disease resistance in sexually propagated crop plants is in a weaker state of development than that for many other traits due largely to changing concepts of host plant resistance, and the development of efficient screening methods. This field clearly merits more careful study. Favret[118,119] already discussed the genetic basis of breeding for disease resistance by induced mutations. With certain of the important mutant traits described previously (semidwarfing, earliness), the available spontaneous sources are few and usually are already known or exploited. Equally important, such mutants often are easier to detect and therefore similar mutants have been readily and repeatedly induced in many genotypes by many investigators.[213,215,263]

Disease resistances, however, are the most important of traits routinely manipulated by crop breeders, who regularly must maintain and work with a wide variety of disease resistance germ-plasm sources. Moreover, genetic variability for disease resistance traits may be hidden unless a compatible pathotype is present in relative abundance. Consequently, the possibility is great that many observed disease resistance variations are due to initially unnoticed outcross and mechanical contaminations, which can and do occur even in the Elite or Breeder Stock of a cultivar. In fact, some contaminants might be difficult to distinguish from mutants, especially when the frequency of variants is low in number. However, Samborski[369] has described methods for distinguishing between induced disease resistant mutants and contaminants. A better understanding of the genetic and physiological nature of disease resistances is needed, as are improvements in mutagen treatment methods able to yield higher frequencies of useful mutations.

Of course, there are truly induced disease resistance mutations in cereals. Their induced mutant origin can be considered certain, because the trait was not found in existing collections, alleles of the trait have been repeatedly induced, or rather special controls have been employed. These mutant disease resistances include resistance to Victoria blight in oats caused by *Helminthosporium victoriae* and resistance to powdery mildew in barley, caused by *Erisiphe graminis* and resistance to downy mildew caused by *Sclerospora graminicola* in pearl millet (*Pennisetum americanum* syn. *Pennisetum typhoides*). Resistance to Victoria blight may really involve modification of a complete locus or pseudo-allelic association rather than pleiotropic relation[123,240,429] between resistance to certain races of crown rust, *Puccinia coronata* and susceptibility to Victoria blight.[220] Oat cultivars, 'Florad' and 'Alamo-X' were isolated by a mass-screening method for resistance to the *H. victoriae* toxin produced in culture.[14,82] These cultivars carry an induced modification of the gene locus, resulting in blight resistance while retaining useful levels of the crown rust resistance. Resistance to downy mildew was induced in the maintainer line of Tift 23A, resulting in the pearl millet hybrid variety NHB5.[185,278] This mutant rectified the male sterile line for an otherwise high yielding hybrid, making possible its continued use in hybrid millet production.[278]

Using tissue culture, Gengenbach and Green[134] and Cassini et al.[81] have isolated mutants resistant to *Helminthosporium maydis* race T corn leaf blight. All of the resistant mutants isolated to date, however, are also fertile not cytosterile as was the objective. Proof was obtained that the mutations occurred in the mitochondrial DNA, demonstrating that the mutants maintained their T cytoplasm. The few mitochondrial DNA alterations, identifiable using electrophoresis, were associated with the change to blight resistance and male fertility (Gengenbach, personal communication 1981). However, because mutagens were not used, it is probable that the range of variation recovered is lower than might yet be possible with finer genetic probes. A number of mutagens, including ethidium bromide[13,237] and azide or its metabolite[210] should probably be tested.[246]

Resistance to yellow mosaic virus disease has been induced in 2 of 8 local mungbean cultivars by gamma radiation treatments to seeds.[380] No resistance to the disease was available in germplasm in Pakistan but resistant germplasm was later reported from India. The 6 mutants obtained varied in degree of resistance or tolerance.

Resistance to powdery mildew in barley was the first induced disease resistance reported in plants, and only after 40 years does it appear that the induced resistance will become widely exploited. This first single instance of the unusual mutant recessive allele *ml-o* was reported in 1940 by Freisleben and Lein,[126] but mutants at the *ml-o* locus have since been repeatedly induced in many laboratories over the world.[143] A spontaneous *ml-o* mutant also was recently found among natural germplasm stocks.[194] The *ml-o* gene is unusual in that it confers resistance to all races of the powdery mildew fungus all over the world. No mildew race virulent on *ml-o* has been found in more than 20 years of study. A second unusual feature of the *ml-o* mutants is the close association of the trait with a non-pathogenic, necrotic spotting condition which breeders found would reduce grain yield by about 10% in the absence of disease. The mutant alleles differ, however, in their associated levels of necrotic spotting.[194,195]

Recently, however, as a result of intensive mutation and genetic breeding studies Jørgensen[194,195] has discovered that the *ml-o* locus has at least three components, i.e. three mutant allele types occur that will genetically recombine at a low frequency to produce mildew susceptible progeny. More important for practical purposes, crosses of some of the low necrotic spotting *ml-o* mutants with modern breeding lines have produced strains which carry *ml-o* resistance without the associated necrotic spotting.[194,195] The selections yield well and thus have variety potential (Jørgensen, personal communication 1981). Breeding work in the Netherlands, in fact, already has led to the release in 1979 of a mildew resistant barley carrying an *ml-o* allele (A. Micke, personal communication 1981). The new Danish barley selections carrying *ml-o* resistance without necrotic spotting are now under advanced testing (Jørgensen, personal communication 1981).

Thus, the practical exploitation of the *ml-o* mutants can be expected to have a significant impact on barley breeding, since mildew is one of the most serious barley diseases worldwide. The success from intensive genetic and breeding efforts to understand and to modify the *ml-o* necrotic spotting complex, establishes a stronger basis for the application of mutation breeding to attain disease resistance in crop plants. However, Gustafsson[143,145] indicates that it has also been possible recently to obtain high yielding and non-necrotic spotting *ml-o* alleles directly from mutagen treatments, suggesting that the mutation process itself can be exploited to obtain the desired alleles if a sufficient number of mutants is produced. Of some 57 mildew resistance mutants isolated over a 5 year period eleven, or 25 per cent of the 44 analysed to date have proved to be *ml-o* alleles.

As appears to be the case also with other traits, the mutability of disease resistance loci is highly dependent on the background genotype even if the genotypes are closely related. Thus, Robbellen and co-workers[358] found that one of three closely related varieties yielded more and different induced mildew resistant mutants than two others, although all three cultivars showed a similar low level non-specific resistance behavior in greenhouse and field tests. Similarly, Yamaguchi and Yamashita[441] obtained three independent *ml-o* locus mutants from 6215 spike progenies of one cultivar cv. Fuji Nijou but only one *ml-o* mutant in 42,000 spike progenies of another cv. Azuma Golden. A third variety cv. Nitta Nijou No. 1 yielded three mildew resistance mutants among 8600 spike progenies, but none were *ml-o*. Yet, the same mutagen treatments were applied to all three varieties and were shown to be effective based on chlorophyll mutant frequencies. Strikingly, the ratio of *ml-o* resistance mutants from Fuji Nijou was 1:165, but only 1:7079 for Azuma Golden.

Thus, the yields of chlorophyll mutants at least as a group provides no indication of the mutability of a genotype for a specific trait, and consequently, the failure to obtain a desired mutant after application of effective mutagen treatments may imply only that the genotype studied does not respond to the particular treatments applied. Therefore, the best recourse is to mutagenize several genotypes, first screening a portion of each population, then concentrating further screening effort on the most promising materials. Better methods are needed for selecting those genotypes most useful for mutation experiments, and surely will be developed as more experience is gained from careful studies.

The disease resistances reported from some mutation studies lack adequate identification, confirmation or replication, and because of the contamination question, it is difficult to assess the validity of reported mutant variation from these studies. However, the results should not be ignored.

Thus, neither the probability of contaminant origin of some reported mutants, nor the hazard of contamination should deter breeders from availing themselves of the potential offered by mutation breeding tools. Some disease resistance types must very rarely occur as are certain unique morphological mutants so far isolated only after mutagen treatments, i.e. semidwarf barleys, as contrasted to erectoid forms,[408] or the tomato mutants with club-shaped leaf hair trichomes or a single stem single fruit cluster.[6,348] These mutants may rarely be found because the desired mutants relate to specific single loci, which perhaps are rarely altered even by the most effective treatments known today, Repeated numbers of mutants for these loci may be recoverable only via extremely efficient screening techniques and from screening very large M_2 populations. In Jute, *Corchorus capsularus* one gamma radiation induced mutant C-36 from the cultivar D-154 was found to have resistance to stem rot disease in artificial inoculation tests. Four other mutants had moderate resistance, while D-154 and the mutant GP were moderately susceptible and one mutant C-7 was susceptible.[160]

The clearly established cases of induced disease resistances thus can serve as indicators of the potential of the method.

(3) *MUTANTS AFFECTING PHYSIOLOGICAL TRAITS*

Although most mutants relate in some way to changes in physiology, certain types of variants are best classified by their physiological alterations, such as the commonly induced earliness or lateness mutants (Table 9.8). Mutants to photoperiod insensitivity also have been obtained in cotton[345], potato[381,411], rice[178], in jute[257,258] and in soybeans[222,223]. In contrast, apparently photosensitive mutants have been induced in sugar cane.[340,418] These mutants do not flower at all under conditions in India and thus are capable of higher yields. Earliness mutants have frequently been induced in a wide range of crop species (Table 9.6). Some induced mutant earliness alleles already have been extensively used in crossbreeding.[262,263] In barley, for example, crossing the early mutant cultivar Mari with an early cross-bred cultivar Otra has led to transgressive segregation for earliness, in the cultivar Eero making it possible to extend barley cultivation even further north than was made possible already with Mari.[209] Moreover, Kivi[208] was able to induce 22 early mutants in an already very early barley variety, 'Pokko' though 350,000 M_2 plants were observed. Two of the 22 mutants proved worthy of advancement in testing beyond M_5 for possible direct use.

Ukai and Yamashita[410] analysed 61 induced early mutants of barley. Four of the mutants carried genes controlling photoperiod response, two of which are allelic to the earliness locus ea_k and three mutants carried genes controlling vernalization effects. They found that the early mutants had varying degrees of changes in response to photoperiod or vernalization. In rice, early ripening mutants sometimes are photoperiod insensitive, and of semidwarf height. Many have become important for successful rice production in colder fringe areas and for the development of varieties for multiple cropping practices in the tropics and sub-tropics. Thus, the mutant cultivar, Nucleorhiza developed in Hungary improves the success of rice production there,[367] whereas numerous other early mutants have markedly extended the germplasm base for earliness breeding in rice.[181] In barley and wheat, vernalization-requiring (winter type) mutants from spring forms are readily obtainable. These mutants lack winter hardiness if the initial variety was not hardy. The reverse, or spring habit, forms may be inducible in some winter habit genotypes. However, mutants of this type need careful examination, since pre-harvest vernalization may occur. Konzak et al.[218] did obtain spring habit, (though still vernaliz-ation responsive) mutants from a vernalization-requiring winter wheat. Late-flowering mutants also may be useful in some cases to extend the ripening beyond a period of inclement weather. Later maturing mutants are in fact common (Konzak, unpublished[92]). A number of mutant cultivars are later maturing than the original (Table 9.8).

Changes in the relative proportion of the vegetative and generative phases in the growth of plants also can be achieved by induced mutation. Kar et al.[198] obtained a mutant HD2237 that ripened 25 days earlier than the original cv. Kalyansona. The vegetative phase of the mutant was shortened from 102 to 70 days, while its generative phase was extended from 42 to 55 days. The grain weight was increased by 15%, possibly a result of the longer grain-filling period. The tillering capacity of the plant was unchanged but tillering was more synchronous and the culm length was about 10 cm shorter. In one test, the mutant yielded 20—30% more than Kalyansona and in other tests it outyielded Kalyansona in all tests at 37 locations in India. Sowing time experiments established that the earliness and thus the relative lengths of the vegetative and generative phases is dependent on sowing time. By late sowing, e.g. in December rather than November, the vegetative phase of Kalyansona was shortened by 3 days whereas that of the mutant was increased by 16 days. In contrast, the generative phase of Kalyansona was shortened by 5 days, but that of the mutant HD2237 was reduced by 18 days.[198,262]

The mutant barley variety Diamant induced by X-radiation in the cv. Valticky.[40] The mutant produces more and longer spikes than Valticky, has a higher thousand grain weight and reduced tiller height. More significant, however, is the changed developmental pattern of the mutant. In the spring, under short day conditions, development of the shoot apex is de-layed, but tillering occurs 10—14 days later than in Valticky but at a comparable tillering stage the shoot apex of the mutant is more differentiated because of its longer juvenile phase, the mutant develops more buds and a better root system. However, between the mutant's gener-ative phase is shortened so that the mutant ripens only 4 days later than Valticky. Even so, the mutant yields about 11% more than Valticky and has a higher harvest index. Diamant has proved to be a valuable parent in cross breeding.[40,41,42] A recent derivative, cv. Trumpf released in the GDR in 1973 outyields the best varieties by 15% and was planted to 70% of the acreage by 1975.[397]

Another mutant with a remarkably altered vegetative phase has been obtained in spi-nach.[151,152] This mutant, induced by EMS treatment of seeds of the cv. Fruremona, had a long lasting vegetative phase, and initially produced only female flowers, but with artificial illumination it was found that 5—10% male flowers could be induced, allowing the production of selfed seeds. The mutant produces more than double the leaf yield of standard cultivars, at a 3 week later harvest, and has high dry matter values as required by the canning industry, as well as four times the iron content due to the later harvest. Moreover, the mutant has a four-fold lower content of oxalic acid.

Physiological variants as a class are probably among the more abundant mutant forms induced. Many mutants involve subtle morphological and physiological changes difficult to classify,[92] (Konzak unpublished). The types and frequencies of these mutants are probably genotype dependent.

Among the more attractive physiological mutant types that might be sought for practical breeding are those which alter reactions to herbicides and fungicides. Mutants of this type would make it possible to gain the benefits of selective chemicals potentially useful in practice

TABLE 9.8. Examples of mutant varieties of seed propagated crops with altered flowering
and/or ripening times

Species and variety	Country and year of release	Characteristics in comparison with parent variety
Allium cepa (onion)		
Brunette	Netherlands 1973	very early, high yield and quality
Arachis hypogaea (groundnut/peanut)		
TG-1	India 1973	late maturing (130 days)
Avena sativa (oat)		
Zelenji	USSR 1976	late ripening, high vegetative production for forage
Brassica juncea (Indian mustard)		
RLM 198	India 1975	5—6 days later ripening, higher yield and oil content
RLM 514	India 1980	early maturity, higher yield, shatter resistant, 11% less erucic acid
Capsicum annuum (pepper)		
Krichimsky ran	Bulgaria 1972	hybrid variety, early maturing high yield
Glycine max (soya)		
Raiden	Japan 1966	25 days earlier ripening
Raiko	Japan 1969	15 days earlier ripening
KEX-2	Korea 1973	11 days earlier ripening, 16% higher yield
Gossypium hirsutum (cotton)		
MCU 7	India 1972	10—15 days earlier ripening, high yield
Rasmia	India 1976	daylength tolerant, high yield, superior quality, wide adaptation
Pusa Ageti	India 1976	short duration, high ginning, Jassid tolerant, fits well in cotton/wheat rotation
Hordeum sativum (barley)		
Haya Shinriki	Japan 1962	very early ripening
Mari	Sweden 1962	8 days earlier, shorter culm
Gamma 4	Japan 1965	2 days earlier, 15 cm shorter culm
Diamant	CSSR 1965	10 days later tillering, 4 days later maturing, shorter culm high yield
Bettina	France 1968	later ripening, shorter culm, lodging resistant
Amagi Nijo 1	Japan 1971	3 days earlier, 10 cm shorter culm
RDB 1	India 1972	early ripening, short culm
Radiation	Korea 1974	early ripening, short culm

Table 9.8 (continued)

Species and variety	Country and year of release	Characteristics in comparison with parent variety
Temp	USSR	2–3 days earlier heading, ripening time unchanged
Markeli 5	Bulgaria 1976	6–8 days earlier maturing, higher productivity, large grain
Ornithopus compressus (serradella)		
Uniserra	Australia 1970	earlier flowering, better pasture plant, good seed yield
Oryza sativa (rice)		
KT 20–74	China 1957	earlier ripening, higher yield
SH 30–21	China 1957	earlier ripening, higher yield
Reimei	Japan 1966	tolerant against low temperatures following early sowing
BPI–121–407	Philippines 1970	early ripening, independent of season, short straw, high tillering erect leaves
Iratom 24	Bangladesh 1971	early ripening, high yield
Iratom 38	Bangladesh 1971	3–4 weeks earlier maturing
IIT48	India 1972	1 week earlier, daylength tolerant
IIT60	India 1972	flowers within 65 days, ripens 1 month earlier, same yield
Nucleoryza	Hungary 1972	ripens 3 weeks earlier, better adapted to cold climate
Kashmir Basmati	Pakistan 1977	ripens 20 days earlier, suitable for growing in Kashmir up to 1500 m elevation (short season)
RD15	Thailand 1978	10 days earlier maturity, better drought tolerance
Yan-Feng-Tsao	People's Rep. China	45 days earlier ripening, 10% higher yield than other early cvs., 14% higher lysine content than common cultivars
Miyanishiki	Japan 1978	early heading, short culm, improved lodging resistance
CNM6	India 1980	earlier maturing 15–23 days
CNM20	India 1980	earlier maturing 10–12 days
CNM25	India 1980	earlier maturing 14–25 days
CNM31	India 1980	earlier maturing 10 days
RD10	Thailand 1981	day length tolerant
Phaseolus vulgaris (common bean)		
Alfa	CSSR 1972	4 days earlier, better disease resistance, 22% higher yield

Table 9.8 (continued)

Species and variety	Country and year of release	Characteristics in comparison with parent variety
Pusa Parvati	India 1970	early, 45 days till maturity, high yield
Pisum sativum (pea)		
Wasata	Poland 1979	early maturity, change of leaflets to tendrils, high yield potential, lodging resistant, suitable for combine harvest
Ricinus communis (castor bean)		
Aruna	India 1969	matures in 120 instead of 270 days, higher yield
Triticum aestivum (wheat)		
Lewis	USA 1964	early ripening, lodging resistant, high yield
Stadler	USA 1964	early ripening, lodging resistant, high yield
Zenkouzi-komugi	Japan 1969	2 days earlier, 15—20 cm shorter culm, 10—15% higher yield
Zea mays (corn)		
CE 200 Hybrid	CSSR 1979	early maturity, high yield, good stalk quality, input to 3-way hybrid by selected inbred mutant Rt 10

Adapted from: A. Micke, "Use of mutation induction to alter the ontogenetic pattern of crop plants", Gamma-Field Symposia No. 18, 'Crop Improvement by Induced Mutation', 1979 and Mutation Breeding Newsletters 13—18, 1979—1980.

but too toxic to the normal crop variety at concentrations sufficient to control weeds or pathogens. Resistance to chemicals is known to occur in nature.[142] In fact, DDT resistant barleys now collected as germplasm, were widely distributed over the Anatolian plateau region of Turkey, long before the resistance was known,[166] and Konzak (unpublished). Low level 2,4-D tolerance occurs in birdsfoot trefoil, *Lotus corniculatus*,[398] and to propanil[387] and metribuzin in soybeans.[26] Moderate level metribuzin also occurs in wheat and barley (Muzik, unpublished 1980).

Genetic tolerance levels available in germplasm of a number of crops is inadequate for safe use of certain highly effective herbicides. However, DDT tolerance is readily induced in barley[423] and results from cell culture studies suggest that herbicide resistance in numerous plants may be readily recovered using cell and tissue culture methods.[24,245,398]

(4) *MUTANTS AFFECTING PROTEIN CONTENT, NUTRITIONAL VALUE OR PROCESSING PROPERTIES*

The general issue of breeding for enhanced protein is considered in Chapter 7. Following the discovery that the grain of certain mutants of maize had a markedly changed amino acid composition[256,300], other mutants with changed protein content and/or composition were isolated from mutagen-treated barley[22,96,98,148], sorghum[18,19] and wheat[94,108,314]. A number of protein mutants of rice and wheat have been reported, and are still being evaluated[71,399,400,402] and (Hermelen, Micke *et al.* personal communication 1981). One high protein wheat mutant was released to commercial growers.[313]

Screening technology for amino acid or protein content changes is not very efficient, and evaluations must deal with significant environmental effects.[187,188,214] In the case of protein content changes, yield/protein relationships must be considered, so that genetic and breeding evaluations may provide the only true identification of protein content mutants. Bhagwat *et al.*[29] have in fact demonstrated that the high protein trait of the Kalyansona mutant TW-1 has a genetic basis and can be transferred to lines that are higher yielding than Kalyansona. This result is notable, in view of the fact that TW-1 itself has a significantly reduced grain yield compared with Kalyansona. However, the grain of TW-1 was also found to have a higher rate of nitrogen accumulation than Kalyansona or Atlas 66.[71] The different ways in which dry matter and nitrogen accumulate in different genotypes suggest greater independence of the two processes than was once thought, whereas other factors, possibly N availability (including efficiency of use) but not competition for the supply of assimilates[30] may limit the realization of high yielding varieties capable of producing high protein grain.[70,71]

It is also of significance that a *T. dicoccoides* wild wheat accession is able to accumulate as high as 43 per cent protein in plump grain when supplied with ample nitrogen in pot experiments.[15] This unusually high protein content for wheat indicates that the particular genotype has no genetic limitation to protein accumulation in the grain. Persson and Karlsson[316] suggest that the high protein content of Hiproly is likely under the control of a polygenic system. Scholz[375] for example was unable to demonstrate recovery of the high protein trait of several lower yielding barley mutants in high yielding progeny of crosses with other varieties or from crosses of high protein mutants.

It is notable that none of the high lysine sources found among natural variations in wheats (except possibly Nap Hal) possess the defective endosperm condition of the opaque mutants of maize and sorghum.[214] An opaque-type or shrunken endosperm type of mutant would likely destroy the very processing properties for which wheat is important as a food because in opaque 2 (O_2) and shrunken endosperm mutants, the proportion of prolamin proteins, vital to the formation of gluten is drastically reduced. Most mutants with significant changes in the proportion of amino acids in proteins have been readily confirmed, but many of the most studied mutants of barley are found to have marked "pleiotropic" effects on yield and other properties.[97,431]

The associated defective (shrunken) endosperm characteristic of most high lysine mutants of barley has proved extremely difficult (perhaps impossible) to separate from the high lysine trait.[97] The exception seems to be for the lys_1 mutant trait of the primitive line 'Hiproly' originating in Ethiopia. High yielding, high lysine recombinants with plump kernels have been obtained by a number of breeders.[409] However, since no cultivar has yet been released, there apparently are still difficulties associated with achieving competitive yield levels even with Hiproly germplasm.[149,316] The breeding, genetics and other analytical results suggest it may not be fruitful to expend much more effort trying to transfer the high lysine trait of the Bomi 1508 and similar mutants to normal endosperm types.[95,211,316] Plarre[323] indicates that high lysine plump endosperm derivatives are obtainable from Hiproly crosses in a wide genetic base. Bansal et al.[23] indicate that plump endosperm derivatives of both Hiproly and Bomi 1508 have been obtained. Similar difficulties have been encountered in transferring the high lysine trait of o_2 and fl_2 mutants of maize to agronomically acceptable, high yielding, hybrids with desirable end user qualities. After considerable effort, more promising modified hard endosperm seed types, still with high lysine are being selected.[412] On the other hand fl_2 maize has been found to be of inferior nutritional value to o_2 maize in feeding trials, showing that biological values may not always coincide with values predicted from chemical analyses.[304]

High threonine levels are produced in the grain of maize mutants isolated by Hibberd and Green[171] from azide-treated callus cultures subjected to inhibition by a combination of the amino acids lysine and threonine. Since threonine is the second limiting amino acid in maize, sorghum, barley and wheat, the method holds promise as a means to isolate threonine over-producing mutants for improving the nutritional quality also of these species. Procedures for selecting mutants resistant to amino acid analogs have been described by Green and Phillips[141] and Widholm[437].

Opaque mutants with characteristics similar to the induced 'hily' mutants of barley and maize have been induced in sorghum and a like allele has been found among primitive Ethiopian cultivars.[19,383] With sorghum, however, many of the problems encountered in transferring the 'hily' (o_2-like) trait to acceptable agronomic types have been avoided by the induction and use in crossbreeding of an EMS mutant P-721 from a high yielding, locally adapted sorghum line P-954114, rather than attempt to transfer the trait from the Ethiopian primitive sorghum germplasms.[17] Axtell et al.[19] indicate that there should be no serious obstacle to the development of high yielding sorghum hybrids with improved protein quality. They suggest that the reduced weight of 'hily' seeds can be compensated readily by an increased seed number to maintain competitive grain yield level in hybrids expressing the mutant trait.

The nutritional quality of rice protein is unusually high for a cereal grain, but generally the levels of protein in rice are exceedingly low. Many possible higher protein induced mutants and selections from land races have been identified.[29,158,162,268,401] Only in a few instances has there seen an indication that both yield and protein content have been improved in the same mutant line.[29,162,268] Tanaka[401] has noted that most of the high protein variants prove to have lower grain yields and often carry one or more phenotypic changes, i.e. smaller seeds, reduced tillering, shorter straw, early heading etc. However, by screening many normal-appearing M_2 plants for improved protein content, Tanaka[400] has been able to recover mutant strains of rice with a 10 to 20 per cent increase in protein productivity per unit area. Most high protein mutants selected in this way have retained the desirable agronomic traits including yield capacity of the initial variety.[157] One mutant was shown to yield more than the initial variety and have higher protein grain under standard fertilization conditions and yield more high protein grain under high fertility.[399]

Progress in breeding using genetic variations available and transferred from wild species had been so disappointing, Coffman and Juliano[85] considered that continued effort to improve rice protein content by breeding at IRRI was not justifiable at present due to the urgency of other problems, such as pest resistance and tolerance to environmental stresses. A major problem common to mutation breeding programs in areas with dynamically changing disease and pest problems is the long time currently required to identify and evaluate mutants for traits that

that show considerable response to environment. Further, it is now becoming clear that candidate protein mutants induced in adapted varieties or lines should be evaluated as soon as possible in crosses with high yielding adapted, disease and pest resistant germplasm stocks. Comparative studies might then be made using the control variety instead of the mutant. Evidence that the high protein trait can be transferred to high yielding lines is the best evidence for the existence of a mutant gene responsible for regulating protein content. Because the environmental component of variation is often great, it seems pointless to evaluate lines under conditions that are not optimal for the expression of heritable differences in protein content.[214,221] Thus, stress conditions should be avoided, and N applications should be optimal for high yield performance, since conditions for maximum yield expression will most likely permit the expression of protein production ability. N levels must be adequate for maximizing yield. When possible, both standard and higher N levels should be used in testing. In the reports cited above, few of the probable mutants has yet been subjected to breeding analyses.

The nutritional quality of fodder also is being improved by the use of mutants. Brown midrib (*bm*) mutants have been found affect the amount of lignin in leaves and stems of maize, resulting in significantly improved digestibility by ruminant animals,[25] and are discussed on page 268. The promise shown by the maize work on brown midrib mutants prompted Porter *et al.*[327] to induce similar mutants in locally adapted grain sorghum lines using diethyl sulfate. Some of the *bmr* mutants proved not to show reductions in lignin percentage, suggesting that several different mutants with similar general phenotype were induced. However, genotypes identified as bmr_6, bmr_{12}, and bmr_{18} appeared promising for further evaluation as germplasm sources for improving the nutritional quality of sorghum plant material. Studies by Fritz *et al.*[127], of progeny from crosses between the *bmr* mutants and 2 grain type and 1 grassy type sorghum (sudan grass), showed that the 3 mutant alleles caused similar reductions in lignin composition of both grain sorghum and sudan grass. The breeding of *bmr* mutant sudan grasses and sorghums can be expected to increase animal production with consequent economic benefits due to the improvements in digestible energy of the *bmr* over normal genotypes.

(5) *MUTANTS AFFECTING CROSS OR SELF-FERTILITY*

Some of the most useful mutants in plants are those which permit greater control of pollination. Hanna and Powell[155,156] have induced partially apomictic mutants of pearl millet (*Pennisetium americanum*) and have identified a mutagen-induced apomictic form in sorghum as well. Hanna and Burton[153] also have hybridized apomictic *P. setaceum* and *P. orientale* with *P. americanum* to obtain apomictic male sterile, interspecific hybrids. The hybrids will be irradiated in an effort to transfer the apomictic trait to a *P. americanum* chromosome. Interestingly, Mujeeb-Kazi[269] has recently obtained apomictic progeny from backcrosses to wheat of *Hordeum vulgare* cv. Manker X *Triticum turgidum* cv. Cocorit, and *H. vulgare* cv. Manker X *T. aestivum* cvs. Bonza and Chinese Spring, barley (2X), durum (4X) and common (2X) wheat hybrids. While the hybrid plants were highly seed sterile, the backcrosses to wheat resulted in recovery of plants exactly like the original F_1. These new hybrids provide unusual new genetic base material for further investigation of the apomixis mechanism and its possible exploitation via the variety of tools, including mutagens, colchicine doubling of chromosomes and callus culture now available for plant genetics and breeding research.

Mutants causing male sterility have become important for hybrid production in a number of species. In ornamentals, as in a number of vegetable crop species, F_1 hybrids are now common. Nuclear genetic male sterility mechanisms are being exploited to eliminate time consuming and costly hand emasculation. Moreover, the marketing of F_1 hybrid seed provides the breeder with complete control of the parental inbreds. The F_1 hybrids may achieve exceptional vigor due to heterosis and usually provide consumers with a more uniform product. In many cases, the heterosis contributes also to seedling vigor, so that fewer F_1 hybrid seeds need be sown to produce a given number of vigorous seedlings. Mechanisms of monogenically controlled male sterility used in commerce vary considerably with the species. In the flower *Tagetes erecta*, F_1 hybrid seed is field-produced using an apetalous mutant form without stamens and with an increased number of pistils. This seed parent is crossed with strains carrying a mutant trait that causes only ray flower formation.[175] In *Antirrhinum*, commercial strains carrying the *def nic* allele isolated by Bauer in 1924.[175,393] This mutant, like ms_5 in the tomato, has fertile pollen, but the pollen usually fails to dehisce. Hand selfings of plants carrying the mutant trait lead to 100% male sterile progeny.[175,352]

In *Begonia*, the 'Cinderella' mutant, used for F_1 hybrid seed production has deformed anthers, while in *Argeratum*, the male sterile mutant used is otherwise normal. Both mutants are recessive, and are maintained by crossing with heterozygote, requiring the removal of fertile plants from rows of the seed parent when F_1 hybrid seed is field produced.

In cereal crops, several uses are being made of nuclear gene male sterile mutants. In barley, an ingeneously devised system involving a trisomic along with a nuclear male sterility gene[335] already has been used in a few commercial hybrids. An improved version is under development which will exploit a specially induced chlorophyll deficient mutant linked in repulsion to the male sterility gene for removing the male fertile portion of the population so that 100% hybrid seed can be produced more simply (Ramage, personal communication 1981). A number of candidate mutants already have been induced.

In barley, one recessive male sterile mutant becomes fertile when gibberellic acid is applied, making it possible to produce 100% male sterile progeny.[200] Unfortunately, the treatment seems not efficient enough to be used in hybrid barley production. In tomato, however, the stamenless mutant *sl* can be converted to a fertile form by gibberellin treatment.[319] The fruit form of the *sl* mutant is defective.[353] Numerous other male sterile mutants have been obtained in tomato, but at the present time few appear to be useful in F_1 hybrid tomato seed production. (Rick, personal communication 1981). However, gene *ms 35* is in use for F_1 hybrid production in France and in Israel.[234,320,321]

Besides their use in the production of hybrid seeds in some species, nuclear gene male steriles are being used effectively to eliminate hand emasculations in plant breeding research programs, and thus to facilitate the development of recurrent selection populations in normally self-pollinated crops. The method, termed male sterile facilitated recurrent selection (MSFRS), has been extensively developed by Ramage[336,337,338,403] in barley and bread wheat. A large number of nuclear mutant male sterility genes are available in barley[172] some of which have been induced. In durum wheat, nuclear genetic male sterile mutants have been induced using radiation by Bozzini.[47] A similar durum wheat male sterile mutant has been induced recently using diethyl sulfate (DES) and is being exploited to develop MSFRS populations by Western Plant Breeders staff (Carleton and Shantz, personal communication 1981). Konzak and Duwayri (unpublished, 1981) also have recently isolated a number of putative male sterile mutants in selections from old Jordanian primitive durum cultivars for use in developing MSFRS populations.

In bread wheat, Driscoll and Barlow[107] have recently induced a simply inherited recessive male sterility mutant located on chromosome 4A. This mutant trait is linked to semidwarfing alleles Rht_1 and Rht_3. The mutant gene, now termed 'Cornerstone', seems particularly useful for MSFRS population breeding of hexaploid wheats.[103,105] Tsunewaki (personal communication 1981) has induced another male sterile mutant ms 6 in Chinese Spring wheat. The mutant appears to be different from 'Cornerstone'. Eight additional recessive male sterile mutants, as well as a male sterile mutant that behaves as a simple dominant were obtained by Frankowiak et al.[125] after EMS treatment of hexaploid bread wheat with Ae squarrosa (T. tauschii) cytoplasm. The dominant mutant FS 6 segregates 1 male sterile to 1 male fertile.[371] The male sterile F_1 progeny are heterozygous, while the male fertile plants are homozygous fertile and thus can be treated as F_1 hybrids with 50% germplasm from the most recent male parent. The other 50% germplasm of the F_1 and later progeny will have been subjected to recombination and any selection applied to the population.

With the MSFRS systems, pedigree information may soon become so complex and the germplasm introduced in the system so diffused that the information value is greatly reduced. Nevertheless, the MSFRS methods facilitate recurrent selection breeding toward advancement of yield, and for any other factors (drought, Al, Mn, salt, etc. tolerances, disease resistances) for which screening is feasible. The MSFRS methods can be exploited most efficiently to complement more 'conventional' methods in which pedigree information is maintained. An MSFRS system employing the nuclear dominant male sterility gene may be preferable for a backcross or modified backcross program because new crosses can be made each generation and the system tends to be more conservative and economical in use of space. Genetic male sterility mutants for use in MSFRS schemes are available or can be readily induced in most self-pollinated plant species.

Cytoplasmic male sterility has certain advantages over some forms of nuclear gene male sterility, especially for hybrid seed production, and in most cases, has been discovered in nature or produced via transferring nuclear material by backcrossing into alien species cytoplasm.[84,201,247,406] Few cases of induced cytoplasmic mutations are known, but Favret and Ryan[121] reported induction of a cytomale sterile mutant of barley. One of the two partially male sterile mutants segregated sterile and male fertile plants. Fertility restorers were not reported. Fertility restoration of the cytomale sterile forms also remains a problem in some cytomale sterile barley species derivatives, although genes for partial fertility restoration have been isolated.[5]

Kinoshita and colleagues[204,205,206,265] have reported induction of cytomale sterile mutants in sugar beets. The features of the cytomale sterility systems among the mutants suggest that a variety of different mechanisms may have been induced. Similarly, Burton and Hanna[77] have induced a cytomale sterile mutant in a pearl millet (Pennisetum americanum) maintainer line. Burnham 41 et al.[72] have described genetic methods for differentiating between types of nuclear gene-cytoplasmic factor interactions mechanisms controlling cytomale sterility and its restoration. As discussed also by Washnok[426] and Burnham et al.[72], cytoplasmic restoration of genetic male sterility is being sought as an alternative to cytomale sterility. In this mechanism, the male sterile stocks would be genetically [N] $msms$, while the male fertile maintainer line would be [cR] $msms$. The F_1 cross with most other genotypes would be male fertile because of the dominant nuclear allele Ms as it would carry the normal [N] cytoplasm introduced with the male sterility ms alleles by the female parent. Mutagens should prove useful for inducing one or the other if not both genetic components of such a system.

Such a system seems feasible for wheat especially since chromosome 1D carries a gene or genes that restore fertility and vigor to hexaploid wheats in *Ae squarrosa* (*T. tauschii*) cyto-plasm. The *Ae squarrosa* cytoplasm causes male sterility and reduced vigor in durum wheat, which does not have the D genome. Both vigor and fertility are restored by the addition of 1D chromosome from hexaploid wheat.[202,247] Normal wheat *T. aestivum* and *T. turgidum durum* cultivars would appear to carry the restorer cytoplasm, and chromosome 1D of *T. aestivum* cultivars tested so far evidently carry the normal allele of the necessary male sterile gene. How-ever, Mukai and Tsunewaki[270] have discovered that when the nucleus of a *T. spelta* line is transferred by backcrossing into *Ae variabilis* or *Ae kotschyi* cytoplasm, the resulting line is male sterile, but nuclei of other hexaploid wheats in the same cytoplasm are male fertile. In the US, Cargill Company has identified a fertile *T. aestivum* strain which does not restore fertility in crosses to lines carrying the *Ae kotschyi* cytoplasm (Curtis, personal communication 1981).

Although the results described above might also be explained in terms of nuclear restorer genes, it is attractive to consider the alternative mechanisms as discussed in detail by Burnham *et al.*[72] and attempt to induce the desired mutants since the use of a cytoplasmic restorer system would simplify the breeding of component lines as well as facilitate hybrid seed pro-duction. Attempts to induce in a (*squarrosa*) *T. aestivum* strain a genetic recessive male sterile mutant gene on chromosome 1D have so far been unsuccessful (Mann, Williams, Sasakuma, personal communication 1981). This male sterile would be restorable by normal *T. aestivum* cultivars, and maintainable by a line homozygous for the 1D *ms* gene but fertile in *T. aestivum* cytoplasm.[125]

As alternative mechanisms for F_1 hybrid production, Driscoll[104] suggests that the 'Corner-stone' nuclear genetic male sterile mutant gene can be exploited for hybrid wheat production using his proposed XYZ system.[105] In the XYZ system lines X, Y also carrying 2 and 1 alien chromosome additions are employed to produce or maintain the hybrid female line Z are all homozygous recessive for a recessive male-sterility gene which is then pollinated by a normal male fertile variety to produce F_1 hybrid seed. The 44 chromosome homozygous alien addition line, homozygous also for the male sterility gene ms is self-reproducible via an Ms allele carried on the alien addition chromosome. The Y plants (Ms/ms + 1' alien Ms) are employed as a male parent to produce and maintain the Z line stock, because the single alien addition chromosome pollen is not competitive with normal 21 chromosome pollen carrying the recessive Ms allele. The Y line is also used as the female parent, employing the X line as pollinator for reproducing or increasing Y stock.[102,103,106] The male sterile line Z, genetically *ms/ms* without an alien addition, is created by crossing line X (*ms/ms* + 1' alien *Ms/Ms*) with normal Ms/ms 21 chromo-some plant, which is easily obtainable, though not in quantity. The homozygous male sterile line Z is pollinated by a normal male fertile variety to produce F_1 hybrid seed.

(iii) Genetic aspects of induced mutants with relation to breeding problems

In principle, induced mutations represent the same kinds of changes in the genes as occur from natural causes. Most induced mutants used in breeding show normal inheritance, and consequently perform in no way different from genetic variations breeders regularly exploit.[212,392] Most induced and spontaneous mutants show recessive inheritance, but more rarely occurring dominant mutants occasionally have been isolated from mutagenized plant populations.[69,125,213] Viable mutants also may show heterosis when crossed.[259,339] Even lethal mutants may contribute to vigor when heterozygous.[146,368] Heterozygosity for lethal or semilethal genes is in fact common in polyploids where it is fixable because of chromosome and gene homoeology.

A reduced frequency of mutant recovery may suggest that the mutant is due to a large deletion, which even so may be modifiable through recombination and selection if the trait is valuable. However, even rather large deletions need not affect viability or productivity.

The greater proportion of mutants isolated from seed treatment in higher plants cannot be due to large deletions, since many different alleles at specific loci have been identified. However, small-sized chromosome segment alterations, base pair deletions, duplications as well as simple base transitions and transversions may be common. The most extensive analyses of induced mutant allelism have been made of barley mutants by Swedish workers.[143,144,145,242] Fine structure analyses of *wx* loci in maize and barley have assumed intra cistron deletions to be the main structural alteration induced, but atypical recombination data for some mutants suggests other causes.[9,303,361] However, the range, frequency, and spectrum of mutations may differ from different mutagens,[305] and for the plant genotype X mutagen interactions.[143,144]

Induced mutant alleles may not be identical to similar alleles isolated from genetic resource collections obtained via plant explorations, or from crossbreeding programs as was once thought. Thus, even if isolated at relatively high frequencies, induced mutants often extend the range of genetic variation known for a species which may include valuable rare forms, not yet known or unlikely to be selected for in natural populations.[143,145] Because so many genes often may be involved in the biochemical genetic basis for many phenotype characters, the concept of homologous series as proposed by Vavilov[413] must be modified, referring to variation as analogous or parallel series.[143] Phenotypically similar mutants may be due to one or more alleles at one or more loci. Thus Wettstein[433] reported 31 gene loci for 33 *albina* mutants examined, whereas 20 gene loci and alleles controlled the *xantha* phenotype. For the virido albina phenotype, 14 gene loci were found from analyses of 20 mutants, and 16 gene loci controlled the *tigrina* and *zonata* phenotype of 27 mutants. Five gene loci with more than one allele were responsible for the phenotype of 49 viridis mutants. In contrast, 22 induced mutants for the *macrolepis* (long awned outer glume) character in barley all proved to be allelic, but the expression of the trait varies considerably in different mutants.[143]

These few examples indicate the very great potential for genetic variability stored within the genomes of plant species, a potential that is being increasingly realized through the induction of mutations using artificial agents. Whether spontaneous or induced, individual mutants may not be exactly identical even if allelic, as has been shown by genetic fine structure analyses, discussed later. Therefore, it can no longer be assumed that an exact copy of a particular mutant can be re-induced readily. Consequently, those mutants shown to have value for breeding or analytical purposes as well as those the investigator believes may have value for breeding should be preserved in germplasm collections.

Numerous induced mutants may be directly usable as varieties, and their mutant genes a preferred germplasm[263]; other induced mutants may represent 'raw' forms, requiring modification by the forces of recombination and selection before their usefulness is realized[213,343] but this situation is not different for induced vs. spontaneous variation.[212,213] For that reason, it is desirable to isolate many mutants of a given type if possible via analysis of large M_2 or M_3 populations to allow for some choice in early selection,[157] and it is desirable also to subject obvious 'raw' mutants to recombination forces via cross or backcrosses in order to 'polish' them or improve their background type as may prove possible.[4,213,343] Closely associated loci affecting different traits sometimes may be deleted, their 'reading frame' shifted or the loci may be jointly modified in other ways (operator-regulator). No detailed analyses are yet available in most mutant systems.

There is now good evidence that mutant traits are modifiable via genetic recombination (possibly involving other loci) in crosses with certain other genotypes. Thus, of the 10 induced *ml-o* allele mutants studied by Jørgensen[193,194,196] several differed notably with regard to the intensity of associated necrotic spotting. With some of the mutants, the necrotic spotting could be eliminated in progeny of crosses with unrelated genotypes[194] (and Jørgensen personal communication 1981). However, it also appears possible to obtain similar mutants without associated negative traits such as the necrotic spotting and lower yield associated with *ml-o* mildew resistance mutants, via the recovery of larger numbers of such mutants in different genetic backgrounds or using different mutagens.[145] Similarly, Tanaka[400] was able to recover high protein mutants without associated yield or other defects by screening large populations of otherwise normal-appearing plants from mutagenized cultures.[157]

In other examples of mutant effects in crossbreeding, Gaul *et al.*[132] and Gaul and Lind[133] discovered that considerable variation in awn length could be released through recombination in crosses between an unrelated genotype and a short awned mutant of barley which showed simple inheritance in most other crosses. Thus, it often may be appropriate to carry out improvement studies with several parents to exploit a 'complex' mutant which carries a desired trait. Such was the case with the *eramosa* mutant of snapdragon, whose distorted flower trait[393] was separated from its dominant main spike character by crossbreeding (F. Scholz, H. Stubbe, personal communication 1981). It may also be advisable in some cases to use a mutant in reciprocal crosses or a male parent in the crosses or backcrosses in case induced negative cytoplasmic changes are present. Some cases of associated possible cytoplasmic damage are already known.[212,213,440] Likewise, considerable modification of the undesirable endosperm characteristics associated with the high lysine opaque o_2 mutant trait in maize has been achieved, providing encouragement that high yielding but normal semi-dent types still with the high lysine o_2 genes can be developed.[412]

Often, the recovery of recombinants without associated undesirable traits may require only screening of a very large segregating population from one or more of several crosses, or sometimes, intensive selection and reselection over several generations from specific crosses.[213,215] Many crosses should be made, involving parents differing in genetic background, and a pedigree method of selection appears essential. However, (MSFRS) male sterile gene or chemical pollen suppressant-facilitated recurrent selection systems merit consideration and might even be more efficient if strong selection pressure is applied toward the desired recombinants in the F_2 or F_3 generation of each cycle.[337]

(iv) Use of mutagens to supplement the genetic variability base for selection

A potentially very effective use of mutagens is to augment genetic variability and recombination by treatments applied to conventionally cross-bred populations. This type of application may have been largely ignored because scientific studies on applications of the procedure are lacking. Acceptable results are very difficult to acquire and to evaluate from studies of mutagen applications to populations that are potentially already highly variable. From a more pragmatic point of view, however, evaluation of this method may have to come from practice over a period of years by numerous investigators. In this context mutagens should be looked upon purely as a means to introduce more genetic variability into a population. If that population already carries some or considerable genetic variability introduced via crossbreeding, it may in principle be so much the better. Thus, the combination of mutation breeding and conventional crossbreeding methods should permit the breeder to attain the benefits of both from the selection efforts followed toward the recovery of improved lines or cultivars.[337,338] In this application, mutagen treatments might be made to pollen of parents or to F_1 or F_2 seeds. Perhaps the most useful applications may be to crosses in which the parents are widely divergent. A 'mutant' variety from mutagenized F_2 populations already has been released.[294]

With the male sterile-facilitated recurrent selection (MSFRS) systems now developing in wheat and barley, it would be meaningful and more feasible to apply mutagens to crossed (F_1) seed since crosses are easier to make when only hand pollination and protection from outcrossing is necessary. Ramage (personal communication 1981) estimates that one person can pollinate and protect enough spikes in one day to yield over 400 g of crossed seed. Evaluation of mutagen treatments might then be possible on a population basis, comparing the variability for various traits recovered from paired populations. It seems to the author that if the mutagen treatments are not so severe as to induce excessive chromosome structural disturbances, there is much to gain and little to lose from the addition of mutagen treatments. Because they induce more mutations and few chromosome aberrations, chemical mutagen treatments, EMS, DES, ENH, MNH (or azide, where effective) are preferable.

In hexaploid wheat, the 7 chromosomes of each of the 3 genomes A, B and D have a high degree of relatedness and compensating ability when corresponding chromosomes are deleted as in nulli-tetrasomic tester stocks. Normally the chromosomes of the A, B and D genomes do not pair with one another, thus are considered as homoeologous. The suppression of pairing among homoeologous chromosomes is largely due to a dominant gene *Ph*, located on chromosome 5B. Thus, in wheat nullisomic 5B (no 5B chromosomes), there is considerable intergenomic chromosome pairing and multivalent chromosome association. Sears[378] induced a mutant deficient for locus *Ph* by X-raying pollen and screening the M_1 plants for pairing mutations. The *ph* locus mutant has somewhat decreased vigor and fertility, and male gametic transmission from the heterozygote is only about 39 per cent. A second mutation with an intermediate level of homoeologous pairing also was induced, but this mutant gene is not on 5B. That either single locus mutant still retains some fertility is probably due to the influence of minor pairing suppressors known to exist on other wheat chromosomes. Wall *et al.*[420,421] also induced pairing suppressor mutants at the *ph* locus in wheat. Use of these mutants by breeders and geneticists is expected to facilitate gene transfers from alien species chromosomes of interspecific and intergeneric crosses.

(v) Uses of induced mutants in analytical studies

While the focus of this chapter has been on the role of induced mutations in the improvement of crops by breeding, there is a multitude of research uses which indirectly, through gains in knowledge of physiological processes, disease and pest resistance mechanisms, etc., result in improved breeding and selection technologies, change breeding strategies and often identify useful breeding materials. Certainly our base of genetic knowledge on crop plants, and on a few species of primarily research interest has grown considerably in recent years, to a large extent because of the rich resources of induced mutants that have become freely available for genetic analyses. Induced mutants can be expected to play an increasing role in genetic analyses of the biosynthetic mechanisms responsible for many important traits, including N and C assimilation and conversion, mineral uptake and metabolism, genetically controlled tolerances to mineral imbalances, and, possibly the most important of all, resistances to diseases and pests.

(vi) Uses in plant physiology and biochemical genetics studies

A recent symposium emphasised the use of mutations as a plant research tool. All too commonly, physiologists especially have focused their investigations on differences between species without realizing that similar differences may occur within a species, and also might be or have been induced.[417] Moreover, while near isogenic and chromosome substitution lines developed for analytical research do markedly reduce the genetic differences between strains, induced mutants may uniquely permit analyses to be focused on much smaller, perhaps truly single gene or gene ultrastructural differences, particularly if efforts are made to exclude secondary mutations. Secondary mutations usually can be excluded from such special stocks by backcrossing them at least once as the male parent to the original genotype and reselecting the mutant type. Thus, there is now an increasing use of mutants in physiological studies. Brunori et al.[70,71] have investigated nitrogen accumulation rates of high protein mutants of bread wheat.

A recessive mutant of tomato btl has a brittle stem. Analyses showed that the mutant is defective in boron metabolism.[68,419] Another tomato mutant has been found to be inefficient in its transport of iron[66,67], whereas a thiamine requiring mutant of tomato has been induced by Langridge and Brock.[228] These few examples indicate the considerable potential for using mutants preferably in the same genetic background and other variants to study plant nutrition as Vose[417] has recently discussed. Using excised roots, Picciurro et al.[322] showed that a radiation-induced mutant had a reduced efficiency for NH_4 uptake compared with its parent variety Capelli.

Treatment with a chemical agent, phenylboric acid has been shown to cause a morphogenetic or phenocopy effect simulating the *lanceolate* leaf form of the *La* tomato mutant. The morphogenetic effect of phenylboric acid and of the lanceolate gene *La* were found reversible by actinomycin D.[249,250,251] These studies may lead to a better understanding both of gene action and of the biological action of actinomycins.

The biosynthesis of plant cuticle waxes is now far better understood through the extensive investigation of 'waxless' *eceriferum* mutants in barley by von Wettstein-Knowles.[435] Associated studies by Lundqvist[242,243] have identified over 1202 *eceriferum* (*cer*) mutant alleles to 69 loci scattered over 1202 eceriferum (cer) mutant alleles to 69 loci scattered over the 6 of the 7 barley chromosomes.[143,145,242] The *cer* mutants show striking specificities with regard to their origin from mutagen treatments, locus *cer-i* mutates abundantly with neutrons, less with X-rays and still less with chemicals, while locus *cer-j* has so far been unaltered by neutrons, but mutates some with X-rays and abundantly with chemicals. The *cer* mutants show a definite organ specificity with regard to the occurrence and amounts of biochemical constituents of the wax.

Gustafsson[143] has analysed a number of earliness mutants in barley, demonstrating the occurrence of 3 groups or types of mutants, including those with (1) *long day (short night) adaptation* — having different graduations of genetically determined earliness in response to different day lengths as well as to the same day length. (2) *short day (long night) adaptation* — having graduations similar to those of (1); and (3) *photoperiod insensitivity* — with development occurring under a wide range of photoperiod conditions. Differences in insensitivity also exist among different mutants. Responses of individual mutants to thermoperiod conditions have not yet been analysed in sufficient detail, but it is already clear that interactions occur among the mutants of each type. Differential thermoperiod response may possibly be the basis for some of the effects of certain mutant alleles.

In barley[351], in maize[8,301,302,303] and in rice[9,10], waxy (wx) mutants are being exploited to learn more about the fine structure of a higher plant gene locus. Crosses among numerous non-complementing alleles have in many cases produced recombinants, indicating that intra-cistron crossing-over has occurred. By assuming the mutants to be due to base deletions, it has been possible to construct genetic fine structure 'maps' of this locus in both barley and maize to show that different wx mutant alleles carry alterations at different points along the read frame structure of the locus.[8,303,361] Amano[8,9] also has demonstrated the occurrence of intermediate (or leaky) wx mutants of both maize and rice, suggesting their origin as missense mutations rather than deletions.

Dosage effects of the ae and wx mutant genes have been investigated by Boyer et al.[45]. Normal maize starch contains about 75 per cent amylopectin and 25 per cent amylose, whereas waxy (wx) mutants may contain 0 per cent amylose. Amylose extender ae also affect the ratio of amylose: amylopectin. High amylose maize hybrids now in production have starch containing about 80 per cent apparent amylose.

Use of physiological mutants in basic aspects of crop improvement can be foreseen. Chapter 8 considers the various current ways in which the nitrogen fixation potential of crop plants might be enhanced. Rennie[350] has described the special role that mutations might have in improving the nitrogen fixation supportive traits (nis) in both legume and non-legume crops. Such potential has been found in rice, Balandreau[20] reporting that N_2-fixation associated with a series of induced mutants of the rice cv. Cesariot varied over a wide range. The mutant with the best N_2-fixing association could fix 60 times that of the poorest, while the best mutant had about 2.5 times the N_2-fixing potential of initial Cesariot.

Mutants have already contributed substantially to our knowledge of basic photosynthetic mechanisms, but in plant breeding there is especial interest in reducing wasteful photorespiration in C-3 crop plants.[298,417] The finding by Zelitch and Day[446] that low photorespiring variants of tobacco exhibited increased net photosynthesis provided encouragement that low photorespiring plants might be induced in other C-3 species.

Genetic effects in plant nutrition are discussed in Chapter 4, but we may note here known variation in genetic tolerance to Al and Mn toxicity (acid soils), Mn-deficiency, Fe-deficiency and toxicity, P-deficiency and salt, especially NaCl, toxicity. In virtually all cases, however, the tolerance levels available in natural resource collections are inadequate for many of the stress levels now present in areas suitable for crop cultivation. Plant breeding effort on these soil-related problems is as yet minimal and should be increased. Little has yet been done to exploit the available screening methods using mutagens to induce increased genetic tolerance to the stresses mentioned. In principle, success should be expected.

Nitrate reductase-deficient mutants in barley and in peas have already revealed new knowledge about the structure of this important enzyme. Unexpectedly, most mutants isolated so far have proved viable and able to grow on nitrate.[122,306] The new results suggest that higher plants have a second, back-up mechanism for reducing nitrate, leading investigations of N metabolism in an unpredicted direction no doubt requiring additional mutants for analysis. Biochemical genetic analyses of the nitrate reductase-deficient mutants are beginning to provide information about the enzyme components, and may prove useful for analyzing the genetic fine structure of the loci coding for the enzyme components (Kleinhofs personal communication 1981).

Besides having practical value, various protein and nutritional quality mutants, have already been used in biochemical and other research leading to an increased understanding of the biosynthetic pathways responsible for the improved traits. Mutants of barley have been used in studies to better understand the biochemical genetics of hordein protein synthesis and transfer to endosperm protein bodies surrounded by endoplasmic reticulum.[431] The hordein deficient mutant Bomi 1508 has a genetic block which affects not only hordein synthesis, but also starch accumulation, and the mutant also has twice the normal level of alpha amylase activity.[431] It is not yet known if the lower starch accumulation is due to catabolism. However, analyses of different mutants and recombinants should provide greater insight into the relationships among the observed phenomena. A considerable number of detailed investigations have been conducted using barley endosperm mutants, with the result that normal protein synthesis in barley is already better known. Amino acid sequencing of some hordein polypeptides has already been done.[374,431]

Brandt and Inverson[49,50] have demonstrated *in vitro* synthesis of barley endosperm proteins on mutant and wild type templates, and were able to isolate and demonstrate the translation of hordein messenger RNA from mutant and normal barley endosperm. Genetic analyses of induced high lysine barley mutants have shown that 3 mutants including the Bomi 1508 mutant are recessive alleles, designated *lys* 3 and located on chromosome 7. Risø mutants 13 and 527, were found to carry recessive high lysine alleles, located on chromosome 6 while the recessive high lysine mutant loci of Risø mutants 16 and 17 are located on chromosome 1. The dominant or semi-dominant high lysine mutant gene of Risø mutant 8 may also be located on chromosome 1.[186]

Amino acid analyses of normal and high lysine double mutants combining the sugary (*su*) and o_2 mutants of maize have shown that preformed lysine in the cell sap of the ear peduncle is highly catabolized in the normal endosperm, but not in the *su* o_2 endosperm.[382] The results indicate that the o_2 mutant trait limits the rate of lysine breakdown in maize endosperm.

Brown leaf midrib mutants of maize have been shown to produce abnormal lignins; some mutants produce less lignin than others, although all have the similar phenotypic expression for which they were selected.[271] Similar mutants have been induced in sorghum.[127,327] Because the low lignin characteristic of these mutants improves digestibility of the plant tissue to ruminant animals, the mutant trait has significant economic value. The mutants also provide a basis for in depth studies of lignin biosynthesis, since many have the same genetic background.

Mutants in barley affecting GA synthesis[174] and response to external GA applications[39] may prove helpful toward understanding not only the GA biosynthetic pathways, but also the role of GA in plant growth, including height expression. Similar mutants appear to have been induced in rice (Kawai and Rutger, personal communication 1981), but so far not in wheat[212], and (Konzak, unpublished).

One type of short straw mutant in barley, the *erectoides* (*ert*) type, has been extensively analyzed by Swedish workers[315] and localized to 29 loci in the barley genome. Most *erectoides* mutants have a reduced spike internode length, hence only a few have proved useful in breeding, and those have been of exceptional value (Micke and Donini in press). However, reduced height mutants with normal or longer spikes in barley, called semidwarfs[7] as in wheat, are readily induced[212,213,408] and these are valuable for breeding because of their more desirable agronomic characteristics. The reduced height mutants of wheat already have significantly extended the base of genetic variability available for research analyses as well as for breeding.[213,215] In both wheat and barley, studies have shown that reduced height mutants generally do not possess the association of short coleoptile length observed for certain GA insensitive spontaneous semidwarf reduced height gene sources[212,213,215,408], and (Ullrich, personal communication 1981) and therefore may prove more useful in breeding varieties adapted to deep seeding. Analyses of the different responses to GA[174] and possible differences in the peroxidase enzyme activity of the reduced height gene sources may extend our understanding of the physiology of height control and expression in plants.

In rice, mature plant height in derived induced mutant lines was not significantly related to laboratory seedling height, and although tall lines as a group showed a higher emergence percentage (seedling vigor) than semidwarf lines, nevertheless semidwarf lines with seedling vigor equal to the tall parent were obtained,[363] indicating that the association of plant height and seedling vigor may involve genetic linkages but is not a pleiotropic relation. Semidwarf height is usually controlled by single recessive genes, and a number of genetically independent mutants have been induced.[347,363]

The potential variety of applications of induced mutants in analytical investigations is considerable, perhaps limited only by the investigator's imagination and ability to obtain a source of mutants, or to induce them in a suitable genetic background.

(vii) Uses in studies on the genetics of host pathogen interactions

Several recent studies on the uniquely non-specific (so far, at least) mildew resistant single gene mutant alleles induced at the *ml-o* locus on chromosome 4 in barley show the locus to be genetically at least tripartite.[192,193,194,195] However, this locus may be even more complex, as analyses of the locus fine structure proceeds through the use of additional *ml-o* mutants.

Röbbelen *et al.*[358] and Einfeld *et al.*[114] have begun an intensive study to investigate the nature of host pathogen relationships using induced mildew resistance mutants of barley induced and evaluated under carefully controlled conditions. Race specificity for induced resistance was demonstrated for 46 induced mutants, indicating mutation from medium grade field resistance to race resistance specificity; adult plant and young plant resistances were observed, as was increased susceptibility. The results suggested that mutations may occur more frequently in regulator than in structurate genes. Yamaguchi and Yamashita[441] also have induced mildew resistance mutants at loci other than *ml-o*, and more recently Röbbelen *et al.*[356] have induced *ml-o* mutants and additional mildew race-specific resistance mutants. They identified allelism for two mutants located close to the *ml-o* locus and allelism for three mutants with partial resistance. The mutants were independent and epistatic over the *Ml-la* locus of the original cv. Bomi. These recent results suggest that an isophenic segment conferring mildew resistance may be located also on chromosome 4. About 25 per cent of all mildew resistance mutants induced in barley over a recent 5 year period by Swedish scientists have proved to be *ml-o* alleles.[143]

Favret and colleagues[118,119,120] have used leaf rust (*Puccinia recondita*) resistant mutants of wheat to learn more about host/pathogen interrelationships with these species. In other studies with mutants from the Argentine cv. 'Sinvalocho', Favret[119] was able to demonstrate that this cultivar carries a duplication of the gene for resistance to leaf rust race group 20. A further gamma radiation induced gene duplication in another mutant, Magnif 96, increased the resistance to leaf rust and improved resistance to stem rust. Borojevic[35,36,37,38] was able to isolate leaf rust (*P. recondita*) resistant variants ranging in reaction from highly to moderately resistant, indicating the presence of both race specific and non-specific resistances in the selections from mutagenized material.

McIntosh[253] studied the genetic interrelations of mildew, stem rust and leaf rust resistances using EMS-induced mutants of mildew resistance gene *Pml* linked in the distal sector of chromosome 7AL of wheat to genes *Sr15* for stem rust resistance and *Lr20* for leaf rust resistance. He found that changes in *Pml* were independent of changes in *Sr15* and *Lr20*, whereas all changes in *Lr20* were associated with changes in *Sr15*. Variation of infection types for *Lr20* and *Sr15* was observed but the relative changes monitored by the two pathogens was not the same. Although the results were interpreted as suggesting that averulence alleles of the two pathogens recognize the same host gene product, it seems possible that the mutants obtained affected both *Sr15* and *Lr20* because they are closely linked. Isolation and analysis of many more mutants would be necessary to differentiate between the two hypotheses. However, as noted above, Favret[119] has observed that the duplication of a leaf rust resistance gene locus confers resistance also to stem rust. Perhaps the two studies are related.

Numerous other materials are available or can readily be induced for similar studies of host-parasite relationships. Konzak[214,215] has recently obtained wheat mutants differing in levels of resistance to stripe rust (*Puccinia striiformis*). These are now being subjected to analysis by plant pathologists, and better methods for analysis of dilatory and non race specific resistances are now available.[217,239] Likewise, Fossati and Bronniman[124] indicate progress in inducing mutants of wheat with greater tolerance to blight caused by *Septoria nodorum*.

The close genetic association of the 'Victoria' type of resistance to crown rust (*P. coronata*) in oats and susceptibility to *Helminthosporium victoriae* blight has already been subjected to preliminary studies,[422,424] but with better mutagens available today, and the extremely efficient screening method available, studies of many more mutations at this locus would be most informative. Since the toxic protein 'victorin' produced by the fungus is now better characterized,[333] parallel mutation studies on the fungus may also prove instructive, especially since the studies by Wallace[424] and Konzak[220] indicate differences in the nature of the *Vb/Pc* locus. Moreover, because screening can probably be done using tissue cultures, it should be possible to isolate many hundreds of mutants for analysis. Rines (personal communication 1981) is now using Vb/Pc:vb/pc oat heterozygotes in mutation studies with tissue cultures.

In maize, Gengenback and Green[134] have selected T. cytoplasm mutants resistant to *H. maydis* pathotoxin and to the fungus. The mutants were selected from tissue cultures without using mutagens. Recent analyses have shown the mutational change to have occurred in mitochondrial DNA (Gengenback, personal communication 1981) but the mutants became male fertile. Application of mutagens might induce more subtle changes permitting the recovery of blight resistant strains that retain the desired T. cytoplasmic male sterility.

SOME GUIDELINES FOR MUTATION BREEDING APPLICATIONS

(i) Vegetatively propagated species

Vegetatively propagated species — successful methods for the induction of mutants in tissues propagated by cuttings, budding, induction of adventitious buds, etc. have been described by Broertjes and van Harten.[62] Since this treatise was published, however, cell culture methods have become more widely used, and where appropriate to the application, should be considered, as described in Chapter 11.

(1) *THE INITIAL MATERIALS*

General guidelines presented by Broertjes and van Harten[62] are appropriate especially with regard to the genetic constitution (if known) of the strain selected for mutation induction. Most often, however, the criterion will be the commercial potential of the initial material, and predicted commercial value of the mutants sought.

(2) *THE V_1 GENERATION*

Exploratory treatments with a sample of the material, will usually be desirable in order to focus a mass scale effort on a more narrow range of mutagen treatments. Even using the most efficient techniques applicable for the species, one will usually not recover as many mutants as might be desirable to allow some choice in selection for propagation.

(3) *POPULATION SIZE*

It usually is desirable to treat as many scions, tissue pieces, etc., as can conveniently be propagated and screened for the desired trait, using a series of treatments following up later in a concentrated effort on a large population from only one or two most effective treatments. However methodologies considered most effective at present differ with the species as do the tissues preferred for treatment.[62]

(4) *PROPAGATION AND EVALUATION*

Tissue culture methods may prove useful for quickly increasing the quantity of mutant material available for evaluation and eventual commercialization. Mutant material is then evaluated by methods followed in breeding by other methods.

(ii) **Seed reproduced species**

(1) *THE INITIAL MATERIALS*

Materials selected for treatment should be as genetically uniform as possible, especially for the type of trait sought by mutation. To be competitive, it is essential that mutation breeding applications be made to materials of the highest performance potential. Thus, in practice, the increase from an advanced generation line or from a Breeder or Elite seed stock grown in isolation may be the best one can do conveniently. However, the breeder must weigh the time, expense, verification requirements, and other factors in deciding how much possible contamination of M_1 material can be tolerated.

Usually, the breeder will have a specific goal. A plan for analysis of the material, i.e. a screening method for the mutant trait(s) being sought, should be developed before starting the experiment. The initial materials selected for mutagen treatment has to be appropriate for the goal(s). Typically, the material selected will have only a single weakness for which a corrective mutation is sought, as for example, a high yielding, disease resistant, good quality line that is too tall for the production environment. Even if mutant gene(s) for further breeding use are sought, the material selected should be adapted, and have high yield potential, since it will prove to be more useful as a parent for future cross-breeding. However, it is most important to conduct the mutation work with a genotype that best suits the goal of the research.

An exception to the general rules concerning specific mutation types or the purity of the initial stock, is the use of mutagens to complement or supplement the variability-creating forces of genetic recombination via treatment of random or selective mating recurrent selection populations, in which crossing is facilitated by the use of genetic male sterility or a chemical pollen suppressant. In this application, the goal is crop improvement, and the origin of the specific variations used is not a concern. However, it is important that mutagen treatments applied are not restrictively damaging viability or fecundity. The benefits, if any, from mutagen treatment might be visible if mutagen treated population samples are compared with non-treated control populations each generation. As mentioned before, this type of application merits more evaluation.

(2) THE M_1 GENERATION

Mutagen treated seed propagated crops should generally be grown first in isolation. Genetic isolation would be sufficient if the treated material were surrounded by a large area of the parent cultivar, another species or even another cultivar generally distinct enough for F_1 hybrids to be distinguishable readily. M_1 treated seeds of cereals and other plants should be sown at a relatively high seeding rates to increase the proportion of progeny resulting from primary shoots. No additional isolation is necessary if controlled pollinations are made, as is usually done with cross pollinated species and male sterile facilitated recurrent selection (MSFRS) populations.

(3) POPULATIONS, EXPECTED MUTATION FREQUENCIES

When seeds are treated, it is generally considered desirable to use several thousand seeds per treatment. Then, if mutant genes for further breeding are sought, one should select for emphasis in M_2 screening those treatments which show moderate to severe injury. If direct mutation breeding applications are intended, screening would be done on those populations showing mild to moderate effects from the mutagen treatments in order to reduce problems from multiple mutations.

Mutation frequencies will be roughly related to the injury (lethality and sterility) observed in the M_1 generation, though the extent of these effects and their correlations vary considerably with the mutagen. Numbers of a specific mutant type sought often are low, perhaps $1/10^4$ to $1/10^6$ M_2 plants or M_2 progenies depending upon the probable number of genes concerned. Any gain above that can be considered good luck. Failure to obtain the desired forms in one trait generally indicates either that the treatments were not sufficiently effective or that the population screened was too low. The frequency of other mutants observed in M_2 is a good indication of treatment effectiveness. A second experiment the next season, improving upon conditions employed in the first might be successful. However, the genetic composition of the material treated can also influence the mutation spectrums, after treatment with a given mutagen.

The original genotype may be an important determinant of mutation frequency, or even whether a mutation can be induced. In the worst case, one genotype may perhaps not mutate at a given locus, because that locus involves a large deletion. Another genotype may not carry the deletion, or may carry an inactivation or frameshift mutation which may be readily mutable. Mutability is a characteristic of the gene locus structure, but may also be affected by the genetic background. Thus, before deciding that a trait is not obtainable by induced mutation, it will be necessary to screen large mutagen-treated populations of different genotypes for that trait. The high mutability of some traits (e.g. plant height/chlorophyll content) actually results because several to many loci are involved. Exceptionally, single locus mutation frequencies as high as 1 per 1000 M_2 progenies can be expected from some loci (as with the *ml-o* mutation in barley[143]. MacKenzie and Martens[254] working with mutagenized oats, failed to obtain a stem rust resistance mutant from screening several very large M_2 populations.

(4) *SCREENING*

The efficiency of mutation breeding, more than any other breeding method is dependent on the effectiveness with which useful genetic variants can be recognized in the M_2 or M_3 generations subjected to screening. Some 'escapes' or environmentally determined variants can be tolerated if the mutation frequency is high, the screened population large, and/or the method of confirmation simple. The mutation breeding approach normally should not be applied without a specific goal, nor applied to problems for which selection methods are inefficient. Thus, semidwarfing and earliness mutants should be obtainable by mutation breeding as long as field screening conditions are reasonably adequate. For earliness, it is essential that screening be done at the time of heading or first flowering, whichever time is most discriminating. Screening for semidwarfs usually can be selected any time after normal height expression is nearly complete. Thin planting of the segregating (usually M_2) population facilitates visual selection of mutants. Screening for disease resistance or chemical toxicity tolerance likewise requires conditions which allow for each differentiation of resistant or tolerant types from large M_2 or M_3 populations.

It generally is better to have an excessively large population with many variants than to have too small a population and far too few variants. The isolation of many similar variants allows the investigator more choice for later selection. Haq[157] has recently reviewed procedures for mutation breeding that have already proved successful for isolation of a number of important agronomic traits in rice. This excellent report provides a current 'state of the art' analysis of mutation breeding practices in which largely field screening and evaluation methods are employed.

ACKNOWLEDGEMENTS

Washington and Agricultural Research Center Projects 1568 and 1570, and Program in Genetics and Cell Biology. Suggestions regarding content, and critical reading of the manuscript by Drs. A. Micke, A.M. van Harten, T. Kawai and B. Donini are deeply appreciated, as are the many hours of manuscript preparation and revision by Ms. Joan Muhs and Ms. Dorothy Larson, Department of Agronomy and Soils and Program in Genetics and Cell Biology.

REFERENCES

1.   Abdullaev, I.K. and Mekhtieva, T.D. Useful mutant forms of strawberry obtained by treatment with chemical mutagesn. *Eksp. Mutagenez. Rast.* **2**, 147–148 (In Russian). *Plant Breeding Abstr.* **47**, No. 1636, 1974.

2.   Abraham, A. Breeding work on tapioca (cassava) and a few other tropical tuber crops. *Tropical Root and Tuber Crops Tomorrow.* D.P. Plucknett (ed.), Vol. 1, 76–78. Univ. Hawaii, Honolulu, Hawaii, 1970.

3.   Adams, M.W.   Plant architecture and physiological efficiency in the field bean. *Potentials of Field Beans and Other Food Legumes in Latin America. Series Seminars No. 2E, 266–278.* Cali, Colombia (Centro Internacional de Agricultura Tropica), 1973.

4.   Adams, M.W.   Plant architecture and yield breeding. *Iowa State J. of Research* **56**, (3), 225–254, 1982.

5.   Ahokas, H.   Cytoplasmic male sterility in barley. *Acta Agriculturae Scandinavica* **29**, 219–224, 1979.

6.   Alexander, L.J., Oakes, G.L. and Jaberg, C.A.   The production of two needed mutations in tomato by irradiation. *Heredity* **62**, 311–315, 1971.

7.   Ali, Mohamed A.M., Okiror, S.O. and Rasmusson, D.C.   Performance of semidwarf barley. *Crop Sci.* **18**, 418–411, 1978.

8.   Amano, E.   Genetic find structure analysis of mutants induced by ethyl methane-sulfonate. *Gamma-Field Symposia No. 11 'Chemical Mutagenesis in Microorganisms and Plants'*, Institute of Radiation Breeding, 43–59, 1972.

9.   Amano, E.   Induction of waxy mutants in rice by reactor radiations. *Annual Report of National Institute of Genetics (Japan)* **29**, 69, 1979.

10.   Amano, E.   Waxy starch mutations in rice. *Annual Report of National Institute of Genetics (Japan)* **26**, 51, 1976.

11.   Anonymous.   *Lupinus cosentinii* Guss. (Sandplain Lupin) cv. "Eregulla" Reg. No. B-7a-2. *J. Austr. Inst. Agr. Sci.* **43**, 87–89, 1977.

12.   Anonymous. *Sugarcane Breeding Institute, Ann. Rep.*, Coimbatore, India, 117, 1966–67.

13.   Ashri, A. and Levy, A.   Sensitivity of developmental stages of peanut (*A. hypogaea*) embryos and ovaries to several chemical mutagen treatments. *Rad. Bot.* **14**, 223–228, 1974.

14.   Atkins, I.M.   Registration of Alamo-x oats, Reg. No. 174. *Crop Sci.* **2**, 531, 1962.

15.   Avivi, L.   High grain protein content in wild tetraploid wheat *Triticum dicoccoides* Korn. *Proc. 5th Int. Wheat Genet. Symp. New Delhi 1978.* 372. Ramanujam, S. (ed.), India Soc. Genetics and Plant Breeding, 1979.

16.   Awan, M.A., Cheema, A.A. and Akbar, M.   Double cropping of rice with an early maturing mutant variety. *Mutation Breeding Newsletter* **13**, 12–13, 1979.

17.   Axtell, J.D.   Breeding for improved nutritional quality. *Plant Breeding II.* K.J. Frey (ed.), 365–432, Iowa State Univ. Press, Ames, 1979.

18.   Axtell, J.D., Oswalt, D.L., Mertz, E.T., Pickett, R.C., Jambunathan, R. and Srinivasan, G.   Components of nutritional quality in grain sorghum. *High-Quality Protein Maize*, 374–385. Dowden, Hutchinson and Ross, Stroudsburg, Pennsylvania, 1974.

19.   Axtell, J.D., VanScoyoc, S.W., Christensen, P.J. and Ejeta, G.   Current status of protein quality improvement in grain sorghum. *Seed Protein Improvement in Cereals and Grain Legumes.* Vol. II. STI/PUB/496, 357–366. IAEA, Vienna, 1979.

20.   Balandreau, J.   Improving N_2-fixation by optimal rice-diazotrophs associations-potential use of induced mutations. *Mutation Breeding Newsletter* **16**, 10–12, 1980.

21.   Balandreau, J. and Knowles, R.   The rhizosphere. *Interactions Between Non-Pathogenic Soil Microorganisms and Plants.* Y.R. Dommergues and S.V. Krupa (eds.), 243–268. Elsevier Scientific Publ. Co. Amsterdam, 1978.

22.   Bansal, H.C.   Induced variability for protein quantity and quality in barley. *Breeding Research in Asia and Oceania. Proc. 2nd General Congress Society for Advancement of Breeding Researches in Asia and Oceania. Session X. Breeding for Protein Nutritive Quality, New Delhi, India.* 560–668; 657–661. Indian Soc. Genet. Plant Breeding, 1974.

23.   Bansal, H.C., Kumar, R., Singh, R.P. and Bhaskaran, S.   Genetic improvement of seed protein in barley. *Proc. 4th Int. Barley Genet. Symp. Edinburgh* (in press).

24.   Barg, R. and Umiel, N.   Selection for herbicide resistance in tissue cultures and phenotypic variation among the resistant mutants. *Production of natural compounds by cell culture methods.* E. Reinhard and A.W. Alfermann (eds.), 337–345. Gesellschaft fur Strahlen- und Umweltforschung mbh, Munich, West Germany, 1978.

25.   Barnes, R.F., Muller, L.D., Bauman, L.F. and Collenbrander, V.F.   *In vitro* dry matter disappearance of brown midrib mutants of maize (*Zea mays* L.). *J. Anim. Sci.* **33**, 881–884, 1971.

26.   Barrentine, W.L., Edwards Jr, C.J. and Hartwig, E.E.   Screening soybeans for tolerance to metribuzin. *Agron. J.* **68**, 351–353, 1976.

27. Bartholomew, D. Vegetable oil fuel. *J. Am. Oil Chemists Soc.* 58, 286–288, 1981.

28. Bauer, R. Westra, an X-ray-induced erect-growing black-currant variety, and its use in breeding. *Polyploidy and Induced Mutations in Plant Breeding.* STI/PUB/359, 13–20, 1974.

29. Bhagwat, S.G., Bhatia, C.R., Gopalakrishna, T., Joshua, D.C., Mitra, R.K., Narahari, P., Pawar, S.E. and Thakare, R.G. Increasing protein production in cereals and grain legumes. *Seed protein improvement in cereals and grain legumes.* Vol. II. STI/PUB/496, 225–236. IAEA, Vienna, 1979.

30. Bhatia, C.R. and Rabson, R. Bioenergetic considerations in cereal breeding for protein improvement. *Science* 194, 418–421, 1976.

31. Blake, T.K. New techniques for evaluating lysine content in hordeins. *Barley Genet. Newsl.* 11, 79–83, 1981. (By permission).

32. Blixt, S. A crossing programme with mutants in peas: Utilization of a gene bank and a computer system. *Induced Mutations in Cross-Breeding.* STI/PUB/447, 21–26. IAEA, Vienna, 1976.

33. Blixt, S. and Vose, P.B. Breeding towards an ideotype — aiming at a moving target? *Contemporary Bases for Crop Breeding.* P.B. Vose and S. Blixt (eds.), Pergamon Press, London and New York, 1983.

34. Blum, A., Gozlan, G. and Mayer, J. The manifestation of dehydration avoidance in wheat breeding germ plasm. *Crop Sci.* 21, 495–499, 1981.

35. Borojevic, K. Evaluating resistance to *Puccinia recondita tritici* in mutant lines selected in wheat after mutagenic treatments. *Radiation Botany* 15, 367–374, 1975.

36. Borojevic, K. Studies on resistance to *Puccinia recondita tritici* in wheat population after mutagen treatments. *Induced Mutations Against Plant Diseases.* IAEA-SM-214/1, 393–401. IAEA, Vienna, 1977.

37. Borojevic, K. Tolerance to *Puccinia recondita tritici* in the population of *Triticum aestivum vulgare* after mutagenic treatments. *Genetika* 8, No. 3, 233–239, 1976a.

38. Borojevic, K. Type of infection, severity and tolerance to *Puccinia recondita tritici* in old mutant lines of wheat. *Induced Mutations for Disease Resistance in Crop Plants.* (FAO/IAEA/SIDA Research Co-ord. Meeting, Ames, Iowa, 1975). IAEA-181, IAEA, Vienna, 41–61, 1976b.

39. Boulger, M.C., Sears, R.G. and Kronstad, W.E. An investigation of the association between dwarfing sources and GA response in barley. *Fourth Int. Barley Genet. Symp., Edinburgh.* (in press).

40. Bouma, J. A new variety of spring barley, "Diamant" in Czechoslovakia. *Induzierte Mutationen und ihre Nutzung.* 177–182. Akademie-Verlag Berlin, 1967.

41. Bouma, J. The spring barley mutant cultivar Diamant, its economic importance and breeding value. *Mutat. Breed. Newsl.* 7, 2–3, 1976.

42. Bouma, J. and Minarik, F. Results of breeding of malting barley in the Czech socialist republic. *Breeding and Productivity of Barley. Proc. Intern. Symp. 1972.* 139–148. Komeriz, Czechoslovakia, 1973.

43. Boyer, C.D. and Preiss, J. Multiple forms of starch branching enzyme of maize: evidence for independent genetic control. *Biochem. Biophys. Res. Comm.* 80, 169–175, 1978.

44. Boyer, C.D. and Preiss, J. Properties of citrate-stimulated starch synthesis catalysed by starch synthase I of developing maize kernels. *Plant Physio.* 64, 1039–1042, 1979.

45. Boyer, C.D., Garwood, D.L. and Shannon, J.C. Interaction of the Amylose-extender and waxy mutants of maize. *Heredity* 67, 209–214, 1976.

46. Boxus, P., Quoirin, M. and Laine, J.M. Large scale propagation of strawberry plants from tissue culture. *Plant Cell, Tissue and Organ Culture.* J. Reinert and Y.P.S. Bajaj (eds.), 130–143. Springer Verlag, Berlin, 1977.

47. Bozzini, A. Radiation-induced male sterility in *durum* wheat. *Polyploidy and Induced Mutations in Plant Breeding.* STI/PUB/359, 23–25. IAEA, Vienna, 1974.

48. Brandt, A. Cloning of double stranded DNA coding for hordein polypeptides. *Carlsberg Res. Comm.* 4, 255–267, 1979.

49. Brandt, A. and Inverson, J. *In vitro* synthesis of barley endosperm proteins on wild type and mutant templates. *Carlsberg Res. Comm.* 41, 312–320, 1976.

50. Brandt, A. and Inverson, J. Isolation and translation of hordein messenger RNA from wild type and mutant endosperms in barley. *Carlsberg Res. Comm.* 43, 451–469, 1978.

51. Breider, H. Uber die zuchterische Auswertung und uber the praktische Verwertung rontgeninduzierter somatische mutationen bei langlebigen und vegetativ vermehrbaren Kulturpflanzen. *Mitt. Kloster-neuberg*, Ser. A. **14**, 165–171, 1964.

52. Brock, R.D. Mutation plant breeding for seed protein improvement. *Seed protein improvement in cereals and grain legumes.* Vol. I. STI/PUB/496, 43–55. IAEA, Vienna, 1979.

53. Broertjes, C. Mutation breeding of autotetraploid *Achimenes* cultivars. *Euphytica* **25**, 297–304, 1976.

54. Broertjes, C. Mutation breeding of *Achimenes. Euphytica* **21**, 48–63, 1972a.

55. Broertjes, C. Mutation breeding of *Streptocarpus. Euphytica* **18**, 333–339, 1969.

56. Broertjes, C. The development of (new) *in vivo* and *in vitro* techniques of significance for mutation breeding of vegetatively propagated crops. *Improvement of vegetatively propagated crops through induced mutation.* TEC/DOC/173, 23–31. IAEA, Vienna, 1975.

57. Broertjes, C. The production of polyploids using the adventitious bud technique. *Polyploidy and Induced Mutations in Plant Breeding.* STI/PUB/359, 29–35. IAEA, Vienna, 1974.

58. Broertjes, C. Use in plant breeding of acute, chronic or fractionated doses of X-rays or fast neutrons as illustrated with leaves of Saintpaulia. PUDOC, Wageningen. *Agr. Res. Rep.* **776**, 74, 1972b.

59. Broertjes, C. and Alkema, H.Y. Mutation breeding of flowerbulbs. *First Int. Symp. on Flowerbulbs.* **2**, 407–411. Noordwijk/Lisse, 1970.

60. Broertjes, C. and Ballego, J.M. Mutation breeding of *Dahlia variabilis. Euphytica* **16**, 171, 1967.

61. Broertjes, C., Haccius, B. and Weidlich, S. Adventitious bud formation on isolated leaves and its significance for mutation breeding. *Euphytica* **17**, 321–344, 1968.

62. Broertjes, C. and Van Harten, A.M. *Application of mutation breeding methods in the improvement of vegetatively propagated crops. An interpretive literature review.* Elsevier Scientific Publ. Co., Amsterdam, 316, 1978.

63. Broertjes, C., Koene, P. and Van Veen, J.W.H. A mutant of a mutant of a . . .: Irradiation of progressive radiation-induced mutants in a mutation-breeding programme with *Chrysanthemum morifolium* Ram. *Euphytica* **29**, 525–530, 1980.

64. Broertjes, C., Roest, S. and Bokelmann, G.S. Mutation breeding of *Chrysanthemum morifolium* Ram. using *in vivo* and *in vitro* adventitious bud techniques. *Euphytica* **25**, 11–19, 1976.

65. Broertjes, C. and Verboom, H. Mutation breeding of *Alstroemeria. Euphytica* **23**, 39–44, 1974.

66. Brown, J.C., Chaney, R.L. and Ambler, J.E. A new tomato mutant inefficient in the transport of iron. *Physiol. Plant.* **25**, 48–53, 1971.

67. Brown, J.C., Clark, R.B. and Jones, W.E. Efficient and inefficient use of phosphous by sorghum. *Soil Sci. Soc. Amer. J.* **41**, 747–750, 1977.

68. Brown, J.C. and Jones, W.E. Differential transport of boron in tomato. *Physiol. Plant.* **25**, 279–282, 1971.

69. Brown, P.D., McKenzie, R.I.H. and Mikaelsen, K. Agronomic, genetic and cytological evaluation of a vigorous new semidwarf oat. *Crop Sci.* **20**, 303–306, 1980.

70. Brunori, A., Micke, A. and Hermelin, T. Dry matter and nitrogen accumulation in the endosperm of wheat mutants. *Induced Mutations – A Tool in Plant Research. Extended Synopses.* STI/PUB/591, 244–245. IAEA, Vienna, 1981.

71. Brunori, A., Axmann, H., Figueroa, A. and Micke, A. Kinetics of nitrogen and dry matter accumulation in the developing seed of some varieties and mutant lines of *Triticum aestivum. Z. Pflanzenzuchtg.* **84**, 201–218, 1980.

72. Burnham, C.R., Phillips, R.L. and Albertson, M.C. Inheritance of male sterility in flax involving nuclear-cytoplasmic interaction, including methods of testing for cytoplasmic restoration. *Crop Sci.* **21**, 659–663, 1981.

73. Burrows, V.D. Use of dwarf oat mutant in cross breeding. *Mutation Breeding Newsletter* **14**, 4, 1979.

74. Burton, G.W. Induced mutations for improving millets, apomictic crop plants and vegetatively propagated grasses. *Induced Mutations for Crop Improvement in Africa.* TEC/DOC/222, 33–40. IAEA, Vienna, 1979.

75. Burton, G.W. Radiation breeding of warm season forage and turf grasses. *Polyploidy and Induced Mutations in Plant Breeding.* STI/PUB/359, 35–39. IAEA, Vienna, 1974.

76. Burton, G. Tifway-2 bermudagrass. *Mutation Breeding Newsletter* **18**, 8–10, 1981.

77. Burton, G.W. and Hanna, W.W. Ethidium bromide-induced cytoplasmic male-sterility in pearl millet. *Crop Sci.* **16**, 731—732, 1976.

78. Burton, G.W. and Hanna, W.W. Performance of mutants induced in sterile turf bermudagrass. *Mutation Breeding Newsletter* **9** (1) 4, 1977.

79. Calhoun, W., Crane, J.M. and Stamp, D.L. Development of a low glucosinolate, high erucic acid rapeseed breeding program. *J. Am. Oil Chem. Soc.* **52**, 363—365, 1975.

80. Campbell, A.I. and Wilson, D. Prospects for the development of disease-resistant temperate fruit plants by mutation induction. *Induced Mutations Against Plant Diseases.* STI/PUB/462. IAEA, Vienna, 215—226, 1977.

81. Cassini, R., Bousquet, J.F., Berville, A. and Cornu, A. *Helminthosporium maydis* et sterilite male cytoplasmique chez le mais. *Am. de Phytopthol.* **7**, 210, 1975.

82. Chapman, W.H., Luke, H.H., Wallace, A.T. and Pfahler, P.L. Florad Oats. *Florida Agr. Exp. Sta. Circ.* S-128, 8, 1961.

83. Choudhary, D.K. and Kaul, B.L. Radiation induced methyl-eugenol deficient mutant of Cymbopogon flexuosus (Nees ex Steud.) *Wats. Proc. Indian Acad. Sci. Sec. B.* **88**, 225—228, 1979.

84. Clayton, E.E. Male sterile tobacco. *J. of Hered.* **41**, 171—175, 1950.

85. Coffman, W.R. and Juliano, B.O. Seed protein improvement in rice. *Seed protein improvement in cereals and grain legumes.* Vol. II. STI/PUB/496, 261—277. IAEA, Vienna, 1979.

86. Conger, B.V. *Cloning agricultural plants via in vitro techniques.* 272. CRC Press, Inc. Boca Raton, Florida, 1981.

87. Crane, M.B. and Lawrence, W.J.C. *The genetics of garden plants.* 86—90. McMillan and Co. London, 1956.

88. Dalrymple, D.G. Development and spread of high-yielding varieties of wheat and rice in less developed nations. *Foreign Agric. Econ. Rep.* No. **95**, 134. USDA in Coop. with USAID, Washington, DC, 1978.

89. Darmodjo, S. and Wirioatmodjo, B. Resistance mutation of sugarcane species (*Saccharum officinarum*) from Top Borer (*Scirpophaga nivella*) as the effect of irradiation with gamma rays. *Lokakarya Pemberantasan Hama Dengan Radiasi* **4**, 10. Jakarta, 1976 (In Indonesian with English summary).

90. Davies, D.R. Creation of new models for crop plants and their use in plant breeding. *Applied Biology* **2**, 87—127, 1977.

91. Davies, D.R. Crop structure and yield in *Pisum. Advances in Legume Science, Vol. 1 of Proc. of Int. Legume Con., Kew, July 31—Aug. 4, 1978.* R.J. Summerfield and A.H. Bunting (eds.). Her Majesty's Stationery Office, London, 637—641, 1980.

92. DeKock, M.J. The actions of gamma irradiation and N-methyl-N-nitrosourea in plants. *Thesis.* Washington State University, Pullman, Washington, 1972.

93. Dellaert, L.M.W. Comparison of selection methods for specified mutants in self-fertilizing crops: theoretical approach. *Seed Protein Improvement in Cereals and Grain Legumes.* Vol. I. STI/PUB/496, 57—74. IAEA, Vienna, 1979.

94. Denic, M. Some characteristics of proteins in mutant lines of hexaploid wheat. *Seed Protein Improvement by Nuclear Techniques.* 365—381. STI/PUB/479. IAEA, Vienna, 1978.

95. Doll, H. Genetic possibilities for improving the nutritional quality of barley protein. *Proc. 4th Int. Barley Genet. Symp. Edinburgh* (in press).

96. Doll. H. Inheritance of the high-lysine character of a barley mutant. *Hereditas* **74**, 293—294, 1973.

97. Doll, H. Storage proteins in cereals. *Genetic Diversity in Plants, Proc. of a Symp. Genetic Control of Diversity in Plants, Lahore, Pakistan, March 1—7, 1976.* A. Muhammed, R. Aksel and R.C. von Borstel (eds.), Plenum Press, NY, 337—346, 1977.

98. Doll, H., Koie, B. and Eggum, B.O. Induced high lysine mutants in barley. *Radiat. Bot.* **14**, 73—80, 1974.

99. Donini, B. Induction and isolation of somatic mutations in vegetatively propagated plants. *Improvement of vegetatively propagated plants through induced mutations.* 35—51. TEC/DOC/173. IAEA, Vienna, 1975.

100. Donini, B. The use of radiations to induce useful mutations in fruit trees. *Improvement of vegetatively propagated plants and tree crops through induced mutations.* 55—67. IAEA-194. IAEA, Vienna, 1976.

101. Dommergues, P. La destinee de la cellule mutee: Consequences dans le cas des plantes a multiplication vegetative et dans le cas des plantes a reproduction sexuee. *Eucarpia Cong. Assoc. Em. Amelior. Plant. (Paris).* 115–139, 1962.

102. Driscoll, C.J. A chromosomal male-sterility system of producing hybrid wheat. *Proc. 4th Int. Wheat Genetics Symp.* E.R. Sears and L.M.S. Sears (eds.), 669–674. Missouri Agri. Exp. Sta., Columbia, 1973.

103. Driscoll, C.J. Induction and use of the 'Cornerstone' male-sterility mutant in wheat. *Proc. 5th Int. Wheat Genetics Symp., New Delhi,* 499–502, 1978.

104. Driscoll, C.J. *Perspectives in chromosome manipulation.* Phil. Trans. R. Soc. Lond. B **292**, 535–546, 1981.

105. Driscoll, C.J. Registration of cornerstone male-sterile wheat germplasm. *Crop Sci.* **17**, 190, 1977.

106. Driscoll, C.J. XYZ system of producing hybrid wheat. *Crop Sci.* **12**, 516–517, 1972.

107. Driscoll, C.J. and Barlow, K.K. Male sterility in plants: Induction, isolation and utilization. *Induced Mutations in Cross-Breeding.* STI/PUB/447, 123–131. IAEA, Vienna, 1976.

108. Dumanovic, J., Denic, M., Jovanovic, C. and Ehrenberg, L. Variation in content and composition of protein in wheat induced by mutation. *Nuclear Techniques for seed protein improvement.* STI/PUB/328, 153–161. IAEA, Vienna, 1973.

109. Durzan, D.J. and Campbell, R.A. Prospects for introduction of traits in forest trees by cell and tissue culture. *N.Z. J. For. Sci.* 4, 261–266, 1974a.

110. Durzan, D.J. and Campbell, R.A. Prospects for the mass production of improved stock of forest trees by cell and tissue culture. *Can. J. For. Res.* 4, 151–174, 1974b.

111. Durzan, D.J. and Lopushanski, S.M. Propagation of American elm via cell suspension cultures. *Can. J. For. Res.* **5**, 273–277, 1975.

112. Duwayri, M.A., Polle, E. and Konzak, C.F. Screening of wheat genotypes for drought tolerance. *Agronomy Abst.* p. 10, 1981.

113. Ehrenberg, L. and Wachtmeister, C.A. Safety precautions in work with mutagenic and carcinogenic chemicals. B.J. Kilbey, M. Legator, W. Nichols and C. Ramel (eds.). *Handbook of Mutagenicity Test Procedures,* 401–418. Elsevier, 1977.

114. Einfeld, E., Abdel-Hafez, A.G., Fuch, W.H., Heitefuss, R. and Robbelen, G. Investigations on resistance of barley against mildew (*Erisiphe graminis*). *Induced Mutations for Disease Resistance in Crop Plants.* (FAO/IAEA/SIDA Research Co-ord. Meeting, Ames, Iowa, 1975). IAEA-181, 81–90. IAEA, Vienna, 1976.

115. El-Tekriti, R.A., Lechtenberg, V.L., Bauman, L.B. and Colenbrander, V.F. Structural composition and *in vitro* dry matter disappearance of brown midrib corn residue. *Crop Sci.* **16**, 387–389, 1976.

116. Epstein, E. Genetic potentials for solving problems of soil mineral stress: adaptation of crops to salinity. *Proc. of Workshop on Plant Adaptation to Mineral Stress in Problem Soils.* M.J. Wright (ed.), 73–82. Cornell Univ. Agr. Exp. Sta. Special Publications, Ithaca, New York, 1977.

117. Epstein, E., Kingsbury, R.W., Norlyn, J.D. and Rush, D.W. Production of food crops and other biomass from seawater culture. *The biosaline concept.* A. Hallaender (ed.), 77–79. Plenum Pub. Co. New York, 1979.

118. Favret, E.A. Basic concepts on induced mutagenesis for disease reaction. *Mutation Breeding for Disease Resistance.* STI/PUB/271, 55–65. IAEA, Vienna, 1971.

119. Favret, E.A. Breeding for disease resistance using induced mutations. *Proc. of an Advisory Group. Induced Mutations in Cross-Breeding.* STI/PUB/447, 95–111. IAEA, Vienna, 1976.

120. Favret, E.A. Different categories of mutations for disease reaction in the host organism. *Proc. of a Panel. Mutation Breeding for Disease Resistance.* IAEA, Vienna, 1970.

121. Favret, E.A. and Ryan, G.S. New useful mutants in plant breeding. *Mutations in Plant Breeding.* STI/PUB/129, 49–61. IAEA, Vienna, 1966.

122. Feenstra, W.J., Jacobsen, E. and De Visser, A.J.C. Isolation of a nitrate reductase deficient mutant of Pea (*Pisum sativum*) and its use in the study of the inhibiting effect of nitrate (NO_3–) on nitrogen fixation. *Induced Mutations – A Tool in Plant Research, Extended Synopses.* STI/PUB/591, 327–332. IAEA, Vienna, 1981.

123. Finkner, V.C. Inheritance of susceptibility to *Helminthosporium victoriae* in crosses involving Victoria and other crown rust resistant oat varieties. *Agron. Jour.* **45**, 404–406, 1953.

124. Fossati, A. and Bronnimann, A. Tolerance to *Septoria nodorum* Berk. in wheat: inheritance and potential in breeding. *Induced Mutations for Disease Resistance in Crop Plants.* IAEA-181, 91–100. IAEA, Vienna, 1976.

125. Frankowiak, J.D., Mann, S.S. and Williams, N.D. A proposal for hybrid wheat utilizing *Aegilops squarrosa* L. cytoplasm. *Crop Sci.* 16, 725–728, 1976.

126. Freisleben, B. and Lein, L. Uber die Auffinding einer mehltau-resistenten Mutante nach Rontgenbestrahlung einer anfalligen reinen Linie von Sommergerste. *Naturwissenschaften* 30, 608, 1942.

127. Fritz, J.O., Cantrell, R.P., Lechtenberg, V.L., Axtell, J.D. and Hertel, J.M. Brown midrib mutants in sudangrass and grain sorghum. *Crop Sci.* 21, 706–709, 1981.

128. Fujita, H. and Nakajima, K. Retreatment of induced mulberry mutants with gamma rays. *Japan. J. Breed.* 22, *Suppl. 1,* 107, 1972.

129. Gangadharan, C. and Misra, R.N. A very early mutant in rice. *Curr. Sci.* 44, 140, 1975.

130. Gangadharan, C., Misra, R.N. and Ghosh, A.K. Improvement of scented rice varieties (Basmati 370) through induced mutation. *Curr. Sci.* 45, 597–598, 1976.

131. Gaul, H. Studies on diplontic selection after X-irradiation of barley seeds. *Effects of Ionizing Radiations on Seeds.* STI/PUB/13, 117–136. IAEA, Vienna, 1961.

132. Gaul, H., Grunewaldt, J. and Hesemann, C.U. Variation of character expression of barley mutants in a changed genetic background. *Mutations in Plant Breeding II,* 77–95. IAEA, Vienna, 1968.

133. Gaul, H. and Lind, V. Variation of the pleiotropy effect in a changed genetic background demonstrated with barley mutants. *Induced Mutations in Cross-Breeding.* STI/PUB/447, 55–69. IAEA, Vienna, 1976.

134. Gengenback, B.G. and Green, C.E. Selection of T-cytoplasm maize callus culture resistant to *Helminthosporium maydis* race T pathotoxin. *Crop Sci.* 15, 645–649, 1975.

135. Gichner, T., Veleminsky, J., Pokorny, V. and Zadrazil. Post-treatment effects in barley seeds treated with mutagenic alkylating compounds. *Polyploidy and Induced Mutations on Plant Breeding.* STI/PUB/359. IAEA, Vienna, 1974.

136. Gladstones, J.S. and Hill, G.D. Selection for economic characters in *Lupinus angustifolius* and *L. digitatus.* 2. Time of flowering. *Aust. J. Agric. Anim. Husb.* 9, 213–220, 1969.

137. Gottschalk, W. Induced mutations in gene-ecological studies. *Induced Mutations – A Tool in Plant Research.* STI/PUB/591, 411–436. IAEA, Vienna, 1981.

138. Gottschalk, W. and Kaul, M.L.H. Investigations on the cooperation of mutated genes. II. Floral structure and seed production of *cochleata* mutants and recombinants of *Pisum sativum.* Ber. *Deutsch. Bot. Ges. Bd.* 85, 513–524, 1974.

139. Gottschalk, W. and Muller, H.P. The reaction of an early-flowering *Pisum* recombinant to environment and genotypic background. *Seed Protein Improvement in Cereals and Grain Legumes.* Vol. I. STI/PUB/496, 259. IAEA, Vienna, 1979.

140. Gottschalk, W. and Patil, S.H. The reaction of *Pisum* mutants to different climatic conditions. *Indian J. Genet. and Pl. Breed.* 31, 403–406, 1971.

141. Green, C.E. and Phillips, R.L. Potential selection system for mutants with increased lysine, threonine and methionine in cereal crops. *Crop Sci.* 14, 827–830, 1974.

142. Gressel, J. and Segel, L.A. The pancity of plants evolving genetic resistance to herbicides: Possible reasons and implications. *J. Theor. Biol.* 75, 349–371, 1978.

143. Gustafsson, A. The genetic analysis of phenotype patterns in barley. *Induced Mutations for Crop Improvement in Africa.* TEC/DOC/222, 41–53. IAEA, Vienna, 1979.

144. Gustafsson, A. The genetic architecture of phenotype patterns in barley. *Proc. of a Study Group Meeting, 7. Induced Mutations and Plant Improvement.* IAEA, Vienna, 1972.

145. Gustafsson, A. and Lundqvist, U. Mutations and parallel variation. *Induced Mutations – A Tool in Plant Research.* STI/PUB/591, 85–110. IAEA, Vienna, 1981.

146. Gustafsson, A., Nybom, N. and Von Wettstein, U. Chlorophyll factors and heterosis in barley. *Hereditas* 36, 383–392, 1950.

147. Hadjichristodoulou, A. and Della, A. Barley mutants with improved protein and grain yield in Cyprus. *Induced Mutations – A Tool in Plant Research.* STI/PUB/591, 245–246. IAEA, Vienna, 1981.

148. Hadjichristodoulou, A. and Della, A. Increase and improvement of protein content of cereals and legumes in Cyprus. *Nuclear techniques for seed protein improvement.* STI/PUB/320, 163–179, IAEA, Vienna, 1973.

149. Hagberg, A., Persson, G., Erman, R., Karlsson, K.E., Tallberg, A.M., Stoy, V., Betholdsson, N.O., Mounla, M. and Johansson, H. The Svalov protein quality breeding program. *Seed Protein Improvement in Cereals and Grain Legumes*, II. STI/PUB/496, 303–313. IAEA, Vienna, 1979.

150. *Handbook of Reactive Chemical Hazards*. Bretherick, L. (ed.), 317, CRC Press, Cleveland, Ohio, 1975.

151. Handke, S. A mutant with a very long lasting vegetative phase in spinach. *Mutation Breeding Newsletter* 18, 11–12, 1981.

152. Handke, S. A mutant with a very long lasting vegetative phase in Spinach. *Proc. EUCARPIA Meeting on leafy Vegetables. Littlehampton (UK)*, 11–14, 1980.

153. Hanna, W.W. and Burton, G.W. Use of mutagens to induce and transfer apomixis in plants. Extended Synopses, IAEA/SM/251, 110. IAEA, Vienna, 1981.

154. Hanna, W.W., Monson, W.G. and Gaines, T.P. IVDMD, total sugars, and lignin measurements on normal and brown midrib (*bmr*) sorghums at various stages of development. *Agron. J.* 73, 1050–1052, 1981.

155. Hanna, W.W. and Powell, J.B. Radiation-induced female-sterile mutant in pearl millet. *J. Hered.* 65, 247–249, 1974.

156. Hanna, W.W. and Powell, J.B. Stubby head, an induced facultative apomict in pearl millet. *Crop Sci.* 13, 726–728, 1973.

157. Haq, M.S. Rice breeding with induced mutation. *Induced Mutations for Crop Improvement in Africa*, 55–66. IAEA/TEC/DOC/222. IAEA, Vienna, 1979.

158. Haq, M.S., Rahman, M.M. and Chowdhury, M.H. Studies of the quality of induced mutants of rice. *Nuclear Techniques for Seed Protein Improvement*. 139–144, 1973.

159. Haq, M.S., Rahman, M.M., Mansur, A. and Islam, R. Breeding for early, high-yielding and disease resistant rice varieties through induced mutations. *Rice Breeding with Induced Mutations III.* IAEA Tech. Report Series 131:35–46, 1971.

160. Haq, M.S., Rahman, M.M., Moa, M.M. and Ahmed, H.U. Disease resistance of some mutants induced by gamma rays. *Induced Mutations for Disease Resistance in Crop Plants*. STI/PUB/388, 150. IAEA, Vienna, 1974a.

161. Haq, M.S., Tin, M.M. and Myint, T. The prospect of breeding red rot resistant sugarcane varieties through induced mutations in Burma. *Induced Mutations for Disease Resistance in Crop Plants*. STI/PUB/388, 150–151. IAEA, Vienna, 1974b.

162. Harn, C., Won, J.L., Park, C.K. and Yoo, J.Y. Mutation breeding for improvement of rice protein. *Nuclear Techniques for Seed Protein Improvement*. STI/PUB/320, 115–137. IAEA, Vienna, 1973.

163. Harte, C. *Antirrhinum majus* L. *Handbook of Genetics, Vol. 2, Plants, Plant Viruses and Protists*, 315–331. Plenum Press, New York, 1974.

164. Harten, A.M. van. Mutation breeding in vegetatively propagated crops with emphasis on contributions from the Netherlands. *Proc. of Eucarpia Conf. Induced Variability in Plant Breeding, Wageningen*, 1981.

165. Harten, A.M. van, Bouter, H. and Broertjes, C. *In vitro* adventitious bud techniques for vegetative propagation and mutation breeding of potato (*Solanum tuberosum* L.) II. Significance for mutation breeding. *Euphytica* 30, 1–8, 1981.

166. Hayes, J.D. Varietal resistance to spray damage in barley. *Nature* (London) 183, 551–552, 1959.

167. Heinz, D.J. Sugar-cane improvement through induced mutations using vegetative propagules and cell culture techniques. *Proc. of a Panel. Induced Mutations in Vegetatively Propagated Plants*. 53–59. IAEA/STI/PUB/339. IAEA, Vienna, 1973.

168. Heinz, D.J. and Mee, G.W.P. Plant differentiation from callus tissue of Saccharum species. *Crop Sci.* 9, 346–348, 1969.

169. Heinz, D.J., Krishnamurthi, M., Nickell, L.G. and Maretzki, A. Cell, tissue and organ culture in sugarcane improvement. *Plant, Cell, Tissue, and Organ Culture*. J. Reinert and Y.P.S. Bajaj (eds.), 3–17. Springer Verlag, Berlin, 1977.

170. Hensz, R.A. Introducing Star Ruby — the new Texas grapefruit. *The Packer*. 913 (Jan. 1, 1972).

171. Hibberd, K.A. and Green, C.E. *Inheritance and expression of lysine plus threonine resistance selected in maize tissue culture*. 79. Proc. Natl. Acad. Sci. USA (in press).

172. Hockett, E.A. and Reid, D.A. Spring and winter genetic male-sterile barley stocks. *Crop Sci.* 21, 655–659, 1981.

173. Holdgate, D.P. Propagation of ornamentals by tissue culture. *Applied and Fundamental Aspects of Plant, Cell, Tissue and Organ Culture*. J. Rinert and Y.P.S. Bajaj (eds.), 18—43. Springer Verlag, Berlin, 1977.

174. Hopp, H.E., Favret, G.C. and Favret, E.A. Study of the control of barley development using dwarf mutants. *Induced Mutations — A Tool in Plant Research. Extended Synopses*. STI/PUB/591, 243. IAEA, Vienna, 1981.

175. Horn, W. F_1-hybrids in floriculture. *Proc. XIXth Inter. Horticultural Congress IV*, 268—277, 1974.

176. Horner, C.E. and Melouk, H.A. Screening, selection and evaluation of irradiation-induced mutants of spearmint for resistance to *Verticillium* wilt. *Proc. of a Symp, Induced mutations Against Plant Diseases*. IAEA/SM/214/20. IAEA, Vienna, 1977.

177. Howland, G.P. and Hart, R.W. Radiation biology of cultured plant cells. *Plant, Cell, Tissue and Organ Culture*. J. Reinert and Y.P.S. Bajaj (eds.). Springer Verlag, Berlin, 1977.

178. Hsieh, S.C. and Chang, T.M. Radiation induced variations in photoperiod-sensitivity, thermosensitivity and the number of days to heading in rice. *Euphytica* 24, 487—496, 1975.

179. *Int. Agency Res. Cancer Monogr.* 1, 125. Lyon, France, 1972.

180. Ishikawa, M. The new soybean varieties 'Raiden' and 'Raiko' induced by Gamma-ray irradiation. *Natn. Reg. Tohoku Exp. Sta. Japan. Res. Rep.* 40, 65, 1970.

181. Ismachin, M. and Mikaelsen, K. Early maturing mutants for rice breeding and their use in cross breeding programs. *Induced Mutations in Cross Breeding*. STI/PUB/447, 199—121. IAEA, Vienna, 1976.

182. Jagathesan, D. Mutation breeding in sugarcane. *Induced Mutations for Crop Improvement in Africa*, 67—82. Tech/Doc/222. IAEA, Vienna, 1979.

183. Jagathesan, D., Balasundaram, N. and Alexander, K.C. Induced mutations for disease resistance in sugarcane. *Induced Mutations for Disease Resistance in Crop Plants*. STI/PUB/388, 151—153. IAEA, Vienna, 1974.

184. Jagathesan, D. and Ratnam, R. A vigorous mutant sugarcane (*Saccharum* sp.) clone co 527. *Theor. Appl. Genet.* 51, 311—313, 1978.

185. Jain, H.K. and Pokhiryal, S.C. Improved pearl millet hybrids. *Mutation Breeding Newsletter* 6, 11—12, 1975.

186. Jensen, J. Chromosomal location of one dominant and four recessive high lysine genes in barley mutants. *Seed Protein Improvement of Cereals and Grain Legumes*. STI/PUB/496, 89—96. IAEA, Vienna, 1979.

187. Johnson, V.A. Wheat protein. *Genetic Diversity in Plants, Proc. of a Symp., Genetic Control of Diversity in Plants, Lahore, Pakistan, March 1—7, 1976*, 371—385. Plenum Press, New York, 1977.

188. Johnson, V.A., Mattern, P.J. and Kuhr, S.L. Genetic improvement of wheat protein. *Seed protein improvement in cereals and grain legumes*. STI/PUB/496, 165—181. IAEA, Vienna, 1979.

189. Jonsson, R. Breeding for improved oil and meal quality in rape (*Brassica napus* L.) and turnip rape (*Brassica campestris* L.). *Hereditas* 87, 205—218, 1977a.

190. Jonsson, R. Erucic acid heredity in rapeseed (*Brassica napus* L. and *Brassica campestris* L.). *Hereditas* 86, 159—170, 1977b.

191. Jørgensen, J.H. An allelic series of mutant genes for powdery-mildew resistance in barley. *Barley Genetics* II, R.A. Nilan (ed.), 475—477. Pullman, Washington, 1971.

192. Jørgensen, J.H. Location of the *ml-o* locus on barley chromosome 4. *Induced Mutations Against Plant Diseases*. STI/PUB/462, 533—549. IAEA, Vienna, 1977a.

193. Jørgensen, J.H. Spectrum of resistance conferred by *ml-o* powdery mildew resistance genes in barley. *Euphytica* 26, 55—62, 1977b.

194. Jørgensen, J.H. Studies of recombination between alleles in the *ml-o* locus of barley and on pleiotropic effects of alleles. *Induced Mutations for Disease Resistance in Crop Plants*. IAEA/Tech/doc/181, 129—140. IAEA, Vienna, 1976.

195. Jørgensen, J.H. and Jensen, P. Inter-allelic recombination in the *ml-o* locus in barley. *Barley Genet. Newsletter* 9, 37—39, 1979. By permission.

196. Jørgensn, J.H. and Mortensen, K. Primary infection by *Erisiphe graminis* f. sp. *hordei* of barley mutants with resistance genes in the *ml-o* locus. *Phytopath* 67, 678—685, 1977.

197. Kaplan, R.W. Chromosomen and Fakformutationsratren in Gerstenkornern bei verscheidenartigen Quellungsbehandlunger oder Kalte Wahrend oder nach bei Rontgen bestrahlung sowie bei Dosis-fractionierung. *Z. Indukt. Abstamm. Vererbungsl.* **83**, 347–382, 1951.

198. Kar, G.N., Chakrabarti, S.N. and Prasad, A.B. Improvement of bread wheat through mutation breeding. *Annual Wheat Newsl.* **24**, 62–63, 1978.

199. Kartha, K.K., Gamborg, O.L., Constabel, F. and Shyluk, J.P. Regeneration of cassava plants from apical meristems. *Plant. Sci. Lett.* **2**, 107–113, 1974.

200. Kasembe, J.N.R. Phenotypic restoration of fertility in a male sterile mutant by treatment with gibberellic acid. *Nature* **215**, 668, 1967.

201. Kihara, H. Cytoplasmic relationships in the *Triticinae*. *Proc. III Int. Wheat Symp.*, Finlay K. Wand (ed.), 125–134, 1968.

202. Kihara, H. Characteristics of *Aegilops squarrosa* cytoplasm. *Proc. IV Int. Wheat Genet. Symp.*, F. Sears and L. Sears (eds.), 351–353. Univ. Missouri Press, Columbia, 1973.

203. Kimber, G. and Sallee, P.J. A hybrid between: *Triticum timopheevii* and *Hordeum bogdanii. Cereal Res. Comm.* **4**, 33–37, 1976.

204. Kinoshita, T. Genetical studies on cytoplasmic male sterility induced by gamma ray irradiation in sugar beets. *Japan. J. Breed* **26**, 256–265, 1976.

205. Kinoshita, T., Takahashi, M. and Mikami, T. Induction of cytoplasmic male sterility by chemical mutagens in sugar beets (a preliminary report). *Zeiken Ziho* **27–28**, 66–71, 1979.

206. Kinoshita, T. and Takahashi, M.E. Induction of cytoplasmic male sterility by gamma ray irradiation in sugar beets. *Jap. J. Breeding* **19**, 445, 1969.

207. Kishore, H., Das, B., Subramanyam, K.N., Chandra, R. and Upadhya, M.D. Use of induced mutations for potato improvement. *Improvement of Vegetatively Propagated Plants Through Induced Mutations.* TEC/DOC/173, 77–83. IAEA, Vienna, 1975.

208. Kivi, E.I. Earliness mutants from NaN$_3$-treated, six-row barley. *Fourth Int. Barley Genet. Symp., Edinburgh*. (in press)

209. Kivi, E.I., Redunden, M. and Varis, E. Use of induced mutations in solving problems of northern crop production. *Polyploidy and Induced Mutations in Plant Breeding.* STI/PUB/359, 187–194. IAEA, Vienna, 1974.

210. Kleinhofs, A., Owais, W.M. and Nilan, R.A. Azide. *Mutation Research* **55**, 165–195, 1978.

211. Knudsen, K.E.B. and Munk, L. The feasibility of high lysine barley breeeding based on economical considerations and the nutritional analysis of the botanical components of Bomi and the 1508 mutant. *Fourth Int. Barley Genet. Symp. Edinburgh.* (in press).

212. Konzak, C.F. A review of semidwarfing gene sources and a description of some new mutants useful for breeding short-stature wheats. *Induced Mutations in Cross-Breeding.* STI/PUB/447, 79–93. IAEA, Vienna, 1976.

213. Konzak, C.F. Evaluation and genetic analysis of semidwarf mutants. I. Wheat. *Proc. of Coord. Research Programme on Evaluation of Mutant Stocks for Semi-Dwarf Plant Type as Cross-Breeding Materials in Cereals. Vienna, March 11, 1981.* (in press).

214. Konzak, C.F. Genetic control of the content, amino acid composition, and processing properties of proteins in wheat. *Adv. Genetics* **19**, 407–582, 1977.

215. Konzak, C.F. Induced mutations for genetic analysis and improvement of wheat. *Induced Mutations – A Tool for Plant Research.* STI/PUB/591, 469–488. IAEA, Vienna, 1981.

216. Konzak, C.F. Induced mutations in host plants for the study of host-parasite interactions. *Plant Pathology-Problems and Progress (Proc. of the Golden Jubilee Symposium of the American Phytopathological Society)*, 202–214. University of Wisconsin Press, 1959.

217. Konzak, C.F., Nilan, R.A. and Kleinhofs, A. Artificial mutagenesis as an aid in overcoming genetic vulnerability of crop plants. *Genetic Diversity in Plants.* Muhammed, A., R. Aksel and R.C. Von Borstel (eds.), 163. Plenum Press, New York, 1977.

218. Konzak, C.F., Nilan, R.A., Froese-Gertzen, Edith E. and Ramirez, I.A. Physical and chemical mutagens in wheat breeding. *Proc. 2nd Int. Wheat Genetics Symp., Lund 1963, Hereditas, Suppl.* **2**, 65–84, 1966.

219. Konzak, C.F., Nilan, R.A., Harle, J.R. and Heiner, R.E. Control of factors affecting the response of plants to mutagens. Fundamental Aspects of Radiosensitivity. *Brookhaven Symposia in Biology* No. **14**, 128–157, 1961.

220. Konzak, C.F. and Randolph, L.R. Radiation and iris breeding. *Bulletin Am. Iris Society*, **142**, 68–74, 1956.

221. Konzak, C.F. and Rubenthaler, G.L. Protein improvement in cereals. I. Wheat. *Induced Mutants for Cereal Grain Protein Improvement*. TECDOC-259, 143–151. IAEA, Vienna, 1982.

222. Koo, F.K.S. Mutation breeding in soybean. *Induced Mutations and Plant Improvement*. STI/PUB/297, 285–292. IAEA, Vienna, 1972.

223. Koo, F.K.S., Cuevas-Ruiz, J. and Guttierrez-Colon, G. Gamma ray induction of mutations in soybeans for environmental adaptation. *Improving Plant Protein by Nuclear Techniques*. STI/PUB/258, 367–377. IAEA, Vienna, 1970.

224. Kreis, M. and Doll, H. Starch and prolamin synthesis in single and double high lysine mutants. *Physiol. Plant.* **48**, 139–143, 1980.

225. Kukimura, H., Ikeda, F., Fujita, H., Maeta, T., Nakajima, K., Katagiri, K., Nakahira, K. and Somegou, M. Genetical, cytological and physiological studies on the induced mutants with special regard to effective methods for obtaining useful mutants in peremial woody plants (II). *Improvement of Vegetatively Propagated Plants and Tree Crops Through Induced Mutations*. TEC/DOC/194, 93–137. IAEA, Vienna, 1976.

226. Kwon, S.H., Im, K.H. and Kim, M.S. A new soybean variety, KEX2, selected from a X-ray irradiated population. *Korean J. Breeding* **5**, 13–16, 1973.

227. Lacey, C.N.D. and Campbell, A.I. Mutation breeding of apple at Long Ashton, UK. *Mutation Breeding Newsletter* **17**, 2–5, 1981.

228. Langridge, J. and Brock, R.D. A thiamine-requiring mutant of the tomato. *Aust. J. Biol. Sci.* **14**, 66–69, 1961.

229. Lapins, K.O. Induced mutations in fruit trees. *Induced Mutations in Vegetatively Propagated Plants*. STI/PUB/339, 1–19. IAEA, Vienna, 1973.

230. Lapins, K.O. Compact Stella sweet cherry introduced. *Mutation Breeding Newsletter* **4**, 18, 1974.

231. Lapins, K.O. The Lambert compact cherry. *Fruit Var. Hostic. Dig.* **19**, 23, 1965.

232. Lapins, K.O. Segregation of compact growth types in certain apple seedling progenies. *Can. J. Plant Sci.* **49**, 765, 1969.

233. Lapins, K.O., Bailey, C.H. and Hough, L.F. Effects of gamma rays on apple and peach leaf buds at different stages of development. I. Survival growth and mutation frequencies. *Radiat. Bot.* **9**, 379–389, 1969.

234. Lapushner, D. and Frankel, R. Practical aspects and the use of male sterility in the production of hybrid tomato seed. *Euphytica* **16**, 300–310, 1967.

235. Lawless, E.W., Ferguson, T.L. and Meiners, A.F. *Guidelines for the Disposal of Small Quantities of Unused Pesticides*. EPA-670/2-75-057, 1975.

236. Lechtenberg, V.L., Muller, L.D., Bauman, L.F., Rhykerd, C.L. and Barnes, R.F. Laboratory and *in vitro* evaluation of inbred and F_2 populations of brown midrib mutants of *Zea mays* L. *Agron. J.* **64**, 657–660, 1972.

237. Levy, A. and Ashri, A. Ethidium bromide – An efficient mutagen in higher plants. *Mutation Research* **28**, 397–404, 1975.

238. Lewis, D. and Crowe, L.K. Structure of the incompatibility gene. IV. Types of mutation in *Prunus avium* L., *Heredity* **8**, 357, 1954.

239. Line, R.F., Allan, R.E. and Konzak, C.F. Identifying and utilizing resistance to *Puccinia striiformis* in wheat. *Induced Mutations for Disease Resistance in Crop Plants*. FAO/IAEA/SIDA Research Coord. Meeting, Ames, Iowa, 1975. IAEA-181, 151–159. IAEA, Vienna, 1976.

240. Litzenberger, S.C. Inheritance of resistance to specific races of crown and stem rust, to *Helminthosporium* blight, and of certain agronomic characters of oats. *Iowa Agr. Exp. Sta. Res. Bul.* **370**, 453–496, 1949.

241. Love, J.E. Somatic mutation induction in Poinsettia and sweet potato. *Mutation Breeding Workshop, Knoxville, Tenn.* M.J. Constantin (ed.). Univ. Tennessee, Knoxville, Tenn., 1972.

242. Lundqvist, U. Locus distribution of induced *eceriferum* mutants in barley. *Barley Genetics* **III**, Gaul, H. (ed.), 162–163. Verlag Karl Thiemig, Munich, 1976.

243. Lundqvist, U. Locus distribution of induced *eceriferum* mutants in barley. *Int. Symp. Experimental Mutagenesis in Plants. Varna, Bulgaria.* Publ. Bulgarian Academy of Sciences, 263–265, 1978.

244. Lyrene, P.M. Tissue culture and mutations in sugarcane. *Sugarcane Breed. Newsl.* **38**, 61–62, 1976.

245. Maliga, P. Resistance mutants and their use in genetic manipulation. *Frontiers of Plant Tissue Culture.* Thorpe, T.A. (ed.), 381–392. Univ. Calgary Press, Calgary, Canada, 1978.

246. Maliga, P., Sidorov, V., Cseplo, A. and Menczel, L. Induced Mutations in advancing in vitro culture techniques. *Induced Mutations — A Tool in Plant Research. Extended Synopses.* STI/PUB/591, 339–352. IAEA, Vienna, 1981.

247. Mann, S.S. Cytoplasmic variability in triticinae. *Proc. IV Int. Wheat Genet. Symp.* Sears, E. and Sears, L. (eds.), 367–373. Univ. Mo. Press, Columbia, 1973.

248. Manual on Mutation Breeding. II. STI/DOC/10/119. IAEA, Vienna, 1977.

249. Mathan, D.S. Morphogenetic effect of phenylboric acid on various leaf-shape mutants in the tomato, duplicating the effect of the *lanceolate* gene. *Z. Vererbungsl.* **97**, 157–165, 1965a.

250. Mathan, D.S. Phenylboric acid, a chemical agent simulating the effect of the *lanceolate* gene in tomato. *Am. J. Bot.* **52**, 185–192, 1965b.

251. Mathan, D.S. Reversing the morphogenetic effect of phenylboric acid and of the *lanceolate* gene with the actinomycin D in the tomato. *Genetics* **57**, 15–23, 1967.

252. McGuire, P.E. and Dvorak, J. High salt tolerance potential in wheat grasses. *Crop Sci.* **21**, 701–705, 1981.

253. McIntosh, R.A. Nature of induced mutations affecting disease reaction in wheat. *Induced Mutations Against Plant Diseases.* STI/PUB/462. IAEA, Vienna, 1977.

254. McKenzie, R.I.H. and Martens, J.W. Breeding for stem rust resistance in oats. *Induced Mutations for Disease Resistance in Crop Plants.* STI/PUB/388, 45–48. IAEA, Vienna, 1974.

255. Merriman, D.T. *Gamma ray induced mutations in the red raspberry cultivar Heritage.* Washington State University, Western Washington Research and Extension Center, Puyallup, Washington. (unpublished report), 1979.

256. Mertz, E.T., Bates, L.S. and Nelson, O.E. Mutant gene that changes protein composition and increases lysine content of maize endosperm. *Crop Sci.* **145**, 279–280, 1964.

257. Mia, M.M., Shaikh, M.A.Q. and Saha, C.S. Genetical, anatomical and yield studies from early maturing jute (*Corchorus capsularis*) mutants. *Ind. J. Exp. Biol.* **14**, 71–72, 1976.

258. Mia, M.M., Shaikh, M.A., Saha, C.S. and Bhuiya, A.D. Gamma-ray induced mutations in jute. *SABRAO Journal* **6**, 175–179, 1974.

259. Micke, A. Improvement of low yielding sweet clover mutants by heterosis breeding. *Induced Mutations in Plants.* STI/PUB/231, 541–550. IAEA, Vienna, 1969.

260. Micke, A. Annex I. List of varieties of vegetatively propagated plants developed using induced mutations. *Induced Mutations in Vegetatively Propagated Plants.* 195–202. IAEA/STI/PUB/339. IAEA, Vienna, 1973.

261. Micke, A. Mutant varieties of crop plants. *Induced Mutations for Crop Improvement in Africa.* IAEA/TEC/DOC/222, 257–265. IAEA, Vienna, 1979a.

262. Micke, A. Use of mutation induction to alter the ontogenetic pattern of crop plants. *Gamma-Field Symp. No. 18, 'Crop Improvement by Induced Mutation',* 1–23, Ohmiya, Japan, 1979b.

263. Micke, A. and Donini, B. Use of induced mutations for the improvement of seed propagated crops. *Proc. Conf. of the EUCARPIA Section Mutations and Polyploidy, Wageningen, 31 Aug. – 4 Sept. 1981. Induced Variability in Plant Breeding.* (in press)

264. Mikaelsen, K., Sajo, Z. and Simon, J. An early maturing mutant: Its value in breeding for disease resistance in rice. *Rice Breeding with Induced Mutations III. IAEA Tech. Reports Ser.* **131**, 97–101. IAEA, Vienna, 1971.

265. Mikami, T., Kinoshita, T. and Takahashi, M. A study on cytoplasmic variation of male sterile mutants in sugar beets (in Japanese with English summary). *Seiken Ziho* **27–28**, 72–80, 1979.

266. Mingwei, Gao. Genetic analysis of the earliness of the early maturing mutants of Indica rice. *Annual Breeding Newsl.* **17**, 9, 1981.

267. Moh, C.C. Does a coffee plant develop from one initial cell in the shoot apex of the embryo? *Rad. Bot.* 1, 97–99, 1961.

268. Monyo, J.H., Sugujama, T. and Kihupi, A.N. Potentially high-yielding and high protein rice in induced mutation breeding. *Seed Protein Improvement in Cereals and Grain Legumes Vol. II.* STI/PUB/ 496, 293–301. IAEA, Vienna, 1979.

269. Mujeeb-Kazi, A. Apomictic progeny derived from intergeneric *Hordium-Triticum* hybrids. *J. Hered.* 72, 284–285, 1981.

270. Mukai, Y. and Tsunewaki, K. Use of the *Kotschyi* and *Variabilis* cytoplasms in hybrid wheat breeding. *Genetic Diversity of the Cytoplasm in Triticum and Aegilops.* Tsunewaki, K. (ed.), 237–250. Japan Soc. for Promotion of Science, 1980.

271. Muller, L.D., Barnes, R.F., Bauman, L.F. and Colenbrander, V.F. Variations in length and other structural components of brown midrib mutants of maize. *Crop Sci.* 11, 413–415, 1971.

272. Muller, L.D., Lechtenberg, V.L., Bauman, L.F., Barnes, R.F. and Phykerd, C.L. *In vivo* evaluation of a brown midrib mutant of *Zea mays* L. *J. Anim. Sci.* 35, 883–889, 1972.

273. Murashige, T. Manipulation of organ initiation in plant tissue cultures. *Bot. Bulletin Academia Sivica* 18, 1–24, 1977.

274. Murray, M.J. Additional observations on mutation breeding to obtain *Verticillium*-resistant strains of peppermint. *Mutation Breeding for Disease Resistance.* STI/PUB/271, 171–195. IAEA, Vienna, 1971.

275. Murray, M.J. Successful use of irradiation breeding to obtain *Verticillium*-resistant strains of peppermint, *Mentha piperita* L. *Induced Mutations in Plants.* STI/PUB/231, 345–371. IAEA, Vienna, 1969.

276. Murray, M.J. and Todd, W.A. Registration of Todd's Mitcham peppermint. *Crop Sci.* 12, 128, 1972.

277. Murray, M.J. and Todd, W.A. Role of mutation breeding in genetic control of plant diseases. *Biology and Control of Soil-borne Plant Pathogens*, Bruehl, G.W. (ed.), 172–176. Am. Phytopath. Soc., St. Paul, Minn., 1975.

278. Murty, B.B. Mutation breeding for resistance to downy mildew and ergot in *Pennisetum* and to *Ascochyta* in chickpea. *Induced Mutations for Disease Resistance in Crop Plants.* Tech/Doc/181, 89–100. IAEA, Vienna, 1976.

279. *Mutation Breeding Newsletter, No. 1.* IAEA, Vienna, 1972.

280. *Mutation Breeding Newsletter, No. 2.* IAEA, Vienna, 1973.

281. *Mutation Breeding Newsletter, No. 3.* IAEA, Vienna, 1974.

282. *Mutation Breeding Newsletter, No. 4.* IAEA, Vienna, 1974.

283. *Mutation Breeding Newsletter, No. 5.* IAEA, Vienna, 1975.

284. *Mutation Breeding Newsletter, No. 6.* IAEA, Vienna, 1975.

285. *Mutation Breeding Newsletter, No. 7*, 12–15. IAEA, Vienna, 1976.

286. *Mutation Breeding Newsletter, No. 8.* IAEA, Vienna, 1976.

287. *Mutation Breeding Newsletter, No. 9.* IAEA, Vienna, 1977.

288. *Mutation Breeding Newsletter, No. 10.* IAEA, Vienna, 1977.

289. *Mutation Breeding Newsletter, No. 11.* IAEA, Vienna, 1978.

290. *Mutation Breeding Newsletter, No. 12.* IAEA, Vienna, 1978.

291. *Mutation Breeding Newsletter, No. 13.* IAEA, Vienna, 1979.

292. *Mutation Breeding Newsletter, No. 14*, 17. IAEA, Vienna, 1979.

293. *Mutation Breeding Newsletter, No. 15*, 14. IAEA, Vienna, 1980.

294. *Mutation Breeding Newsletter, No. 16.* IAEA, Vienna, 1980.

295. *Mutation Breeding Newsletter, No. 17*, 20. IAEA, Vienna, 1981.

296. *Mutation Breeding Newsletter, No. 18*, 17. IAEA, Vienna, 1981.

297. Nakajima, K. Induction of useful mutations of mulberry and roses by gamma rays. *Induced Mutations in Vegetatively Propagated Plants.* STI/PUB/339, 105–116. IAEA, Vienna, 1973.

298. Nasyrov, Y.S. Genetic control of photosynthesis and improving of crop productivity. *Ann. Rev. Plant Physiol.* 29, 215–237, 1978.

299. Nayar, G.G. Improving tapioca by mutation breeding. *J. Root Crops* 1, 55–58, 1975.

300. Nelson, O.E. Improvement of plant protein quality. *Improving Plant Protein by Nuclear Techniques.* STI/PUB/258, 43–49. IAEA, Vienna, 1970.

301. Nelson, O.E. The waxy locus in maize. I. Intralocus recombination frequency estimates by pollen and by conventional analysis. *Genetics* 47, 737–742, 1962.

302. Nelson, O.E. The waxy locus in maize. II. The location of the controlling element alleles. *Genetics* 60, 507–524, 1968.

303. Nelson, O.E. The waxy locus in maize. III. Effect of structural heterozygosity on intragenic recombination and flanking marker assortment. *Genetics* 79, 31–44, 1975.

304. Nelson, O.E. and Mertz, E.T. Nutritive value of floury-2 maize. *Nuclear Techniques for Seed Protein Improvement.* IAEA/STI/PUB/320, 321–328. IAEA, Vienna, 1973.

305. Nilan, R.A. Induced mutations and winter barley improvement. *Induced Mutations and Plant Improvement.* STI/PUB/297, 349–351. IAEA, Vienna, 1972.

306. Nilan, R.A., Kleinhofs, A. and Warner, R.L. Use of induced mutants of genes controlling nitrate reductase, starch deposition, and anthocyanin synthesis in barley. *Induced Mutations – A Tool in Plant Research. Extended Synopses.* STI/PUB/591, 183–200. IAEA, Vienna, 1981.

307. Nilan, R.A., Konzak, C.F., Wagner, J. and Legault, R.R. Effectiveness and efficiency of radiations for inducing genetic and cytogenetic changes. *The Use of Induced Mutations in Plant Breeding (supplement to Radiation Botany, Vol. 5) (Proc. of a Symp., Rome, 1964). FAO/IAEA, Rome,* 71–89. Pergamon Press, 1965.

308. Nuffer, M.G., Jones, L. and Zuber, M.S. The mutants of maize. *Crop Sci.* 74. Soc. Am., Madison, Wisconsin, 1968.

309. Oberle, G.D. The occurrence of spur type trees in seedling apple populations. *Fruit Var. Hort. Digest* 19, 43, 1965.

310. Ono, S. Studies on the improvement of the components of essential oil of genus *Mentha* by radiation. *Gamma Field Symp. No. 18, 'Crop Improvement by Induced Mutation',* 97–113. Ohmiya, Japan, 1979.

311. Oono, K. Test tube breeding of rice by tissue culture. *Tropical Agric. Res. Ser.* 11, 109–124, 1978.

312. Pakendorf, K.W. Selection for low alkaloid mutants in *Lupinus mutabilis. Induced Mutations – A Tool in Plant Research. Extended Synopses.* STI/PUB/591, 233–247. IAEA, Vienna, 1981.

313. Parodi, P.C. and Nebreda, I.M. Mutation breeding to increase protein content in wheat. *Seed Protein Improvement by Nuclear Techniques.* IAEA/STI/PUB/479, 33–39. IAEA, Vienna, 1978.

314. Parodi, P.C. and Nebreda, I.A. Protein and yield response of six wheat (*Triticum* spp.) genotypes to gamma radiation. *Seed Protein Improvement in Cereals and Grain Legumes* II. IAEA/STI/PUB/ 496, 201–209. IAEA, Vienna, 1979.

315. Persson, G. and Hagberg, A. Induced variation in a quantitative character in barley. Morphology and cytogenetics of *erectoides* mutants. *Hereditas* 61, 115–178, 1969.

316. Persson, G. and Karlsson, K.E. Progress in Breeding for improved nutritive value in barley. *Cereal Res. Comm.* 5, 169–179, 1977.

317. Peterson, C.L., Auld, D.L., Thomas, V.W., Withers, R.V., Smith, S.M. and Bettis, B.L. Vegetable oil as an agricultural fuel for the Pacific Northwest. *Univ. Idaho Agr. Expt. Sta. Bulletin 598.* 8, 1981a.

318. Peterson, C.L., Wagner, G.L. and Auld, D.L. Performance testing of vegetable oil substitutes for diesel fuel. *Am. Soc. Agr. Eng. Tech. Paper* 81–3578. Chicago, Illinois, 1981b.

319. Phatak, S.C., Wittwer, S.H., Honma, S. and Bukovac, M.J. Gibberellin induced pollen development in a stameless tomato mutant. *Nature* 209, 635–636, 1966.

320. Philouze, J. Etude de deux genes de sterilite male chez la Tomate: ms-32 et ms-35. *Ann. Amelior. Plantes* 19, 443–457, 1969.

321. Philouze, J. Utilisation pratique d'une lignee male-sterile ms-35 chez la Tomate. *Ann. Amelior. Plantes* 24, 129–144, 1974.

322. Picciurro, G., Ferrandi, L., Bonifoti, R. and Bracciocurti, G. Uptake of ^{15}N-labelled NH_4^+ in excised roots of a *durum* wheat mutant line compared with *durum* and bread wheat. *Isotopes in Plant Nutrition and Physiology Proc.,* 511. IAEA, Vienna, 1967.

323. Plarre, W.K.F. Developing new recombinants by crossing Hiproly X complicated composite genotypes. *Proc. 4th Int. Barley Symp. Edinburgh.* (in press).

324. Pohlheim, F. Induced mutations for investigation of histogenetic processes as the basis for optimal mutant selection. *Induced Mutations — A Tool in Plant Research.* STI/PUB/591, 489—495, 1981.

325. Polle, E., Konzak, C.F. and Kittrick, J.A. Visual detection of aluminum tolerance levels in wheat by hematoxylin staining of seedling roots. *Crop Sci.* **18**, 823—827, 1978.

326. Porsche, W. Results of X-irradiation in breeding of *Lupinus albus. Induzierte Mutationen und ihre Nutzung.* 241—244. Akademie-Verlag, Berlin, 1967.

327. Porter, K.S., Axtell, J.D., Lechtenberg, V.L. and Colenbrander, V.F. Phenotype, fiber composition, and *in vitro* dry matter disappearance of chemically induced brown midrib (*bmr*) mutants of sorghum. *Crop Sci.* **18**, 205—208, 1978.

328. Powell, J.B. Induced mutations in highly heterozygous vegetatively propagated grasses. *Induced Mutations in Cross Breeding.* STI/PUB/447, 219—224. IAEA, Vienna, 1976.

329. Powell, J.B. and Murray, M.J. The vegetative bud mutation technique for breeding improved *Poa pratensis* L. *Proc. XIII Int. Grass Cong. Leipsig, GDR, 18—17 May 1977.*

330. Powell, J.B. and Toler, R.W. Induced mutants in 'Floratam' St. Augustine grass. *Crop Sci.* **21**, 644—646, 1980.

331. Powers, P.W. *How to dispose of toxic substances and industrial wastes.* Noyes Data Corp., Park Ridge, N.J., 1976.

332. Preiss, J. and Boyer, C.D. Evidence for independent genetic control of the multiple forms of maize endosperm branching enzymes and starch syntheses. *Mechanism of Saccharide Polymerization and Depolymerization.* Marshall, J.J. (ed.). Academic Press, New York, 1979.

333. Pringle, R.B. Comparative biochemistry of the phytopathogenic fungus *Helminthosporium* XVI. The production of victoximine by *H. sativum* and *H. victoriae. Can. J. Biochem.* **54**, 783—787, 1976.

334. Quellet, C.E. and Pomerleau, R. Recherches sur la resistance de l'orme d'amerique au Cerato cystis ulmi. *Can. J. Bot.* **43**, 85—96, 1965.

335. Ramage, R.T. Balanced tertiary trisomics for use in hybrid seed production. *Crop Sci.* **5**, 177, 1965.

336. Ramage, R.T. Comments about the use of male-sterile facilitated recurrent selection. *Barley Newsletter* **24**, 52—53, 1981.

337. Ramage, R.T. Male sterile facilitated recurrent selection and happy homes. *Fourth Regional Winter Cereal Workshop Barley, Vol.* **II**, 92—98. Amman, Jordan, 1977a.

338. Ramage, R.T. Varietal improvement of wheat through male sterile facilitated recurrent selection. *Food and Fertilizer Technology Center Tech. Bulletin* **37**, 1—6, 1977b.

339. Ramirez, A., Allan, R.E., Konzak, C.F. and Becker, W.A. Combining ability of winter wheat. *Induced Mutations in Plants.* STI/PUB/231, 445—455. IAEA, Vienna, 1969.

340. Rao, P.S. Effects of flowering on yield and quality of sugar cane. *Exp. Agric.* **13**, 381—387, 1977.

341. Rao, P.S. Mutation breeding for non-flowering in sugar cane. *Mutation Breeding Newsletter* **3**, 9, 1974.

342. Rao, T.J., Srinivasan, K.V. and Alexander, K.C. A red rot resistant mutant of sugar-cane induced by gamma radiation. *Proc. Indian Acad. Sci., Sect. B.* **4**, 224—230, 1966.

343. Rasmusson, D.C., Banttari, E.E. and Lambert, J.W. Registration of M21 and M22 semidwarf barley germplasm. (Reg. No. GP 13 and GP 14). *Crop Sci.* **13**, 777, 1973.

344. Raut, R.N. and Jain, H.K. A daylength insensitive mutant of cotton. *Mutation Breeding Newsletter* **8**, 2—3, 1976.

345. Raut, R.N., Jain, H.K. and Panwar, R.S. Radiation-induced photoinsensitive mutants of cotton. *Current Science* **14**, 383—384, 1971.

346. Raut, R.N. and Panwar, R.S. Two mutants of cotton with wide adaptability suitable for cultivation in different cotton zones of India. *Mutation Breeding Newsletter* **16**, 6, 1980.

347. Reddy, G.M. and Padma, A. Some induced dwarfing genes non-allelic to Deo-Geo-Woo-Gen in rice variety Tellakattera. *Theor. Appl. Genet.* **47**, 115—118, 1976.

348. Reeves, A.F., II. Tomato trichomes and mutations affecting their development. *Am. J. of Botany* **64**, 186—189, 1977.

349. Reinert, J.A., Toler, R.W., Bruton, B.D. and Busey, P. Retention of resistance by mutants of 'Flora-tam' St. Augustine grass to the southern chinch bug and St. Augustine decline. *Crop Sci.* **21**, 464–466, 1978.

350. Rennie, R.J. Potential use of induced mutations to improve symbioses of crop plants with N_2– fix-ing bacteria. *Induced Mutations – A Tool in Crop Plant Research.* STI/PUB/591, 293–321. IAEA, Vienna, 1981.

351. Rennie, R.J. and Larson, R. Dinitrogen fixation associated with disomic chromosome substitution lines of spring wheat. *Can. J. Bot.* **57**, 2771–2775, 1979.

352. Rick, C.M. Genetics and development of nine male-sterile tomato mutants. *Hilgardia* **18**, 599–633, 1948.

353. Rick, C.M. The tomato. *Handbook of Genetics, Vol.* **2**, King, R.C. (ed.), 247–280. Plenum Press, New York, 1975.

354. Rick, C.M. Tomato *Lycopersicon esculentum (Solanaceae). Evolution of Crop Plants.* Simmonds, N.W. (ed.), 268–287. Longman, London and New York, 1976.

355. Röbbelen, G. Selection for oil quality in rapeseed. *The Way Ahead in Plant Breeding. Proc. Sixty Cong. Eucarpia.* Lupton, F.G.H., Jenkins, G. and Johnson, R. (eds.), 215–219. Cambridge Univ. Press, 1972.

356. Röbbelen, G., Kwon, H.J. and Heun, M. Inheritance and pathotype specificity of mildew resistance of induced barley mutants, 87. *Fourth Int. Barley Genet. Symp., Edinburgh.* (in press).

357. Röbbelen, G. and Nitsch, A. Genetical and physiological investigations on mutants for polyenic fatty acids in rape seed *Brassica napus* L. I. Selection and description of new mutants. *Z. Planzen-zuecht.* **75**, 93–105, 1975.

358. Röbbelen, G.P., Abdel-Hafez, A.G. and Reinhold, M. Use of mutants to study host/pathogen rela-tions. *Induced Mutations Against Plant Diseases.* IAEA/SM/214/6. IAEA, Vienna, 1977.

359. Roest, S. and Bokelmann, G.S. *In vitro* adventitious bud techniques for vegetative propagation and mutation breeding of potato (*Solanum tuberosum* L.). 1. Vegetative propagation *in vitro* through adventitious shoot formation. *Potato Res.* **23**, 167–181, 1980.

360. Roest, S. and Bokelmann, G.S. Vegetative propagation of *Solanum tuberosum* L. *in vitro. Potato Res.* **19**, 173–178, 1976.

361. Rosichan, J.L. Genetic fine structure analysis of the *waxy* locus in barley (*Hordeum vulgare*). *Master of Science Thesis* 39. Washington State University, 1979.

362. Roy, K. and Jana, M.K. EMS induced early mutation in rice. *Current Sci.* **42**, 101–102, 1973.

363. Rutger, J.N. *Use of induced and spontaneous mutants in rice genetics and breeding. FAO/IAEA re-search coordinating meeting on evaluation of mutant stocks for semi-dwarf type as cross breed-ing materials in cereals. IAEA, Vienna, 2–6 March, 1981.* (in press).

364. Rutger, J.N., Peterson, M.L., Carnahan, H.L. and Brandon, D.M. Registration of mutant rice variety 'M-101'. *Mutation Breeding Newsletter* **15**, 5–6, 1980.

365. Rybakov, M.N. *Variability of garden strawberry clones induced by gamma rays and ethylene imine.* Timirjazevsk. S-kh. Akad. IXV (2), 36–44, 1966. (In Russian and English summary).

366. Sagawa, Y. and Mehlquist, G.A.L. The mechanism responsible for some X-ray induced changes in flower color of the carnation *Dianthus caryophyllus. Amer. J. Bot.* **44**, 397–403, 1957.

367. Sajo, Z. and Simon, J. A mutant variety in Hungarian rice productions. *Mutation Breeding Newsletter* **8**, 4–6, 1976.

368. Salerno, J.C., Hopp, H.E. and Favret, E.A. Persistence of lethal genes in maize populations due to natural selections. *Induced Mutations – A Tool in Crop Plant Research.* STI/PUB/591, 81, 1981.

369. Samborski, D.J. Mutants with improved disease resistance: Their recognition, identification and dis-tinction from outcrosses. *Induced Mutations Against Plant Diseases.* IAEA/SM/214/43. IAEA, Vienna, 1977.

370. Sano, Y., Fujii, T., Iyama, S., Hirota, Y. and Komagata, K. Nitrogen fixation in the rhizosphere of cultivated and wild rice strains. *Crop Sci.* **21**, 758–761, 1981.

371. Sadakuma, T., Maan, S.S. and Williams, N.D. EMS-induced male sterile mutants in euplasmic and alloplasmic common wheat. *Crop Sci.* **18**, 850–853, 1978.

372. Sato, H. Rice breeding with induced mutants in Japan. *Mutation Breeding Newsletter* **15**, 2–4, 1980.

373. Scarascia Venezian, M.E. and Esposito Seu, M. Fioritura, epoca di fruitificazione e produzione di piante di fragole (varieta Elite Climax, Cambridge e Huxley) sottoposte ad irraggiamento cronico gamma. *Agric. Ital.*, XI (7), 3—14, 1965.

374. Schmitt, J.M. and Svendson, I. Amino acid sequences of hordein polypeptides. *Carlsberg Res. Comm.* 45, 143—148, 1980.

375. Scholz, F. Experience and opinions on using induced mutants in cross-breeding. *Induced Mutations in Cross-Breeding.* STI/PUB/447, 5—19. IAEA, Vienna, 1976.

376. Schwarzbach, E. Response to selection for virulence against the *ml-o* based mildew resistance in barley, not fitting the gene-for-gene hypothesis. *Barley Genet. Newsletter* 9, 85—88, 1979.

377. Scott, K.G. and Campbell, A.I. Variation in self compatability in Cox clones. *Long Ashton Res. Sta. Report* 23, 1970.

378. Sears, E.R. An induced mutant with homoeologous pairing in common wheat. *Can. J. Genet. Cytol.* 19, 585—593, 1977.

379. Shaikh, M.A.Q., Ahmed, Z.U., Majid, M.A., Kau, A.D. and Mia, M.M. *Mutation Breeding Newsletter* 16, 1—3, 1980.

380. Shakoor, A., Ahsanul-Haq, M., Sadiq, M. and Sarwar, G. Induction of resistance to yellow mosaic virus in mungbean through induced mutations. *Induced Mutations Against Plant Diseases.* IAEA/SM/214/37. IAEA, Vienna, 1977.

381. Shephard, J.F., Bidney, D. and Shahin, E. Potato protoplasts in crop improvement. *Science* 208, 17—24, 1980.

382. Silva, W.J. and Arruda, P. Evidence for the genetic control of lysine catabolism in maize endosperm. *Phytochemistry* 18, 1803—1805, 1979.

383. Singh, R. and Axtell, J.D. High lysine mutant gene (*hl*) that improves protein quality and biological value of grain sorghum. *Crop Sci.* 13, 535—539, 1973.

384. Singh, R. and Iyer, C.P.A. Chemical mutagenesis in pineapples (*Ananas comosus*). *Proc. Int. Hortic. Cong.* 19 (1A), 108, 1974.

385. Slate, G.L. *Lilies for American gardens.* 258. Charles Scribner's Sons. New York, London, 1947.

386. Slein, M.W. and Sansone, E.B. *Degradation of chemical carcinogens.* Van Nostrand Reinhold Co., New York, 1980.

387. Smith, R.J., Jr. and Caviness, C.E. Differential response of soybean cultivars to propanil. *Weed Sci.* 21, 279—281, 1973.

388. Snoad, B. and Hedley, C.L. The potential for redesigning the pea crop using spontaneous and induced mutations. *Induced Mutations — A Tool in Plant Research. Extended Synopses.* STI/PUB/591, 111—126. IAEA, Vienna, 1981.

389. Soost, R.K. and Cameron, R.K. Citrus. *The Citrus Industry, Vol.* I. Webber, H.J. and Bachelor, L.D. (eds.), 507—540. Univ. of Calif. Press, Berkeley, California, 1975.

390. Srinivasachar, D., Seetharam, A. and Malik, R.S. Combination of the three characters (high oil content, high iodine value and high yield) in a single variety of linseed *Linum usitatissimum* L. obtained by mutation breeding. *Curr. Sci.* 41, 169—171, 1972.

391. Strobel, G.A., Steiner, G.W. and Byther, R. Deficiency of toxin binding protein activity in mutants of sugar cane clone H54-755 as it relates to disease resistance. *Biochem. Genet.* 13, 557—565, 1975.

392. Stubbe, H. Considerations of the genetical and evolutionary aspects of some mutants of *Hordeum, Glycine, Lycopersicon* and *Antirrhinum.* Cold Spring Harb. Symp. Quant. Biol. 24, 31—40, 1959.

393. Stubbe, H. *Genetik und Zytologie von Antirrhinum L. sect. Antirrhinum.* (In German). Veb Gustav Fischer Verlag Jena, 1966.

394. Stubbe, H. Mutanten der Kulturtomate *Lycopersicon esculentum* Miller VI. *Kulturpflanze* 19, 185—230, 1972a.

395. Stubbe, H. Mutanten der Wildtomate *Lycopersicon pimpinellifolium* (Jusl.) Mill. IV. *Kulturpflanze* 19, 231—263, 1972b.

396. Stubbe, H. New mutants of *Antirrhinum majus* L. *Kulturpflanze* 22, 189—213, 1974.

397. Stubbe, H., Scholz, F. and Zacharias, M. Induzierte Mutationen und Pflanzanzuchtung, Wissenschaft und Fortschritt 26, 13—17, 1976.

398. Swanson, E.B. and Tomes, D.T. *In vitro* responses of tolerant and susceptible lines of *Lotus corniculatus* L. to 2,4-D. *Crop Sci.* **20**, 792–795, 1980.

399. Tanaka, S. Difference in nitrogen absorption between a radiation-induced high protein rice mutant and its original variety. *Seed Protein Improvement by Nuclear Techniques.* STI/PUB/479, 199–210. IAEA, Vienna, 1978.

400. Tanaka, S. Induction of mutations in protein content of rice. *Evaluation of Seed Protein Alterations by Mutation Breeding.* STI/PUB/426, 139–140. IAEA, Vienna, 1976.

401. Tanaka, S. Varietal differences in protein content of rice. *Nuclear Techniques for Seed Protein Improvement.* STI/PUB/320, 107–113. IAEA, Vienna, 1973.

402. Tanaka, S. and Takagi, Y. Protein content of rice mutants. *Improving Plant Protein by Nuclear Techniques.* STI/PUB/258, 55–62. IAEA, Vienna, 1970.

403. Thompson, R.K. and Shantz, K.C. Registration of MSFRS wheat germplasm composite corsses A and B. *Crop Sci.* **18**, 698, 1978.

404. Todd, W.A., Green, R.J., Jr. and Horner, C.E. Registration of Murray Mitcham peppermint. *Crop Sci.* **17**, 188, 1977.

405. *Toxic and hazardous industrial chemicals safety manual.* 646. International Technical Information Institute, 1980.

406. Tsunewaki, K. (ed.). Genetic diversity of the cytoplasm in *Triticum* and *Aegilops. Japan So.c for Promotion of Sci.* **290**, 1980.

407. Ullrich, S.E. and Eslick, R.F. Lysine and protein characterization of induced shrunken endosperm mutants of barley. *Crop Sci.* **18**, 963–966, 1978.

408. Ullrich, S.E., Nilan, R.A. and Bacaltchuk, B. Evaluation and genetic analysis of semi-dwarf mutants. II. Barley. *Proc. of Coord. Research Programme on Evaluation of Mutant Stocks for Semi-Dwarf Plant Type as Cross-Breeding Materials in Cereals. IAEA, Vienna, March 11, 1981.* (in press).

409. Ullrich, S.E., Blake, T.K., Kleinhofs, A., Coon, C.N. and Nilan, R.A. Protein improvement in cereals. II. Barley. *Induced Mutations for Cereal Grain Protein Improvement.* IAEA-TECDOC-259, 153–160. IAEA, Vienna, 1982.

410. Ukai, Y. and Yamashita, A. Early mutants of barley induced by ionizing radiations and chemicals. *Fourth Int. Barley Genet. Symp., Edinburgh.* (in press).

411. Upadhya, M.D., Dayal, T.R., Dev, B., Chaudhri, V.P., Sharda, R.T. and Chandra, R. Chemical mutagenesis for day-neutral mutations in potato. *Polyploidy and Induced Mutations in Plant Breeding.* STI/PUB/359, 379–383. IAEA, Vienna, 1974.

412. Vasal, S.K., Villegas, E. and Bauer, R. Present status of breeding quality protein maize. *Seed Protein Improvement in Cereals and Grain Legumes, Vol.* **II**, 127–150. STI/PUB/496, 127–150. IAEA, Vienna, 1979.

413. Vavilov, N.I. The law of homologous series in variation. *J. Genet.* **12**, 47–89, 1922.

414. Vasudevan, K.N., Nair, S.G., Jos, J.S. and Magoon, M.L. Radiation-induced mutations in cassava. *Indian J. Hortic.* **24**, 95–98, 1967.

415. Vincent, J.M. Potential for enhancing biological nitrogen fixation. Chapter 8. *Contemporary Basis for Crop Breeding.* Vose, P.B. and Blixt, S. (eds.). Pergamon Press, Oxford, 1983 [This Book.]

416. Visser, T. Methods and results of mutation breeding in deciduous fruits, with special reference to the induction of compact and fruit mutations in apple. *Induced Mutations in Vegetatively Propagated Plants.* STI/PUB/339, 21–33. IAEA, Vienna, 1973.

417. Vose, P.B. Potential use of induced mutants in crop plant physiology studies. *Induced Mutations — A Tool in Plant Research.* STI/PUB/591, 159–181. IAEA, Vienna, 1981.

418. Walker, D.I.T. and Sisodia, N.S. Induction of a non-flowering mutant in sugar cane. *Crop Sci.* **9**, 551–551, 1969.

419. Wall, A.M., Riley, R. and Chapman, V. Wheat mutants permitting homoeologous chromosome pairing. *Genet. Res.* **18**, 311–328, 1971a.

420. Wall, A.M., Riley, R. and Gale, M.D. The position of a locus on chromosome 5B of *Triticum aestivum* affecting homoeologous meiotic pairing. *Genet. Res.* **18**, 329–339, 1971b.

421. Wall, J.R. and Andrus, C.F. The inheritance and physiology of boron response in tomato. *Am. J. Bot.* **49**, 758–762, 1962.

422. Wallace, A.T. Increasing the effectiveness of ionizing radiations in inducing mutations at the vital locus controlling resistance to the fungus *Helminthosporium victoriae* in oats. *The Use of Induced Mutations in Plant Breeding. Radiation Botany Supplement* 5, 237–250, 1965.

423. Wallace, A.T., Rillo, F.O. and Browning, R.M. DDT resistance in barley, *Hordeum vulgare* L. I. Induced mutants for resistance. *Radiat. Bot.* 8, 381, 1968.

424. Wallace, A.T., Singh, R.M. and Browning, R.M. Induced mutations at specific loci in higher plants. III. Mutation response and spectrum of mutations at the *Vb* locus in *Avena byzantina* C. Kock. *Induced Mutations and Their Utilization.* 47–57. Akademie Verlag Berlin, 1967.

425. Walther, F. and Preil, W. Mutants tolerant to low temperature conditions induced in suspension culture as a source for improvement of *Euphorbia pulcherrima* Woold. ex Klotsch. *Induced Mutations − A Tool in Plant Research.* STI/PUB/591, 399–405. IAEA, Vienna, 1981.

426. Washnok, R.F. Cytoplasmic restoration of *ms*-sterility. *Maize Genet. Coop. Newsl.* 46, 25–27, 1972.

427. Watanabe, I. and Cabrera, D.R. Nitrogen fixation associated with the rice plant grown in water culture. *Appl. Eviron. Microbiol.* 37, 373–378, 1979.

428. Watkins, P.A.C. Strawberry meristem culture. *John Innes Inst. Ann. Rep.* 66, 18–19, 1975.

429. Welsh, J.N., Peturson, B. and Machacek, J.E. Associated inheritance of reaction to races of crown rust, *Puccinia coronata avenae* Erikss., and to Victoria blight, *Helminthosporium victoriae* M. and M., in oats. *Canad. J. Bot.* 32, 55–68, 1954.

430. Wenzel, G., Schieder, O., Prezewozny, T., Sopory, S.K. and Melchers, G. Comparison of single cell culture derived *Solanum tuberosum* L. plants and a model for their application in breeding programs. *Theor. App. Genet.* 55, 49–55, 1979.

431. Wettstein, D. Biochemical and molecular genetics in the improvement of malting barley and brewers yeast. *Proc. 17th Congress, Berlin (West), European Brewery Convention, 1979.* 587–629. European Brewery Convention (E.BC.), Rotterdam, 1979.

432. Wettstein, D., Jende-Strid, B., Ahrenst-Larsen, B. and Sorensen, J.A. Biochemical mutant in barley renders chemical stabilization of beer superfluous. *Carlsberg Res. Commun.* 42, 341–351, 1977.

433. Wettstein, D., Henningsen, K.W., Boynton, J.E., Kannangara, G.C. and Nielsen, O.F. The genic control of chloroplast development in barley. *Autonomy and Biogenesis of Mitochondria and Chloroplasts.* 205–233. North Holland, 1971.

434. Wettstein-Knowles, P. Genetics and biosynthesis of plant epicuticular waxes. *Advances in the Biochemistry and Physiology of Plant Lipids.* Appelqvist, L.A. and Liljenberg, C. (eds.). 1–26. Elsevier/North Holland Biomedical Press, 1979.

435. Wettstein-Knowles, P. The molecular phenotypes of the *eceriferum* mutants. *Barley Genetics II.* Nilan, R.A. (ed.), 146–193. Washington State Univ. Press. Pullman, Washington, 1971.

436. Wettstein-Knowles, P. and Søgaard, B. The *cer-cqu* region in Barley: gene cluster or multifunctional gene. *Carlsberg Res. Commun.* 45, 125–141, 1980.

437. Widholm, J. Selection and characterization of amino acid analog resistant plant cell cultures. *Crop Sci.* 17, 597–600, 1977.

438. Wilson, D. Breeding for morphological and physiological traits. *Plant Breeding II.* Frey, K.J. (ed.), 233–290. Iowa State Univ. Press, Ames, 1979.

439. Wilson, L.A. Stimulation of adventitious bud production in detached sweet potato leaves by high levels of nitrogen supply. *Euphytica* 22, 324–326, 1973.

440. Worland, A.J., Law, C.N. and Shakoor, A. The genetical analysis of an induced height mutant in wheat. *Heredity* 45, 61–71, 1980.

441. Yamaguchi, I. and Yamashita, A. Induced mutation of two-rowed barley resistant to powdery mildew *Erysiphe graminis* f. sp. *hordei.* I. Comparison of effects of gamma rays and ethylene imine in induction of resistant mutation. *Jap. J. Breed.* 29, 217–227, 1979.

442. Yamakawa, K. and Sparrow, A.H. Correlation of interphase chromosome volume and reduction of viable seed set by chronic irradiation of 21 cultivated plants during reproductive stages. *Radiat. Bot.* 5, 557–566, 1965.

443. Yonezawa, K. and Yamagata, H. Optimum mutation rate and optimum dose for practical mutation breeding. *Euphytica* 26, 413–414, 1977.

444. Zacharias, M. The yields of early ripening soybean mutants in relation to climatic conditions. *Induzierte Mutationen und ihre Nutzung.* 245–250. Akademie-Verlag Berlin, 1967.

445. Zagaja, S.W. and Przybla, A. Compact type mutants in apple and sour cherries. *Improvement of Vegetatively Propagated Plants and Tree Crops Through Induced Mutations.* TEC/DOC/194, 171–184. IAEA, Vienna, 1976.

446. Zelitch, I. and Day, P.R. The effect on net photosynthesis of pedigree selection for low and high rates of photorespiration in tobacco. *Plant Physiol.* **52**, 33–37, 1973.

447. Zhukov, O.S. and Gavrilina, L.V. The role of mutants in breeding sour cherry varieties with an apomictic system of reproduction and self-fertility. *Induced Mutations — A Tool in Plant Research.* STI/PUB/591, 155–156. Extended Synopses. IAEA, Vienna, 1981.

CHAPTER 10

APPLICATION OF COMPUTERS TO GENEBANKS AND BREEDING PROGRAMMES

Stig Blixt
Weibullsholm Plant Breeding Institute,
Landskrona, Sweden

INTRODUCTION

A characteristic of both genebank and plant breeding information is the need to store large quantities of data obtained from different sources, different places and different years, and used repeatedly for different purposes over a long period of time. In plant breeding the average data, such as from a yield trial, may be in active use for two decades; genebank information for half a century, or more. The final cost of such information depends on the initial cost of collecting the data but even more on the longevity and utility. It is therefore essential that the information is in a form of data so collected and with such a content that degeneration is minimized and utility maximized. Computers with disk or tape storage facilities are excellent tools to store, retrieve and process such data.

The presentation in this chapter is intentionally limited to subjects and details of which the author has first hand personal experience from the construction of systems now in operation. A number of references are given for background reading, but they are not cited specifically in the text.

293

GENEBANK VERSUS PLANT BREEDING INFORMATION

It is useful in organizing data and writing processing programmes to distinguish two categories of information.

MANAGEMENT INFORMATION:

related to identification, organization of experiments and data management, with elements such as accession number, origin of material, plot size, etc. As a rule, these are constant, over a certain period of time or over a certain amount of material, or both.

RESULT INFORMATION:

related to results of observations, measurements, etc., such as yield data, disease resistance, etc. In contrast to management information a new set of data is, as a rule, produced every time material is handled. This type of information is therefore updated much more frequently.

Comparing genebank and plant breeding with respect to these categories it is found that the management information is very similar and the result information almost identical with respect to the elements contained. Therefore, genebank and plant breeding information problems, with or without computer application, can advantageously be treated jointly, both being part of the same process of utilization of genetic variation in man-controlled biological production. From these and other considerations, it seems sometimes to have been concluded that the plant breeder is the only valid user of genebank material. But such a conclusion would be an oversimplification. More often than not the character has a long way to go, via the geneticist, the physiologist, the pathologist, etc., before it is actually ready to be employed in a breeding programme. It is therefore essential that the information collected by genebanks is designed for the whole range of plant sciences.

Having noted the similarities one important difference, relating perhaps more to usage than content, needs emphasizing. In the case of the genebank the information is a perpetual integrated vital part of the material stored for the benefit of the users, breeders being but one, though the most important category. In breeding, on the other hand, the information must be seen as a disposable tool for obtaining new varieties. In fact, obtaining, saving and storing information is a costly process competing for the resources in breeding projects. Observation followed by immediate discarding of unsuitable plant material, leaving for harvest only the smallest possible fraction of successful genotypes would come close to the ideal.

Thus, breeders can normally be expected to provide information only on characters in materials of immediate interest in on-going breeding projects. With this as the only source of information, genebank material could be very unevenly covered. The genebanks are, however, by definition expected to provide material and information that widens the choice for the breeder; to provide new information and new material or at least new information on old material. It is therefore important for future successful utilization of genebanks that the process of information collection is systematized so that the material becomes broadly and evenly described. With the number of accessions in all crops all over the world to-day probably over a million, this is an enormous task even if a large number of these accessions are redundants, i.e. unwanted duplicates.

To try to put the task in some perspective, the story of the pea collection at the Weibullsholm Plant Breeding Institute can briefly be told: initiated about 1917; systematized and characterization begun in 1930; computerized in 1973. From a beginning of less than a dozen characters observed in a dozen accessions it grew to the present state of over 400 characters characterized in over 3000 accessions. The first observations made 60 years ago, regarding the gene z, are still in use: pure genetic information does not degenerate. As new information was added the utility of the material steadily increased.

The lesson is that if one takes a longer perspective, characterization and evaluation of slowly degenerating genetic information can be a continuous process which, in the way of biology, does not really have an end. What is pressing today is, therefore, not so much as to finish a task but to get started in the right direction.

Against this has to be balanced the fact that the utility of a genebank is low when information is missing or scanty; in fact, a genebank has probably to reach a certain level of information before it makes an impact. It is therefore understandable that those providing funding for genebanks want to see this level reached as soon as possible. One temptation to avoid, however, is to invest in quickly collected but low level rapidly degenerating information, making the task unpleasantly paralleling that of ancient Sisyphus.

In pursuing the right direction for genebanks, it might be as well to start by considering the fact that collecting and storing information costs time and money, and further, that information degenerates and becomes obsolete after longer or shorter periods of time. The final cost of genebank information can and must, therefore, be related to initial cost of collecting it, and its longevity, content, and utility. The longevity is closely related to the genetic part of the content, as information of an entirely genetic nature degenerates slowest, if at all. Much of such information, e.g. discrete morphological traits and disease resistance, are simply inherited, often monogenic and can be obtained by simple and inexpensive methods of observation. Also information on certain quantitative traits, polygenically inherited, which are easily measured and have a high heritability can be obtained at reasonable cost in the first place. With initial low cost and long degeneration time the cost per unit information per unit time becomes small. The collection of such information, i.e. **characterization**, should therefore be the first priority and responsibility of genebanks.

In contrast to characterization can be put **evaluation**, a term often used for the collection of information on standard characters of plant breeding, variety testing and marketing; such as yield, straw stiffness, earliness, etc. Several of these characters have a very varying genetic content, often with a very high degree of genotype-environment interaction. Such information needs, to be useful, to be objectively measured and provided with such statistics as means, standard deviations and heritability. These statistics are also required for the large proportion of accessions in genebanks, notably from cross-pollinated species, that must be assumed to be as a rule polymorphic and heterozygous.

The costs of obtaining and storing this type of information is already high initially. Because of the great interdependence of these characters with such factors as agrotechnological, commercial and others, the information degenerates rather rapidly. If in addition the information is taken as gradings, in order to save initial time and costs, the degeneration may become very rapid. The costs per unit information per unit time may then become high to very high.

The evaluation type of information is normally obtained from specific breeding projects with specific breeding aims, or from variety test trials or similar projects which have a more limited timespan. It seems therefore natural that evaluation should come from plant breeders, national variety testing institutions or other similar institutions, feeding back the information to genebanks for maintenance; anyway it should not be considered a first-hand responsibility of genebanks.

Grading poses a problem in genebank information systems. The method is initially cheap and quick and, provided it is done by highly qualified personnel, also informative. But, because of rapid degeneration with distance in time and space from the grader, it has no place in characterization although perhaps in evaluation, as such data can be expected to degenerate anyway. The information system should in such cases, however, always be designed to store objectively-obtained information as well as gradings of the same character. When relating the low initial cost of grading to cost of storing, longevity and utility the overall cost may be found to be unpleasantly high.

In contrast to genebank information, plant breeding information should from the outset be expected to be degenerating and incomplete. The emphasis here is on the few selected positive genotypes, the new varieties to come and their performance; therefore the material of interest changes continuously. An information system for plant breeding purpose will thus emphasize the processing part, improving the capacity of selection in larger materials and aiming at proving and presenting the superiority of each new result. Storing data in a form suitable for statistical treatment is therefore essential in this area, while at the same time grading is here an efficient and important tool in the selection work.

A more research-directed use of computers in breeding work will, however, undoubtedly ask for more information being stored on early generation selection material.

In the case of e.g. analysing material for suitability as parents in crossing programmes, breeders should be eager to include computerized genebank information in their own computer processing, thus complementing as far as possible the inherent unevenness of breeding information. This will be facilitated if genebank and breeding information is largely compatible.

To summarize, genebank and plant breeding information is in essence the same, and therefore has great advantages for both fields in trying to make information systems and data as compatible as possible without losing their own particular identity and aims. To make genebank material available to all categories of users, it is important that the information on the material is as specific as possible. As genebank material is of indefinite interest, the information should as far as possible be non-degenerating, i.e. mainly genetic. To make this possible may not be all that demanding, once it is realized and accepted that it must take its time. Valuable genetic information already exists in various forms at different institutes. Much certainly can be achieved by developing close collaboration with existing institutions in all disciplines of plant science and as a first step collect and systematize this information and integrate it in the genebank collections.

COMPUTER AS STORAGE MEDIUM

Modern computers, as with modern cars, are definitely not all the same. It is, however, highly unlikely that one will find a make that will function very badly or not at all. The problem is therefore not which computer but rather that of finding the right combination of computer make, size and service agreement, with a programing language and a system solution that constitutes the optimal choice for the particular location, crops, people and problem.

There is one criterion, however, for suitability of computer hardware for breeding and genebank purposes that is rather self-evident: the need for inexpensive and reliable disk storage. A solution including disk drives with one or several removable disks allows the rather high cost of the drive to be distributed over a large number of disks in use. Magnetic tape has the disadvantage of longer access time but has an advantage as exchange medium.

The most applied, and also most time consuming, processes in handling this kind of information are data input, sorting and printing. The amount of arithmetical operations, on the other
hand, applied to the standard statistical treatments used in breeding and genebanks is, from the
computer point of view, negligible.

The print-out, the presentation of data and the processing of results, are often critical for
having full and volunteer utilization of the material by non-computer specialist users whether
breeder, genebank curator, or whoever. In a choice between high speed and more readable print
it is usually advisable to choose the latter as far as time permits.

Computer applications to plant breeding and genebank work have progressed comparatively
slowly in the last decade despite considerable investments. One reason has probably been the
lack of systems originally and particularly designed for the purpose. The efforts to take general
systems originally designed for administrative or other purposes and to use them with or without modifications have not been entirely successful. This is hardly surprising considering the
sometimes very specific nature of the biological part of the problem.

Another reason was certainly an underestimation of the importance of accessibility in this
type of work. Accessibility in terms of space and time will often turn out to be the factor that
determines whether a computerized system will be a success or a failure. The mini-computers,
which today are very far from small computers, located on-site, multiprogramable so as to
allow terminals for simultaneous data input have therefore drastically altered the situation. This
does not mean that big computers are now out of the picture, rather the contrary. The mini-
computer seems to present the necessary on-site cheap data storage and data management
capacity, capable of taking care of most − or even all − of the data processing. In addition, it
can secondarily be linked up with a big computer when there is a need for big core capacity
and also to other micro-computers that may take care of most of the data input and, if convenient, output.

The transfer of all information from paper to computer storage media may be a formidable
task for an institution with some decades of history, and quite a few breeding institutes can
look back on a century or more. Most genebanks are fortunately younger. The process may be
eased by following certain pathways now at least partly tried out, and avoid others for the same
reason. It is, for instance, advisable when considering computerization of an established plant
breeding programme to start at the end, i.e. to define what data must be available to perform
the analyses and present the results really needed. The beginning of the computerization may
therefore be to store data from the end process of the breeding work: the results from the yield
trials. Once these are stored on disk or tape, such a process as multiple classification variance
analysis, over years, locations and varieties becomes a simple matter of routine, Search processes for pair-wise comparisons of breeding lines and varieties covering different locations and
years can easily be applied; tables and lists can be printed out rapidly and repeatedly. From
here, the step to a system which takes care of the entire information process is not too great.
Storing data at the end of the trial may easily be modified into storing data from the very lay-
out of the same trials.

What has been accomplished in such a process is only transferring the information storage
from one medium, paper, to another, a computer memory, be it disk or tape. This transfer
can, though it may not always have been the case, be done with minimum disturbance to a
working system. The gains of the transfer process alone can be considerable in terms of better
organization of data and work, in addition to the raw time-saving from electronic processing

and print-out.

With regard to genebanks the matter may be more simply stated but more difficult to implement. The most important is probably to get a simple sequential database organization, open-
ended in terms of number and kinds of descriptors to be used as well as to number of accessions
that can be handled.

THE COMPUTER IN MODEL STUDIES

The computer is a very efficient information storage medium and calculating device. It
could, however, be put to more sophisticated uses, such as model studies in plant breeding. As a
first step the 'Existing ideotype' may be analysed. The information needed for this analysis

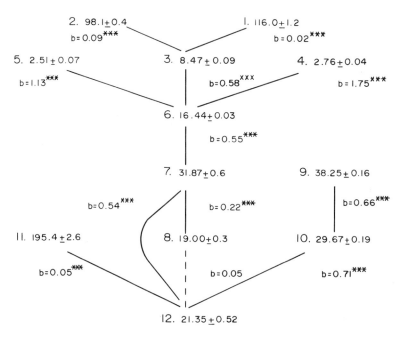

Fig. 10.1. Average value of character ± standard error and regression coefficient, b,
with significance of correlation indicated (* ～ 0.05, ** ～ 0.01, *** ～ 0.001).
1000 pea lines investigated.

should be obtained from the genebanks which are supposed, in due course, to cover the width
of genetic variation in crop plants.

In 1977 an attempt was made at establishing such an 'Existing ideotype' by analysing the
information in the pea collection at Weibullsholm Plant Breeding Institute. Simple linear correlations were calculated between 20 characters in 1000 lines from the collection of spontaneous
genetic variation. Eleven of these characters were found correlated in such a way (Fig. 10.1) as
to suggest the relation shown in Fig. 10.2. These characters are thus most probably components
of yield in the pea.

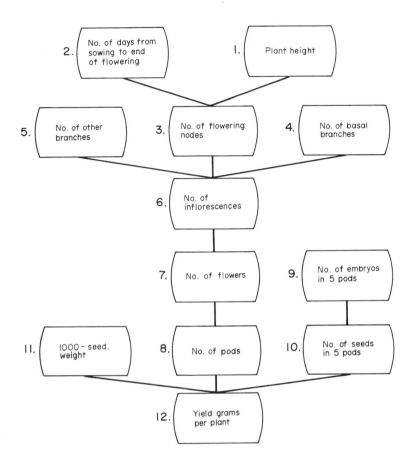

Fig. 10.2. Eleven components of yield in the pea from standard list of characters.

As indicated in Table 10.1 the genetics of some of these characters are partly known, enough to show that the characters 1 to 11 used are each composite. Plant height for instance is certainly a composite, or a mixture in this case, of number of internodes and length of internodes. Each of these characters is again a composite of a number of characters and at this level the characters may be monogenically inherited. Here, however, the information is presently far too scanty and the genes analysed and isolated too little studied. Up till now, the investigation and isolation of such genes have been almost totally chance occurrences, there has been no systematic approach.

Other standard characters, such as number of flowers or pods per plant are obviously not well defined. They are functions of real gene-determined ones, such as number of flowering nodes, number of flowers per inflorescence, number of branches. Plus in the case of number of pods per plant a tremendous environmental influence.

Though this experiment in model building was extremely simple it has shown that it is now possible to start such work. It is clear, however, that for further work the material of the collection with extreme variation in background genotype is not the right material. But from

TABLE 10.1. Characters of the pea related to yield presented in Figures 10.1 and 10.2

No.	Character	Genes involved
1.	Plant height in cm, measured at the beginning of maturity	Length of internodes: *Coe, Cob, Cona, Cry, H, La, Le, Lm, Min, Prae* Number of internodes: *Mie, Mier, Mine, Miu*
2.	Number of days from sowing to the end of flowering	*E, Hr, Lf, Sn*
3.	Number of flowering nodes on the main stem	
4.	Number of basal branches, at the beginning of maturity	*Fr, Fru*
5.	Number of other branches, at the beginning of maturity	*Ram*
6.	Number of inflorescences on the plant	
7.	Number of flowers on the plant	*Fn, Fna*
8.	Number of pods on the plant	
9.	Number of embryos in 5 pods	*Bt?*
10.	Number of seeds in 5 pods	
11.	Thousand grain weight	*Sg-1, Sg-2*
12.	Yield, in grams of seeds per plant	

the variation in the collections it is possible to produce a series of near-isogenic lines, as seems to have been done already by some researchers in peas. This would allow the production of crosses segregating for a limited number of known genes to be studied in detail.

It can hardly be expected that plant breeders could indulge in such basic studies at this stage. Genebanks should, however, both have the material needed and the capacity to guarantee the maintenance of the near-isogenic material produced. One of the major practical obstacles in producing this type of material was until now the lack of such a guarantee of continuity. Each new effort usually had to start again from scratch. As a matter of fact computers, making it possible to handle the complex information fundamental to such work, are probably about to create quite a new situation. Computer based model studies in plant physiology have actually been operating, with varying success, for some time. Perhaps the necessary base also for those models is such a genetically defined material. There must be definite advantages in being able to control the genetic effects and distinguish them from effects of other experimentally or otherwise administered parameters.

CHARACTER VERSUS DESCRIPTOR

Plant breeding and genebank programs are living plants, growing or resting. In order to systematize and analyse these plants they are described. It is these descriptions of actual characters in the plants which are stored and processed in the computer. **Descriptor** is now widely accepted as the computer term for the character and **descriptor state** for the quantity or quality of the character observed in the plant.

Much effort has been spent discussing uniformity in naming and spelling of descriptors. Incidentally, an extremely simple manner of making descriptors uniform in the computer would be to use a number. It has the same spelling and format in every language. The print-out of this number in text can then be left to the user. Much effort has also been spent discussing descriptor states and much of this discussion has been limited to the use of different scales or gradings. This may have partly a historical explanation: gradings and scales, preferably one-digit ones, were easy and economical to handle in the early stone age of computers; and the term descriptor perhaps made it too easy to have a discussion getting too remote from the character in the plant.

Therefore, discussion now needs to centre around the content of the descriptor: the definition of the character in the plant and particularly its heritability. It is, after all, the genetic part of the character that can be utilized in a breeding program as recombination of DNA. Progress in computer application to breeding and genebanks beyond the stage of simple storage and calculating, therefore needs the clearest possible definition of characters in the plant, and to define characters as the smallest possible heritable units.

Yield, for instance, is a very useful parameter but in this sense it is not a character. What has to be defined and assigned descriptors are first of all the heritable characters, the genes which when known, are the components of yield and the expression of which results in the yield. The descriptor 'Yield' and other similar ones describing the result of complex interactions are equally interesting but the difference should be made distinct.

It is clear that with the present state of genetic knowledge of most crop plants this cannot be done instantly. It should be equally clear, however, that in time it will become possible once the work has begun.

DESCRIPTORS AND DESCRIPTOR STATES

It is very useful in organizing information, as well as in programing, to distinguish between descriptors of different kinds, for different use and of different content. It is thus quite useful to code descriptors as to kind, e.g. letter codes, number codes, gene-symbols, integer numbers, decimal numbers, strings of text, etc.

From the point of use and content, several categories and types can be distinguished.

PASSPORT DESCRIPTORS,
used to store information concerning identity and history of the material.

MANAGEMENT DESCRIPTORS,
used for information relating to the handling of data or for organizing experiments and plant material. Passport and management descriptors may jointly be said to store management information in contrast to result information, i.e. the following category.

CHARACTER DESCRIPTORS,
used for information regarding specific characters. These may again be divided with respect to content. For qualitative characters, determined by one (to a few) genes, these can be used:

GENETIC DESCRIPTORS,

which as a descriptor states the alleles of a known gene. This information is totally utilizable in a breeding program and it may from this point of view be considered the ideal form of descriptor as it describes directly the genetics of the character.

QUALITATIVE DESCRIPTORS,

for cases not genetically analysed but where the characters are distinct and clearly recognizable. These can be represented by equally clear and distinct descriptor states expressed for example, by a numerical scale of integers. The utility in breeding programs is roughly inversely correlated with the number of genes involved in determining the character or the descriptor state.

The difference in genetic descriptors is the number of descriptor states: for a genetic descriptor this is equal to the number of alleles; for the qualitative descriptor equal to the number of distinguishable phenotypes as determined by the number of genes and alleles involved. From the information point of view this is of some importance and can be illustrated by an example from Table 10.1.

Assuming (for the sake of simplicity) only two alleles for each gene, plant height stored as gene-symbols needs 13 descriptors and together 26 states; storing it as a qualitative character takes one descriptor with 8192 states; storing it therefore as a quantitative descriptor, all existing genetic information is lost. Therefore genetic representation requires more descriptors but also preserves most of the information and the information degenerates slowest. Qualitative descriptors are about equal when up to 4 genes are involved. After this the character must be broken down if the genetic information is to be preserved.

For quantitative characters, polygenically inherited, these may be used:

METRIC DESCRIPTORS,

with the absolute value of an objectively measured character as the descriptor state. The utility in breeding will be about correlated with the genetic content of the information; therefore heritability should be added.

GRADING DESCRIPTORS,

the descriptor states being a scale of the 1 to 9 kind or similar. When the descriptor state is determined by subjective grading the utility is more than dependent on the user.

DATABASES

Computer storage media, disks or tape, are usually organized in sectors. Data are stored in these sectors using logical records. A **logical record** is organized by a series of descriptors, a **descriptor list**. The descriptors 'Variety', 'Replication', 'Grain yield per plot', 'Water content of grain' constitute such a descriptor list that could be used to organize a logical record for a (very simple) yield trial. A series of logical records makes up a file and one or many files can constitute a database. In the logical record the descriptor state of a given descriptor is stored in the place and in the format as defined in the descriptor list.

There are two main ways of storing the descriptor state, fixed or free format (Fig. 10.3).

In fixed format, a certain space is reserved for each descriptor to store the datum of the descriptor state. In free format the datum is occupying only the actual space taken; thus, no datum should occupy no space, though in reality usually at least one byte is used for a signal. Sometimes a mixed format can be used, i.e. with one (or several) logical record in free format and another (or others) in fixed. The general rule is that in fixed format the computer can recognize the datum from an address given as a sector or logical record number and find it without searching. Within the logical record the address is given in the descriptor list. In free format the computer reads through the data and knows its position by counting records and descriptors. Therefore, fixed format is as a rule easier to organize and to search, free format takes the least space in storage.

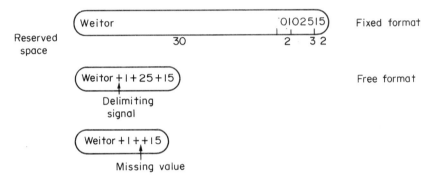

Fig. 10.3. Organization of a logical record.

The kind of database presented here is a very simple one. It can be made much more complicated, or sophisticated. To enter into more complicated forms needs, however, very strong motivation. If the databases are kept in this simple, sequential form the problems in data management and data exchange are minimized. When there is a need to rearrange the data in other formats, e.g. matrix or table formats, this can always be done by specific processing, building a temporary working database in any specific format from the basic sequential one.

THE NGB/WBP SYSTEM

As an example of computer application the system used at present by the Nordic Genebank (NGB) for genebank purposes and at Weibullsholm Plant Breeding Institute (WBP) for plant breeding purposes, will be presented. It is written in BASIC for a 16 K Wang 2200 mini-computer with one disk drive for one fixed and one removable disk. It handles the entire information process, such as planning of field experiments, printing of sowing lists, labels, field books, statistical treatments of results and storage and distribution details. The system is a fixed format system. A standard subroutine has been developed that can transfer it into a free format one if databases need to be stored with minimum space requirement. The software package for the database management is five small programs identical in NGB and WBP; the processing programs are partly different due to the differences in application and need. For minicomputers smaller than 16 K core the programs may be sectioned.

Three parallel files store the main descriptor list, which contains all descriptors used by the system. From this, and only from this, descriptors can be selected freely to organize logical records for a database.

One file stores the **Descriptor Table** (.DT), a second the **Descriptor Name** (.DN) and a third various explanatory information, such as, e.g. the format of decimal numbers of the **Descriptor State** (.DS). The unique identification of a descriptor is the descriptor number. Changes in format or spelling in .DN or .DS will therefore not affect the data search or other processes inside the computer.

The .DT holds all information necessary for dimensioning and organizing the data in a logical record (Fig. 10.4). Positions 7−10 not used in the main descriptor table are entered by computer when creating a database. Only .DT is normally used in processing the data; .DN and.DS are operative mainly during input of data and output of results.

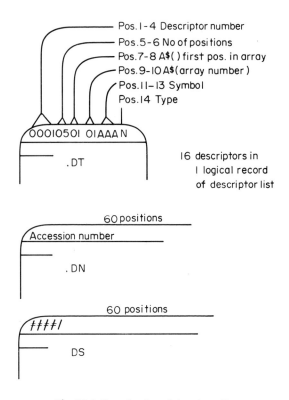

Fig. 10.4. Organization of descriptor lists.

The files .DT, .DN and .DS can be any number of logical records, each containing 16 descriptors. Each such logical record can be loaded and processed separately. In the Wang equipment used, the logical record of .DT has been dimensioned to exactly one sector on a disk, .DN and .DS to four sectors each. Direct access to any given descriptor is possible by means of a converting table in processing programs and the DSKIP statement, thereby diminishing search times to a great extent.

There are only two restrictions on the free arrangement of descriptors: in genebank use Accession Number must be the first descriptor in a .DT of a database containing this descriptor; and in both versions a .DT must always end with the End Signal Descriptor.

Already existing databases can be rearranged, diminished or enlarged by adding or deleting descriptors.

There are five programs for database management. The operator works interactively. Questions and choices appearing on the screen are to be answered by given alternatives.

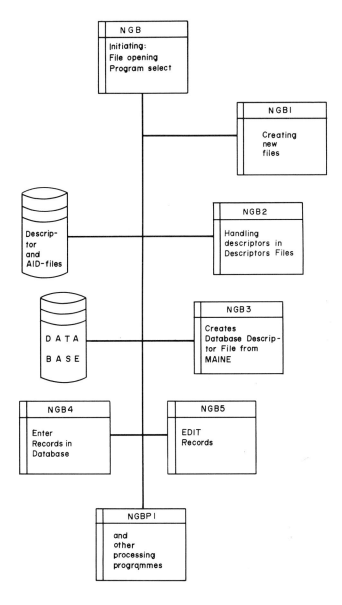

Fig. 10.5. Organization of data management.

NGB

(or WBP) program will initiate the system by opening the appropriate files and databases and link in the requested program.

NGB1

will be linked in when a new database or other file is to be created.

NGB2

handles descriptor lists, editing, rearranging, listing, etc.

NGB3

is used for creating descriptor lists for databases from the main descriptor list, MAINE. MAINE stands in the system (Fig. 10.5) for MAIN (descriptor list) (English version). To adapt the system to another language amounts to translating the MAINE and the program text printed in lists or on screen. Using parallel lists different language versions can be run.

NGB4

enters records in all databases created by the system.

NGB5

edits and changes records of databases. Edit means here that data already in the database are produced on the screen and can be operated on by special EDIT-functions and then restored; it is not necessary to rewrite the whole record or even the single datum.

Database

in Fig. 10.5 stands for any database. In genebank use, one crop, existing collection or collecting trip as a rule constitutes one database. In a plant breeding version, a single trial is stored in one file and all the files of a crop form the database (an SRD, see below). A small steer-file is then used to open the different files under program control. When analysing, e.g. performance of varieties over locations and years, it is enough to sort the steer-file.

AID-

files are a number of files, specific databases, used for storing any kind of information used by the system but not part of an accession database, e.g. the decoded forms of codes used, such as addresses, literature references, field books, etc.

In the system versions now operating more kinds of databases are needed for genebank use and more specific processing and output programs are used in breeding. The organization of NGB and WBP is shown in Fig. 10.6. Only the part to the right of the dotted line is used for the WBP version.

CROP INVENTORY DATABASES

store data from inventories of varieties and advanced breeding material. It is intended to be utilized for two main purposes: for a genebank to store the specific type of information of interest for this type of material from a region or country; also for breeders for obtaining genealogical records. The bases from crop inventories are normally seed-catalogues, variety lists and breeders field books. No plant material is involved.

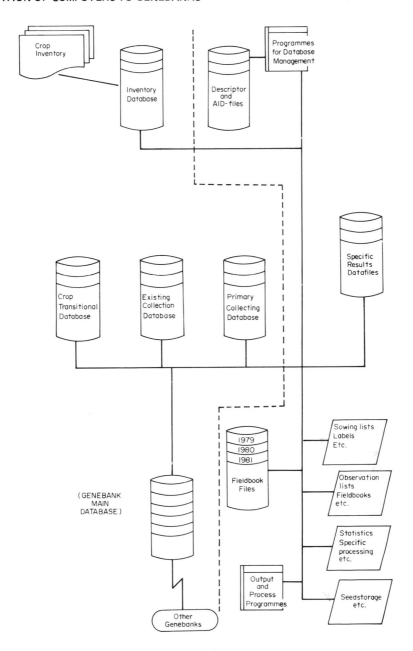

Fig. 10.6. Organization of NGB/WBP system.

CROP TRANSITIONAL DATABASES

store data on a crop basis, i.e. all accessions on, for example, wheat, will form a wheat database. A CTD may operate on very specific institute or individual descriptors. The designation 'Transitional' was given to indicate a more or less temporary function. It is intended, but not yet implemented, that the accessions with information may enter a Genebank Main Database (see below), following checking of plant material and information, characterization and evaluation, and necessary conformation of data format with international standards.

EXISTING COLLECTION DATABASE
is used to store information on specific collections entering a genebank. It may either be tempo-rary, i.e. the material is, following checking and characterization, partly discarded and partly transferred to a CTD; or permanent when the specific collection is to form its own subunit in the genebank.

PRIMARY COLLECTING DATABASES
store the data from collecting expeditions, with full details on geographical, ecological, edaphical and other similar information. It is visualized, though not yet implemented, that a PCD will hold the information on the original collection of seeds, plants etc. When selection of specific genotypes from the original collection have been made for plant breeding purposes such selections may be given their own accession numbers and transferred to a CTD. Only part of the original passport information and a reference to the PCD may then need to be included in the CTD. A similar procedure can be followed when the material is considered to have drifted too far from the original collection to be preserved as such.

SPECIFIC RESULT DATAFILES
It has often been a problem to store in genebank databases evaluation data from several years and sites. This system provides a solution with SRD, where such data are stored in chrono-logical order with passport descriptors defining the kind of material. In the genebank version a reference and/or sector address is made in the accession; in the breeding version a multi-file SRD is the main database, where all results from field trials are located.

THE GENEBANK MAIN DATABASE
is for the time being purely hypothetical. After international genebank standards have been universally adopted, it is assumed that CTD, in which information can be stored in a site-specific form, may be replaced by a GMD. Here information will be stored in a form directly exchangeable on an, as yet hypothetical, global information exchange network.

FIELD BOOK FILES
store the management information on the material. The organization may, of course, differ greatly in details from crop to crop and user to user. In this version are included such de-scriptors as field number of the year/growing season, field number of the previous year/season, origin of material, either in text or through reference to other databases, purpose of cultivation, generation or similar of cross progeny, etc. This file guides the processing programs handling print-out of labels, sowing lists, field plans, etc. It also provides information on geneaology of breeding material, etc.

THE OUTPUT AND PROCESSING PROGRAM PACKAGE
The following programs are examples from those presently in operation:
   Planning of field experiments (peas)
   Print-out of sowing lists (peas)

Print-out of labels (peas)

Print-out of pocket size field-book (peas)

Print-out of full-size field-book, observation lists (peas)

Processing of trial results including statistical treatments such as
averages, variance, relative values, correlations (all crops)

Print-out of tables of results (all crops)

Handling of seed storage (peas)

Print-out of index seminum (peas)

These processing programs are all written in BASIC and tailored to the needs and wishes of the individual users; they are both crop- and user-specific. The WBP version prints, e.g. in Swedish language. Though the systems could certainly be used by others straight away in many cases, certain modifications should be made, as the best use of the computer is usually obtained when it is instructed to work specifically for one user. It is, however, also possible to utilize already available standard software instead of such tailored processing programs. It is only necessary to produce a program interface connecting the standard program as a module with the NGB/WBP database. When the hardware permits, the standard program used can then also be in another programing language.

REFERENCES

1. Andrews, D.J., Hardwick, R.C. and Hardaker, J.M. A computer based retrieval system for plant breeding material. *Euphytica* **27**, 849—853, 1978.

2. *A Report on Data Retrieval/Report Generator Programs.* International Board for Plant Genetic Resources. Report of meeting in Rome, 8—9 November 1979 (Only for Committee's information: briefly presents PRINTGEN, EASYTRIEVE, FILETAB, SPRINT1, DATAMAN, SYSSIF, ANSWER2).

3. Blixt, S. A crossing programme with mutants in peas. Utilization of a genebank and a computer system. *Proc. Symp. Induced mutations in cross breeding,* 21—36. IAEA, Vienna, 1976.

4. Blixt, S. The genesymbols of *Pisum. Pisum Newsletter Suppl.* **9**, 2—59, 1977.

5. Blixt, S. Problems relating to pea breeding. *Agri Hortique Genetica* **36**, 56—87, 1978 (presenting a computerized model study of the pea).

6. Blixt, S. Descriptive list of genes for *Pisum. Pisum Newsletter* **10**, 80—101, 1978.

7. Blixt, S. *A genebank information system.* Technical paper presented at the First meeting of the Governing Board, FAO/UNDP European Cooperative Programme for Conservation and Exchange of Plant Genetic Resources, 15—18 December 1980. Geneva, 1980.

8. Brennan, J.P.M., Ross, R. and Williams, J.T. (eds.). *Computers in Botanical Collections.* Plenum Press, London, 1975.

9. Brill, R.C. The TAXIR primer. *Occ. Paper No. .1.* Inst. Arct. and Alp. Res. Univ. Colorado, Boulder, 1971.

10. Carmer, S.G. Computer programs for the construction of experimental layouts. *Agron. Journ.* **57**, 312—313, 1965.

11. Curtis, B., Roberts, B.J., Johnston, D.R., Kleim, D.L., Gile, D.E., Larson, R.C. and Lawson, J.W. Computerization of breeding program. *Annual Wheat Newsletter* **21**, 85, 1975.

12. Erskine, W. and Williams, J.T. The principles, problems and responsibilities of the preliminary evaluation of genetic resources samples of seed-propagated crops. *Plant Genetic Resources Newsletter* **41**, 19—33, 1980.

13. Johnson, M.W. The J.I.I./F.F.R.I. computer based system for *Pisum. Rothamsted Experimental Station Computer Department Newsletter.* Dec. 1977, 13—14, 1977.

14.   Rathjen, A.J. and Lamacraft, R.R. The use of computers for information management in plant breed-
       ing. *Euphytica* **21**, 502–506, 1972.
15.   Schicke, P. and O'Swath, J.   Verrechnung von Mehrklassen- Befalls-bonitierungen. *Zeitschrift für
       Pflanzenkrankheiten und Pflanzenschutz* **84**, (1), 1–17, 1977 (Statistical treatment of grading
       results according to central limit theorem by Townsend and Heuberger 1943; with 31 references
       relating to problems of grading in disease resistance).
16.   Technical aspects of Information Management and Means of Communication in Plant Genetic Re-
       sources Work for a future Utilization of Genetic Material in Plant Breeding. *Proceedings of a
       Colloqium held in Braunschweig on 7/8th Nov. 1978. 1979.*
17.   Townsend, G.R. and Heuberger, J.W. Methods for estimating losses caused by diseases in fungicidal
       experiments. *Pl. Dis. Reptr.* **27**, 340–343, 1943.
18.   Whan, B.R. and Buzza, G.C. A flexible computer system for constructing plant breeding trials. *J.
       Natn. Inst. Agric. Bot.,* **14**, 262–271, 1977.

CHAPTER 11

PLANT TISSUE CULTURE: RAPID PROPAGATION, INDUCED MUTATIONS, AND THE POTENTIAL ROLE OF PROTOPLAST TECHNIQUES

Marcel Cailloux
Department of Biological Sciences,
University of Montreal,
Montreal, Canada

THE RAPID PROPAGATION OF PLANTS

A little more than a decade ago, the Rockefeller Foundation sponsored a conference on 'Crop Improvement Through Tissue Culture'. It had then been felt[179] that regeneration of plants from callus and cell cultures had been accomplished with enough species to consider that it could be done with all plants. The present report, in some way, gauges the progress that has been accomplished since. It deals with the potential that has been exploited, the modification of outlook stemming from basic and applied research, and tries to see if some general conclusions can be drawn. This is not to say that no general article has been written on the applications of plant tissue culture. In particular, articles written by Murashige[169,170,171] are of interest.

A far reaching revolution in horticultural practice is in the making. It started in the laboratory and involved plant parts grown in test tubes. The speed at which this revolution will bring practical results depends on how much a happy wedding between the test tube and down to earth practices is possible. This revolution really started in 1964 when the late Dr. G. Morel of INRA, Versailles, France, proved that orchids could be propagated rapidly through meristem culture[163,164]. This was like a bomb in the orchid industry, where true-to-type propagation was a major problem. Soon tissue culture laboratories sprung up the world over in the orchid industry. This fact, more than any other, popularized tissue culture for the rapid propagation of plants, although very successful results had already been obtained in carnations, dahlias and potatoes[163,164]. Now over 125 species and varieties of orchids are being successfully propagated by tissue culture[201].

After the firm establishment of tissue culture in the orchid industry came the development of the propagation of ornamentals by tissue culture. This field of endeavor developed rapidly because ornamentals usually root easily, and growers soon found out that plants that propagated easily are good candidates for easy cultivation by tissue culture: their successful cultivation requires a minimum of experimentation to achieve practical results. Holdgate[102] has given a list of ornamentals with demonstrated potential for clonal multiplication through tissue culture, and deals with the economics of propagation of ornamentals. The list of species that are presently propagated commercially by tissue culture is not known with certainty because commercial nurseries do not publish their successes and tend to keep their techniques for themselves. However, the number of species that are actually being cultivated is impressive.

Perspectives of *in vitro* culture of tropical ornamentals, have been given by Miller and Murashige[157] and by Kunisaki[133]. For horticulturists, it is heartening to see more and more scientific articles dealing specifically with the rapid propagation of plants, a facet which is most important for the commercial feasibility of tissue culture. For example Seabrooke and Cumming[213] and Start and Cumming[225] who have worked out the rapid propagation of *Saintpaulia*. In addition, means for the rapid propagation of daffodils[213], daylilies[156], and carnation[65] have been worked out. Lilies have proved more difficult to achieve success, but a solution is in the making, because Tarayama and Misawa[235] have succeeded in obtaining differentiation with bulbscales. The list could be lengthened indefinitely.

A very interesting method having potential for the rapid propagation of plants has been developed recently. It consists of mechanically separating cells of a tissue, each cell having the potential of regenerating whole plants. Thus a method of homogenisation has been worked out for the Staghorn fern and for the Hare's-foot fern[49]. On the other hand, Schwenk[212] had devised a technique by which cells from cotyledons of soybean are mechanically isolated and then cultivated. This method of mechanical separation seems to have the possibility of being applied to a great number of plants. That is not to say that plants had not been previously regenerated from cell suspensions, in fact, it had been done quite often. However, the technique of obtaining the cells was indirect and brought about a great risk of obtaining mutations. First, a friable callus had to be generated, usually by use of artificial auxin 2,4-D, which is now known to cause frequent variations in chromosome number following abnormal cell divisions. Cells from this callus were liberated by mechanical agitation in a liquid culture medium and the isolated cells or cell clumps could then be cultivated to obtain complete plants. It has been shown that this method cannot be used to obtain consistently true-to-type plants (clones), but has the potential for giving rise to mutations, some of which may possess economic importance. This will be discussed later.

The development of practical uses of tissue culture first centered around house plants and ornamentals because these plants could be cultivated and multiplied *in vitro* with a minimum of research: their culture was relatively easy. With herbaceous and woody crop plants it has been much more difficult, but progress is steadily being made. For example, among the herbaceous crop plants that have recently been propagated by tissue culture, are the eggplant[120], rhubarb[241], white cabbage[242], and red cabbage[200]. A solution to the culture of red pepper is likely, since Gunay and Rao[84] have been able to regenerate it from hypocotyl and cotyledon explants.

Plant regeneration from calli of legumes has long proven to be very difficult, but a solution is in sight. Walker et al.[243] obtained regeneration with callus tissue of *Medicago sativa*. They even investigated the smallest cellular aggregate that is morphogenetically competent in their system and found it to be 105 μm. Contrary to the general case with legumes, Mukhopadhyay and Bhojwani[167] readily obtained shoot formation in *Lathyrus sativus* from cells derived from apical meristems, but without root formation. However, from the known factors that bring about root formation, it should be possible to obtain whole plants.

It is now known that plant regeneration through calli is not the best means of obtaining clonal propagation, and this is also true for legumes: genetic variations may appear. Thus, the culture of apical meristems of legumes has been attempted. Bajaj and Dhandju[8] have had success with various cultivars of *Cicer arietinum*, *Lens esculentum*, *Pisum sativum*, *Phaseolus aureus* and *Phaseolus mungo*. This last work was carried out in India where, during the last decade, legumes (pulses) have had almost static yields due to enormous losses caused by various pathogens, especially viruses. Fortunately the culture of true meristems is also a method of obtaining virus-free plants. At the same time, Martin et al.[143] investigated among other aspects, the rooting of meristem cultures of the field bean. Thus the propagation of legumes through culture of apical meristems is well under way.

Cereals and grasses have been grown in tissue culture mainly to exploit the formation of new mutations. In this case, one proceeds from the formation of a callus which is then induced to regenerate into complete plants. For a review on plant regeneration in cereals and grasses, see Green[80]. In general it has been difficult to achieve plant regeneration in this group. Early work was done in rice[234] and on sugar cane[93], then on sorghum[145]. More recently Green and Philips[81] found that calluses derived from immature embryos of maize possess higher morphogenic potential than those of mature parts and older embryos, while Mascarenhas et al.[144] demonstrated organ differentiation in maize, wheat and rice, while Dudits et al.[61] were successful in obtaining complete plants of wheat from rachis and shoot callus, later confirmed by Chin and Scott.[42] Orton[189] obtained regeneration from calli of *Hordeum vulgare*, *Hordeum jubatum* and their interspecific hybrid. Brar et al.[28] summarized what was known on tissue culture of corn and sorghum, while Dunstan et al.[62] obtained plantlet production from cultured tissues of *Sorghum bicolor*.

The rapid propagation of small fruits began to be achieved in the early 70's. For example, Boxus[26] propagated strawberries from meristem tips in order to get virus-free plants, and the strawberry is now being propagated commercially in Belgium, France, and Italy to supply virus-free stocks. The *in vitro* propagation of blackberry was worked out by Broome and Zimmerman.[32] Kiss and Zatyko[130] have worked in Hungary on *Rubus* species with Zatyko[248] obtaining complete red currant plants from adventitious embryos induced *in vitro* via calli. Success has been achieved with grapes, and Bernard and Mur[13] obtained buds from *Vitis vinifera* grown in tissue culture, which is a pre-requisite for the rapid propagation of that plant. Favre and Grenan[70] obtained *in vitro* production of tendrils, flowers and berries of the grape.

Until the mid 70's, woody plants proved difficult to propagate by tissue culture, except for those which rooted easily in nature. Rooting seemed to be the main problem and this was so much so that Murashige[169] stated that major breakthroughs were needed in order to obtain plant regeneration from woody plants. He probably meant plant regeneration from calli, which is more difficult to achieve than plant regeneration from shoot tips (micro cuttings) or apical meristems. However, it now seems that no major breakthrough is actually needed, but only more intelligent use of principles; like the application of the knowledge that root initiation and root growth require different auxin concentrations; that certain new auxin conjugates might not be so readily oxidized as IAA and still retain better morphogenic potential than NAA or IBA or other artificial auxins; that in some cases phenols have a protecting effect on auxins and act synergistically to promote root initiation[95]; that controlling sucrose is an important factor; etc.

For the rooting of micro-cuttings (the name given to the *in vitro* culture of stem sections), recent research has also proved that no major breakthroughs were needed to get results. For example, James and Thurborn[112] showed that with difficult-to-root apple cultivars like M-9, phloroglucinol, a phenol normally found in apple leaves and other parts of apple trees, is all that is needed to induce rooting. The use of phenols and other special factors will be dealt with later in more detail.

Among fruit trees, regeneration of pear plants from shoot meristem-tips has been achieved by Lane[136]. In this case, apical dominance was reduced by removing 0.25 cm of the tip from a 5 cm shoot, which was then placed horizontally or inverted. This appeared to have reduced apical dominance and helped the production of a great number of shoots. Tissue culture may also revolutionize the production of peach.[218] Rootstocks of plum and cherry have also been propagated.[117] Propagation of the *Citrus* group has proven difficult to achieve, although work has been done on grape-fruit[14] and general aspects have been discussed by Altman and Goren.[2] Considering forest trees, in *Eucalyptus*, Sita[217] achieved plantlet production from cotyledonary callus, while shoot callus obtained from one year old plants did not differentiate. Among coniferous trees, Boulay[23] has described the rapid propagation of *Sequoia sempervirens* while Janson and Bornman[113] obtained *in vitro* phyllomorphic regeneration and shoot buds and shoots in *Picea abies*. It was noted that while phyllomorphic regeneration is possible, many problems in relation to shoot isolation, elongation and rooting remain. Minocha[158] obtained calli and adventitious shoot formation with excised embryos of white pine, *Pinus strobus*. Arnod and von Eriksson[5] worked with resting vegetative buds of Norway spruce, which were induced to form adventitious bud primordia on medium containing cytokinin, and after transfer of the induced buds to medium lacking cytokinin, adventitious buds developed. They arose from meristems formed *de novo* in the needle primordia, and it is interesting to note that no differences were found in the ability to form adventitious buds among buds collected from trees ranging from 5 to 50 years old. Coleman and Thorpe[45] obtained plantlet formation with the western red cedar, *Thuya plicata*, opening up possibilities for the rapid propagation of this tree. With the Douglas fir, adventitious bud formation was obtained by Cheng (1975) while Campbell and Durzan[35] described the vegetative propagation of *Picea glauca*. Bonga[21] obtained organogenesis in cultures of embryogenic shoots of the balsam fir. The general results with conifers show that no big breakthrough is needed to obtain rapid propagation eventually. All growth factors seem to be known, it is therefore a matter of using them in the right manner.

Work has also been done recently on some woody plants in the tropics. For example Litz and Conover[139] have described *in vitro* propagation of *Papaya*. Nair *et al.*[173] have shown regeneration of shoot apical meristems of cassava (*Manihot esculenta*), a method which eventually can lead to the rapid propagation of this plant.

For aspects of woody plants propagation prior to 1978, the reader is advised to read Lawson[137]; for forest trees in particular, Bonga[22]; and for coffee, Monaco et al.[159]

THE CLONING OF PLANTS AND THE PROBLEM OF PLANT VARIABILITY

The object of cloning is to provide strictly identical plants and horticulturists specializing in plant propagation want these in great numbers. Theoretically, plant tissue culture offers great promise of obtaining just this, but in practice, this has not always happened. In some cases, but only in some cases, abnormal plants appeared in alarming numbers and this tarnished considerably the theoretical image of what plant tissue culture could offer. Let us consider what happened and see how it can be avoided.

(i) Tissue culture

The 'tissue culture' encompasses in fact a great many techniques, which have in common only the use of nutrient media and of being carried out *in vitro*. Thus, plant parts as diverse as buds, apical meristems, leaves, petioles, embryos, stem segments, pith, ovaries, seeds, root tips, cambium, intercalary meristems, epidermis, etc. can be used as starting tissue, and each can be cultivated by different methods depending on the objective. On top of that, single cells derived from any plant part can be cultivated. They may be obtained in some cases directly from plant parts by mechanical means, but more often they have been obtained from calli, themselves generated by a variety of means and from a widely diverse variety of plant organs. These single cells, cultured mostly in liquid media, may either be whole cells — that is with their cell wall present — or with cell walls removed by enzyme treatment, in which case one deals with what are called protoplasts. With many of these types of plant tissue culture, whole plants have been regenerated. Depending on the donor tissue, not all of these having the same regenerative capacity, it being influenced by the family, the genus, species or cultivar. Thus in an impressive number of cases, they can be used potentially for the multiplication of plants. However, if the aim is the propagation of plants that must all be genetically identical, some basic principles must be followed, otherwise genetic variability may show up and destroy the very purpose of the endeavor.

First of all, it must be understood that in a normal plant all of its cells do not necessarily possess the same chromosome complement. In the variety of tissues that comprise a plant, only the meristems keep the faculty of dividing without bringing about irregularities in the chromosome composition. Even then there might be rare exceptions.[89] As the cells differentiate they may start to lose their capacity of keeping intact their chromosome complement, and the first trend observed is towards polyploidy. For this reason, as far back as 1966 Murashige and Nakano[171] proposed tissue culture as a potential tool to obtain polyploid plants. Then aneuploidy may appear. Mehra and Mehra[151] give a list starting from 1958 of authors who detected unstable chromosomal constitution in plants.

Different plant parts do not respond in the same way in bringing about variable chromosome constitution. Chand and Roy[38] studied this aspect with *Nigella sativa*. They found that calli from leaf segments had a tendency to maintain the normal chromosome number, while stem calli showed much polyploidy and aneuploidy. The maximum frequency of tetraploid mitosis occurred in seed calli. The hormonal composition of the culture medium has also a strong influence. Vanzulli et al.[238] studied the influence of 22 hormonal combinations on the caryological stability of *Datura innoxia* calli.

Secondly, as a means of rapidly propagating plants, many workers have induced callusing of plant parts and further induced these calluses to regenerate whole plants. The method seems attractive, because when plant regeneration is possible with a certain type of callus usually one can induce the formation of many buds on a single callus, and each bud can form a whole plant upon transferring to a rooting medium. Thus a great many plants can be obtained rapidly by this method.

It has been known for a long time that calli are composed of cells of heterogenous chromosomal composition. For example, Torrey[237] made a study on the progressive changes that occur in calli with regard to chromosome composition, and stated that the longer the period of subculture (in this case on yeast extract plus 2,4-D) the greater the range of abnormal nuclear behaviour in the callus tissue, with increasing frequency of aneuploid mitoses, variations in ploidy which included odd multiples of the basic chromosome numbers (3n, 5n) and aneuploids around these chromosome numbers. The development of a high degree of polyploidy is accompanied by a corresponding loss of cells at the diploid level. Concurrently, during the prolonged period of subculture there is a progressive loss of organ-forming capacity. After several years of continuous sub-culture, some callus tissues formed 'normal' tetraploid roots and, still later, the calli lost completely the capacity to initiate roots. This loss was paralleled by increasing abnormalities in the chromosomal constitution, including higher chromosome numbers and greater frequency of aneuploidy. Thus, within the framework of experimental work, tetraploid roots were obtained. But much more variation can be obtained when propagation in great numbers is attempted. It is not impossible that in some cases aneuploid cells may not lose completely the potential for plant regeneration, and that mutations follow.

As more refined techniques of culture are being developed, it is observed more and more frequently that **somatic** embryos arise directly from the surface of calli. These embryos, sometimes called embryoids, can be cultured and give rise to complete plants. Somatic embryo formation from callus cells was first observed in the carrot by Steward *et al.*[226] and immediately afterwards by Reinert[203]. Since then, the phenomenon of somatic embryogenesis has been demonstrated in tissue and cell culture of many plants. For example, Halperin and Wetherell[85] obtained some from the wild carrot, Staritsky[224], Söndahl and Sharp[222] and Nassuth *et al.*[175] from coffee, Bonga from *Arceuthobium pusilum*, Jeleska[114] from pumpkin, Button *et al.*[34] from the 'Shamouti' orange, Krul and Worley[132] from 'Seyval', a French grape; Pareek and Chandra[191] from cauliflower, Zatyko[248] from the red currant, and Chang and Hsing[39] from ginseng, etc. Since the calli from which these embryos arise may contain cells with abnormal chromosome number, the question arises: will somatic embryos form from these cells? This, to my knowledge, has not been studied and one can only speculate on the subject. It is known that diploid cells normally keep the highest potential for plant regeneration, this is followed by polyploid cells, and finally the possibility arises that some aneuploid cells may be capable of plant regeneration. This could also be the case for the initiation of somatic embryos. If this is true, the possibility of mutations arising from the cultivation of somatic embryos may materialize. Thus, if clonal multiplication is desired, it is advisable not to use the technique of callus culture prior to bud or embryo formation.

Cell suspensions have also been used for the propagation of plants. At first sight they offer an attractive system for the initiation of huge populations, since each cell has theoretically the potential of giving rise either to a somatic embryo or to a callus which itself can give rise to many plants. Let us consider this system from the point of view of the horticulturist who needs a strictly clonal propagation, meaning that no mutation should appear.

There are two general methods through which cell (that is cells with their cell wall) suspension cultures can be originated. The oldest one consists of obtaining initially a friable callus, which is then placed in a liquid culture medium and agitated gently so as to loosen-up individual cells or cell clumps. A friable callus is obtained by the judicious choice of hormones. Usually 2,4-D is the best. From what has been said previously, it is now clear that any system derived from a callus is potentially a source of abnormal cells, which may under favourable conditions give rise to mutated plants. Thus a cell suspension obtained from calli is not usable for obtaining a pure clone.

The second method is only beginning to be developed. It consists of mechanically separating cells from a tissue, as for example colyledonary tissue[212], and cultivating these cells. Cell cultures are capable of giving rise to somatic embryos. Thus if mechanically obtained cells were cultured in such a way as to give rise directly to embryos (this has been done by Constabel et al.[48] but not from cells obtained mechanically), the chances of obtaining pure clones would be theoretically perfect. But this has not yet been achieved, for example Schwenk[212] used his mechanically obtained cells to induce callus formation.

Another technique for multiplying plants is to use meristem tips derived from buds. By meristem tips, or apical meristem, is meant a piece of tissue no more than 0.5 mm long taken from the meristematic dome region and including at most three leaf primordia. At the beginning of the culture the technique does not seem to be rapid, because the development of such a small piece of tissue takes time and some of the meristems do not develop at all, particularly when the explant is too small. But as soon as this meristem has developed into a small stem with buds the propagation rate becomes rapid: each small stem or bud produced has the potential of producing an entire plant with axillary stems and the possible multiplication rate becomes geometric. This technique shows the following advantages: the plants derived from meristems are true to type and, as an additional advantage, may be rid of virus diseases that infected the parent plants. However, one must be careful to adjust the hormone composition of the nutrient medium so as to avoid callusing of the stem and the production of adventitious buds which arise from such calli. For example, the use of 2ip with the apple system[227] will induce abnormal proliferation of stem cells (presumably cambial cells) which produce white spots on the surface of the stem or a swelling of the base of the stem from which adventitious buds may arise.

The multiplication of plants through such buds is not guaranteed to give rise to true-to-type plants. On the other hand, if the hormonal composition of the medium is adjusted in such a way as to induce the formation and development of only axillary buds, one may be assured of the best clonal propagation possible. It might not be the most rapid, but will avoid to the greatest degree the production of variants. When there is production of a callus at the base of the stem and this callus produced adventitious buds, only the axillary buds arising from the stem should be used as explants. This technique is used by Boxus (personal communication) for the rapid propagation of virus-free strawberry plants. Then mutations are avoided. For house plants, where unwanted mutations are easily spotted at an early stage, commercial propagation can be, and actually is, commonly carried out through simultaneous adventitious and axillary bud formation.

Another technique which offers great promise for the horticulturist is the cultivation of shoot tips. In this technique one uses, instead of only the meristematic dome, the whole apical bud together with a portion of the stem which can be 1 to 2 cm long. This procedure does not require the use of a dissecting microscope and is more rapid to start with. However it requires the use of virus-free stock, and furthermore, making the stem aseptic is more difficult. Virus-free stock is obtained through traditional methods, whereas aseptic material can be obtained by resorting to several tricks like using greenhouse grown plants instead of outdoor grown plants which have heavier surface infection, dipping in 80% alcohol or a detergent prior to disinfection with hypochlorite, etc.

In order to be assured that no mutations, variations etc. appear in the process of multiplication one must be careful, as in the case of apical meristems, that multiplication is assured through the formation and development of axillary buds, and not from adventitious buds arising from callused regions. This is clearly shown by the work of Wasaka[245] who experimented with pineapple, even though he did not make a clear distinction between meristematic and non-meristematic tissues. In this case, individuals re-differentiated from syncarps and from 'slips' produced a very high frequency of variant plants. The variation index was so high from syncarps that no individual of the original phenotype showed up among 104 re-differentiated plants. In contrast, only two individuals out of 29 initiated from crowns were variants. The crown evidently included the meristematic region and produced less variation.

(ii) Transferring to soil

In classical tissue culture, plantlets obtained *in vitro* have also been rooted *in vitro* and mainly on agar. Once a plantlet has rooted, one of the main problems is the transfer to soil. If the proper technique is not employed, some 20% or more of the plants may be lost in the process. The cause of loss of plants may be twofold. (1) The plantlets may lose excessive amounts of water because they have not developed systems appropriate for the uptake and conservation of water, (2) roots formed in agar are not fit to grow in soil.

Let us consider first the case of conservation of water. Up to now plantlets have been transferred to soil under a mist or a tent in a greenhouse. One of the first things that one can observe is that roots that have been grown in agar are not functional when washed from the agar and transplanted into soil. Because of that, new roots have to be grown by the plantlet in order to be able to absorb water from the soil. As a consequence, during a period of days, the plant has to rely solely on the humidity absorbed by the leaves from the moist air. Furthermore, the roots grown in agar are not resistant to the soil organisms that gradually invade the previously sterile soil. In order to solve this problem, some have tried to root *in vitro* the plantlets on a substrate other than agar. They have used mixtures of vermiculite, perlite and sphagnum moss etc. impregnated with a nutrient solution. But good contact between the artificial soil and the rootless plantlet is difficult to achieve consistently, resulting in a poor rooting percentage. Also the handling of the artificial soil is somewhat messy and requires extra labour. The tendency now is for commercial growers to root the plantlets directly in sterile soil under mist. This saves one *in vitro* step, thereby saving time, labour and money. The roots that develop in these conditions are more vigorous and more resistant to soil bacteria, thus giving rise to healthier and more resistant plants.

Aside from the aspect of roots and rooting, plantlets that have been grown *in vitro* differ widely from plants grown in the field, particularly in two respects: wax formation on the leaves and xylem regeneration. These aspects have been examined in cauliflower[82], while Sutter and Langhans[231] have studied epicuticular wax formation in carnation. Carnations grown in the field or in the greenhouse are glaucous due to a dense cover of tiny wax rods. Those grown from shoot-tip cultures *in vitro* were mostly non-glaucous and no structured wax was seen on

non-glaucous plantlets. The low survival rate of carnation was explained by their lack of epi-cuticular wax structure which results in excess dessication when they are transferred from *in vitro* conditions to the greenhouse. The cuticle itself was not examined, but it is very possible that *in vitro* grown plants have a thinner and less waxy composition compared to their counter-part grown normally. Considering the fact[231] that non-glaucous plants did not develop epicuticular wax in 2½ weeks in the greenhouse, one must accept that the plantlet must grow new leaves in order to be able to possess this protective coating. Thus rootless plantlets trans-ferred to soil must achieve two things in order to be able to survive in normal air, (a) they must make new roots, (b) they must grow new leaves to be better adapted to dry air stress.
dry air stress.

When new roots are being formed, another aspect has to be considered. The conducting elements of the new root must bridge with the existing elements in the stem. This may take a few days to a week in order to be functional.

SPECIAL CULTURE FACTORS

Most tissue cultures are now started on standard media, with auxins and cytokinins at various concentrations. When this does not work, special factors are tried. Here the intention is not to deal in detail with all special factors that can be used, but to mention only some that have been used in very recent years.

(i) Phenolic compounds

These compounds have long been known to act synergistically with auxins to promote adventitious root formation. Hess[95] studied this system in mung cuttings, and the protective effect of phenols on auxin level was examined by Gaspar *et al.*[73] in relation to their effect on initiation of root primordia and the subsequent growth of the roots. They used chlorogenic and ferulic acids.

Phloridzin is a major phenolic compound in apple trees, and it represents 3 to 7% of the dry weight of leaves. It is also important in the *Rosacae*. Two of the phloridzin breakdown pro-ducts, phloroglucin and phloretic acid have been used by Jones and Hatfield[116] to improve rooting of the Malling 7 apple rootstock which rooted poorly with auxins. James and Thur-born[112] also used phloroglucinol to root the difficult Malling 9 rootstock, and found a syner-gistic effect of indole butyric acid and phloroglucinol on rooting.

With the peach, other phenolic compounds have been used to induce rooting: quercitin and rutin, which are 0-diphenols. Rutin was found to be the most effective of the two com-pounds.[165]

Bud formation and shoot growth can also be enhanced by phenols. Jones[115] used phloridzin and phloroglucinol with the Malling 7 apple. He found that the two phenolic compounds markedly stimulated shoot growth, internode extension, leaf production and expansion. Like-wise Mosella *et al.*[165] used phloridzin and phloroglucinol for the development of buds. They found that these gave a better response than quercitin and rutin which, on the other hand, are more effective in rooting. James[111] found that phloroglucinol favours adventitious root form-ation in *Rubus* and *Fragaria* grown *in vitro*, demonstrating that the growth promoting effect of phloroglucinol is not specific to Malus species.

The beginning of an explanation of the beneficial effect of phenols on plantlets of apple explants grown *in vitro* from meristem tips has been found (Phan and Cailloux, unpublished). Calli as well as plantlets derived from meristem tips were completely devoid of phloridzin, contrary to the twigs and buds of the normal tree. Instead, they contained low levels of acidic phenolic compounds, free hydroxycinnamic acids or 'acidic' esters. Thus both the callus and the young plants grown from meristem tips have lost their capacity for synthesizing phloridzin. It has also been found that the *in vitro* cultures benefit from phloridzin when they have been in contact with the added phloridzin or phloroglucinol for not more than a few weeks. Thereafter the cultures decline if they are left continuously in contact with these phenols. A possible explanation for the phenomenon is that one or more of the enzymes responsible for the synthesis of these phenols is/are inductive: phloroglucinol or phloridzin is needed to stimulate them. Once they have been stimulated to synthesize the phenols, the exogenous phloridzin or phloroglucinol must be omitted from the culture medium. Otherwise the concentration of these probably gets too high and the cultures wither slowly.

Phloroglucinol has also been used recently in the culture of *Chincona ledgerianum*[106]. Previous attempts to culture apical meristems of this plant failed, being accompanied by browning and eventual necrosis, and treatment with anti-oxidants like ascorbic or citric acids had no effect. The stimulation to rapid growth was both significant and persistent and, although oxidative browning of explant cut surfaces was still present, it appeared to have little influence on growth.

(ii) Special hormones

Ordinarily the known natural and artificial auxins and cytokinins suffice for most plant tissue cultures. However, in special instances the plant material may not respond to these, and one may try some of the other plant hormones, especially gibberellic acid, and in some instances, ethylene.

Gibberellic acid has been tried in many instances without success in attempts either to improve morphogenesis or to enhance shoot growth. But, Wochok and Sluis[246] have used it successfully to promote shoot multiplication and elongation in *Atriplex*. It works also on morphogenesis in cassava, and it seems likely that BA and GA_3 together are necessary for complete morphogenesis in cassava.[125]

Ethylene has not been generally used in tissue culture. However it may have some applications. Goren *et al.*[77] have used it to induce callus formation in Citrus bud cultures.

Indole acetic acid (IAA), the natural auxin, is destroyed rapidly by many tissues, thereby limiting its use in tissue culture. In such cases, it is ordinarily replaced by the artificial auxins naphthalene acetic acid (NAA) or 2,4-dichlorophenoxy acetic acid (2,4-D). In certain cases, the latter may induce undesirable side-effects such as callus formation, polyploidy, aneuploidy or chromosome damage[237], or suppression of organized growth. In quest of reducing the rate at which IAA is oxidized, Hangarter *et al.*[86] have studied auxin conjugates which mimic bound auxins that are the natural reservoir from which a plant draws to obtain free IAA at a gradual rate. Conjugates of IAA are protected from peroxidative destruction and can act as a reserve source of IAA. They are probably hydrolysed slowly, releasing a constant low steady state concentration of free IAA. It was found that IAA-L-alanine and IAA glycine are good auxin sources for the production of callus, while the commercially available auxins are not nearly as good. They also found that while IAA normally favours root development, a number of IAA conjugates favour shoot development instead. They even have been able to regenerate shoots from tomato roots. They also have obtained friable callus and finely divided suspension cultures from explants of field corn seedlings using IAA-L-alanine. It remains to be seen if chromosome stability is better than with 2,4-D, opening the possibility of obtaining clonal multiplication without mutations through callus and cell suspension cultures.

Mecoprop [2(4-chloro-2-methyl) phenoxy propionic acid], an analogue of 2,4-D, has been shown to possess auxin activity even greater than 2,4-D in some bioassays. Experiments were carried out to test its capacity to induce callus[208], and it was found that it produced a greater proportion of 'good' calli with wheat and rye, and that it is also better for supporting the callus growth. There is an indication that it is more effective in cereal callus culture.

Plant juice, extracts of yeast, tomato, grape, banana, coconut milk etc. have been used extensively in the past because they contain beneficial growth factors. In modern tissue culture they are used less and less, mainly for two reasons: (1) their action can be replaced almost universally by cytokinins (2) their growth promoting action is not equal from batch to batch. But in certain cases they may be useful. For example, Bapat and Rao[9] had to resort to coconut milk in order to achieve regeneration of plantlets from shoot apical meristems of morning glory (*Pharbitis nil*).

(iii) Sugars

In most cases, sucrose is used successfully in tissue culture to replace photosynthesis in the test tube. But sucrose also controls differentiation and has for example some effect on the formation of xylem elements. Very recently, Maretzki and Hiraki[142] showed that in sugarcane sucrose promotes root formation, on plantlets regenerated from calli.

(iv) Amino acids

Calli of certain plants seem not to be able to synthesize certain of their amino acids. For example, Bhojwani and Hayward[15] grew wheat calli on B_5 medium. The calli showed intense rooting, but no shoot differentiation. By adding to the culture medium 2000 mg/l of asparagine and placing the cultures in the dark, 4 or 6 cultures differentiated multiple shoots. Since the differentiation was only sporadic, they felt that all the bud formation factors had not been found.

(v) Mechanically isolated cells

Isolated cells have been obtained by dissociating the cells from a friable callus. As said previously, the induction of callus and its subsequent growth produces polyploidy and cells with varying chromosome numbers. If these chromosome variations are unwanted, a good alternative is to mechanically isolate cells. Mechanically isolated cells from soybean were obtained as early as 1953, but they did not enlarge nor divide. Schwenk[212] has devised a method for mechanically isolating cells from cotyledons of soybean. Cotyledonary cells possess a better growing capacity and a much higher potential for regeneration of plantlets than cells from the leaves, so that much more success in growing them is expected. Free cells were obtained by placing small pieces of soybean cotyledons in a 25 × 300 mm test tube possessing internally six small protuberances and placing this in a Lab-Line Supermixer running at top speed for one minute. The vortex causes the cotyledon pieces to impinge on the indentations on the side of the wall of the test tube and causes the cells to dislodge from the mass. However, in this case, the cells were not grown as cell suspensions but were induced to form calli which produced vigorous roots.

(vi) The regenerative capacity of calli obtained from different plant parts

When calli are wanted, particularly for the production of mutants, the first requirement is that the calli be able to regenerate whole plants. Since calli can be produced from various parts of a plant, a question arises: which plant part is best to start a callus? Research workers are finding more and more that a huge difference exists in the regenerative capacity between a callus from cotyledons, for example, and a callus obtained from the pith or from leaves. The former possesses a much higher regenerative capacity.

Morel[162] noted growing evidence to suggest that calli derived from growing apical meristems possess higher morphogenic potential than those from mature parts. Sita[217], working with eucalyptus, found that callus culture of hypcotyls and of cotyledons differentiated into plantlets, while leaf and shoot callus obtained from one year old plants do not differentiate. Wheat, corn, rice, barley, sorghum and oats are the cereals most resistant to callus formation and regeneration. Green and Philips[81], working with maize found that even with the same plant part, immature embryos possess a higher morphogenic potential than those from older embryos. Later, Orton[188] found that, for barley, immature embryos give rise to the most vigorous and easily regenerated callus.

With respect to different plant parts, it is interesting to note that besides their regenerative capacity, they may be typed according to their specific cellular determinants. Stems, leaves, pistils, and anthers, have been examined by Raff *et al.*[199] who found that they possess different antigens which are still expressed in callus cells after four subcultures. The organ specific antigens are associated with the protein rather than that of the polysaccharides.

(vii) Inverted position

By placing the explant in a position other than vertical, one may cause changes in the hormonal distribution and translocation within the explant. This change in hormonal distribution affects morphogenesis, and sometimes in the right direction. For example, Abbott and Whiteley[1] obtained more consistent root formation with *Malus sylvestris*, 'Cox's Orange Pippin', when the explant was placed in an inverted position. In normal position they obtained 30% rooting, while in inverted position the rooting percentage rose to 80%. Placing the explant horizontally or vertically may also reduce apical dominance. Shoot orientation influences shoot multiplication. With pear, Lane[136] showed that if the tip (0.25 cm) is removed from a 5 cm pear shoot, which is then placed horizontally or inverted in a 5 μM benzyladenine medium, most leaf axillary buds grow and form shoots.

(viii) Temperature

The influence of temperature on organogenesis has not often been investigated. Usually, experiments are carried out at laboratory temperature. However, rooting, e.g. of apple never reaches 100% and every single factor which improves rooting is welcome. The finding by Lane[135] that temperatures below 28°C decrease rooting efficiency of apple might lead to improvement. The best temperature for growing potato calli is 28°C. Probably the tissue culture of tropical plants should be carried out at higher temperatures. For plants of temperate climates, it is possible that fluctuating temperatures within certain limits might prove beneficial. In certain cases, dormancy must be lifted by a cold treatment.

POTENTIAL FOR OBTAINING MUTANTS

In contrast to plant propagators who do not want mutations to appear among their plants, plant breeders welcome them, because some of them may possess desirable traits. Plant tissue culture offers a great variety of means by which nuclear and protoplasmic inheritance can be altered. However, it is only in extremely rare cases that this potential of tissue culture has been exploited with commercial success. The techniques for inducing genetic changes and for selection of the desirable variants have to be perfected a good deal before practical routine procedures are established. The science and art of plant tissue culture is so new that the necessary time to perfect it for these specific aims has not been sufficient. Tissue culture could drastically shorten the time necessary for developing new varieties by conventional breeding and selection programmes.

Here, the term mutation is being used in its widest sense: any change in chromosomal or cytoplasmic composition arising by artificial means, as opposed to conventional breeding by cross pollination. Tissue culture can bring about changes in chromosomal or cytoplasmic composition, or both, by diverse means.

(i) The use of calli

When plants are regenerated from calli or isolated cells, it is well known that when moved from the stabilizing environment of the intact organism into the alien environment of the culture vessel, tissues and cells display more nuclear irregularity, including both chromosomal and gene mutations.[50,192,229,230]

Polyploidy is the most common phenomenon which occurs in callus culture. This polyploidy arises from two distinct causes. When the initial explant is taken from a plant part other than a meristem or a cambium, this explant may already contain polyploid cells.[171] These cells multiply together with the others in the callus and may differentiate into somatic embryos or buds which usually can be developed into whole plants. Secondly, *in vitro* culture conditions, particularly the type of phytohormones used and their relative concentration, may play a significant role in bringing about polyploidy. The fact that polyploidy is induced by tissue culture of calli was recognized for some years and some scientists, e.g. Nitsch *et al.*[184], took advantage of this fact to produce diploid tobacco plants from haploid calli. Horak *et al.*[103] produced polyploid plants from tissue cultures of *Brassica oleracera* and they later[104] produced polyploid plants from stem pith explants of diploid marrow stem of kale. Thus polyploid plants may originate either from polyploid cells already present in the original plant, or from cells which become polyploid during callus formation and growth.[104]

Tissue culture of calli also brings other changes in chromosomal composition. In many cases, these changes are caused by the use of 2,4-D, a hormone used to bring about callusing and maintain the callused state. Different species react differently and one cannot set a general rule. For example, Roy[206] made a comparative study of chromosomal variations in the callus tissues of *Allium tuberosum* and of *Allium cepa*. For *Allium tuberosum* calli, obtained with 1 mg/l of 2,4-D, it was found that during the first passage cells showed mostly the normal chromosome number (2n = 32), while a few cells revealed aneuploidy (21–31). In the second passage, triploidy, early separation and scattering of chromosomes showed up. The proportion of aneuploid cells to normal cells was higher in the third passage (3 months old). Fragments of chromosomes and micronuclei were also noted. In the fourth passage, the occurrence of micronuclei was very high. In certain cells, 4 to 5 chromosomes were lagging, whereas under the same conditions, in *Allium cepa* there was mostly polyploidy.

As chromosome composition becomes aberrant, it seems that the regenerative capacity diminishes greatly[206,207], in other words, plant regeneration selects for cells with specific chromosomal constitution[186]. The regenerative capacity manifests itself at the highest frequency with diploid cells, then come the poplyploid ones. As the cells become aneuploid, they can still regenerate complete plants[207], but the regenerative capacity becomes lower. The frequency with which calli give rise to aneuploid plants is not generally known, because caryological studies are not always made with mutants. For example, Beach and Smith[10] induced callus formation of carnations with 11 μM of naphthalene acetic acid, 10 μM of 2,4-D and kinetin. From the calli were obtained, in addition to morphologically normal, self fertile plants, crimson plants that were both male and female sterile and which showed definite physical abnormalities in regard to leaf shape and flower head structure. A cytological study of these abnormal plants was not done, but one may wonder if they were aneuploids or ones with chromosome abnormalities. In some cases, cytogenetic studies were made of variants, such as that of Heinz and Mee[93] with sugarcane, but it is a pity that in so many cases, only a morphological description of the variations is made. Thus a genetic basis for the building-up of variants is not yet established.

One may conclude that plant regeneration from calli can become a tool for the novel expression of the genetic pool already present in a plant. While in many cases plant regeneration from calli still proves to be difficult, some progress is steadily being made. Among recent work, one may cite Greef and Jacobs[79] who report the establishment of a habituated callus line of sugarbeet with a high regeneration capacity; Beach and Smith[10] who regenerated plants from callus of red and crimson clover (morphogenesis has always been difficult to achieve with legumes); Tilquin[236] who obtained plant regeneration from stem callus of cassava, and Orton[189] who studied chromosomal variability in regenerated plants of *Hordeum*.

(ii) Adventitious buds

One may also obtain genetic variability by inducing the production of adventitious buds arising from callused portions of stems. No one has ever made a specific study of this aspect, but it may very well explain the variability obtained sometimes when apical meristems and shoot tips are cultivated. Depending on species, and even cultivars, callusing is very easily obtained with kinetin, BA or 2ip. As stated previously, buds arising from calli may be polyploid in composition and even aneuploid. This feature could be exploited systematically in order to obtain genetic variability.

(iii) Cell cultures

Cell cultures can and have been used as a means of obtaining variant strains, particularly cell cultures initiated from calli. The more often the calli and the cell suspensions have been subcultured, the more variation that can be expected. On top of that, variability is further enhanced if the donor tissue is a chimera. For example, the research done on sugarcane by Heinz *et al.*[91] can be cited. "Mature parenchyma tissue of clone H 50-7209 was used to produce a callus. Material from this original callus was subsequently grown as a suspension culture. Several years later, single cells were isolated from the suspension, and a callus produced from each of them by the nurse culture technique[166]. Of five isolated individual cells, each produced tissue differing in appearance. One was white, two beige, and two yellow. The two beige ones grew at different rates, as did the two yellow ones. It appeared that there were five different clones".

Through cytogenetic studies it was found that each of these subclones had different chromo-some numbers.[92] It was established that the suspension culture from which the explants were made contained cells of varying chromosome numbers. Later, in field-grown plants of the parent variety, it was discovered that individual plants are not made up of cells of uniform chromosome number. In other words, this particular variety exists as a 'chromosomal mosaic'. From this work it can be seen that a great range of mutations can be expected by exploiting the cell suspension technique. The nurse culture technique developed by Muir et al.[166] does not necessarily have to be used. Many natural and induced variations can be detected and isolated through plating of cell suspensions. The variant may arise through mutation or by epigenic change (the expression of genes that were repressed). In either case, the new characteristics can be retained over long periods if propagated clonally.

Sugarcane breeders have probably been the most successful in utilizing cell culture for breed-ing purposes. For example, they have been able to develop cultivars that resist eyespot disease caused by *Helminthosporium sacchari* and the downy mildew caused by *Sclerospora sacchari* as well as viruses like the sugarcane mosaic and Fiji viruses, by utilizing cell cultures[178]. How-ever, regeneration of whole plants from single cells has not been achieved for all species of plants. Still more basic research has to be done to attain this goal, but a good deal of practical work can presently be done with those species and clones for which the procedure of obtaining whole plants from single cells has been worked out.

(iv) The use of mutagens

Mutagens in connection with tissue culture have not been used extensively, but they offer a powerful tool to enhance the variability already cited and could be used routinely for the pro-duction of new cultivars.

Chemical mutagens have been used in connection with sugarcane breeding programmes. As early as 1969, Mee et al.[150] reported on their effects on cell suspensions. Heinz and Mee[93] have studied particularly methyl methanesulphonate at 50 mg/l and ethyl methanesulphonate at 50 mg/l. At that concentration they found it lethal to certain clones and non-lethal to others, so that the dosage has to be worked out for each clone. Malepszy et al.[141] used ethyl methane sulphonate and 5-bromo-2-desoxyuridine to bring about a considerable broadening of the range of alkaloid content of *Nicotiana sylvestris*. Colchicine has been used by Lin et al.[138] as an aid to sugarcane breeding. He obtained aneuploid cells and plants by treatment of cell suspension cultures. The same chemical has been used by Chen and Goedenkallemeyn[41] to obtain tetra-ploid plants from diploid daylily callus.

Radiation biology of cultured cells has been studied quite extensively.[105] However, radi-ation has been tried to a very limited and timid extent in connection with practical induction of mutations *in vitro* for use by horticulturists. Only the possibilities of the technique have been shown. For example, Nitsch et al.[182,184] demonstrated that mutant plants can be obtained by 1.5 to 3 Krad doses of γ irradiation to microspores or plantlets in tobacco anthers. Leaf variation, albinism, flower colour and petal shape variations were obtained. Devreux and Saccardo[53], also working with haploid tobacco, claim to have obtained a high frequency of mutations with 1 Krad of X-rays. Note that using haploid cells permits direct selection for re-cessive mutations. Eriksson[66] reported having obtained a high propensity for anthocyanin production with UV-irradiated *Haplopappus* culture. In connection with UV, Howland and Hart[105] state that "UV mutagenesis, because of its specificity, availability, convenience, and relative safety, should find wide application in studies with isolated plant cells. More is known about the molecular effects of UV on DNA (i.e. pyrimidine dimer induction) and about the repair of these DNA lesions than for any other agent. Mutation frequencies may be enhanced or suppressed by careful selection of the experimental conditions".

Very often experimental work is directed only to evaluate radiosensitivity of the cultured cells or tissues. This type of work is reported by Howland and Hart[105] in some detail, and research on these lines will in the long run benefit plant breeders but the practical aspects are long coming. For example, Heinz[90] tried 2.5 and 8 Krads of gamma radiation on clones of sugarcane, and found that two clones were killed while two others survived, showing clonal differences in the sensitivity to gamma radiations.

(v) The incorporation of foreign DNA

Incorporation of new genes in bacteria has been achieved with success in a variety of cases, but genetic engineering has proven to be much more difficult with vegetative parts of adult plants, calli and single cells. Here, there is the problem of the DNA being adsorbed by cell walls, by the nuclei or simply dissolved in the cytoplasm.

The one successful experiment by which genes from a bacterium have been transplanted into the genome of a plant of horticultural interest is in the case of the tomato. Even then, the particular genes that have been transferred are of no interest to horticulturists. In this case (Doy *et al.*[56,57,58]), tomato calli were treated with phage material carrying the bacterial genes for galactose-1-phosphate uridyl transferase. Normally, tomato calli die or grow very slowly on media containing galactose as a carbon source. Following treatment with the phage material, normal growth was induced in some calli on galactose media. The same was true on lactose media following a treatment with phages carrying the bacterial gene for B-galactosidase. Working concurrently, Johnson *et al.*[119] did the same general type of experiment, this time with suspension cultures of *Acer pseudoplatanus*. The bacteriophage or its DNA was taken up by the plant cells and *lac* genes were expressed leading to enzymes which confer the capacity to assimilate lactose. Thus the possibility of gene transfer in a plant of horticultural interest has been demonstrated.

Some genes from bacteria could be extremely important for higher plants. For example, molecular nitrogen fixing genes would be of immense value to crop plants, because they could then fix atmospheric nitrogen and would have reduced requirement for nitrogen fertilizers.

DNA extracted from higher plants, instead of from bacteria, may be very useful in creating planned mutations. This, for example, has been done with *Petunia*. The transfer of genetic material from one pure line to the other for flower colour (anthocyanin synthesis) was apparently obtained by treating ovaries[101], seeds[97], or seedlings[96,97,98,99,100], with appropriate DNA. Treatment of a white-flowering pure line with the DNA of a red-flowering anthocyanin-synthethising pure line resulted in a genetically stable correction for anthocyanin synthesis in 0.06% of the treated plants. This kind of genetic exchange technique has some potential for horticulturists, but up till now it has been confined to the research laboratory, and shows no immediate sign of being adaptable for practical purposes.

THE SELECTION OF MUTANTS

When mutants are sought which relate to the mature morphology or physiological response of the plant, then there will be no alternative to propagating and growing material to maturity, for screening by conventional methods. However, when the mutants sought relate to disease resistance or tolerance to some toxicity which acts at a cellular level, then the possibility arises of selecting very early. In this way it is possible to take advantage of the potential variation present in tissue cultures where, particularly with cell suspensions and protoplasts, there are hundreds of thousands of small individuals which cannot possibly be grown to maturity in order to detect useful mutants. Thus early screening methods must be devised. Some have been found that may be of some use to plant breeders, but progress in this field has not been great.

(i) Haploidy and the selection of mutants

Most mutations are recessive in nature, and they are often unable to manifest themselves in the duplicate unmutated gene. Haploid plants simplify the detection of mutants because there are no masking effects of modifications in recessive genes by their dominant counterparts.[170] Haploid tissues can be obtained either by anther culture or pollen culture, and the production and application of haploid plants is considered at length in Chapter 12.

The first culture of anthers was done by Guha and Maheshwari[83] with *Datura*. The pollen in these anthers was stimulated to produce embryos from which plants could be obtained. Then Bourgin and Nitsch[25] produced haploid plants from anther culture of *Nicotiana*. However, this technique suffers from one main disadvantage: the developed plants may not originate from the pollen only, but also from various other parts of the anther. The result of this is that one obtains a mixed population with various levels of ploidy. For example, Durr and Fleck[64] obtained 40% of haploids and 60% of diploids with tobacco anthers, and screening procedures are not easy. Despite this difficulty, Nakamura *et al.*[174] improved the fluecured tobacco variety MC 1610 by using this type of culture. They obtained disease-resistant tobacco varieties in 2 years, instead of the usual 6 with traditional methods.

Another culture has been successful in a diversity of plants. For example, Niizeki and Oomo[177] did it with rice in 1968 and, according to Melchers[152], the Chinese have been able to produce improved rice varieties by hybridizing plants that originated in anther culture. McComb[149] writes that in China particular efforts are being made with anther culture for plant breeding purposes. Anther culture has also been done with wheat[190], barley[40], belladona[202], potato[223], *Solanum surattense*[216] and *Arabidopsis thaliana*[211].

Haploid plants have been obtained from isolated pollen from a very few plants like tobacco[183,204], *Datura*[181], and *Petunia*[209], but as the technique develops many advantages over the culture of anthers will be found: the number of plants obtained is higher, and the time to produce them is shorter. For example, it is now possible, at least with *Nicotiana* and *Datura*, to start with a microspore, and by doubling the chromosome number of the microspore, collect seeds from the homozygous diploid five months later.[180]

Protoplasts from haploid plants can also be used for the large scale propagation of haploid plants and the induction of mutations. Facciotti and Pilet[68] have obtained somatic embryos and plants from *Nicotiana sylvestris* protoplasts.

(ii) Selection for resistance to disease

Selection for disease resistance is a complex affair because of the great variety of causes of disease. For a general article on the subject, read Brettell and Ingram[31]. Two main methods will be described here: (1) exposure of the cultured cells to a particular parasite (2) exposure of the cultured cells to a toxin produced by the parasite. The selection can be done either on cell suspensions or on calli.

In certain cases, one can try to grow the parasite on calli of the plant under study (the parasite will grow on certain calli, while not on others). Plants regenerated from the resistant calli may show resistance to the growth of the parasite. In this respect Ingram and Robertson[109] showed that cultured tissues of susceptible and resistant varieties of *Solanum tuberosum* exhibit the same response as intact plants to infection by the potato blight *Phytophtora infestans.* Also, Helgeson *et al.*[94] showed that callus cultures derived from tobacco plants resistant to black shank disease would not support growth of the offending fungus, unlike cultures from sensitive plants which did support its growth. Ingram[107] discussed the general aspects of biotrophic parasites in tissue culture. It must be noted[108] that there are several examples in which there appears to be little correlation between intact plants and the tissue cultures derived from them, in their response to fungal parasites. For example, this is so between *Brassica* species and the disease caused by *Peronospora parasitica.* It is also the fact for the interaction between sugar beet and *Peronospora farinosa.* Thus the method needs to be tried beforehand in order to know if it is effective or not. Also, in many cases the parasite is difficult to grow on calli and some research has to be done in order to achieve this goal, e.g. it is only recently that Ando *et al.*[4] succeeded in growing *Puccinis horiana*, white rust, on calli of chrysanthemum.

In the case of exposing single cells or calli to a host-specific toxin the technique involves presenting the cells or calli to a just-sublethal concentration of toxin produced by the parasite under study.

The technique of selection of single cells by a toxin analog has been used by Carlson[36] with methionine sulfoximine (MSO) which resembles the tabtoxin produced by the bacterial pathogen causing the wildfire disease of tobacco. It will elicit a response from tobacco leaves identical to the true toxin produced by *Pseudomones tabaci.* For this research, haploid plants of tobacco were initiated from anther culture and single cells and protoplasts from these were next treated with the mutagen ethyl methyl sulfonate. They were then cultured in the presence of MSO. Most of the calli obtained from the surviving cells could not grow with MSO, but three calli succeeded in growing and gave resistant diploid plants.

A toxin in conjunction with calli has been used with success[74] to select *Zea mays* calli resistant to the *Helminthosporium maydis* Race T patho-toxin of the southern corn leaf blight. Selection was done on sublethal concentrations of the toxin using callus carrying the Texas male-sterile cytoplasm which is susceptible to growth inhibition. Apparently the selection has been accomplished for genetically altered mitochondria. Brettell *et al.*[30] used the same type of system to select for maize resistant to *Drechslera maydis* T-toxin.

The toxin does not necessarily have to be pure to permit screening for disease resistance. For example, Behnke[11] obtained general resistance to late blight of *Solanum tuberosum* from callus resistant to culture filtrates of *Phytophtora infestans.* This bypasses the time-consuming and costly process of purifying the toxin and may very well be used as a general technique.

(iii) Physical factors in selection for mutants

One of the important physical factors in the selection for mutants is temperature, and this has been used to select for cold resistance. Dix and Street[54] for example, used cell lines of mutagen-treated and untreated *Nicotiana sylvestris* and selected them for enhanced resistance to chilling injury by exposing them during 21 days at $-3°C$ to $+5°C$. Some of the lines retained cold resistance after growing at $24°C$ for 28 days.

Tomato is very susceptible to injury at chilling temperatures, whereas the potato is relatively chilling resistant. Smillie *et al.*[220] fused tomato with potato protoplasts and found that the chilling resistance of all four tomato-potato hybrids they obtained was intermediate between the tomato and the potato.

PROTOPLASTS AND PLANT IMPROVEMENT

Protoplasts are plant cells which have been artificially rid of their cell walls. These naked cells, because of their lack of cell walls, exhibit special properties such as capacity of fusion with other protoplasts, intake of organelles, intake of bacteria and foreign DNA which may, in the future, be of great use in developing new crop varieties.

(i) General

Protoplasts, as such, have been known for a long time. As long ago as 1892, Klerker[131] obtained some by mechanical means. He first plasmolysed tissues and then cut them into thin slices to liberate the cell wall-entrapped plasmolysed cells. He called these protoplasts. The yield was relatively poor, and furthermore, he obtained a mixture of protoplasts, sub-protoplasts (the protoplasts that have been sectioned during the slicing process, plus those that had been created by plasmolysis itself) and liberated vacuoles. Many workers tried to improve the technique, but without much success. Then, 68 years later, Cocking[43] developed a radically different technique. Instead of cutting the cell walls, he digested them by means of a cellulase. He first plasmolysed the cells of root tips of tomato seedlings, and then subjected the root tips to the action of *Myrothecium* cellulase. The advantages of this technique were immediately realized: (1) large quantities of protoplasts could be obtained easily (2) the plasmolysis did not have to be as severe, imposing less strain on the isolated cells (3) sub-protoplasts, which are amputated cells, did not contaminate the protoplast suspension.

A decade later, successful experiments taking advantage of some properties of protoplasts and having practical overtones began to be published. For example, Power, Cummins and Cocking[196] were the first to fuse interspecific protoplasts in a reproducible way. They fused oat and maize protoplasts, using 0.25 M sodium nitrate. Carlson *et al.*[37] then not only succeeded in fusing *Nicotiana glauca* with *Nicotiana langsdorfii*, but obtained plants from the fusion products. They claimed that the parasexually produced hybrids were identical in every way, including chromosome number (2n = 24 + 18 = 42) to the sexually produced hybrids. This success greatly attracted the attention of plant breeders, in view of the vast possibilities of the technique, and has given a powerful impetus to research in this field. So much so that complete plants have been obtained from enzymatically isolated protoplasts of tobacco, carrot and petunia. Bajaj[6] summarised the prospects for protoplast use as follows:

(1)   the regeneration of somatic hybrids from fused protoplasts of sexually incompatible or distantly related plants,

(2)   the introduction of nitrogen-fixing bacteria and blue-green algae into non-legumes, specially the cereals,

(3)   the induction of disease resistance in a crop by the incorporation of a selective genome into the protoplast,

(4)   the induction and ease of detection of mutations in haploid protoplasts,

(5)   the transplantation of 'foreign chloroplasts' into crop plants with inefficient photosynthetic system,   .

(6)   culture of protoplasts to raise clones for vegetative multiplication and also for genetic diversity,

(7)   insertion of a part or complete genome by transgenosis, cell modification, and nuclear transplantation could replace or at least help to modernize some of the conventional practices used for plant improvement.

Much work has to be done in order that they may become reality.

One important step was the discovery that polyethylene glycol (PEG) is a better fusing agent than sodium nitrate. This discovery, as shown by the fusion of *Daucus carota* protoplasts[244] and by the high frequency of intergeneric fusion of protoplasts[123], has greatly enhanced the possibility of obtaining somatic hybridization, and PEG is the main fusion agent used today.

Protoplasts from dicotyledons have proved to be relatively easy to obtain and cultivate. Those from monocotyledons have been more difficult to achieve. They have been subject to spontaneous lysis or premature senescence. It was only in 1975 that Brenneman and Galston[29] succeeded in obtaining calli with protoplasts of oat, and the following year that Deka and Sen[52] obtained differentiation of calli originated from protoplasts of rice. In 1977 Potrykus *et al.*[193] obtained calli from protoplasts of corn, and Gamborg *et al.*[72] obtained whole plants of sorghum, the latter success being due in great part to the use of immature embryos instead of leaves as a source of protoplasts.

Each protoplast, if successfully cultured, can develop into several plants through the production of buds on calli plus eventual rooting (plant regeneration) and sometimes the formation of somatic embryos. Since the number of protoplasts obtained from maceration of a tissue can be fantastic, a huge number of plants can theoretically be obtained from a single explant. This fact can permit the full exploitation of a rare mutation through cloning.

The following is a cross-section of the work that has been done in the regeneration of whole plants from protoplasts. The first time whole plants were obtained from enzymatically isolated protoplasts is only a decade ago, when Nitsch and Okyama[185] obtained protoplasts from haploid tobacco and grew them to full haploid plants. The same year Takebe *et al.*[233] realized the feat with diploid tobacco. Then followed the regeneration of petunia plants by Durand *et al.*[63], and fully flowered diploid tobacco plants by Okyama and Nitsch[187]. Bui and Mackenzie[33] obtained growth and differentiation from protoplasts of asparagus, and Kartha *et al.*[126] were successful in getting rape plants. The following year Vardi *et al.*[239] were able to regenerate orange plantlets through the formation of somatic embryos. As mentioned Gamborg *et al.*[72] obtained whole plants of sorghum. Investigations within the family of *Solanaceae* have been extensive. For example, *Solanum dulcamara* was successfully regenerated by Binding and Nehls[17] while Facciotti and Pilet[68] obtained a high yield of haploid *Datura innoxia, Petunia hybrida, Nicotiana tabacum* and *Nicotiana sylvestris* through the formation of somatic embryos, which had rarely been obtained in the *Nicotiana* genus. Lörz and Potrykus[140] regenerated plants from *Atropa belladona* and *Hyoscyanus* protoplasts and Bourgin *et al.*[24] showed the slight differences needed to cultivate several *Nicotiana* species.

Recently Kao and Michayluk[123] succeeded in the difficult task of regenerating whole plants from legume protoplasts with alfalfa, but they found that the ability to form somatic embryos and to regenerate into plants varied considerably from plant to plant in the same cultivar. They attributed this to the heterogeneity of the population. Vasil and Vasil[240] extended the culture of cereal protoplasts to *Pennisetum americanum* while Xuan and Menczel[247] improved protoplast culture and plant regeneration from protoplast-derived callus in *Arabidopsis thaliana*. Shahin and Shepard[215] were only able to obtain shoot formation from mesophyll protoplasts of cassava. It can be seen that plant regeneration from protoplasts, although started barely a decade ago, is being achieved in a wider and wider range of plants. The field is now open for practical applications.

(ii) Somatic hybridization

Somatic hybridization is the result of the fusion of two somatic cells of different species, genera, or family, in contrast to normal hybridization which is achieved through fecundation of flowers of related species. The fusion of somatic cells has been rendered possible by the use of protoplasts and fusing agents. This possibility opens up vast new avenues, as theoretically it permits crosses that are not possible in nature due to incompatibility resulting in either non-fecundation, or abortion of the fecundated ovules or of the embryo. Somatic hybridization and its significance in crop improvement has been emphasized by Cocking[44] and by Bajaj[6], while the subject has been generally covered by Gamborg et al.[71], Constabel[46], Bajaj[7], Melchers[154].

(iii) Interspecific fusion of protoplasts

Obtaining a whole plant from interspecific fusion of protoplasts was first achieved by Carlson et al.[37]. They fused protoplasts of *Nicotiana glauca* with those of *Nicotiana langsdorfii* and were able to select for and develop the hybrids into whole plants. In 1976 Power et al.[198] were successful in creating a hybrid between *Petunia hybrida* and *Petunia parodii*, while Smith et al.[221] studied interspecific somatic hybridization in *Nicotiana*. Up till then somatic hybridization had given results only in the family of *Solanaceae*, but Dudits et al.[59] were able to regenerate plants from a somatic cross between *Daucus carota* and *Daucus capillifolius*. Work continued within the *Solanaceae* by hybridizing *Datura innoxia* with *Datura discolor*[210], while Nagao[172] obtained hybrids between *Nicotiana tabacum* and *Nicotiana glutinosa* as well as between *Nicotiana tabacum* and *Nicotiana alata*. Brar et al.[27] in Canada, tried fusing *Sorghum* and corn protoplasts, and obtained calli but no regeneration of plants was achieved.

One may conclude that interspecific somatic hybridization can be successfully achieved, but that so far it has been done mainly within the *Solanaceae*. It will most probably be done in the relatively near future with many other families.

(iv) Intergeneric fusion of protoplasts

Intergeneric fusion of protoplasts offers greater possibilities to the plant breeder because, theoretically, it can produce crosses that cannot be had by cross pollination. In the mid-seventies many intergeneric fusions were tried and, at first, one was content to obtain a few cell divisions of the fusion products. For example, Kao et al.[122] obtained fusion and cell division with corn + soybean, pea + soybean and barley + soybean. Until 1976, only a few reports concerning the successful culture of intergeneric heterokaryons were published and they were obtained only when soybean protoplasts were involved. Then Gosch and Reinert[78] produced heterokaryons from carrot and petunia protoplasts which were induced to regenerate plantlets. The same year Dudits et al.[60] obtained fusion and division of carrot + barley protoplasts, but no plantlets. By 1978, Constabel[47] was able to give a list of 24 cases where viable intergeneric somatic hybrids had been obtained but still, in almost all cases, hybrid plants could only be generated when the species involved could be crossed by sexual methods. Then Melchers et al.[155] created hybrids between potato and tomato, hybrids that cannot be achieved by sexual means, but the plants were malformed, sterile and grew slowly except when grafted. Then Dudits et al.[59] fused protoplasts from an albino mutant of carrot with those of normal *Aegopodium podagraria*, to produce a green hybrid having *Ae. podagraria* markers for roots and abnormal inflorescences. These had a low growth rate and could be maintained only on sterile culture. Also Gleba and Hoffman[76] were able to obtain plants from the fusion products of *Arabidopsis thaliana* and *Brassica campestris*, which are from taxonomically distant tribes. They fused colourless *A. thaliana* callus-derived protoplasts with green ones from the leaves of *B. campestris* to obtain green hybrids. Some of the hybrids, had a full complement of chromosomes and showed characters of both species. They were called by the authors 'Arabido-brassica', but the plants did not flower. Other hybrids obtained had lost the *Brassica* chromosomes and exhibited a full spectrum of forms, from morphologically perfect flowering types to severely malformed shoot-like or leaf-like teratomas. They were called by the authors 'asymetric hybrids'.

In summary, up till now some isolated successes have been obtained in regenerating plants resulting from the fusion of protoplasts from plants that cannot be crossed sexually. But in most cases, the plants exhibited malformations, grew slowly or were sterile. Perhaps one could obtain better structurally equipped plants by using haploid protoplasts and by fusing again protoplasts from these somatic crosses with one of the parents?

(v) The segregation of chromosomes

When a heterokaryon is produced as the result of protoplast fusion, the nuclei still keep a certain autonomy which is reflected by the division cycle. If the heterokaryon has been created by fusing closely related species, there is a good chance that the cell division cycles are synchronous, resulting in a normal distribution of chromosomes in the daughter cells. On the other hand, if the cell division cycle is different, as can happen between unrelated plants, then one can witness a segregation of chromosomes of one of the parents while they get stranded in the cytoplasm. This feature is not restricted to the fusion of protoplasts but has been found also in sexual hybrids. For example Subrahmanyan and Kasha[228] observed a selective chromosome elimination in crosses from as closely related species as *Hordeum vulgare* and *Hordeum bulbosum*. They attributed this phenomenon to differences in mitotic cycle times. In the case of protoplasts, this mechanism may not be the only one involved, but this has yet to be cleared up.

Chromosome segregation has been observed by Kao[121] in somatic hybrids between soybean and *Nicotiana glauca*. The hybrids were able to retain all of the soybean chromosomes but only a few of tobacco after a long period of culture. The chromosome behaviour was not synchronous in the first few cell generations and the tobacco chromosomes had a tendency to stick together and break into pieces. After 6 months of culturing some of the *Nicotiana glauca* chromosomes were still retained, and were then in synchrony with soybean chromosomes during cell division. Binding and Nehls[18] observed the same pattern with fusion products of *Vicia faba* and *Petunia hybrida*. In cells containing only a few *Vicia* chromosomes, these tended to become eliminated from the nucleus and were found in the cytoplasm. Segregation was also observed by Dudits et al.[59] in their somatic cross between *Aegopodium podagraria* and *Daucus carota*. They found that the hybrid possessed only the carrot chromosomes (2n = 18) + (2n = 42) = (2n = 18).

Segregation of chromosomes, however, does not necessarily happen in all intergeneric cases. For example, Gleba and Hoffman[76] observed in their somatic cross between *Arabidopsis thaliana* and *Brassica campestris*, in four out of six lines investigated, no specific elimination of the genetic material of one of the species during 15 months of unorganized cell growth. As for the two others, they did observe chromosome segregation, the *Brassica* chromosomes having a tendency to be eliminated. Thus, in the same fusion experiment, some of the fusion products may retain all their chromosomes while others will not. Gleba and Hoffman[76] have used the term 'asymetric hybrid' to describe a hybrid in which one of the parents is represented by a less than full complement of chromosomes.

(vi) The transfer of cytoplasmic inheritance

In traditional crosses, during pollen fertilization, only the nucleus is incorporated in the oosphere, but no cytoplasm. Thus only the female cytoplasm is inherited. On the other hand, when a somatic cross is obtained by fusion of protoplasts, the protoplasm of the two parents are mixed and the resulting hybrid cells contain self replicating organelles of both parents. This cytoplasmic make-up may be retained for a certain time, but may eventually be lost. For example, Power et al.[197] fused *Petunia* and *Parthenocissus* protoplasts. The calli obtained were shown to possess the chromosomes of *Parthenocissus* only, yet exhibited isoperoxydases of both parents. The calli retained isoperoxydases associated with both species for approximately one year, but thereafter the *Petunia*-specific isoperoxydases were progressively lost.

Dudits et al.[59] were able to transfer chloroplasts from *Aegopodium podagraris* to an albino carrot mutant by fusing protoplasts of the two. The hybrid possessed only the carrot chromosomes but had gained structurally normal chloroplasts. However, the total amount of chlorophyll and carotenoids was significantly lower than in the leaves of normal plants.

Plant breeders will benefit from the fact that male sterility can be transmitted through fusion of protoplasts because, in some cases, male sterility is controlled by cytoplasmic factors such as mitochondria, and the cytoplasm of both parents is mixed during the fusion process. For example Belliard et al.[12] produced a somatic hybrid between the tobacco variety Techne, which is male-sterile, and the variety 'Xanthi' which has normal flowers. The most modified hybrids were fertile Techne and sterile Xanthi. The 'cybrids', that is, the cytoplasmic hybrids, retained only one nucleus belonging to one of the parents and had only one or the other of the two parental chloroplastic DNAs. Izhar and Power[110] were able to create cybrids by fusing leaf mesophyll protoplasts of a cytoplasmic sterile *Petunia* with leaf mesophyll protoplasts of a fertile *Petunia*. Some of the hybrids obtained combined the cytoplasm of the male-sterile line with the genome of the fertile line.

When no segregation of the chromosomes of the donor plant occurs naturally, cell division of the donor plant can be eliminated by treatment with X-rays. This treatment blocks cell division of the donor plant and thus only the cytoplasm is transmitted. This technique has been used by Zelcer et al.[249]. They obtained male-sterile *Nicotiana sylvestris* by fusing X-ray treated protoplasts of a male sterile cultivar of *Nicotiana tabacum* with non-irradiated ones of *Nicotiana sylvestris*. Thus male sterility has been transmitted in more than one system. This shows that the technique of protoplast fusion could in the future be of considerable assistance to plant breeders.

(vii) The incorporation of exogenous genetic material

Exogenous genetic material can be incorporated into protoplasts via fusion with organelles, microorganisms, or DNA. The subject has been reviewed by Giles[75] and by Binding[16]. Two recent cases have effectively achieved incorporation. Landgren and Bonnett[134] incorporated chloroplasts into albino leaf protoplasts of tobacco and while green plants were obtained, they were pale green, and none was capable of autotrophic growth. Syono et al.[232], instead of incorporating microorganisms directly into protoplasts chose to use protoplasts already containing bacteria and fused them with non-infected protoplasts. They fused legume pea root nodule protoplasts with tobacco mesophyll protoplasts in an attempt to overcome the difficulty presented by the infection barrier, and to establish a new symbiotic relationship.

Plasmids have been used with success for the genetic engineering of bacteria. Rörsch and Schilperoort[205] described the possible advantages of using these for genetic manipulations of higher plants. Davey et al.[51] actually used the technique in a model system to impart new properties to Petunia protoplasts. They succeeded in transforming the cells by giving them the characteristics of crown-gall tumors, such as continued proliferation on hormone-free media, tumor formation when grafted on to host plants, octopine synthesis and lysopine dehydrogenase activity. Thus they have proved the possibility of using Ti (tumor-inducing) plasmids as vectors to modify crop plants by the introduction of isolated genes.

Cucumber mosaic virus and virus RNA have been incorporated into cucumber protoplasts by the use of polyethylene glycol[147,148]. The further elaboration of this method might be very useful in horticulture.

(viii) The selection of somatic hybrids and of mutants

Methods of selection must be found in order to select the desired somatic hybrids or the proper mutants in protoplast culture. The minimum plating density in order to cause protoplasts to divide is 10^3 cells per ml. Assuming that 95% of these will form cell colonies, and eventually calli which can be regenerated to form whole plants, one can easily understand that selection procedures must be employed at an early stage to recover the few hybrids or mutants which may exist in this huge population. The work of Melchers and Labib[154] gives an idea of the small quantity of hybrid material that can be recovered from a somatic hybridization experiment, even where a high frequency of fusion is obtained: an efficient selection procedure gave a recovery index of one for every 200,000 calli developing from protoplasts.

A relatively recent article of Power and Cocking[195] reviewed selection systems for somatic hybrids. Carlson et al.[37] utilized a system based on known nutritional requirements. While it is effective, it requires prior knowledge of the nutritional requirements of the somatic hybrid, which is not available when an entirely new hybrid is wanted. Melchers and Labib[154] used light-sensitive mutants as a basis for selection, since chlorophyll defects are not uncommon and are easily detected, but this approach is rather limiting in its practical possibilities.

Naturally occurring differential drug sensitivities was exploited by Power and Cocking[195]. Also Power et al.[194] utilized differences in sensitivity to actinomycin D in the selection for somatic hybrids of Petunia hybrida + Petunia parodii. Along similar lines it is possible to take advantage of the tolerance to relatively high concentrations of growth regulators, probably resulting from intergeneric complementation.[195] Zilkah and Ezra[250] discussed herbicide screening, while Nehls[176] evaluated the use of metabolic inhibitors. A technique which may hold promise in selecting for disease resistance has been used by Matern et al.[146]. They used pathotoxin on populations of potato protoplasts. Physical factors may be also used. Thus Smillie et al.[220] exploited chilling resistance to isolate tomato and potato somatic hybrids. Harms and Potrykus[87] used a novel approach, and investigated the enrichment of the number of fused cells by the utilization of iso-osmotic gradients before plating. Thus many techniques are being developed.

(ix) Special procedures and special factors

A number of special techniques have been used in the past three years.

The fact that dimethylsulfoxide (DMSO) increases the efficiency of PEG-mediated fusion of human cells led Haydu et al.[88] to investigate this compound in order to check if it has the same properties with plant protoplasts. They found indeed that DMSO increases the frequency of plant protoplast fusion induced by PEG. It was also known that, with human cells, a rise of temperature enhances fusion and this was confirmed by Keller and Melchers[128] for protoplasts. No kinetic analysis of the reaction had been done, but this is now available through the work of Senda et al.[214]. They worked with protoplasts of Rauwolfia serpentina at temperatures of 2–45°C and were able to develop a mathematical expression between the extent of fusion and temperature. However, no parallel physiological studies were carried out and no indication is given as to the effect of higher temperatures on the subsequent survival of the fused protoplasts, and their capacity to divide. Thus it seems that the highest temperature compatible with a protoplast system should be determined in each case, in order to induce the highest level of fusion.

Protoplasts from leaves of cereals often do not undergo sustained mitotic division when cultured on a wide range of media. They suffer from a decrease in the capacity for incorporating precursors of protein and nucleic acids, chlorophyll breakdown, and abnormal levels of hydrolases such as RNase and Protease which bring about lysis. This gives the impression that they are subject to premature senescence. Altman et al.[3] were able to combat this in oats by the use of polyamines such as spermine, spermidine and cadaverine. In the same laboratory, Kaur-Sawhney et al.[127] showed that these polyamines could reverse the process of senescence and favour cell division by inducing DNA synthesis.

Donn[55] used special procedures to induce cell divisions in leaf protoplasts of Vicia narbonensis which cannot be cultivated under standard conditions. Pre-culture of leaf fragments on agar medium yielded protoplasts that gave only 1–3% cell division, but addition of asparagine, glutamine and serine enhanced cell division to 25%. Farmer and Lee[69] found L-glutamine supplement helpful in obtaining viable cultures of protoplasts of Durum wheat. Using Douglas fir protoplasts derived from cotyledons, Kirby and Chen[129] found that in absence of glutamine, only two cycles of cell division occurred, but addition of 2 mM glutamine stimulated cell division to proceed for several cycles.

Binding and Nehls[19] succeeded in obtaining better plating efficiency, with the low densities of Vicia faba protoplasts required for their survival, by co-culturing them with those of Petunia hybrida. Co-culture of protoplasts which divide easily with protoplasts that do not has been achieved also by Evans.[67] By this method, he succeeded in obtaining whole plants from protoplasts of four species of Nicotiana: N. langdorfii, N. repanda, N. nesophiala and N. stocktonii which are valuable for somatic hybridization experiments.

Kirby and Chen[129] pre-conditioned cotyledons of Douglas fir on a solid nutrient medium with 15 μM BAP plus 500 nM NAA for 8–14 days before submitting them to cell wall-degrading enzymes. In this way they obtained a two-fold increase in the number of protoplasts obtainable with fresh cotyledons. Preconditioning in a liquid medium was done by Shahin and Shepard[215] to obtain leaf protoplasts of cassava. They were floated for 48 h on a dilute water solution containing 1 mM $CaCl_2$, 1 mM NH_4NO_3, 1 ppm NAA and 5 ppm BAP.

Using 0.4 M or 0.5 M glucose instead of mannitol in connection with Gamborg's B_5 medium, increased plating efficiency and permitted plant regeneration from *Arabidopsis thaliana*[247]. Brar *et al.*[27] obtained improved cell division with the fusion products of sorghum and corn protoplasts, in 8—10 day old liquid cultures, by adding 0.2 ml of protoplast culture medium containing 2.5% sucrose at 3 day intervals. Specific sugar alcohols were found to be indispensable for protoplast division of cassava by Shahin and Shepard.[215] They used D-mannitol, myo-inositol, sorbitol and xylitol, all at 0.025 M in sucrose at 0.2 M as the osmotic stabilizer.

Thus it seems, that starting from the classical work with tobacco and petunia, special procedures must be used in many cases to obtain (1) more protoplasts (2) protoplasts that divide (3) high frequency fusion. In general, these special procedures include: pre-treatment of the tissues; addition of aminoacids or of polyamines; use of special sugars; and use of DMSO and high temperatures in connection with PEG to obtain fusion with higher frequency.

CONCLUSION

To close this chapter, one can note that enormous progress has been achieved in tissue culture during the past ten years. Tissue culture has different roles to play in crop science, because in the case of field crops it will primarily be a means of obtaining new varieties with improved attributes. Most of this application has still to come, but in horticulture applications in propagation are now routine. From the point of view of the horticulturist, a decade of expectant waiting may have seemed long, but many problems had to be solved, and the main ones have indeed been solved. In most cases, it requires only matters of detail to achieve success.

That plant tissue culture is a huge success in the field of plant propagation is manifest by the setting up of laboratories throughout the world for the commercial propagation of plants. In the USA only, there are unofficially about 200 such commercial laboratories. In France, Italy, England, Denmark, Belgium, etc., numbers of huge laboratories are already in production. In France and Belgium, certified stock is accepted only if it has been obtained through tissue culture. Thus, at least this sector in plant tissue culture is in full commercial expansion.

The feasibility of inducing useful mutations in tissue culture has been demonstrated and has only to be expanded to give practical results, while in protoplast culture very little or nothing has yet been done commercially, but progress is now so rapid in this field that we will surely not have to wait another decade to find practical applications. However, one must be aware that all tissue culture work has to be done in a laboratory and this requires techniques that seem very far from down-to-earth present practice. Horticulture and agriculture will profit by this new type of venture by team work between scientists and skilled technicians, even though this osmosis may seem difficult to achieve.

Abbreviations: see page 372.

REFERENCES

1.   Abbott, A.J. and Whiteley, E. Culture of *Malus* tissues *in vitro*. 1. Multiplication of apple plants from isolated shoot apices. *Scientia Hort.* **4**, 183—189, 1976.
2.   Altman, A. and Goren, R. Development of citrus bud explants in culture. *J. Amer. Soc. Hort. Sci.* **103**, 120—123, 1978.
3.   Altman, A., Kaur-Sawhney, R. and Galston, A.W. Stabilization of oat leaf protoplasts through polyamine-mediated inhibition of senescence. *Plant Physiol.* **60**, 570—574, 1977.

4. Ando, K., Katsuya, K. and Sato, S. Production of aerial mycelium and teliospores of *Puccinia horiana* in chrysanthemum callus cultures. *Can. J. Bot.* **57**, 2162–2166, 1979.

5. Arnold, S. and Von Eriksson, T. Induction of advantitious buds on buds of Norway spruce grown *in vitro. Physiol. Plantarum* **45**, 29–35, 1979.

6. Bajaj, Y.P.S. Potentials of protoplast culture work in agriculture. *Euphytica* **23**, 633–649, 1974.

7. Bajaj, Y.P.S. Protoplast isolation, culture and somatic hybridization. *Plant Cell, Tissue and Organ Culture,* 467–496. Reinert, J. and Bajaj, Y.P.S. (eds.). Springer-Verlag, Berlin, 1977.

8. Bajaj, Y.P.S. and Dhanju, M.S. Regeneration of Plants from Apical Meristem tips of some Legumes. *Current Science* **48**, 906–907, 1979.

9. Bapat, V.A. and Rao, P.S. Shoot apical meristem culture of *Pharbitis nil. Plant Sci. Lett.* **10**, 327–334, 1977.

10. Beach, K.H. and Smith, R.R. Plant regeneration from callus of red and crimson clover. *Plant Sci. Lett.* **16**, 231–237, 1979.

11. Behnke, M. General Resistance to Late Blight of *Solanum tuberosum* Plants Regenerated from Callus Resistant to Culture Filtrates of *Phytophtora infestans. Theoret. Applied Genet.* **56**, 151–153, 1980.

12. Belliard, G., Pelletier, G., Vedel, F. and Quetier, F. Morphological Characteristics and Chloroplast DNA Distribution in Different Cytoplasmic Parasexual Hybrids of *Nicotiana tabacum. Molec. Gen. Genet.* **165**, 231–237, 1978.

13. Bernard, A.C. and Mur, G. *Vitis vinifera* buds grown *in vitro. Ann. Amelior. Plantes* **29**, 311–325, 1979.

14. Bhansali, R. and Arya, H.C. Differentiation in explants of *Citrus paradisi* Macf. (Grape fruit) growing in culture. *Indian J. Expt. Biol.* **16**, 409–410, 1978.

15. Bhojwani, S.S. and Hayward, C. Some observations and Comments on Tissue Culture of Wheat. *Z. Pflanzenphysiol.* **85**, 341–347, 1977.

16. Binding, H. Subprotoplasts and organelle transplantation. *Plant, Cell and Tissue Culture. Principles and applications,* 789–805. Sharp, W.R., Paddock, E.F. and Raghavan, V. (eds.). Columbus Ohio State Press, 1979.

17. Binding, H. and Nehls, R. Regeneration of isolated protoplasts to plants in *Solanum dulcamara* L. *Z. Pflanzenphysiol.* **85**, 279–280, 1977.

18. Binding, H. and Nehls, R. Somatic cell hybridization of *Vicia faba + Petunia hybrida. Molec. Gen. Genet.* **164**, 137–143, 1978a.

19. Binding, H. and Nehls, R. Regeneration of Isolated Protoplasts of *Vicia faba* L. *Z. Pflanzenphysiol.* **88**, 327–332, 1978b.

20. Bonga, J.M. Formation of holdfasts, callus, embryoids and haustorial cells in the *in vitro* cultures of dwarf mistletoe *Arceuthobuim pusillum. Phytomorphology* **21**, 140–143, 1971.

21. Bonga, J.M. Organogenesis in *in vitro* cultures of embryogenic shoots of *Abies balsamea* (balsam fir). *In Vitro* **13**, 41–48, 1977a.

22. Bonga, J.M. Application of Tissue Culture in Forestry. *Plant Cell, Tissue, and Organ culture,* 93–108. Reinert, J. and Bajaj, Y.P.S. (eds.). Springer-Verlag, Berlin, 1977b.

23. Boulay, M. Multiplication et clonage rapide de *Sequoia sempervirens* par la culture *in vitro.* Association Foret-Cellulose. *Etudes et Recherches* **12**, 49–55, 1979.

24. Bourgin, J.–P., Chupeau, Y. and Missonier, C. Plant regeneration from mesophyll protoplasts of several *Nicotiana* species. *Physiologia Plant.* **45**, 288–292, 1979.

25. Bourgin, J.–P., and Nitsch, J.P. Obtention de *Nicotiana* haploides à partir d´étamines cultivées *in vitro. Ann. Physiol. Végétale* **9**, 377–382, 1967.

26. Boxus, P. The production of strawberry plants by *in vitro* micropropagation. *J. Hort. Sci.* **49**, 209–210, 1974.

27. Brar, D.S., Rambold, S., Constabel, F. and Gamborg, O.L. Isolation, Fusion and Culture of *Sorghum* and Corn Protoplasts. *Z. Pflanzenphysiol.* **96**, 269–275, 1980.

28. Brar, D.S., Rambold, S., Gamborg, O. and Constabel, F. Tissue Culture of Corn and *Sorghum. Z. Pflanzenphysiol.* **95**, 377–389, 1979.

29.   Brenneman, F.N. and Galston, A.W. Experiments on the cultivation of protoplasts and calli of agri-
      culturally important plants. I. Oat (*Avena sativa*). *Biochem. Physiol. Pflanzen* **168**, 453–471,
      1975.

30.   Brettell, R.I.S., Goddard, B.V.D. and Ingram, D.S. Selection of Tms-cytoplasm maize tissue cultures
      resistant to *Drechslera maydis* T-toxin. *Maydica* **24**, 203–215, 1979.

31.   Brettell, R.I.S. and Ingram, D.S. Tissue culture in the production of novel disease-resistant crop
      plants. *Biol. Rev.* **54**, 329–345, 1979.

32.   Broome, O.C. and Zimmerman, R.H. *In vitro* propagation of Blackberry, *Hort. Science* **13**, 151–153,
      1978.

33.   Bui, D.H. and Mackenzie, I.A. The division of protoplasts from *Asparagus officinalis* L. and their
      growth and differentiation. *Protoplasma* **78**, 215–221, 1973.

34.   Button, J., Kochba, J. and Bornman, C.H. Fine structure of an embryoid development from embryo-
      genic ovular callus of Shamouti orange. *J. Exp. Bot.* **25**, 446–457, 1974.

35.   Campbell, R.A. and Durzan, D.J. Vegetative propagation of *Picea glauca* by tissue culture. *Can. J.
      For. Res.* **6**, 240–243, 1976.

36.   Carlson, P.S. Methionine-sulfoximine-resistant mutants of tobacco. *Science* **180**, 1366–1368, 1973.

37.   Carlson, P.S., Smith, H.H. and Dearing, R.D. Parasexual interspecific plant hybridization. *Proc. Natl.
      Acad. Sci. USA* **69**, 2292–2294, 1972.

38.   Chand, S. and Roy, S.C. Study of Callus Tissues from Different Parts of *Nigella sativa* (*Ranuncu-
      laceae*). *Experientia* **36**, 305–306, 1980.

39.   Chang, W. and Hsing, Y. *In vitro* flowering of embryoids derived from mature root callus of ginseng
      (*Panax ginseng*). *Nature* **284** (5754), 341–342, 1980.

40.   Chapman, D. Haploid *Hordeum* plants from anthers *in vitro*. *Z. Pflanzenzeucht.* **69**, 142–155, 1973.

41.   Chen, C.H. and Goedenkallemeyn, Y.C. *In virto* Induction of Tetraploid Plants from Colchicine-
      Treated Deploid Daylily Callus. *Euphytica* **28**, 705–711, 1979.

42.   Chin, J.C. and Scott, K.J. Studies on the formation of roots and shoots in wheat callus cultures. *Ann.
      Bot.* **41**, 473–481, 1977.

43.   Cocking, E.C. A method for the isolation of plant protoplasts and vacuoles. *Nature* **187**, 962–963,
      1960.

44.   Cocking, E.C. Plant Cell Modification: Problems and Perspectives. *Protoplastes et fusion de cellules
      somatiques. Coll. Int. CNRS (France)*. N° 212, 327–337, 1973.

45.   Coleman, W.K. and Thorpe, T.A. *In vitro* culture of western red cedar (*Thuja plicata* Donn.). Plant-
      let formation. *Bot. Gaz.* **138**, 298–304, 1977.

46.   Constabel, F. Somatic hybridization in higher plants. *In Vitro* **12**, 743–748, 1976.

47.   Constabel, F. Development of Protoplast Fusion Products, Heterokaryocytes, and Hybrid Cells.
      Frontiers of Plant Tissue Culture 1978. Thorpe, T.A. (ed.). International Assn. for Plant Tissue
      Culture. Univ. of Calgary, Alberta, Canada, 1978.

48.   Constabel, F., Miller, R.A. and Gamborg, O.L. Histological studies on embryos produced from cell
      cultures of *Bromus inermis*. *Can. J. Bot.* **49**, 1415–1417, 1971.

49.   Cooke, R.C. Homogenisation as an aid in tissue culture propagation of *Platycerium* and *Davallia*.
      Hort. Science **14**, 21–22, 1979.

50.   D'Amato, F. The problem of genetic stability in plant tissue and cell cultures. *Crop resources for
      today and tomorrow*. Frankel, O. and Hawkes, J.G. (eds.). Univ. Press, Cambridge, U.K., 1975.

51.   Davey, M.R., Cocking, E.C., Freeman, J., Pearce, N. and Tudor, I. Transformation of *Petunia* proto-
      plasts by isolated *agrobacterium* plasmids. *Plant Sci. Lett.* **18**, 307–313, 1980.

52.   Deka, P.C. and Sen, S.N. Differentiation of calli originated from isolated protoplasts of rice (*Oryza
      sativa*) through plating technique. *Molec. Gen. Genet.* **145**, 239–243, 1976.

53.   Devreux, M. and Saccardo, F. Mutazioni sperimentali osservate su piante aploidi di tabacco ottenute
      per colture in vitro de antere irradiate. *Atti Ass. Genet. Ital.* **16**, 69–71, 1971.

54.   Dix, P.J. and Street, H.E. Selection of plant cell lines with enhanced chilling resistance. *Ann. Bot.*
      (London) **40**, 903–910, 1976.

55. Donn, G.   Cell Division and Callus Regeneration from Leaf Protoplasts of *Vicia narbonensis*.   *Z. Pflanzenphysiol.* **86**, 65—75, 1978.

56. Doy, C.H., Gresshof, P.M. and Rolfe, B.G.   Biological and molecular evidence for the transgenosis from bacteria to plant cells. *Proc. Natl. Acad. Sci. USA* **70**, 723—726, 1973a.

57. Doy, C.H., Gresshoff, P.M. and Rolfe, B.G.   Time course of phenotypic expression of *E. coli* gene Z following transgenosis in haploid *Lycopersicum esculentum* cells. *Nature New Biol.* **244**, 90—91, 1973b.

58. Doy, C.H., Gresshoff, P.M. and Rolfe, B.   Transgenosis of bacterial genes from *Escherichia coli* to cultures of haploid *Lycopersicum esculentum* and haploid *Arabidopsis thaliana* plant cells. *The Biochemistry of Gene Expression in Higher Organisms*. Pollack, T.K. and Lee, J.W. (eds.). Sidney: Bosh Co., 1973c.

59. Dudits, D., Hadlaczky, G.Y., Bajszar, G.Y., Koncz, C.S., Lazar, G. and Horvath, G.   Plant Regeneration from intergeneric cell hybrids. *Plant Sci. Lett.* **15**, 101—112, 1979.

60. Dudits, D., Kao, K.N., Constabel, F. and Gamborg, O.L.   Fusion of carrot and barley protoplasts and division of heterokaryocytes. *Can. J. Genet. Cytol.* **18**, 263—269, 1976.

61. Dudits, D., Nemet, G. and Haydu, Z.   Study of callus growth and organ formation in wheat (*Triticum aestivum*) tissue cultures. *Can. J. Bot.* **53**, 957—963, 1975.

62. Dunstan, D.I., Short, K.C. and Dhaliwal, E.T.   Further Studies of Plantlet Production from Cultured Tissues of *Sorghum bicolor*. *Protoplasma* **101**, 355—363, 1979.

63. Durand, J., Potrykus, J. and Donn, G.   Plantes issues de protoplastes de *Petunia*.   *Z. Pflanzenphysiol.* **69**, 26—34, 1972.

64. Durr, A. and Fleck, J.   Production of haploid plants of *Nicotiana langsdorfii*.   *Plant Sci. Lett.* **18**, 75—79, 1980.

65. Earle, E. and Langhans, R.   Carnation propagation from shoot tips cultured in liquid medium. *Hort. Science* **10**, 608—610, 1975.

66. Eriksson, T.   Cell cultures of *Haplopappus gracilis* as testing material for radiomimetic compounds. *Hereditas* **57**, 127—148, 1967.

67. Evans, D.A.   Chromosome stability of plants regenerated from mesophyll protoplasts of *Nicotiana* species. *Z. Planzenphysiol.* **95**, 459—465, 1979.

68. Facciotti, D. and Pilet, P.—E.   Plants and embryoids from haploid *Nicotiana sylvestris* protoplasts. *Plant Sci. Lett.* **15**, 1—6, 1979.

69. Farmer, I. and Lee, P.E.   Culture of protoplasts derived from Ramsey Durum wheat. *Plant Sci. Lett.* **10**, 141—145, 1977.

70. Favre, J.M. and Grenan, S.   *In vitro* production of tendrils, flowers and berries on Grape. *Ann. Amelior. Plantes.* **29** (3), 247—253, 1979.

71. Gamborg, O.L., Constabel, F., Fowke, L.C., Kao, K.N., Ohyama, K., Kartha, K.K. and Pelcher, L.E.   Protoplast and cell culture methods in somatic hybridization in higher plants. *Can. J. Genet. Cytol.* **16**, 737—750, 1974.

72. Gamborg, O.L., Shyluk, J.P., Brar, D.S. and Constabel, F.   Morphogenesis and plant regeneration from callus of immature embryos of *Sorghum*. *Plant Sci. Lett.* **10**, 67—74, 1977.

73. Gaspar, T., Smith, D. and Thorpe, T.   Arguments supplémentaires en faveur d'une variation inverse du niveau auxinique engogène au cours des deux premières phases de la rhizogénèse. *C.R. Acad. Sci. Paris* **D285**, 327—330, 1977.

74. Gengenbac, B.G. and Green, C.E.   Selection of T-cytoplasm maize callus cultures resistant to *Helminthosporium maydis* race T pathotoxin. *Crop. Sci.* **15**, 645—649, 1975.

75. Giles, K.L.   The Uptake of Organelles and Microorganisms by Plant Protoplasts; Old Ideas but New Horizons. *Frontiers of Plant Tissue Culture 1978*. Thorpe, T.A. (ed.). International Assn. for Plant Tissue Culture. Univ. of Calgary, Alberta, Canada, 1978.

76. Gleba, Y.Yu. and Hoffman, F.   *Arabidobrassica*: Plant-Genome Engineering by Protoplast Fusion. *Naturwissenschaften* **66**, 547—554, 1979.

77. Goren, R., Altman, A. and Giladi, I.   The Role of Ethylene in ABA-Induced Callus Formation in Citrus Bud Cultures (Abstract). *Israel J. Bot.* **28**, 57, 1980.

78. Gosch, G. and Reinert, T. Continuous Division of Heterokaryons from *Daucus carota* and *Petunia* hybrida protoplasts. *Naturwissenschaften* **63**, 534, 1976.

79. Greef, W. de and Jacobs, M. *In vitro* culture of the sugarbeet: description of a cell line with high regeneration capacity. *Plant Sci. Lett.* **17**, 55–61, 1979.

80. Green, C.E. *In Vitro* Plant Regeneration in Cereals and Grasses. *Frontiers of Plant Tissue Culture.* Thorpe, T.A. (ed.). International Assn. of Plant Tissue Culture. Univ. of Calgary. Alberta, Canada, 411–419, 1978.

81. Green, C.E. and Philips, R.L. Plant Regeneration from tissue cultures of maize. *Crop. Sci.* **15**, 417–421, 1975.

82. Grout, B. and Aston, M. Transplanting of cauliflower plants regenerated from meristem culture. I. Water loss and water transfer related to changes in leaf wax and to xylem regeneration. *Hort. Res.* **17**, 1–7, 1977.

83. Guha, S. and Maheshwari, S.C. *In vitro* production of embryos from anthers of *Datura. Nature* **204**, 497, 1964.

84. Gunay, A.L. and Rao, P.S. *In vitro* plant regeneration from hypocotyl and cotyledon explants of red pepper (*Capsicum*). *Plant Sci. Lett.* **11**, 365–372, 1978.

85. Halperin, W. and Wetherell, D.F. Adventive embryony in tissue cultures of the wild carrot, *Daucus carota. Am. J. Bot.* **51**, 274–283, 1964.

86. Hangarter, R.P., Peterson, M.D. and Good, N.E. Biological Activities of Indoleacetylamino Acids and Their Use as Auxins in Tissue Culture. *Plant Physiol.* **65**, 761–767, 1980.

87. Harms, Ch.T. and Potrykus, I. Enrichment for heterokaryocytes by the use of iso-osmotic density gradients after plant protoplast fusion. *Theor. Appl. Genet.* **53**, 49–55, 1978.

88. Haydu, Zs., Lazar, G. and Dudits, D. *Increased frequency of polyethylene glycol induced protoplast fusion by dimethylsulfoxide, Plant Sci. Lett.* **10**, 357–360, 1977.

89. Hebert, L.–P. Recherches cytogéographiques et cytotaxonomiques sur des espèces méditerranéennes et centre-européennes du genre *Sedum* L. (*Crassulaceae* D.C.). *Thesis*. Faculty of Science, Univ. of Neuchatel (Switzerland), 1980.

90. Heinz, D.J. Sugarcane improvement through induced mutations using vegetative propagules and cell culture techniques. FAO/IAEA Panel Proc. *Induced Mutations in Vegetatively Propagated Plants.* PL 501/5, 53–59, 1973.

91. Heinz, D.J., Krisnamurthi, M., Nickell, L.G. and Maretzki, A. Cell, Tissue and Organ Culture in Sugarcane Improvement. *Plant Cell, Tissue, and Organ Culture,* 3–16. Reinert, J. and Bajaj, Y.P.S. (eds.). Springer-Verlag, Berlin, 1977.

92. Heinz, D.J. and Mee, G.W.P. Plant differentiation from callus tissue of *Saccharum* species. *Crop Sci.* **9**, 346–348, 1969.

93. Heinz, D.J. and Mee, G.W.P. Morphologic, Cytogenetic and enzymatic variation in *Saccharum* species hybrid clones derived from callus tissue. *Amer. J. Bot.* **58**, 257–262, 1971.

94. Helgeson, J.P., Kemp, J.D., Haberlach, G.T. and Maxwell, D.P. A tissue culture system for studying disease resistance; the black shank disease in tobacco callus cultures. *Phytopathology* **62**, 1439–1443, 1972.

95. Hess, C.E. Characterization of rooting co-factors extracted from *Hedera Helix* L. and *Hibiscus rosa-sinensis* L. *Proc. XVI Int. Hort. Congr. Brussels* **4**, 382–386, 1964.

96. Hess, D. Versuche zur Transformation an höheren Pflanzen: Induktion und konstante Weitergabe der Anthocyansynthese bei *Petunia hybrida. Z. Pflanzenphysiol.* **60**, 348–353, 1969a.

97. Hess, D. Versuche zur Transformation an höheren Pflanzen: Wiederholung der Anthocyaninduktion bei Petunia und erste Charakterizierung des transformierenden Prinzips. *Z. Pflanzenphysiol.* **61**, 286–298, 1969b.

98. Hess, D. Versuche zur Transformation an höheren Pflanzen: Genetische Charakterisierung einiger mutablich transformierter Pflanzen. *Z. Pflanzenphysiol.* **63**, 31–43, 1970a.

99. Hess, D. Versuche zur Transformation an höheren Pflanzen: Mögliche Transplantation eines Gens für Blattform bei *Petunia hybrida. Z. Pflanzenphysiol.* **63**, 461–467, 1970b.

100. Hess, D. Beseitigung der transformierenden aktivität durch DNase. *Naturwissenschaften* **58**, 366, 1971.

101. Hess, D. Transformationsversuche an höheren Pflanzen: Untersuchungen zur Realisation des Exo-somen-Modells der Transformation bei *Petunia hybrida*. *Z. Pflanzenphysiol.* **68**, 432–440, 1973.
102. Holdgate, D. Propagation of ornamentals by tissue culture. *Applied and fundamental aspects of plant cell, tissue and organ culture*, 18–43. Reinert, J. and Bajaj, Y.P.S. (eds.). Springer-Verlag, Berlin, 1977.
103. Horak, J., Landa, Z. and Lustinek, J. Production of polyploid plants from tissue cultures of *Brassica oleracera* L. *Phyton* **28**, 7–10, 1971.
104. Horak, J., Luslinek, J., Mesicek, J., Kaminek, M. and Polackova, D. Regeneration of diploid and poly-ploid plants from the stem pith explants of diploid marrow stem kale (*Brassica oleracea* L.). *Ann. Bot.* **39** (161), 571–577, 1975.
105. Howland, G.P. and Hart, R.W. Radiation Biology of Cultured Plant Cells. *Plant Cell, Tissue and Organ Culture*, 731–756. Reinert, J. and Bajaj, Y.P.S. (eds.). Springer-Verlag, Berlin, 1977.
106. Hunter, C.S. *In vitro* culture of *Chincona ledgerianum* L. *J. Hort. Sci.* **54**, 111–114, 1979.
107. Ingram, D.S. Growth of biotrophic parasites in tissue culture. *Encyclopedia of Plant Physiology, New Series, Vol.* 4, 743–759. Heitefuss, R. and Williams, P.H. (eds.). Springer-Verlag, Berlin, 1976.
108. Ingram, D.S. Applications in plant pathology. *Plant Tissue and Cell Culture.* 463–500. Street, H.E. (ed.). Blackwell, Oxford, 1977.
109. Ingram, D.S. and Robertson, N.F. Interaction between *Phytophtora* infestans and tissue cultures of *Solanum tuberosum. J. Gen. Microbiol.* **40**, 431–437, 1965.
110. Izhar, S. and Power, J.B. Somatic hybridization in *Petunia*: a male sterile cytoplasmic hybrid. *Plant Sci. Lett.* **14**, 49–55, 1979.
111. James, D.J. The role of auxins and phloroglucinol in adventitious root formation in *Rubus* and *Fragaria* grown *in vitro. Hort. Sci.* **54**, 273–277, 1979.
112. James, D.J. and Thurborn, I.J. Rapid *in vitro* rooting of the apple rootstock M.9. *J. Hort. Sci.* **54**, 309–311, 1979.
113. Jansson, E. and Bornman, C.H. *In vitro* phyllomorphic regeneration of shoot buds and shoots in *Picea abies. Physiol. Plant* **49**, 105–111, 1980.
114. Jeleska, S. Embryogenesis and organogenesis in pumpkin explants. *Physiol. Plant.* **31**, 257–261, 1974.
115. Jones, O.P. Effect of phoridzin and phloroglucinol on apple shoots. *Nature* **262**, 392–393, 1976.
116. Jones, O.P. and Hatfield, S.G.S. Root initiation in apple shoots cultured *in vitro* with auxins and phenolic compounds. *J. Hort. Sci.* **51**, 495–499, 1976.
117. Jones, O.P. and Hopgood, M.E. The successful propagation *in vitro* of two rootstocks of *Prunus*: the plum rootstock 'Pixy' (*P. insititia*) and the cherry rootstock F12/1 (*P. avium*). *J. Hort. Sci.* **54**, 63–66, 1979.
118. Jones, O.P., Hopgood, M.E. and O'Farrell, D. Propagation *in vitro* of M.26 apple rootstocks. *J. Hort. Sci.* **52**, 235–238, 1977.
119. Johnson, C.B., Grierson, D. and Smith, H. Expression of λplac 5 DNA in cultured cells of a higher plant. *Nature New Biol.* **244**, 105–107, 1973.
120. Kamat, M.G. and Rao, P.S. Vegetative multiplication of eggplants (*Solanum melongena*) using culture techniques. *Plant Sci. Lett.* **13**, 57–65, 1978.
121. Kao, K.N. Chromosomal behaviour in somatic hybrids of soybean – *Nicotiana glauca. Mol. Gen. Genet.* **150**, 225–230, 1977.
122. Kao, K.N., Constabel, F., Michayluk, M.R. and Gamborg, O.L. Plant protoplast fusion and growth of intergeneric hybrid cells. *Planta (Berlin)* **120**, 215–227, 1974.
123. Kao, K.N. and Michayluk, M.R. A method for high frequency intergeneric fusion of plant protoplasts. *Planta (Berlin)* **115**, 355–367, 1974.
124. Kao, K.N. and Michayluk, M.R. Plant Regeneration from Mesophyll Protoplasts of Alfalfa. *Z. Pflanzenphysiol.* **96**, 135–141, 1980.
125. Kartha, K.K., Gamborg, O.L., Constabel, F. and Shyluk, J.P. Regeneration of cassava plants from apical meristems. *Plant Sci. Lett.* **2**, 107–113, 1974.

126. Kartha, K.K., Michayluk, M.R., Kao, K.N., Gamborg, O.L. and Constabel, F. Callus formation and plant regeneration from mesophyll protoplasts of rape plants (*Brassica napus* L. cv. Zephyr). *Plant Sci. Lett.* **3**, 265–271, 1974.

127. Kaur-Sawhney, R., Flores, H.E. and Galston, A.W. Polyamine-induced DNA Synthesis and Mitosis in Oat Leaf Protoplasts. *Plant Physiol.* **65**, 368–371, 1980.

128. Keller, W.A. and Melchers, G. The effect of high pH and calcium on tobacco leaf protoplast fusion. *Z. Naturforsch.* **28c**, 737–741, 1974.

129. Kirby, E.G., and Chen, Tsai-Ying. Colony formation from protoplasts derived from Douglas fir cotyledons. *Plant Sci. Lett.* **14**, 145–154, 1979.

130. Kiss, F. and Zatyko, J. Vegetative propagation of Rubus species *in vitro*. (In Hungarian). *Bot. Közlm.* **65**, 65–68, 1978.

131. Klerker, J. AF. Eine Methode zur Isolierung lebender Protoplasten. *Svenka. Vet-Acad. Forb.* Stockholm **9**, 463–471, 1892.

132. Krul, W. and Worley, J.F. Formation of adventitious embryos in callus cultures of 'Seyval' – a French grape. *Hort. Science* **12**, 411. (Abstract), 1977.

133. Kunisaki, J.T. Tissue culture of tropical ornamental plants. *Hort. Sci.* **12**, 141–142, 1977.

134. Landgren, C.R. and Bonnett, H.T. The culture of albino tobacco protoplasts treated with polyethylene glycol to induce chloroplast incorporation. *Plant. Sci. Lett.* **16**, 15–22, 1979.

135. Lane, W.D. Regeneration of apple plants from shoot meristem-tips. *Plant Sci. Lett.* **13**, 281–285, 1978.

136. Lane, W.D. Regeneration of pear plants from shoot meristem-tips. *Plant Sci. Lett.* **16**, 337–342, 1979.

137. Lawson, L.W. Morphogenesis in Clonal Propagation of Woody Plants. *Frontiers of Plant Tissue Culture 1978.* 419–427. Thorpe, T.A. (ed.). International Assn. of Plant Tissue Culture. Univ. of Calgary, Alberta, Canada, 1978.

138. Lin, M.C., Shang, K.C., Chen, W.H. and Shin, S.C. Tissue and cell culture as aids to sugarcane breeding. III. Aneuploid cells and plants induced by treatment of cell suspension cultures with colchicine. *Proc. Int. Soc. Sugarcane Technol.* (Brazil) **16**, 29–41, 1977.

139. Litz, R.E. and Conover, R.A. *In vitro* propagation of Papaya. *Hort. Science* **13**, 241–242, 1978.

140. Lörz, H. and Potrykus, I. Regeneration of plants from mesophyll protoplasts of *Atropa belladona*. *Experientia* **35**, 313–314, 1979.

141. Malepszy, S., Corduan, G. and Przybecki, Z. Variability in the level of alkaloids in *Nicotiana sylvestris* plants after mutagenesis *in vitro*. *Bull. Acad. Pol. Sci. ser. Sci. Biol.* **25**, 737–740, 1977.

142. Maretzki, A. and Hiraki, P. Sucrose Promotion of Root Formation in Plantlets Regenerated from Callus of *Saccharum-Spp.* *Phyton* **38**, 85, 1980.

143. Martin, C., Carre, M. and Duc, G. Studies on field bean tissue cultures (*vicia faba* L.). Rooting of stem cuttings, callus culture, meristem culture. *Ann. Amelior. Plantes* **29**, 277–289, 1979.

144. Mascarenhas, A.F., Patha, M., Hendee, R.R., Ghugale, D.D. and Jagannathan, V. Tissue culture of maize, wheat, rice and sorghum. IV. Studies of organ differentiation in tissue culture of maize, wheat and rice. *Indian J. Exp. Biol.* **13**, 116–119, 1975.

145. Masteller, V.J. and Holden, D.J. The growth of an organ formation from callus tissue of sorghum. *Plant Physiol.* **45**, 362–364, 1970.

146. Matern, V., Strobel, G. and Shepard, J. Reaction to pathotoxin in a potato population derived from mesophyll protoplasts. *Proc. Nat. Acad. Sci.* (USA) **75**, 4935–4949, 1978.

147. Maule, A.J., Boulton, M.I. and Wood, K.R. An Improved Method for the Infection of Cucumber Leaf Protoplasts with Cucumber Mosaic Virus. *Phytopathol. Zeit.* **97**, 118–127, 1980a.

148. Maule, A.J., Boulton, M.I., Edmunds, C. and Wood, K.R. Polyethylene Glycol-mediated Infection of Cucumber Protoplasts by Cucumber Mosaic Virus and Virsu RNA. *J. Gen. Virol.* **47**, 199–209, 1980b.

149. McComb, J.A. Use of Tissue Culture, Particularly Anther Culture, for Plant Breeding and Propagation in China. *J. Austral. Inst. Agric. Sci.* **45**, 187–192, 1979.

150. Mee, G.W.P., Nickell, L.G. and Heinz, D.J. *Chemical mutagens – their effects on cells in suspension cultures.* 7–8. Ann. Rep. Exp. Sta., Hawaiian Sugar Planter's Ass., 1969.

151. Mehra, A. and Mehra, P.N. Organogenesis and plantlet formation *in vitro* in almond. *Bot. Gaz.* **135**, 61–73, 1974.

152. Melchers, G. *Plant Res. and Dev.* **5**, 86–110, 1977.

153. Melchers, G. Plant hybrids by fusion of protoplasts. *Proc. Symp. on Plant Tissue Culture.* Science Press. Peking, 1978.

154. Melchers, G. and Labib, G. Somatic hybridization of plants by fusion of protoplasts. I. Selection of light resistant hybrids of 'haploid' light sensitive varieties of tobacco. *Mol. Gen. Genet.* **135**, 277–294, 1974.

155. Melchers, G., Sacristan, M.D. and Holder, A.A. Somatic hybrid plants of potato and tomato regenerated from fused protoplasts. *Carlsburg Res. Commun.* **43**, 203–218, 1978.

156. Meyer, M.M. Propagation of Daylilies by tissue culture. *Hort. Science* **11**, 485–487, 1976.

157. Miller, L.R. and Murashige, T. Tissue culture propagation of tropical foliage plants. *In Vitro* **12**, 797–813, 1977.

158. Minocha, S.C. Callus and adventitious shoot formation in excised embryos of white pine (*Pinus strobus*). *Can. J. Bot.* **58**, 366–371, 1980.

159. Monaco, L.C., Sondahl, M.R., Carvalho, A., Crocomo, O.J. and Sharp, W.R. Applications of Tissue Culture in the Improvement of Coffee. *Plant Cell, Tissue and Organ Culture.* 109–126. Reinert, J. and Bajaj, Y.P.S. (eds.). Springer-Verlag, Berlin, 1977.

160. Morel, G. Tissue culture – new means of clonal propagation in orchids. *Am. Orchid. Soc. Bull.* **33**, 473–478, 1964.

161. Morel, G. Clonal propagation of orchids by meristem culture. *Cymbidium Soc. News.* **20**, 3–11, 1965.

162. Morel, G.M. Morphogenesis of stem apical meristem cultivated *in vitro*: Application to clonal propagation. *Morphogenesis in Plant Cell, Tissue and Organ Cultures.* 5–12. Internat. Symp. Report, Dept. Bot. Univ. Delhi, 1971.

163. Morel, G. and Martin, C. Guérison de Dahlias atteints d'une maladie à virus. *Compt. Rend. Acad. Sci.* Paris **D235**, 1324–1325, 1952.

164. Morel, G. and Martin, C. Guérison des plantes atteintes de maladies à virus par culture de méristèmes apicaux. *Report 14th Intern. Hort. Congr. Netherlands.* 303–310, 1955.

165. Mosella, L., Macheix, C.H. and Machelix, J.J. *In vitro* Shoot Development of the Peach Tree (*Prunus persica* Batsch.): Influence of Certain Phenolic Compounds. *C.R. Acad. Sci. Paris* **D289**, 567, 1979.

166. Muir, W.H., Hildebrandt, A.C. and Riker, A.J. Plant tissue cultures produced from single isolated cells. *Science* **119**, 877–878, 1954.

167. Mukhopadhyay, A. and Bhojwani, S.S. Shoot-Bud Differentiation in Tissue Cultures of Leguminous Plants. *Z. Pflanzenphysiol.* **88**, 263–268, 1978.

168. Murashige, T. Plant propagation through tissue cultures. *Ann. Rev. Plant Physiol.* **25**, 135–165, 1974.

169. Murashige, T. Current status of plant cell and organ culture. *Hort. Sci.* **12**, 127–130, 1977.

170. Murashige, T. The impact of Plant Tissue Culture on Agriculture. *Frontiers of Plant Tissue Culture 1978.* 15–27. Thorpe, T.A. (ed.). International Assn. for Plant Tissue Culture. Univ. of Calgary, Alberta, Canada, 1978.

171. Murashige, T. and Nakano, R. Tissue Culture as a Potential Tool in Obtaining Polyploid Plants. *J. Hered.* **57**, 114–118, 1966.

172. Nagao, T. Somatic Hybridization by Fusion of Protoplasts. 2. Combination of *Nicotiana tabacum* and *N. glutinosa* and of *N. tabacum* and *N. alata*. *Crop Sci.* **48**, 385–393, 1979.

173. Nair, N.G., Kartha, K.K. and Gamborg, O.L. Effect of Growth Regulators on Plant Regeneration from Shoot Apical Meristems of Cassava (*Manihot esculenta* Crantz) and on the culture of Interodes *in vitro*. *Z. Pflanzenphysiol.* **95**, 51–57, 1979.

174. Nakamura, A., Yamada, T., Kadotani, N. and Itagaki, T. Improvement of fluecured tobacco variety MC 1610 by means of haploid breeding method and investigations on some problems of this method. *Proc. Haploids in Higher Plants-Advances and Potential.* 277–278. Kasha, K.J. (ed.). Univ. Guelph., 1974.

175. Nassuth, A., Wormer, T.M., Bouman, F. and Staritsky, G. The histogenesis of callus of *Coffea cenaphora* stem explants and the discovery of early embryoid initiation. *Acta Bot. Neerl.* **29,** 49, 1980.

176. Nehls, R. The use of metabolic inhibitors for the selection of fusion products of higher plant protoplasts. *Mol. Gen. Genet.* **166,** 117–118, 1978.

177. Niizeki, H. and Oomo, K. Induction of haploid rice plant from anther culture. *Japan Acad. Proc.* **44,** 554–557, 1968.

178. Nickell, L.G. Crop improvement in Sugarcane: Studies using *in vitro* methods. *Crop Sci.* **17,** 717–719, 1977.

179. Nickell, L.G. and Torrey, J.G. Crop improvement through plant cell and tissue culture. *Science* **166,** 1068–1069, 1969.

180. Nitsch, C. Culture of isolated microspores. *Plant Cell, Tissue and Organ Culture.* 268–278. Reinert, J. and Bajaj, Y.P.S. (eds.). Springer-Verlag, Berlin, 1977.

181. Nitsch, C. and Norreel, B. Effect d'un choc thermique sur le pouvoir embryogène du pollen de *Datura innoxia* cultivé dans l'anthère ou isolé de l'anthère. *C.R. Acad. Sc. Paris* **D276,** 303–306, 1973.

182. Nitsch, J.P. Haploid plants from pollen. *Z. Pflanzenzüchtung* **67,** 3–18, 1972.

183. Nitsch, J.P., Nitsch, C. and Hamon, S. Production de *Nicotiana* diploides à partir de cals haploides cultivés *in vitro.* *C. R. Acad. Sci. Paris* **D269,** 1275–1278, 1969.

184. Nitsch, J.P., Nitsch, C. and Pereau-Leroy, P. Obtention de mutants à partir de *Nicotiana* haploides issus de pollen. *C.R. Acad. Sci. Paris* **D269,** 1650–1652, 1969.

185. Nitsch, J.P. and Okyama, K. Obtention de plantes à partir de protoplastes haploides cultivés *in vitro.* *C.R. Acad. Sci. Paris* **D273,** 801–804, 1971.

186. Novak, F.J. and Vystot, B. Karyology of callus cultures derived from *Nicotiana tabacum* haploids and ploidy of regenerants. *Z. Planzenzüchtung* **75,** 62–70, 1975.

187. Okyama, K. and Nitsch, J.P. Flowering haploid plants obtained from protoplasts of tobacco leaves. *Plant Cell Physiol.* **13,** 229–236, 1972.

188. Orton, T.J. Quantitative Analysis of Growth and Regeneration from Tissue Cultures of *Hordeum vulgare, H. jubatum* and their Interspecific Hybrid. *Envir. Exper. Bot.* **19,** 319–337, 1979.

189. Orton, T.J. Chromosomal Variability in Tissue Cultures and Regenerated Plants of *Hordeum. Theor. Appl. Genet.* **56,** 101–113, 1980.

190. Ouyang, T.–W., Hu, H., Chuang, C.–C. and Tseng, C.C. Induction of pollen plants from anthers of *Tricitum aestivum* L. cultured *in vitro. Sci. Sin.* **16,** 79–95, 1973.

191. Pareek, L.K. and Chandra, N. Somatic embryogenesis in leaf callus from cauliflower (*Brassica oleracea* var. *botrytis*). *Plant Sci. Lett.* **11,** 311–316, 1978.

192. Partanen, C.R. Cytological behavior of plant tissues *in vitro* as a reflection of potentialities *in vivo. Proceedings of the International Congress of Plant Tissue Culture.* 463–471. White, P.R. and Grove, A.R. (eds.). McCutchan, Berkeley, 1965.

193. Potrykus, I., Harms, C.T., Lörz, H. and Thomas, E. Callus formation from stem protoplasts of corn (*Zea mays* L.). *Molec. Gen. Genet.* **156,** 347–350, 1977.

194. Power, J.B., Berry, S.F., Frearson, E.M. and Cocking, E.C. Selection procedures for the production of interspecies somatic hybrids of *Petunia hybrida* and *Petunia parodii.* I. Nutrient media and drug sensitivity complementation selection. *Plant Sci. Lett.* **10,** 1–6, 1977.

195. Power, J.B. and Cocking, E.C. Selection Systems for somatic hybrids. *Plant Cell, Tissue and Organ Culture.* 497–505. Reinert, J. and Bajaj, Y.P.S. (eds.). Springer-Verlag, Berlin, 1977.

196. Power, J.B., Cummins, S.E. and Cocking, E.C. Fusion of isolated plant protoplasts. *Nature* (London) **225,** 1016–1018, 1970.

197. Power, J.B., Frearson, E.M., Hayward, C. and Cocking, E.C. Some consequence of the fusion and selective culture of Petunia and Parthenocissus protoplasts. *Plant Sci. Lett.* **5,** 197–207, 1975.

198. Power, J.B., Frearson, E.M., Hayward, C., George, D., Evans, P.K., Berry, S.F. and Cocking, E.C. Somatic hybridisation of *Petunia hybrida* and *P. parodii. Nature* **263,** 500–502, 1976.

199. Raff, J.W., Hutchinson, J.F., Knox, R.B. and Clarke, A. Cell recognition: Antigenic determinants of plant organs and their cultured callus cells. *Differentiation* **12,** 179–186, 1979.

200. Prino-Millo, W. and Harada, H. Morphogenesis and plant propagation from leaf tissue of red cabbage (*Brassica oleracera* cv. tete de négre). *C.R. Acad. Sci. Paris* **D280**, 2845–2848, 1975.

201. Rao, A.N. Tissue culture in the Orchid industry. *Applied and Fundamental Aspects of Plant Cell, Tissue and Organ Culture.* 44–69. Reinert, J. and Bajaj, Y.P.S. (eds.). Springer-Verlag, Berlin, 1977.

202. Rashid, A. and Street, H.E. The development of haploid embryoids from anther cultures of *Atropa belladonna* L. *Planta* **113**, 263–270, 1973.

203. Reinert, J. Uber die kontrolle der morphonese und die induktion von adventivombryonen an gewebe-kulturen aus carotten. *Planta* **53**, 318–333, 1959.

204. Reinert, J., Bajaj, Y.P.S. and Heberle, E. Induction of haploid tobacco plants from isolated pollen. *Protoplasma* **84**, 191–196, 1975.

205. Rörsch, A. and Schilperoort, R.A. *Agrobacterium tumefaciens* plasmids: potential vectors for genetic engineering of plants. *Genetic Engineering.* 189–195. Boyer, H.W. and Nicosia, S. (eds.). Elsevier/North Holland. Biomedical Press, 1978.

206. Roy, S.C. Chromosomal Variations in the Callus Tissues of *Allium tuberosom* and *A. cepa. Protoplasma* **102**, 171–177, 1980.

207. Sacristan, M.D. and Melchers, G. The kariological analysis of plants regenerated from tumorous and other callus cultures of tobacco. *Mol. Gen. Genet.* **105**, 317–333, 1969.

208. Sanchez de Jimenez, E. and Murillo, E. Mecoprop as a growth factor in wheat and rye tissue cultures. *Can. J. Bot.* **57**, 1479–1483, 1979.

209. Sangwan, R.S. and Norreel, B. Induction of plants from pollen grains of Petunia cultures *in vitro. Nature* **257** (5523), 222–223, 1975.

210. Schieder, O. Somatic hybrids of *Datura innoxia* Mill + *Datura discolor* Bernh, and of *Datura innoxia* Mill + *Datura strammonium* var. *tatula* L. *Mol. Gen. Genet.* **162**, 113–119, 1978.

211. Scholl, R.L. and Amos, J.A. Isolation of Doubled-Haploid Plants Through Anther Culture in *Arabidopsis thaliana. Z. Pflanzenphysiol.* **96**, 407–415, 1980.

212. Schwenk, F.W. Callus formation from mechanically isolated soybean cotyledonary cells. *Plant Sci. Lett.* **17**, 437–442, 1980.

213. Seabrook, J.E.A. and Cumming, B.G. Tissue culture – a new way to propagate Daffodils. *The Daffodil Journal* **13**, 16–18, 1976.

214. Senda, M., Morikawa, H., Katagi, H., Takada, T. and Yamada, Y. Effect of Temperature on membrane fluidity and Protoplast Fusion. *Theor. Applied Genet.* **57**, 33–37, 1980.

215. Shahin, E.A. and Shepard, J.F. Cassava mesophyll protoplasts: isolation, proliferation and shoot formation. *Plant Sci. Lett.* **17**, 459–465, 1980.

216. Sinha, S., Roy, R.P. and Jha, K.K. Callus formation and shoot bud differentiation in anther culture of *Solanum surattense. Can. J. Bot.* **57**, 2524–2528, 1979.

217. Sita, G.L. Morphogenesis and plant regeneration from cotyledonary cultures of *Eucalyptus. Plant Sci. Lett.* **14**, 63–68, 1979.

218. Skirvin, R.M. and Chu, M.C. Tissue culture may revolutionize the production of peach shoots. *Illinois Research, University of Illinois Agric. Exp. Sta.* **19**, 18–19, 1977.

219. Skoog, F. and Miller, C.O. Chemical regulation of growth and organ formation in plant tissue cultures *in vitro. Symp. Soc. Exp. Biol.* **9**, 118–131, 1956.

220. Smillie, R.M., Melchers, G. and Wettstein, D. von. Chilling resistance of somatic hybrids of tomato and potato. *Carlsberg Res. Commun.* **44**, 127–132, 1979.

221. Smith, H.H., Kao, K.N. and Combatti, N.C. Interspecific hybridization by protoplast fusion in *Nicotiana. J. Hered.* **67**, 123–128, 1976.

222. Söndahl, M.R. and Sharp, W.R. High frequency induction of somatic embryos in cultures leaf explants of *Coffea arabica. Z. Pflanzenphysiol.* **81**, 395–408, 1977.

223. Sopory, S.K., Jacobson, E. and Wenzel, G. Production of monohaploid embryoids and plantlets in cultures anthers of *Solanum tuberosum. Plant Sci. Lett.* **12**, 47–54, 1978.

224. Staritsky, G. Embryoid formation in callus tissues of coffee. *Acta Bot. Neerl.* **19**, 502–507, 1970.

225. Start, N.D. and Cumming, B.G. *In vitro* propagation of *Saintpaulia jonantha. Hort. Science* **11**(3), 204–206, 1976.

226. Steward, F.C., Mape, M.O. and Mears, C. Growth and organized development of cultured cells. II. Organization in cultures grown from freely suspended cells. *Am. J. Bot.* **45**, 705–708, 1958.

227. Strahlheim, E. and Cailloux, M. La propagation *in vitro* des cultivars Ottawa-3 et Malling-9 du pommier (*Malus domestica*) à partir de méristèmes apicaux. *Thesis*. Département des Sciences Biologiques. Université de Montréal, Canada, 1980.

228. Subrahmanyam, N.C. and Kasha, K.J. Selective chromosome elimination during haploid formation in barley following interspecific hybridisation. *Chromosoma. Berl.* **42**, 111–125, 1973.

229. Sunderland, N. Nuclear cytology. *Plant Tissue and Cell Culture*. 161–190. Street, H.E. (ed.). Blackwell, Oxford.

230. Sunderland, N. Nuclear cytology. *Plant cell and tissue culture. Vol. 2.* 177–206. Street, H.E. (ed.). Univ. of California Press, Berkeley, Calif., 1977.

231. Sutter, E. and Langhans, R.W. Epicuticular Wax Formation on Carnation Plantlets Regenerated from Shoot-Tip Culture. *J. Amer. Soc. Hort. Sci.* **104**, 493–496, 1979.

232. Syono, K., Nagata, T., Susuki, M., Kajita, S. and Matsui, C. Fusion of Pea Root Nodule Protoplasts with Tobacco Mesophyll Protoplasts. *Z. Pflanzenphysiol.* **95**, 449–457, 1979.

233. Takebe, I., Labib, G. and Melchers, G. Regeneration of whole plants from mesophyll protoplasts of tobacco. *Naturwissenschaften* **58**, 318–320, 1971.

234. Tamura, S. Shoot formation in calli originated from rice embryo. *Japan Acad. Proc.* **44**, 544–548, 1968.

235. Tarayama, S. and Misawa, M. Differentiation in *Lilium* bulbscales grown *in vitro*. Effects of activated charcoal, physiological age of bulbs and sucrose concentration on differentiation and scale leaf formation *in vitro. Physiol. Plantarum* **48**, 121–126, 1980.

236. Tilquin, J.P. Plant regeneration from stem callus of cassava. *Can. J. Bot.* **57**, 1761–1763, 1979.

237. Torrey, J.G. Morphogenesis in relation to chromosomal constitution in long term plant tissue cultures. *Physiol. Plant.* **20**, 265–275, 1967.

238. Vanzulli, L., Magnein, E. and Olivi, L. Caryological stability of *Datura innoxia* calli analysed by cytophotometry for 22 hormonal combinations. *Plant Sci. Lett.* **17**, 181–193, 1980.

239. Vardi, A., Spiegel-Roy, P. and Galun, E. Citrus cell culture: isolation of protoplasts, plating densities, effect of mutagens and regeneration of embryos. *Plant Sci. Lett.* **4**, 231–236, 1975.

240. Vasil, V. and Vasil, I.K. Isolation and Culture of Cereal Protoplasts. Part 2: Embryogenesis and Plantlet Formation from Protoplasts of *Plennisetum americanum. Theor. Appl. Genet.* **56**, 97–101, 1980.

241. Walkey, D.G.A. and Matthews, K.A. Rapid clonal propagation of rhubarb (*Rheum rhapondicum* L.) from meristem tips in tissue culture. *Plant Sci. Lett.* **14**, 287–290, 1979.

242. Walkey, D.G.A., Neely, H.A. and Crisp, P. Rapid propagation of white cabbage by tissue culture. *Scientia Horticulture* **12**, 99–109, 1980.

243. Walker, K.A., Wendeln, M.L. and Jaworski, E.G. Organogenesis in callus tissue of *Medicago sativa*. The temporal separation of induction processes from differentiation processes. *Plant Sci. Lett.* **16**, 23–30, 1979.

244. Wallin, A., Glimelius, K. and Eriksson, T. The induction of aggregation and fusion of Daucus *carota protoplasts* by polyethylene glycol. *Z. Pflanzenphysiol.* **74**, 64–80, 1974.

245. Wasaka, K. Variation in the plants differentiated from the tissue culture of pineapple. *Japan J. Breed.* **29**, 13–22, 1979.

246. Wochok, Z.S. and Sluis, C.J. Gibberellic acid promotes *Atriplex* shoot multiplication and elongation. *Plant Sci. Lett.* **17**, 363–369, 1980.

247. Xuan, L.T. and Menczel, L. Improved Protoplast Culture and Plant Regeneration from Protoplast-Derived Callus in *Arabidopsis Thaliana. Z. Pflanzenphysiol.* **96**, 77–81, 1980.

248. Zatyko, J.M. Complete red currant (*Ribes rubrum* L.) plants from adventive embryos induced *in vitro. Current Science* **48**, 456–457, 1979.

249. Zelcer, A., Aviv, D. and Galun, E. Interspecific Transfer of Cytoplasmic Male Sterility by Fusion between Protoplasts of Normal *Nicotiana sylvestris* and X-Ray Irradiated Protoplasts of Male-Sterile *N. Tabacum. Z. Pflanzenphysiol.* **90**, S, 397–407, 1978.

250. Zilkah, S. and Ezra, G. Herbicide Action, Resistance and Screening in Cultures vs Plants. *Frontiers of Plant Tissue Culture 1978.* 427–437. Thorpe, T.A. (ed.). International Assn. of Plant Tissue Culture. Univ. of Calgary, Alberta, Canada, 1978.
Bibliography completed June 1980.

CHAPTER 12

PRODUCTION AND APPLICATION OF HAPLOID PLANTS

W.R. Sharp, S.M. Reed and D.A. Evans

DNA Plant Technology Corporation, 2611 Branch
Pike, Cinnaminson, NJ 08077, U.S.A.

INTRODUCTION

The haploid condition of any organism, tissue or cell refers to the chromosome number of the normal gametes of the species.[30] Depending on the ploidy level in somatic cells of the species considered, haploid individuals are called **monohaploid** (1x), **dihaploids** (2x), or **polyhaploids** (nx). The first cytogenetically analyzed haploid plant was reported in *Datura stramonium*.[19] Soon after this report spontaneous haploids were also reported in *Nicotiana tabacum* and *Triticum compactum*. Spontaneous haploids occur in many species, but usually at low frequencies. Haploid plants are recovered via parthenogenesis, androgenesis, chromosome elimination or by using tissue culture methods.

Haploid production using tissue culture became a reality following work with *Datura innoxia* anther culture.[73] The potential for high frequency production of haploid plants in Angiosperms was suggested by Sunderland and Wicks[207] who estimated that 2,000 haploid plantlets could be obtained per tobacco anther assuming that only 5% of the ca. 40,000 pollen grains underwent embryogenesis. This method of haploid production is of interest due to the simplistic protocol for high frequency induction,[133] as compared to the low frequency of spontaneous haploids. Under optimal conditions, induction frequencies close to 100% have been obtained in anther cultures of *Datura innoxia* and *Nicotiana tabacum* (see Table 13.6). The induction of *in vitro* androgenesis occurs within the first 24 hours in *D. innoxia* and following a few days in *N. tabacum*, with recovery of haploid plants within three weeks.

Haploid plants and tissues offer an important tool in studies of cytogenetics, gene action, experimental embryogenesis, evolution, and plant breeding. Haploids are important in plant breeding because of decreasing the time for obtaining homozygous cultivars following diploidization, and facilitating genetic analysis. For example, using haploid analysis Nakamura et al.[144] confirmed that leaf shape in flue-cured tobacco is controlled by two genes and Nakata & Kurihara[146] found F_1 hybrids from crosses between resistant and susceptible varieties of tobacco to wildfire and tobacco mosaic virus segregated into 1:1 ratios.

As haploids contain a single copy of functional genes, they are extremely useful for cellular genetics. Induced recessive genetic changes produced by mutagenesis or transfer of genetic material can be detected in haploid cells, while the presence of recessive genes remains masked in diploid or polyploid cells. As evidence suggests that auxotrophic and other metabolic mutants are recessive, haploid cells may be necessary to study this class of mutants in higher plants.

OBTAINING HAPLOID PLANTS

Conventional breeding methods

The technique of anther and pollen culture has yielded the greatest number of haploid plants, but haploids have been obtained in many plant species from a range of other methods. With few exceptions, the yield of haploids obtained with most of the conventional techniques has not been great enough to be of practical use in a breeding program. Nevertheless, these haploids have been of interest, particularly for use in genetic and physiological studies of the consequences of haploidy. Since the purpose of this review is to describe the applications of haploidy to plant breeding, discussion of conventional methods will be limited to those techniques other than anther culture that have produced haploids at a sufficient frequency to be used in conjunction with a conventional breeding program.

Haploids have been obtained, either spontaneously or through the use of an induction treatment, in over 125 species representing 23 families.[111,116,124,158] Spontaneous haploids have arisen through monoembryonic or polyembryonic parthenogenesis[124], androgenesis[33] and semigamy[217]. Commonly used induction techniques include parthenogenesis or chromosome elimination following interspecific hybridization[88,217], crosses to an inducer stock[110] and treatment with various physical and chemical agents.[124]

The best described conventional system for producing a high frequency of haploids is the Bulbosum Method in barley. Kasha and Kao[103] reported that a high frequency of barley haploids was obtained in the progeny of crosses of cultivated barley, *Hordeum vulgare*, with *H. bulbosum*. Upon investigation, it was determined that haploidy is the result of selective elimination of *H. bulbosum* chromosomes in the developing embryos of the interspecific hybrid *H. vulgare* × *H. bulbosum* and its reciprocal cross.[9,199] Since hybrid endosperm development ceases at an early stage, embryo culture of the hybrid is necessary for haploid production. The basic procedure for production of barley haploids through the Bulbosum Method involves the following: (1) emasculation of *H. vulgare* followed by pollination with fresh pollen of *H. bulbosum* (2) application of 75 ppm GA_3 (one drop/floret) for 2−3 days following pollination

(3) excision and culture of embryos at 14–16 days after pollination, and (4) transfer of differentiated embryos to light.[102] The frequency of haploids among the cultured embryos has ranged from 11 to 68%.[102]

The genetic basis for chromosome elimination in barley interspecific hybrids has been of great interest. Genomic balance is known to play a role in determining whether haploids or hybrids are obtained. In crosses between diploid and tetraploid forms of both *H. vulgare* and *H. bulbosum*, it was determined that only the ratio of one *H. vulgare* genome to two *H. bulbosum* genomes produced hybrids. Haploids were obtained whenever this ratio was disturbed.[102,117] It was later determined, using trisomic stocks of *H. vulgare*, that this 1:2 ratio is only important for chromosome 2 and 3.[8,85,105] At least 3 major factors, one on each arm of chromosome 2 and one on the short arm of chromosome 3 of *H. vulgare*, control elimination of the *H. bulbosum* chromosomes in the *H. vulgare–H. bulbosum* hybrids.[105]

Haploids obtained through the Bulbosum Method have been used extensively in several barley breeding programs. Kasha and Reinbergs[104] reported data from six of these breeding efforts. While the most commonly employed haploid breeding method involves the extraction and chromosome doubling of haploids from F_1 hybrids, haploid extraction has occasionally been delayed until a more advanced generation. Breeding studies have indicated that the doubled haploid technique is as effective as the more conventional breeding schemes, such as pedigree, bulk and single seed descent methods, in the production of superior barley genotypes.[165,194] Computer simulation studies have determined that the doubled haploid method might be most effective for improving quantitative traits in barley.[176]

The biggest advantage of the doubled haploid technique over other breeding schemes is the shorter time needed to produce a new variety. The barley variety 'Mingo', developed by the Bulbosum Method, was released by Ciba-Geigy Seeds, Canada only 5 years after the breeding program was initiated.[84] This represents a tremendous time savings over conventional breeding techniques, where the average time for development of a new variety is 12 years.[104]

Chromosome elimination following interspecific hybridization is not limited to the *H. vulgare–H. bulbosum* hybrids. Several additional *Hordeum* interspecific crosses give rise to haploids, apparently through chromosome elimination.[102,196,197,198] A pattern of selective chromosome elimination in *Nicotiana* interspecific hybrids has been well documented; however, aneuploids rather than haploids are produced from this system.[79,80,137] The most recently elucidated example of chromosome elimination following wide hybridization was reported in wheat. It was found that haploids of wheat are produced in *Triticum aestivum* X *H. bulbosum* hybrids through elimination of *H. bulbosum* chromosomes in the hybrid embryos.[7,193] Although haploid production is presently limited to certain genotypes of *T. aestivum*, these reports open the possibilities of a more widespread use of the Bulbosum Method.

Haploids are also obtained in potato (*Solanum tuberosum* subsp. *tuberosum*) following interspecific hybridization, but through a different mechanism than that found in *Hordeum*. Pollination of *S. tuberosum* with *S. phureja* was found to induce haploid parthenogenesis in *S. tuberosum*.[88] Because *S. tuberosum* is an autotetraploid (2n = 4x = 48), the resulting plants are dihaploids (2n = 2x = 24), rather than monoploids as found in *Hordeum*. Recently this work has been extended to include the production of true haploids (n = x = 12) following hybridization of *S. tuberosum* dihaploids with *S. phureja*.[23,97] Monoploids have also been obtained through anther culture of dihaploid potato lines.[61]

In order to facilitate identification of potato haploids, marker stocks of *S. tuberosum* and *S. phureja* are utilized. For example, in the work by Hougas and coworkers[90], non-pigmented (pppp) tetraploid clones of *S. tuberosum* were crossed to pigmented (PP) *S. phureja* lines. The non-pigmented progeny, which could easily be identified by a lack of anthocyanin development in hypocotyls of the seedlings, represented potential haploids. The remaining pigmented progeny were of hybrid origin.

Various factors have been found to influence the frequency of haploid parthenogenesis following interspecific hybridization in *Solanum*. Five to ten times more fruit per pollination can be obtained by allowing pollinated fruit to mature on stems that have been removed from the plant and placed in a bottle of water rather than on intact plants.[179] Selection of a superior seed parent was found to increase the number of haploids from 0.1 to 4.1%.[89] Some studies have indicated that particular *Solanum tuberosum* genotypes favor parthenogenesis and that the characteristic is dominant.[139] It has also been proposed that a high frequency of haploids is related to a low number of recessive lethal genes in the seed parent.[89]

The selection of a pollinator has a large effect on haploid parthenogenesis frequency. While several *Solanum* species are capable of inducing parthenogenesis in *S. tuberosum*, *S. phureja* is the most effective.[64,90] Selection among *S. phureja* lines improved haploid frequency from 0.3 to 40.4%.[89] Superior pollinator lines have been found to have a high frequency of pollen tubes containing a single male gamete.[140] The trait of superior haploid production is thought to be recessive and is controlled by either one or a few genes.[65] An interaction between seed-parent and pollinator has also been proposed.[62]

Several uses for *S. tuberosum* haploids have been proposed.[167] Interspecific gene transfer between wild diploid *Solanum* species and tetraploid *S. tuberosum* is limited by ploidy barriers to hybridization. Crosses between dihaploid *S. tuberosum* and these diploid *Solanum* species would allow a flow of desirable genes into *S. tuberosum*. Chase[32] has proposed carrying out intensive inbreeding and selection on dihaploids, followed by the resynthesis of the tetraploid. The production of tetraploids with dominant genes for disease and insect resistance in the duplex state through use of dihaploids has also been suggested.[91] Finally, chromosome mapping is facilitated by the use of dihaploids. Using crosses between tetraploid nulliplex and heterozygous dihaploid individuals, the distance between the P locus and the centromere was determined to be 13 map units.[135] It is anticipated that additional genes can be mapped as appropriate heterozygous diploid and nulliplex tetraploid marker stocks are generated. Haploid parthenogenesis has also been observed in *Nicotiana tabacum* following pollination by *N. africana*[25] resulting in maternal *N. tabacum* haploids.

Another system that has yielded a high frequency of haploids is the semigametic production of haploids in cotton. Semigamy occurs when a microspore nucleus enters an egg cell but nuclear fusion does not occur. Both nuclei divide independently, thereby producing an embryo that is sectored for maternal and paternal tissue. Semigamy was first observed in Pima cotton (*Gossypium barbadense*) in a line that originated as a doubled haploid.[217] This doubled haploid was obtained from a spontaneous haploid found growing in a commercial field. While the frequency of spontaneous haploidy in *G. barbadense* is normally less than 1%,[163] over 30% of the progeny of the original semigametic doubled haploid were haploid.[218] The occurrence of semigamy was verified by the appearance of chimeral plants among the progeny of crosses between marker stocks.[218] Chromosome counts of these chimeral plants revealed both haploid and diploid chromosome numbers among the maternal tissue. Only haploid tissue was observed in sectors possessing the paternal genotype. Therefore, only those plants containing paternal sectors suitable for chromosome doubling are considered useful for the production of doubled haploid lines.[220,221]

Semigamy in Pima cotton appears to be conditioned by single dominant gene.[221] Thus, incorporation of the trait into a large number of genotypes is a relatively simple procedure. Once the gene for semigamy is incorporated into a breeding line, homozygosity can be obtained at any step of a breeding program.[219] In a comparison of the yield of three inbred Pima cotton commercial varieties and three doubled haploid lines, it was determined that lint yields of the doubled haploids were as stable as those of the varieties from which they were developed.[60,222] Therefore, the doubled haploid method seems capable of producing cotton varieties comparable to those derived by conventional methods.

Use of an inducer stock is the basis for one method of haploid production in *Zea mays*. Corn haploids are produced from crosses with lines carrying the 'indeterminate gametophyte' (ig) mutation.[110] When use of ig is coupled with use of the R^{nj} allele, a dominant gene that produces anthocyanin pigmentation in the kernel, haploids can easily be identified among the seed of a cross. For example, the paternal haploid progeny arising from the cross $IgigR^{nj}R^{nj}$ X $Igigr^{g}r^{g}$ are identified by the lack of kernel anthocyanin pigmentation. In the same manner, maternal haploids produced from a cross such as $Igigr^{g}r^{g}$ X $IgigR^{nj}R^{nj}$ can be selected. Using ig inducer stocks, the frequency of haploid production can be as high as 3%. Corn haploids are primarily utilized for the development of doubled haploid lines, which are used in breeding programs for identifying superior parents for hybrid production.[158]

There are other species in which conventionally derived haploids have been utilized in plant breeding. The tomato variety 'Marglobe' was developed from a spontaneous haploid[141], as were varieties of *Brassica napus*[214] and *Pelargonium*.[45] Monosomic lines have been extracted from *Avena*[150] and *Triticum*[186] spontaneous haploids and used in breeding efforts. Spontaneous haploids in *Capsicum*[168] and haploidy induced through interploid crosses in *Medicago*[16,17] have been utilized for breeding on the haploid level. Unfortunately, the usefulness of haploidy in many species has been diminished by a low frequency of haploid production. However, development of *in vitro* techniques for induction of haploidy, as will be discussed in the next section, has widened the range of species for which haploidy can now be put to use.

Anther culture

The feasibility of obtaining haploid tissue, embryos, and plantlets in higher plants with high frequency using tissue culture was first reported by Guha and Maheshwari.[73] Mature anthers of *Datura innoxia* were cultured in Nitsch's medium and embryo formation was observed only when 5, 15, or 30% of coconut water (CW) or 1 μM kinetin (KIN) was added to the culture medium. A multicellular globular mass of cells was observed to grow within the exine wall. Subsequently, the mass of cells was liberated and eventually organized into normal or abnormal (with multiple cotyledons) embryos. Cytological investigation of recovered embryos and plantlets confirmed the haploid chromosome number of n = 12.[74,75]

Mature haploid plants were also recovered in large quantities from cultured *Nicotiana* sp. anthers.[22] The haploid condition was confirmed by chromosome counts and sterility of haploid flowers. Formation of plantlets occurred one month following culture of anthers containing uninucleate microspores. The haploid embryos followed the normal embryogenic process with globular, heart, and torpedo stages observed. Haploid *Nicotiana* plants are smaller than comparable diploid plants and an abundant number of flowers are frequently produced. Atypical morphological features are usually observed *in vitro* but new growth following transfer to pots becomes phenotypically normal.[202]

Sunderland and Wicks[207] determined the best developmental stage for anther culture of

Nicotiana tabacum to obtain plantlet formation. After testing buds at different developmental stages, they concluded that the most rapid response was achieved using (a) petals of 8—16 mm with uninucleated pollen grains or microspores, then (b) petals of 17—20 mm with microspores undergoing the first pollen grain mitosis. They concluded that the first pollen grain mitosis was the critical transitional point for plantlet formation in *N. tabacum.* Unfortunately, there is no single optimum microsporogenic stage for cultured anthers, even for closely related species. First pollen grain mitosis seems to be the ideal for *Nicotiana tabacum* and *Datura innoxia,* whereas the tetrad stage (late meiosis) proved ideal for embryogenic development in *Arabidopsis thaliana* and *Lycopersicum esculentum.*[69,70] In barley, early microsporogenesis was the best stage[37], while in wheat, premicrosporogenesis was optimum.[162]

During normal microsporogenesis in *N. tabacum,* the vacuole pushes the nucleus to one end of the spore prior to the first pollen mitosis. The first pollen mitosis is an asymmetrical division (quantal mitosis) resulting in two unequal cells. The generative cell retains only a small portion of the microspore cytoplasm which is partitioned from the vegetative cell when the cell plate fuses with the intine. The vegetative cell, in contrast to the generative cell, has a large nucleus which stains lightly and has a large, well defined nucleolus (indicative of an active transcriptional process). Later the generative nucleus is detached from the intine and is retained inside the vegetative cell (bicellular condition).

When microspores are set in culture, three routes of development have been described by Sunderland.[203,204] In Route A, as occurs in *N. tabacum,* embryos are formed from the vegetative cell while the generative cell degenerates soon after or just following one or two mitotic divisions. Route B, as for *D. innoxia,* results in the absence of an asymmetrical mitosis and leads to the formation of two equal, diffuse vegetative-type nuclei. Route B can occur in anthers cultured at any stage prior to pollen mitosis. The frequency of Routes A and B vary considerably between species and even from one anther to another within the same plant. Both routes of microspore division have been observed in *Triticum aestivum, Triticale,* and *Hordeum vulgare.*[37,162,201,224,225] There are some suggestions that the generative cell can follow an organogenesis pattern of development as reported by Sharp *et al.*[187] and Raghavan.[171] Haploid plants are not recovered in Route C in which the generative and vegetative nuclei fuse prior to nuclear division.

TECHNIQUES

In *Datura* and *Nicotiana* spp. only sucrose (2%) in agar was necessary for induction of embryogenesis up to the globular stage with subsequent development occurring only when inorganic salts are added to the medium. Nitsch[156] reported the role of iron as a key element in the complete development of embryos. Iron-deficiency in higher plants has been associated with an abnormally high level of arginine as demonstrated by suppression of embryo development when arginine was added to the culture medium of *Nicotiana* anthers.

The reaction of anthers to different culture media varies widely both between and within species. For example, the *japonica* subspecies of *Oryza sativa* gives rise to haploid callus in the presence of relatively low auxin concentrations[149], whereas the *indica* subspecies can undergo no haploid callus formation with even high exogenous auxin concentrations.[76] Exogenous auxin can readily induce callus formation from the anther wall as in the case of *Datura innoxia.*[75] In general, the ideal induction is achieved with the anther wall remaining quiescent. If both somatic and gametophytic tissues develop callus, attention should be given to the morphological characters and texture of the calluses, and chromosome number of developing callus should be monitored.

Induction of the anther is dependent on the plant age and growth environment[53,57,174]. Anther response is greater at the beginning of the flowering period than at the end[208] and there are differences between seasonal plantings of greenhouse plants. The percent of embryogenesis in *Datura innoxia* decreased from 19.7% to 0% as anthers were cultured between June and October.[154] It is claimed that short days are more proper for tobacco anther development than long days.[53]

Pretreatment of anthers before placing in culture has been shown to enhance the frequency of anther induction. A cold shock treatment (3–5°C for 48–72 h) resulted in an increased number of embryos per anther in *Nicotiana* and *Datura*.[152,153] Inflorescences of *Oryza sativa* have been treated with 2-chloroethyl phosphonic acid for 48 h at 10°C to increase the anther response.[226]

TABLE 12.1. Mineral salt concentrations of culture media most frequently used for regeneration of haploid plants *in vitro*

	Medium			
	MS	B5	White	Nitsch & Nitsch[155]
Macronutrients (mM)				
NH_4NO_3	20.6	–	–	12.5
KNO_3	18.8	25	0.8	9.9
$CaCl_2 \cdot 2H_2O$	3.0	1.0	–	–
$MgSO_4 \cdot 7H_2O$	1.5	1.0	2.9	0.23
KH_2PO_4	1.25	–	–	2.2
$(NH_4)_2SO_4$	–	1.0	–	–
$NaH_2PO_4 \cdot H_2O$	–	1.1	0.12	–
$Ca(NO_3)_2 \cdot 4H_2O$	–	–	0.12	2.12
KCl	–	–	0.9	0.9
Na_2SO_4	–	–	1.4	–
Micronutrients (μM)				
KI	5	4.5	4.5	4.8
H_3BO_3	100	50	25	26
$MnSO_4 \cdot 4H_2O$	100	–	31.4	–
$MnSO_4 \cdot H_2O$	–	60	–	29
$ZnSO_4 \cdot 7H_2O$	30	7	10.5	9.3
$Na_2MoO_4 \cdot 2H_2O$	1.0	1.0	–	–
MoO_3	–	–	trace	–
$CuSO_4 \cdot 5H_2O$	0.1	0.1	0.001	–
$CoCl_2 \cdot 6H_2O$	0.1	0.1	–	–
$Fe_2(SO_4)_3$	–	–	6.3	–
Fe-EDTA chelate	100	100	–	87

Data on the techniques resulting in successful anther culture experiments based on published literature has been summarized in Tables 12.2–12.9. For each reported species, haploid plants have been recovered either via androgenesis or by plant regeneration from microspore-derived haploid cells following callus formation. Forty-three of 67 species (64%) undergo androgenesis while 24 of 67 species (36%) produce haploid plants following callus formation. Taxonomic differences are apparent between these two distinct developmental pathways. Of all species that are capable of undergoing androgenesis, 77% (33 of 44 species) are *Solanaceae*. In addition, the majority (87%) of the Solanaceous species that are capable of producing microspore-derived haploid plants undergo direct embryogenesis. On the other hand, the majority (75%) of *Gramineae* capable of producing haploid plants undergo callus-mediated regeneration. Consequently, the possibility of obtaining plants from cultured anthers as well as the developmental route for plant production varies between species.

The formulations for the common culture media used for haploid plant regeneration are listed in Table 12.1. MS medium is used most frequently for both androgenesis and callus-mediated plant regeneration (Tables 12.5 and 12.9). Nitsch medium is commonly used for androgenesis (Table 12.9) and also to initiate callus formation from cultured anthers capable of regeneration (Table 12.5). Each medium contains similar nutrients, while White's medium contains the lowest concentrations of inorganic nutrients.

The two developmental pathways for obtaining haploid plants are summarized separately because the culture medium for the callus-mediated mode of regeneration is usually different from that for androgenesis. Species capable of haploid plant regeneration following callus formation are summarized in Table 12.2, and of the 24 species listed there are 9 *Gramineae* (37.5%) and 5 *Solanaceae* (20.8%). For most species, anthers are cultured at earlier (*Arabidopsis*) or later (*Brassica*) stages of development. All 9 Graminaceous species are cultured at the uninucleate stage. The percentage of responding anthers varies both within and between species. Optimum frequencies for responsive species vary between less than 1% to 60% of cultured anthers.

For each species, callus was induced on a primary culture medium followed by transfer to a secondary medium to achieve plant regeneration (Table 12.2). In some species the anthers were subcultured onto a new culture medium, while in other species only the hormone concentration was latered. A survey of hormones used for callus initiation and plant regeneration of these 24 species is summarized in Table 12.3. For callus initiation 2,4-D, KIN, then NAA are included in most primary culture media, while CW, then 6BA and IAA are also frequently used for callus formation. For plant regeneration KIN, NAA, then IAA are used most frequently. Regeneration occurs in hormoneless medium for 6 of 24 species (25%) following callus formation. While KIN, a cytokinin, is often used for callus initiation, in each case it is used with either 2,4-D or NAA (Table 12.4). Consequently, a cytokinin is not used alone for callus formation. 2,4-D may also be used alone (3 species) or with other auxins or cytokinins for callus production. On the other hand, KIN is used most frequently for plant regeneration on secondary media. KIN is usually combined with NAA or IAA for plant regeneration (Table 12.4). The MS culture medium is used most frequently for both primary and secondary culture media (Table 12.5), with Nitsch, Miller, then B5 media used less frequently. The mineral salts were changed to achieve regeneration for only 4 species.

Representative species capable of androgenesis are summarized in Table 12.6. Species from 15 genera undergo androgenesis. In many cases, numerous species within a genus are androgenetic (Table 12.10). Seven of the 15 genera are *Solanaceae*, as a total of 34 Solanaceous species are capable of haploid embryogenesis. As with species capable of callus mediated regeneration, anthers are usually cultured in the uninucleate stage of development. *Hyoscyamus, Paeonia* and *Saintpaulia* each are cultured at pollen mitosis. The frequency of embryogenic anthers varies between 0.1% (*Secale cereale*) to 100% (*Nicotiana tabacum*). Among these 15 genera, Nitsch (40%) and MS (40%) media are used most frequently (Table 12.9), while Blaydes, N6, and Bourgin and Nitsch media are used less frequently. Of the species listed in Table 12.6 all except two species require a growth regulator for haploid embryogenesis. KIN and CW are used most frequently suggesting that a cytokinin is necessary to induce androgenesis in a wide range of species (Table 12.7). A cytokinin is added to 10 of the 15 species in Table 12.6. Auxins are used less frequently (Table 12.7) and when used, are usually used in combination with a cytokinin (Table 12.8). A cytokinin is used alone in 5 of 15 species (33.3%). When used, CW is usually added alone to culture media (Table 12.8).

Of the 15 genera that contain at least 1 species capable of androgenesis, 6 genera contain more than one androgenic species (Table 12.10). These genera include *Brassica* (2 species), *Petunia* (2), *Hyoscyamus* (4), *Datura* (6), *Solanum* (9), and *Nicotiana* (12). Methodology for regeneration has varied between species. Culture stage is usually consistent between species, although at least one *Nicotiana* species can be identified that was cultured at the pollen mitosis, the uninucleate, and the binucleate stages (Table 12.10). Different media and growth regulators are used for the 2 *Brassica* species while all 6 *Datura* species are cultured in Nitsch medium. Variation in growth regulator types and concentrations are particularly striking. Three *Datura* species require a cytokinin (CW) while 3 *Datura* species regenerate on hormoneless medium (Table 12.10). In *Nicotiana*, 7 of 12 species require no growth regulators, while 5 of 12 species require an auxin (IAA) and a cytokinin (KIN). All 9 *Solanum* species require an auxin, either IAA or NAA (Table 12.10). Frequency of induction varies between species. Only 0.4% of cultured anthers of *N. sanderae* undergo androgenesis, while nearly 100% of *N. tabacum* anthers undergo androgenesis. Similarly in *Datura*, a range of 3% responsive anthers for *D. metel* to 80% for *D. innoxia* is observed, even though both species contain the same chromosome number.

Isolated pollen culture

Anthers represent a mixed population of somatic diploid and haploid cells. Consequently, difficulty can arise using anther culture to obtain haploid plants if anther walls regenerate or if clumped microspores fuse together. To overcome this difficulty, the culture of isolated microspores has been successfully used in a few species. Binding[13] squashed *Petunia hybrida* anthers in a nutrient solution, filtered on a 160 μm mesh stainless steel mesh and cultured the isolated pollen grains in Petri dishes with 0.6% v/v agar. Sharp *et al.*[188] used a nurse culture technique for pollen of *Lycopersicon esculentum* and observed proliferation after 14 days. Pollen nurse culture consists of culturing a small number of isolated microspores on the top of a filter paper disc which is placed over an intact anther or mass of proliferating callus. In a similar way, Pelletier[166] grew *Nicotiana* spp. microspores on nurse tissues of *Petunia hybrida*. Debergh & Nitsch[47] working with tomato, reported that the isolation of microspores in a liquid medium via filtration was more suitable than isolation by centrifugation. By using this procedure, cell mortality was reduced from 6.3% to only 8.3%. Additionally, Debergh & Nitsch[47] demonstrated the use of a water extract of cultured anthers to enhance the development of isolated microspores.

The culture of isolated microspores has not been quite as successful as the culture of anthers

TABLE 12.2. Characteristics of plant species capable of regeneration of haploid plants following callus proliferation

Species	Embryos vs. Callus	Stage	1° Medium	2° Medium	Culture Environment	Maximum %	n=	References
Aegilops caudata × umbellulata	C	Uninucleate	Miller + 9.1 μM 2,4-D	Miller	25°C	1%	7	Kamata and Sakamoto[100]
Arabidopsis thaliana	C	Meiosis	B5 + 34.3 μM IAA / 9.1 μM 2,4-D / 4.0 μM KIN	Blaydes + 2.7 μM NAA / 46.7 μM KIN	25°C	NA	5	Gresshoff and Doy[70]
Asparagus officinalis	C	Uninucleate	MS + 1.0 μM NAA / 1.0 μM 6BA	MS + 0.9 μM 6BA	26°C dark	NA	10	Hondelman and Wilberg[87]
Brassica oleracea	C	Binucleate	Nitsch + 10% CW / 7.4 μM KIN / 4.5 μM 2,4-D	Nitsch + 4.7 KIN + 5.4 NAA	19–26°C	6.7%	9	Kameya and Hinata[101]
Capsicum annuum	E/C	Uninucleate	MS + 4.7 μM KIN / 4.5 μM 2,4-D	MS + 4.7 μM KIN / 5.4 μM NAA	25–30°C	4%E 25%C	12	Wang et al.[227]
Digitalis purpurea	C	Uninucleate	MS + 22.6 μM 2,4-D	Nitsch	25–30°C 1500–5000 lux 2–12 days 4°C	8%	28	Corduan and Spinx[44]
Fragaria × ananassa	C	Uninucleate	B5 + 10.7 μM NAA / 23.2 μM KIN	B5 + 16.7 μM NAA / 23.2 μM KIN	4°C pretreat 16 h light	NA	28	Rosati et al.[178]
Hevea brasiliensis	C	Uninucleate	MS + 9.3 μM KIN / 9.1 μM 2,4-D	MS, MB + 1–4.7 μM KIN / 0.2 μM 2,4-D	25–29°C	4%	18	Cheng-hua et al.[34]
Hordeum vulgare	C	Uninucleate	MS + 5.7 μM IAA / 4.4 μM 6BA / 10% CW	MS + 5.7 μM IAA / 4.4 μM 6BA	22–27°C	11–30%	14	Clapham[38]
Ipomoea batatas	C	Uninucleate	MS + 11.4 μM IAA / 9.1 μM 2,4-D / 9.3 μM KIN	MS + 5.7 μM IAA / 18.6 μM KIN	25–27°C 16 h	NA	45	Tsay and Tseng[216]
Lilium longiflorum	C	Uninucleate	MS	MS	25°C	NA	12	Sharp et al.[187]
Lolium multiflorum	C	Uninucleate	MS (12% suc) + 5.4 μM NAA / 15% CW	MS + 5.7 μM IAA	25°C darkness	10%	14	Pagniez and Demarly[164]
Nicotiana langsdorfii	C	3 mm	MS + 4.5 μM 2,4-D / 4.4 μM 6BA	MS	4°C for 2 days	20%	9	Durr and Fleck[54]

Table 12.2 continued

Species	Embryos vs. Callus	Stage	1° Medium	2° Medium	Culture Environment	Maximum %	n=	References
Oryza sativa japonica	C	Uninucleate	MS + 5 μM NAA (or) 1–5 μM 2,4-D	MS + 26.9 μM NAA 4.5–22.6 μM 2,4-D	28°C dark	9%	12	Niizeki and Oona[149]
Pelargonium hortorum	C	Uninucleate	White + 10.7–13.4 μM NAA 11.6 μM KIN	MS + 10.7 μM NAA 11.6 KIN	28°C 16 h day	60%	18	Abo El-Nil and Hildebrandt[1]
Setaria italica	C	Uninucleate	Miller + 4.5 μM 2,4-D 9.3 μM KIN 3 g/1 YE	Miller + 11.4 μM IAA 9.3–18.6 μM KIN	28°C dark	NA	9	Ban et al.[6]
Setaria melongena	C	Uninucleate	Bourgin + Nitsch + 15% CW	Bourgin + Nitsch + 22.8 μM IAA 9.3 μM KIN	dark	NA	9	Ban et al.[6]
Solanum surattiense	C	Uninucleate	MS + 9.9 μM 2,4-D 10.2 μM NAA 10.2 μM KIN	MS + 23.2 μM KIN 15% CW	25°C	5%	12	Sinha et al.[191]
S. terracosum	C	Uninucleate	Nitsch 10.7 μM NAA 9.1 μM 2,4-D 9.3 μM KIN	Nitsch	20°C 16 h day	1.4%	12	Irikura and Sakaguchi[94]
Triticale	C	Uninucleate	MS + 9.1–22.6 μM 2,4-D 4.7 μM KIN 15% CW	MS + 1.1 μM NAA 2.9 μM IAA 4.7 μM KIN	25–30°C	17%	28	Wang et al.[224]
Triticum aestivum	C	Uninucleate	MS + 9.1 μM 2,4-D 14.0 μM KIN	MS + 1.1 μM IAA 0.1 μM KIN	25–30°C 9–11 h day	6%	21	Wang et al.[224]
Vitis vinifera X *V. rupestris*	C	Uninucleate	Nitsch + 5 μM 2,4-D 1 μM 6BA	Nitsch	cold treat 27°C dark	40%	19	Rajasekaran and Mullins[172]
Trifolium alexandrium	C	3mm	MS + 5.4 μM NAA 0.5 μM 2,4-D 0.5 μM 2ip	MS + 2.7 μM NAA 2.3 μM KIN	26°C dark	16%	8	Mokhtarzadeh and Constantin[138]
Bromus inermis	C	Uninucleate	MS + 4.7 μM KIN 2.9 μM IAA 9.1 μM 2,4-D	MS + 9.3 μM KIN 2.3 μM IAA	22–25°C	1.3%	14	Zenkteler[236]

which may be due to the stage of microsporogenesis at the time of isolation and placing in culture medium. The stage of development determines the polarity of the first haploid division[195]. The formation of plantlets from *Nicotiana* spp. microspores following the culture of anthers on a very simple medium containing only sucrose (2%) and mineral salts was demonstrated by Nitsch.[156]

The percentage of embryo formation from isolated pollen grains of *Datura innoxia* was improved by exposing buds to a cold temperature trauma (3°C during 48 h) prior to culture[154]. 53.3% of the treated buds gave rise to embryos compared to 19.7% of the control buds when liquid culture media were supplemented with anther extracts. Cold treatment changes the orientation of the spindle at the time of the first pollen grain mitosis. Instead of the typical oblique orientation, the spindle orientation during the first pollen grain mitosis was responsible for the induction of embryo formation. Sunderland[205] presents a different explanation for the chilling effect in which presumably, the trauma allows a greater rate of survival of the pollen in excised buds than occurs in cultures at ambient temperatures. A similar cold treatment (5°C for 62 h) was given to flower buds of *Nicotiana* spp. which gave 40—50% embryos compared with 21% of the control.[152] The most effective way to apply this cold treatment was found to detach the flower buds and place them in Petri dishes with the pedicel immersed in water.[47]

Another type of trauma described to increase the frequency of androgenesis from isolated microspores was excision of the proximal end of an inflorescence prior to pollen culture. In wheat, this resulted in an increase in the frequency of microspores undergoing mitotic division and subsequent haploid plant development.

The limited development of isolated microspores in culture and the enhancement of growth with the nurse culture technique clearly indicate that endogenous factors found in anthers supply essential nutrients for androgenesis demonstrated a requirement for exogenous serine using extracts of *Datura innoxia*. Based on analysis of anther extracts, a completely synthetic medium was devised for the development of isolated microspores of *Datura* and *Nicotiana* spp.[151] This medium contains mineral salts and glutamine (800 mg/l), L-serine (100 mg/l) and inositol (5 g/l).

Sunderland[205] described a slightly different culture method for haploid production whereby anthers are floated on liquid media in Petri dishes. A chilling pretreatment for 8 days at 8°C is recommended for the anthers prior to culture. In *N. tabacum*, the float cultured anthers open and liberate microspores into the liquid medium. Successful development of the released microspores requires the conditioning of the medium by anther tissues. The anther-containing liquid medium is growth limiting unless glutamine, serine, and inositol are included in the medium.[151,152]

CHROMOSOME DOUBLING

Techniques to produce doubled haploids

Much of the interest in haploids stems from the fact that haploids can be utilized to develop completely homozygous diploid lines. Development of these diploid, or doubled haploid, lines depends on reliable techniques for doubling chromosome number. Several chromosome doubling techniques have been utilized in the production of doubled haploids. Many of these methods were developed for obtaining polyploids from diploids. Other techniques, particularly the *in vitro* chromosome doubling procedures, were developed specifically for use with haploid material. The most commonly used procedures double chromosome number either through the use of colchicine or through endomitosis following callus tissue formation. Colchicine and other mitotic poisons induce a process known as C-mitosis. C-mitosis is characterized by in-activation of the spindle, increased chromosome contraction and delayed division of the centro-meres. The final result of C-mitosis is the formation of a restitution nucleus.[177] Endomitosis, on the other hand, is a natural occurrence in certain plant tissues, particularly callus tissue. During endomitosis, the nuclear envelope remains intact, the spindle apparatus does not form and the metaphase plate is eliminated. Subsequently, a restitution nucleus is formed.[98,177]

Colchicine, an alkaloid extract of the autumn-flowering crocus, is used in a variety of chromosome doubling techniques. It is most frequently applied as an aqueous solution, but it can also be dissolved in weak alcohol[48] or 10% glycerin with water[31] or applied as a lanolin or agar paste.[48,51] Optimum concentration of colchicine varies among species. Examples of some commonly used concentrations are 0.05% for *Zea mays*[33], 0.1% for *Hordeum vulgare*[102], 0.25% for *Solanum tuberosum*[118], 0.4% for *Nicotiana tabacum*[24] and 0.25–0.5% for *Lyco-persicon esculentum*[55].

Colchicine is frequently applied to verified haploid seeds[33], seedlings[211], plantlets[24] or shoot tips[18,48,136,148,168]. Seeds, seedlings, and plantlets are usually soaked in the colchi-cine solution for a few hours. Apical buds may also be soaked, but more often drops of colchi-cine solution are applied to the meristem at regular intervals. Lanolin or agar pastes are often utilized for treatment of meristems.[98]

Colchicine has also been applied to *in vitro* cell or tissue cultures. Meristem cultures of monohaploid *Solanum verucosum* were initiated in a medium containing 0.5% colchicine.[229] After 48 hours, the buds were transferred to a colchicine-free medium. A high frequency of diploids was found among the resulting plantlets. Haploid cell cultures of *Atropa bella-donna*[175], *Saccharum officinarum*[81] and *Nicotiana tabacum*[66] have also been diploidized by the addition of colchicine to the suspension culture medium for a specified length of time.

Additions of various substances to the colchicine solution have been reported to increase doubling frequency. In barley, a solution of 0.1% colchicine and 2–4% dimethyl sulfoxide (DMSO) applied to seedlings at the 2–3 leaf stage, yields 56% plants with doubled sectors, com-pared to 37% when colchicine alone is used.[200] Addition of 10 mg/l GA_3 to the colchicine-DMSO solution further increases frequency of doubling in *Hordeum* up to 72%.[211]

Other C-mitotic agents, though successful, have had limited use for chromosome doubling. Nitrous oxide, applied at 90 p.s.i. for 24 hours to excised floral heads of diploid *Trifolium*, pro-duces a high frequency of tetraploids.[209] Nitrous oxide is normally applied to the inflorescence after pollination, so that doubling occurs at the time of fertilization or during the first zygotic division.[161] Podophyllin from *Podophyllus peltatum* and vincaleublastine from *Vinca rosea* also act as C-mitotic agents, but are not used routinely with any plant species.[67,98]

Tissue-wounding, and subsequent callus formation, is one of the earliest techniques used to double chromosome number. Callus is usually induced by decapitation of the shoot tip and removal of all axillary buds. Shoots that develop from the callus frequently have an increased ploidy level. This technique has been successful in producing tetraploids from diplid forms of *Solanum*[99], *Lycopersicon*[122], and *Nicotiana*[68]. In *Nicotiana*, callus formation can be induced by application of 1% indole acetic acid in lanolin to the cut surface.[68]

In vitro techniques have been developed to take advantage of chromosome doubling via endomitosis in callus tissue. These techniques have been most widely used in *Nicotiana*. Kasperbauer and Collins[107] obtained diploids from callus cultures of haploid *Nicotiana* at a frequency of up to 33%. The highest frequency of diploids was obtained when leaf midveins from mature leaves were used as the explant material. Midveins from leaves that were not fully expanded produced only haploid plants, while stem pith yielded a high frequency of aneuploid plants.

Diploids have also been regenerated from callus cultures initiated from stems of 7−8 week old *Nicotiana* haploid plantlets obtained directly from anther culture.[114] Callus cultures have also been used to resynthesize tetraploids from dihaploid *Solanum tuberosum* leaf tissue.[96]

Uses of doubled haploids

Since a doubled haploid represents complete homozygosity, pure lines can be produced by extracting haploids from segregating material without the 6−7 generations of backcrossing normally needed to stabilize a line.[192] If haploids are obtained from an F_1 plant, it is theoretically possible to have the full range of genotypes normally represented in an F_2 generation present in a homozygous form. In practice, however, haploid extraction is sometimes delayed until a more advanced generation, in order to allow more recombination to occur.[104]

Several theoretical studies aimed toward comparing use of the doubled haploid breeding method to conventional breeding methods have been reported. The probability of obtaining the desired genotype in the F_2 population of a self-pollinating species was compared for doubled haploid versus diploid methods. The advantage of the doubled haploid over the diploid procedure was determined to be 2^n, where n is the number of segregating factors.[147] Griffing[71] compared the use of doubled haploids to regular diploids in a recurrent selection program. He concluded that selection based on doubled haploids was up to six times more effective than selection utilizing regular diploids. In another study, response to mass selection utilizing doubled haploids was found to be 14 times faster than with regular mass selection.[35]

Much of the work concerning application of doubled haploid technique to varietal development has been reported for *Nicotiana* and *Hordeum*. As mentioned in an earlier section, release of the barley variety 'Mingo', developed through use of doubled haploids, represented a time saving of 7 years over conventional breeding methods.[84,104] At least six major barley breeding programs are now utilizing doubled haploids.[104]

Many of the breeding efforts in *Nicotiana* have concerned disease resistance. An improvement in resistance to three diseases was obtained in the Japanese flue-cured tobacco variety MC 1610 through selection among doubled haploids extracted from MC 1610 and crosses of MC 1610 to two other commercial varieties.[144,145] Resistance to blue mold (*Peronospora tabacina*) and TMV has been transferred from wild *Nicotiana* species to commercial varieties of *N. tabacum* following the use of the doubled haploid procedure.[228] High yielding, disease resistant lines have been selected from doubled haploids obtained from an F_1 hybrid.[29]

Other breeding efforts in tobacco have been concerned with improving agronomic and chemical qualities. Three varieties with improved yield and quality, developed from doubled haploids, have been released by the Chinese Academy of Agricultural Sciences.[2,3] High-yielding lines of flue-cured varieties have also been obtained from haploids in a Bulgarian tobacco breeding program.[5] Finally, true breeding lines of burley tobacco representing four levels of alkaloid production, were obtained with the use of the doubled haploid procedure.[42]

Doubled haploids have been used in the breeding of several species other than *Hordeum* and *Nicotiana*, though not to the extent that they have been utilized with these two crop species. Doubled haploids have a unique application in *Asparagus officinalis* breeding. Male asparagus plants, with the sex chromosome constitution of XY, are higher yielding and less fibrous than female XX plants. Through anther culture and subsequent chromosome doubling, 'supermale' YY plants can be produced. When these plants are crossed to female plants for hybrid production, only males are present among the F_1 plants.[50,210] Doubled haploid techniques are being utilized in *Brassica* for breeding for lowered glucosinolate content.[158,230] Varietal development in rice[4,158,234] and peppers[168] also currently involve doubled haploids. In corn, doubled haploids are being utilized in a process known as analytical breeding. Because of the decreased variability of the doubled haploids in comparison to the regular diploid lines, doubled haploids of inbred lines are used for evaluation of the lines for use as parents of F_1 hybrids. Following the analysis, F_1 hybrids can be constructed with more confidence.[158]

Doubled haploids can also be used in breeding programs for studying the inheritance of qualitative traits. Because haploids and doubled haploids exhibit the gametic genotype, ratios obtained in genetic analysis are less complex if doubled haploids rather than diploids are used for the analysis.[39] Doubled haploids have also been utilized for the genetic analysis of several quantitative traits in tobacco. In one study, biparental progenies of doubled haploid plants were used to estimate genetic variance for morphological, agronomic and chemical traits in a population of burley tobacco.[121] Procedures for using doubled haploids to estimate additive and additive X additive genetic variances have also been outlined.[36]

HAPLOID PROTOPLASTS

Isolation and culture of protoplasts

Protoplasts are essential for successful mutant isolation, somatic hybridization, and gene transfer experiments in higher plants.[185] The requirement for gene expression in genetically altered plant cells implies that haploid protoplasts, with a single copy of each gene, would be extremely valuable in each of these types of experiments. While haploid protoplasts have been isolated directly from pollen tetrads[12], more often haploid protoplasts have been isolated from

TABLE 12.3. Effective concentrations and types of growth regulators used in the primary and secondary culture medium for the induction of haploid plant development through callus formation

Growth Regulator	1° Culture Medium		2° Culture Medium	
	Frequency of Use	Effective Concentration	Frequency of Use	Effective Concentration
IAA	16.7%	2.9–34.3 μM	33.3%	1.1–22.8 μM
2,4-D	70.8%	0.5–22.6 μM	8.3%	0.2–22.6 μM
NAA	33.3%	1.0–13.4 μM	33.3%	1.1–26.9 μM
KIN	54.2%	4.0–23.2 μM	58.3%	0.1–46.7 μM
6BA	16.7%	1.0– 4.4 μM	8.3%	0.9– 6.4 μM
2ip	4.2%	0.05 μM		
CW	16.7%	10–15%	4.2%	15%
YE	4.2%	0.1%	0%	—
None	4.2%	—	25%	—

TABLE 12.4. Frequency of specific growth regulators or combinations of regulators
used in primary and secondary culture medium for haploid plant development
following callus formation

Growth Regulator	1° Medium	2° Medium
No regulator(s)	4.2%	25%
2,4-D	12.5%	—
2,4-D + NAA	—	4.2%
2,4-D + 6BA	8.3%	—
2,4-D + KIN	12.5%	4.2%
2,4-D + NAA + KIN	8.3%	—
2,4-D + NAA + 2ip	4.2%	—
2,4-D + IAA + KIN	12.5%	—
2,4-D E KIN + CW	8.3%	—
2,4-D + KIN + YE	4.2%	—
IAA	—	4.2%
IAA + KIN	—	20.8%
IAA + 6BA + CW	4.2%	—
IAA + NAA + KIN	—	4.2%
IAA + 6BA	—	4.2%
NAA	4.2%	—
NAA + KIN	8.3%	25.0%
NAA + CW	4.2%	—
NAA + 6BA	4.2%	—
KIN + CW	—	4.2%
6BA	—	4.2%
CW	4.2%	—

TABLE 12.5. Frequency of specific tissue culture media formulations used during primary and secondary culture for the occurrence of haploid plant development following callus formation

Culture Medium	1° Culture Medium	2° Culture Medium
Miller	8.3%	8.3%
B5	8.3%	4.2%
MS	62.5%	62.5%
Nitsch	12.5%	16.7%
White	4.2%	
Bourgin & Nitsch	4.2%	4.2%
Blaydes	0.0%	4.2%

somatic tissue. In most cases, the protoplasts were isolated from leaf mesophyll cells of haploid plants produced via anther culture. Plant regeneration from haploid protoplasts has been observed in ten plant species (Table 12.11). All of the species regenerated from protoplasts, except *Brassica napus*, are Solanaceous plants. Plants have also been regenerated from protoplasts isolated from a cell suspension culture of *Datura innoxia*[63]. Consequently, the technology exists for regeneration of haploid protoplasts that could be used in genetic engineering experiments.

Protoplasts can be isolated using various concentrations of cellulases and pectinases. The most commonly used cellulases were Onozuka R10 and SS (Kinki Yakult Mfg.) at 0.1–3.0% concentration and Driselase (Kyowa Hakko, Kogyo Co., Ltd.) at 0.03–1.5% concentration. Macerozyme (Kinki Yakult Mfg.) and Meicellase (Meiji Seika Kaisha Ltd.) have also been used successfully. Enzymes were usually dissolved in mannitol at 0.25–0.6 M for osmotic stability. In each case, following removal of enzyme solution, protoplasts were cultured in a high osmotic medium prior to transfer to the plant regeneration medium. Mannitol at 0.3–0.6 M has been used most frequently in the protoplast culture medium, but both sorbitol[59] and 0.38 M glucose[21] have been used. A wider range of hormones were used for protoplast culture than for plant regeneration. For 6 of the 10 cultured species 2,4-D is added to the culture medium, while only 1 of 10 species (*Brassica napus*) contains 2,4-D in the regeneration medium. Caboche[26] has defined nutritional requirements of haploid protoplasts that permit culture of protoplasts in a simple defined growth medium. Procedures for plant regeneration from haploid protoplasts are summarized in Table 12.11. In most cases, high concentrations of a cytokinin, 6BA or KIN, are required to achieve plant regeneration.

The chromosome number of plants regenerated from haploid protoplasts is quite variable. As with the chromosome instability of plants obtained from cultured anthers, this may reflect that diploid cells that arise spontaneously may have a decided selective advantage over haploid cells. Nonetheless, haploid plants have been recovered from haploid protoplasts of *N. tabacum*, *Solanum tuberosum*, *Datura innoxia*, *B. napus*, *Hyoscyamus muticus*, and *P. hybrida*. In each of the other species regenerated from protoplasts (Table 12.11), no haploid plants were recovered despite the use of haploid protoplasts. A mixture of haploid and diploid plants were recovered from mesophyll protoplasts of *Solanum tuberosum* and *Datura innoxia*, while only diploid plants were regenerated from haploid leaf protoplasts of *N. sylvestris*, *Datura metel*,

TABLE 12.6. Plant genera capable of androgenesis *in vitro*

Species	Stage	Culture Medium	Growth Regulators	Culture Environment	Maximum %	n=	References
Atropa belladonna	Uninucleate	MS	11.4 IAA 18.6 KIN	NA	18%	36	Zenkteler[235]
Aesculus hippocastanum	Uninucleate	MS	4.7 KIN 4.5 2,4-D	28°C	53%	20	Radojevic[169]
Brassica napus	Uninucleate	MS	0.5 2,4-D 10% CW	25–27°C dark	60%	19	Thomas and Wenzel[212]
Datura innoxia	Uninucleate	Nitsch	15–30% CW	25–30°C	80%	12	Engvild et al.[57]
Festuca arundinacea	Uninucleate	MS	9.1 2,4-D	cold treat 5°C 22°C in dark	NA	21	Kasperbauer et al.[106]
Hyoscyamus muticus	Pollen Mitosis	Nitsch	15% CW or 0.9 6BA	16 h day	47%	14	Wernicke et al.[233]
Nicotiana tabacum	Uninucleate	Bourgin & Nitsch	none	25–28°C	100%	24	Nitsch[151]
Oryza sativa indica	Uninucleate	Blaydes	15% CW	25°C dark	11–26%	12	Guha-Mukherjee[77]
Paeonia hybrida	Pollen Mitosis	MS	0.5 NOA	25°C	40%	5	Sunderland[204]
Petunia hybrida	Uninucleate	Nitsch	15% CW	27°C	12.3%	7	Malhotra and Maheshwari[125]
Saintpaulia ionantha	Pollen Mitosis	Blaydes	none	27°C	18%	14	Hughes et al.[92]
Scopolia physaloides	Uninucleate	Nitsch	15% CW	26°C 12 h light	3%	12	Wernicke and Kohlenbach[232]
Secale cereale	Uninucleate	Nitsch	1.1 2,4-D	25°C dark	0.14%	14	Wenzel and Thomas[231]
Solanum bulbocastanum	Uninucleate	MS	11.4 IAA 9.3 KIN	20°C	32%	12	Irikura[93]
Zea mays	Uninucleate	N6, 15% suc	4.7 KIN	16 h day 25–29°C	3%	10	Shu-hua et al.[190]

TABLE 12.7. Frequency of specific growth regulators and effective concentration ranges
used in the promotion of androgenesis

Growth Regulator	Frequency of Use	Effective Concentration Range
IAA	13.3% (2/15)	11.4 μM
2,4-D	26.7% (4/15)	1.1–9.1 μM
NOA	6.7% (1/15)	0.5 μM
KIN	26.7% (4/15)	4.7–18.6 μM
6BA	6.7% (1/15)	0.9 μM
CW	40.0% (6/15)	10–30 %
No Regulators	13.3% (2/15)	

TABLE 12.8. Frequency of specific growth regulators or combinations of
regulators in the promotion of *in vitro* androgenesis

Growth Regulator or Combination Thereof	Frequency of Use
No Regulator(s)	13.3%
2,4-D	13.3%
2,4-D + KIN	6.7%
2,4-D + CW	6.7%
IAA + KIN	13.3%
NOA	6.7%
KIN	6.7%
6BA	6.7%
CW	33.3%

TABLE 12.9. Frequency of specific tissue culture medium formulations
used in the promotion of *in vitro* androgenesis

Culture Medium	Frequency of Use
MS	40%
Nitsch	40%
Blaydes	13.3%
Bourgin & Nitsch	6.7%
N6	6.7%

TABLE 12.10. Differences in interspecies requirements for androgenesis *in vitro*

Species	Stage	Culture Medium	Growth Regulators	Culture Environment	Maximum %	n=	References
Brassica campestris	Uninucleate	B5	0.5 NAA 0.5 2,4-D	25°C 35°C for 1 day	15%	5	Keller and Armstrong[108]
B. napus	Uninucleate	MS	0.5 2,4-D 10% CW	25–27°C dark	60%	19	Thomas and Wenzel[212]
Datura innoxia	Uninucleate	Nitsch	15–30% CW	25–30°C	80%	12	Engvild et al.[57]
D. metel	Uninucleate	Nitsch	none	27°C	3%	12	Iyer and Raina[95]
D. meteloides	Uninucleate	Nitsch	none	22–28°C	NA	12	Nitsch[157]
D. muricata	Uninucleate	Nitsch	none	22–28°C	NA	12	Nitsch[157]
D. stramonium	Uninucleate	Nitsch	15–30% CW	25–30°C	NA	12	Guha and Maheshwari[73]
D. wrightii	Pollen Mitosis	Nitsch	10–20% CW	NA	20%	12	Kohlenbach and Geier[115]
Hyoscyamus albus	Uninucleate	Nitsch & Nitsch	none	26°C	NA	17	Raghavan[170]
H. pusillus							
Hyoscyamus muticus	Pollen Mitosis	Nitsch	15% CW or 0.9 6BA	16 h day	47%	14	Wernicke et al.[233]
H. niger	Uninucleate	Nitsch & Nitsch	none	28°C 16 h day	10%	17	Corduan[43]
Nicotiana alata	Uninucleate	Nitsch	none	24–30°C	NA	9	Nitsch[156]
N. attenuata	Pollen Mitosis	Nitsch	none	25°C	10%	12	Collins and Sunderland[40]
N. clevelandii	Uninucleate	Nitsch	none	25–32°C	2%	24	Vyskot and Novak[223]
N. glutinosa	Binucleate	MS	5.7 IAA 2.0 KIN	21–26°C	2%	12	Tomes and Collins[215]
N. knightiana	Binucleate	MS	5.7 IAA 2.0 KIN	21–26°C	3%	12	Tomes and Collins[215]
N. otophora	Binucleate	MS	none	25°C	NA	12	Collins et al.[41]
N. paniculata	Binucleate	MS	5.7 IAA 2.0 KIN	21–26°C	25%	12	Tomes and Collins[215]
N. raimondii	Pollen Mitosis	Nitsch	none	25°C	30%	12	Collins and Sunderland[40]

Table 12.10 continued

Species	Stage	Culture Medium	Growth Regulators	Culture Environment	Maximum %	n=	References
N. rustica	Binucleate	MS	5.7 IAA 2.0 KIN	21–26°C	13%	24	Tomes and Collins[215]
N. sanderae	Uninucleate	Nitsch	none	25–32°C	0.4%	9	Vyskot and Novak[223]
N. sylvestris	Pollen Mitosis	MS	5.7 IAA 2.0 KIN	21–26°C	20%	12	Tomes and Collins[215]
N. tabacum	Uninucleate	Bourgin & Nitsch	none	25–28°C	100%	24	Nitsch[152]
P. axillarix × *P. bybrida*	Pollen Mitosis	MS	4.4 6BA 0.5 NAA	23–24°C 16 h day	1.2%	7	Raquin and Pilet[173]
P. bybrida	Uninucleate	Nitsch	15% CW	27°C	12.3%	7	Malhotra and Maheshwari[125]
Solanum bulbocastanum	Uninucleate	MS	11.4 IAA 9.3 KIN	20°C	32%	12	Irikura[93]
S. demisseim	Uninucleate	Nitsch & Nitsch	0.6 IAA	16 h day 20°C	2%	36	Irikura[93]
S. fendleri	Uninucleate	Nitsch & Nitsch	10.7 NAA 3 g AC	16 h day 20°C	2%	24	Irikura[93]
S. bjertingii	Uninucleate	Nitsch & Nitsch	11.4 IAA 8.8 6BA	16 h day 20°C	2%	24	Irikura[93]
S. pbureja	Uninucleate	MS	10.7 NAA 9.3 KIN	16 h day 20°C	18%	12	Irikura[93]
S. polytricbon	Uninucleate	Nitsch & Nitsch	10.7 NAA	16 h day 20°C	16%	24	Irikura[93]
S. stenotomum	Uninucleate	MS	3 g AC 11.4 IAA 3 g AC 9.5 KIN	20°C 16 h day	18%	12	Irikura[93]
S. stoloniferum	Pollen Mitosis	MS	10.7 NAA 3 g AC	20°C 16 h day	2%	24	Irikura[93]
S. tubrosum	Pollen Mitosis	MS	0.05 NAA	25°C	13S	24	Dunwell and Sunderland[53]

TABLE 12.11. Protocol for plant regeneration from haploid protoplasts

Species	Isolation	Culture Medium	Regeneration Medium	Regeneration Hormones in μM	Protoplast Source	References
Datura innoxia n = 12	1% Macerozyme 3% Onozuka SS 0.5 M Mannitol	Durand (1973)	B5	2.2 6BA	leaf	Schieder[182]
D. metel *D. meteloides* n = 12	1% Macerozyme 3% Onozuka SS 0.5 M Mannitol	Durand (1973)	V47	2.2 6BA 8.1 NAA	leaf	Schieder[183]
Hyoscyamus muticus n = 14	0.5% Macerozyme 0.25% Cellulase 0.25 M Mannitol	Nagata & Takebe (1970)	MS	1.0 6BA 0.3 NAA	leaf	Wernicke et al.[233]
Nicotiana alata n = 9	0.02% Macerozyme 0.1% Onozuka R10 0.05% Driselase 0.38 M Mannitol	Kao et al. (1974) Medium 3	T$_0$	4.4 6BA 2.9 IAA	leaf	Bourgin and Missonier[21]
N. sylvestris n = 12	0.3% Onozuka %10 0.03% Driselase 0.15% Pectinase 0.33 M Sorbitol	Murashige & Skoog (1962)	MS	0.2 KIN	leaf	Facciotti and Pilet[59]
N. tabacum n = 24	1% Potassium dextran sulfate 0.5% Pectinase 0.6 M Mannitol	Ohyama & Nitsch (1972)	Med. 1	4.4 6BA 0.5 2,4-D	leaf	Ohyama and Nitsch[160]
Petunia hybrida n = 14	1% Macerozyme 3% Cellulase 0.6 M Mannitol	Durand (1973)	NT	5.0 6BA 2.0 NAA	leaf	Binding[14]

Table 12.11 continued

Species	Isolation	Culture Medium	Regeneration Medium	Regeneration Hormones in μM	Protoplast Source	References
Solanum tuberosum n = 24	1% Macerozyme 2% Meicellase 0.5 M Mannitol	Kao & Michayluk (1974)	MS	15 KIN 5.0 IAA	leaf	Binding et al.[15]
Brassica napus n = 19	1% PATE* 0.5% Onozuka R10 0.5 M Mannitol	Nitsch & Nitsch (1969)	MS	1.0 6BA 0.5 2,4-D	leaf	Thomas et al.[213]
Datura innoxia n = 12	1.5% Driselase 0.5% Pectinase 0.5 M Mannitol	Durand (1973)	B5	NA	cell culture	Furner et al.[63]

* PATE = pectine acid transeliminase.

D. meteloides, and haploid suspension culture protoplasts of D. innoxia. Plants regenerated from leaf protoplasts of haploid N. alata contained the 2n, 3n, and 4n chromosome number. As each experiment was initiated with haploid protoplasts, the culture medium or culture conditions used may have influenced the ploidy of the regenerated plants. Cytokinin concentration has been related to recovery of phenotypically abnormal plants from protoplasts.[189] Furner et al.[63] examined protoplast-derived callus, suspension cultures, and shoots of Datura innoxia. Haploid cells were observed in regenerated callus and suspension cultures but not in the regenerated shoots. Either haploid cells do not regenerate as well as cells of higher ploidy, or a component of the regeneration procedure results in the production of aneuploid and polyploid cells. It should be noted, though, that genetic alterations of haploid protoplasts would still be expressed in regenerated doubled haploid plants.

Techniques of genetic modification

Haploid cells and protoplasts have been used extensively for mutant isolation. Haploid cells are preferred for the isolation of mutations as recessive genes can be expressed in cells with single copies of genes. The only auxotrophic[27,142] and temperature sensitive[128] mutants isolated to date have been recovered from haploid cell lines. Unfortunately most variants isolated from haploid cell lines have not been genetically characterized. Chromosome instability has been reported in plants regenerated from variant cell lines.[129] More often variant haploid cell lines have been isolated by selecting for resistance to toxic chemicals

TABLE 12.12. Variant cell lines isolated from haploid cell cultures

Mutant	Plant Species	Plants Regenerated	Genetic Basis	References
1. auxotrophic				
lysine	N. tabacum	yes	semi-dominant	Carlson[27]
calcium pantothenate	D. innoxia	no	—	Savage et al.[181]
2. temperature sensitive				
lethal	N. tabacum	yes	—	Malmberg[128]
3. resistance				
streptomycin	N. tabacum	yes	cytoplasmic	Maliga et al.[126]
bromodeoxyuridine	N. tabacum	yes	semi-dominant	Maliga et al.[127]
valine	N. tabacum	yes	semi-dominant recessive	Bourgin[20]
chlorate	N. tabacum	yes	—	Muller and Grafe[142]
aminopterin	D. innoxia	yes	—	Mastrangelo and Smith[130]
methionine sulfoximine	N. tabacum	yes	semi-dominant	Carlson[28]
glycine hydroxamate	N. tabacum	yes	—	Lawyer et al.[119]
isonicotinic acid hydrazide	N. tabacum	yes	dominant	Berlyn[11]
chlorate	D. innoxia	no	—	King et al.[112]
sodium chloride	N. sylvestris	no	—	Dix and Street[49]
streptomycin	N. sylvestris	yes	recessive	Maliga et al.[127]
chloramphenicol	N. sylvestris	yes	—	Maliga et al.[127]

(Table 12.12). In the cases classified to date, recessive and dominant single gene nuclear mutants as well as cytoplasmic mutations have been identified and genetically characterized.

Bourgin[20] has used haploid protoplasts of *N. tabacum* in a direct protoplast selection for mutations. A mutagenized population of protoplasts was cultured directly in a medium containing a lethal concentration of valine. Resistant colonies were recovered and regenerated to plants and subsequently characterized as single gene mutations.

Haploid protoplasts have been used extensively in protoplast fusion with lower plants such as *Physcomitrella*[72], but have not been used as extensively for protoplast fusion in higher plants. Melchers and Labib[134] fused mesophyll protoplasts of two haploid albino, light sensitive mutants of *N. tabacum* and recovered, following genetic complementation, normal green intraspecies fusion products. Although chromosome instability followed protoplast fusion, resulting in numerous aneuploid plants, some normal diploid plants were recovered. Consequently, haploid protoplasts have been used in numerous cellular genetic experiments.

Numerous attempts have been made to induce transformation in higher plants. Early reports included attempts to inject DNA into seeds[82] or into isolated pollen grains.[83] Exogenous DNA may be either plant DNA[86] or bacterial DNA.[120] The success of whole plant experiments has been critically reviewed.[113] More recently interest has accumulated in transforming cultured plant cells[143] and plant protoplasts.[123] Successful transfer of DNA may require using naturally infective DNA such as *Agrobacterium* DNA or encapsulation in liposomes to afford protection to exogenous DNA.[131] As these techniques develop, as in other areas of cell manipulation, the value of using haploid cells will be evident.

GENETIC STABILITY

Plants regenerated from cultured anthers are often not haploid. This reflects, in part, that the anther is a mixed population of haploid and diploid cells, and that often plants are regenerated from the somatic tissue of the anther. Even when regeneration occurs from microspores, variation in chromosome number is often observed among regenerated plants. The range of chromosome numbers reflects both endomitosis and fusion of microspore nuclei.[206] In addition, as polyploidy and aneuploidy are quite common in plant callus tissue[46], additional factors associated with *in vitro* culture contribute to variation of the chromosome number of plants regenerated from cultured anthers. As callus is associated with chromosome instability, anthers undergoing embryogenic mode of development may have less chromosomal aberrations than anthers that first produce callus. Due to the multiple causes of chromosomal variation, it is preferable to use a genetic marker in the heterozygous condition when culturing anthers, so that diploid regenerated plants that are microspore-derived can be identified.[133] Use of hybrid explants facilitated identification of doubled haploids in *Arabidopsis* by using isoenzyme markers.[184]

The genetic stability of regenerated plants is dependent on the species cultured. In *Nicotiana tabacum*, most plants recovered from cultured anthers are haploid.[107] On the other hand, plants obtained from *Brassica campestris* (Keller *et al.*[109]) and *Arabidopsis thaliana*[184] are only diploid or polyploid with no haploids produced. Successfully cultured anthers of most species have resulted in recovery of a range of ploidies. These reports include recovery of haploid to triploid plants in *Atropa belladonna*[235], haploid to tetraploid plants in *Hordeum vulgare*[38], haploid to pentaploid plants in *Oryza sativa*[149] and haploid to hexaploid plants in *Datura innoxia.*[57]

McComb and McComb[132] examined cultured anthers of *Nicotiana sylvestris* in detail. A range of haploid to tetraploid plants were recovered with the appearance of 5% mixaploid plants. The mixaploid class included plants with n shoots and 2n roots and 2n shoots and n roots as well as a number of plants with a mixture of n and 2n cells within a root or shoot. This chromosomal variation suggests the occurrence of continuous endomitosis within mature developing plants. As a similar phenomenon has not been detected for 2n to 4n *N. sylvestris*, this work suggests that the n chromosome level is unstable in *N. sylvestris*.

Numerous chemical, physical, and developmental factors may influence the chromosome number of plants regenerated from cultured anthers. Guo[78] demonstrated that addition of kinetin increased chromosome number of callus derived from haploid tissue of a *Nicotiana* hybrid. In addition, alterations in photoperiod were correlated with altered chromosome number.[78] Engvild[56,57] has demonstrated a relationship between plantlet ploidy and developmental stage of floral buds in *Datura innoxia* and *Nicotiana tabacum*. Cultured flower buds containing late binucleate microspores result in recovery of a greater number of polyploid plants.

While in most cases haploid plants recovered from cultured anthers or pollen are maintained as haploid plants, stable haploid tissue maintained *in vitro* would be quite useful for basic studies in cellular genetics. Unfortunately, most haploid cultures maintained as callus cultures do not remain haploid. Chromosome variation, particularly polyploidy, has been reported in long-term callus cultures of *Pelargonium, Crepis,* and *Nicotiana*[10,159,180]. In addition to numerical changes, structural alterations of chromosomes were observed in callus cultures of *Crepis capillaris*[180]. Chromosomally stable haploid suspension cultures have been reported for a number of species, e.g. *Datura innoxia*[58], resulting in recovery of haploid plants. In most cases, though, when cells are cultured under strong selective pressure, such as for mutant isolation, the plants regenerated are not haploid (e.g. Malmberg[129]).

CONCLUDING REMARKS

While haploid plants have been successfully recovered from cultured anthers of a number of plant species, reproducable success has been limited to a few Graminaceous and Solanaceous species. Despite limitations in the success of obtaining haploids, techniques have been developed to apply haploids to conventional breeding programs. Doubled haploids have been used to release new varieties of barley and tobacco. In barley, most haploids are produced by chromosome elimination following interspecific hybridization; while in tobacco, haploids are produced via anther culture. Additionally, haploid cells, with single gene copies, have unique application to somatic cell genetics. Haploid cells have been used extensively in mutant isolation and protoplast culture. As techniques are successfully extended to more crop species, it is likely that additional varieties will be released using doubled haploid lines.

Abbreviations:

cytokinins:

	KIN =	kinetin

auxins:

	6BA =	6 benzyladenine (benzyl amino purine);
IAA	= indole acetic acid;	2ip = 2 isopentenyl adenine;
2,4-D	= 2,4-dichlorophenoxyacetic acid;	*growth additives:*
NAA	= napthaleneacetic acid;	CW = coconut water;
NOA	= β-napthoxyacetic acid;	YE = yeast extract

REFERENCES

1. Abo El-Nil, M.M. and Hildebrandt, A.C. Differentiation of virus-symptomless geranium plants from callus. *Plant Disease Reporter* **55**, 1017–1020, 1971.

2. Anonymous. The evaluation of the progenies of the pollen plants of tobacco. *Acta Genet. Sinica* **1**, 26–39, 1974a.

3. Anonymous. Success of breeding the new tobacco cultivar 'Tan-Yuh No. 1'. *Acta Bot. Sinica.* **16**, 300–303, 1974b.

4. Anonymous. Studies on anther culture *in vitro* in *Oryza sativa* subsp. Shien II. The role of anther culture in purification and selection of rice cultivar. *Acta Genet. Sinica* **3**, 57–60, 1976.

5. Atanassov, A., Pamukov, I., Kunev, K. and Nedeltcheva, S. Results from the application of haploids in tobacco breeding. *CORESTA Information Bulletin, International Tobacco Scientific Symposium. Sofia, 1978.* No. 1978 – Special, p. 82, 1978.

6. Ban, Y., Kokuba, T. and Miyaji, Y. Production of haploid plant by culture of *Setaria italica. Bull. Fac. Agric. Kagoshima Univ.* **21**, 77–81, 1971.

7. Barclay, I.R. High frequency haploid production in wheat (*Triticum aestivum*) by chromosome elimination. *Nature* **255**, 410–411, 1975.

8. Barclay, I.R., Shepherd, K.W. and Sparrow, B.H. Control of chromosome elimination in *Hordeum vulgare–H. bulbosum* hybrids. *Barley Genet. Newsletter* **2**, 22–24, 1972.

9. Bennett, M.D., Finch, R.A. and Barclay, I.R. The time, rate mechanism of chromosome elimination in *Hordeum* hybrids. *Chromosoma* **54**, 175–200, 1976.

10. Bennici, A. Cytological analysis of roots, shoots, and plant regenerated from suspension and solid *in vitro* cultures of haploid *Pelargonium. Z. Pflanzenzuchtg.* **72**, 199–205, 1974.

11. Berlyn, M.B. Isolation and characterization of isonicotinic acid hydrazide-resistant mutants of *Nicotiana tabacum. Theoret. Appl. Genet.* **58**, 19–26, 1980.

12. Bhojwani, S.S. and Cocking, E.C. Isolation of protoplasts from pollen tetrads. *Nature New Biol.* **239**, 29–30, 1972.

13. Binding, H. Nuclear and cell Divisions in isolated pollen of *Petunia hybrida* in agar suspension cultures. *Nature* **237**, 283–285, 1972.

14. Binding, H. Regeneration von haploiden und diploiden Pflanzen aus Protoplasten von *Petunia hybrida* L. *Z. Pflanzenphysiol.* **74**, 325–356, 1974.

15. Binding, H., Nehls, R., Schieder, O., Sopory, S.K. and Wenzel, G. Regeneration of mesophyll protoplasts isolated from dihaploid clones of *Solanum tuberosum. Physiol. Plant.* **43**, 52–54, 1978.

16. Bingham, E.T. and Binek, A. Comparative morphology of haploids from cultivated alfalfa, *Medicago sativa* L. *Crop Sci.* **9**, 749–751, 1969.

17. Bingham, E.T. and Dunbier, M.W. Current research with haploids and haploidy in alfalfa, *Medicago sativa. Haploids in Higher Plants.* 274. Kasha, K. (ed.). Univ. of Guelph, Guelph, Canada, 1974.

18. Blakeslee, A.F. and Avery, A.G. Methods of inducing doubling of chromosomes in plants by treatment with colchicine. *J. Hered.* **28**, 393–411, 1937.

19. Blakeslee, A.F., Belling, J., Farnham, M.E. and Bergner, A.D. A haploid mutant in the jimson weed, *Datura stramonium. Science* **55**, 646–647, 1922.

20. Bourgin, J.P. Valine-resistant plants from *in vitro* selected tobacco cells. *Molec. Gen. Genet.* **161**, 225–230, 1978.

21. Bourgin, J.P. and Missonier, C. Culture of haploid mesophyll protoplants from *Nicotiana alata. Z. Pflanzenphysiol.* **87**, 55–64, 1978.

22. Bourgin, J.P. and Nitsch, J.P. Obtention de *Nicotiana* haploides as partir d'etamines cultivees *in vitro. Ann. Physiol. Veg.* **9**(4), 377–382, 1967.

23. Breukelen, E.W.M. van, Ramanna, M.S. and Hermsen, J.G.Th. Parthenogenetic monohaploids (2n = x = 12) from *S. tuberosum* L. and *S. verrucosum* Schlechtd. and the production of homozygous potato diploids. *Euphytica* **26**, 263–271, 1977.

24.   Burk, L.G., Gwynn, G.R. and Chaplin, J.R. Diploidized haploids from aseptically cultured anthers of
        Nicotiana tabcum. J. Hered. **63**, 355−360, 1972.
25.   Burk, L.G., Gerstel, D.U. and Wernsman, E.A. Maternal haploids of *Nicotiana tabacum* L. from seed.
        Science **206**, 585, 1979.
26.   Caboche, M. Nutritional requirements of protoplast-derived, haploid tobacco cells grown at low cell
        densities in liquid medium. *Planta* **149**, 7−18, 1980.
27.   Carlson, P.S. Induction and isolation of auxotrophic mutants in somatic cell cultures of *Nicotiana
        tabacum. Science* **168**, 487−489, 1970.
28.   Carlson, P.S. Methionine sulfoximine-resistant mutants of tobacco. *Science* **180**, 1366−1368, 1973.
29.   Chaplin, J.F. Genetic manipulation for tailoring the tobacco plant to meet the requirements of the
        grower, manufacturer and consumer. *CORESTA Information Bulletin, International Tobacco
        Scientific Symposium, Sofia 1978*, No. 1978 − Special, 17−32, 1978.
30.   Chase, S.S. Monoploids in maize. *Heterosis*, Gowen, J.W. (ed.), Iowa State College Press, Ames,
        1952a.
31.   Chase, S.S. Production of homozygous diploids of maize from monoploids. *Agron. J.* **44**, 263−267,
        1952b.
32.   Chase, S.S. Analytic breeding in *Solanum tuberosum* L. − a scheme utilizing parthenotes and other
        diploid stocks. *Can. J. Genet. Cytol.* **5**, 359−363, 1963.
33.   Chase, S.S. Monoploids and monoploid-derivatives of maize (*Zea mays* L.). *Bot. Rev.* **35**, 117−167,
        1969.
34.   Cheng-hua, C., Fa-tsu, C., Chang-fa, C., Chuan-hua, W., Shi-jie, C., Hsu-en, H., Hsiao-hui, O., Yung-tao,
        H. and Tsun-min, L. Obtaining pollen plants of *Hevea brasiliensis* Muell. Arg. *Proceedings of
        Symposium on Plant Tissue Culture*, 11−22. Science Press, Peking, 1978.
35.   Choo, T.M. and Kannenberg, L.W. The efficiency of using doubled haploids in a recurrent selection
        programme in a diploid, cross-fertilizing species. *Can. J. Genet. Cytol.* **20**, 505−511, 1978.
36.   Choo, T.M., Christie, B.R. and Reinbergs, E. Doubled haploids for estimating genetic variances and a
        scheme for population improvement in self-pollinating crops. *Theor. Appl. Genet.* **54**, 267−
        271, 1979.
37.   Clapham, D. *In vitro* development of callus from the pollen of *Lolium* and *Hordeum. Z. Pflanzen-
        zuchtg.* **65**, 285−292, 1971.
38.   Clapham, D. Haploid *Hordeum* plants from anthers *in vitro. A. Pflanzenzuchtg.* **69**, 142−155, 1973.
39.   Collins, G.B. and Legg, P.D. Recent advances in the genetic applications of haploidy in *Nicotiana.
        The Plant Genome*, 197−213. Davies, D.R. and Hopewood, D.A. (eds.), The John Innes Charity,
        Norwich, England, 1980.
40.   Collins, G.B. and Sunderland, N. Pollen-derived haploids of *Nicotiana knightiana, N. raimondii*, and
        N. attenuata. J. Experimental Botany **25**, 1030−1039, 1974.
41.   Collins, G.B., Legg, P.D. and Kasperbauer, M.J. Chromosome numbers in anther-derived haploids of
        two *Nicotiana* species. *Heredity* **63**, 113−118, 1972.
42.   Collins, G.B., Legg, P.D. and Kasperbauer, M.J. Use of anther-derived haploids in *Nicotiana*. I. Isola-
        tion of breeding lines differing in total alkaloid content. *Crop Sci.* **14**, 77−80, 1974.
43.   Corduan, G. Regeneration of anther-derived plants of *Hyoscyamus niger* L. *Planta* **127**, 27−36, 1975.
44.   Corduan, G. and Spinx, C. Haploid callus and regeneration of plants from anthers of *Digitalis pur-
        purea* L. *Planta* **124**, 1−11, 1975.
45.   Daker, M.G. Cytological studies on a haploid cultivar of *Pelargonium* and its colchicine-induced di-
        ploids. *Chromosoma* **21**, 250−271, 1967.
46.   D'Amato, F. Chromosome number variation in cultured cells and regenerated plants. *Frontiers of
        Plant Tissue Culture 1978*, 287−296. Thorpe, R.A. (ed.). Univ. of Calgary, Calgary, Canada,
        1978.
47.   Debergh, P. and Nitsch, C. Premiers resultats sur la culture *in vitro* de grains de pollen isoles chez la
        tomate. *C.R. Acad. Sci. Paris* **276**, 1281−1284, 1973.
48.   Dermen, H. Colchicine polyploidy and technique. *Bot. Rev.* **6**, 599−635, 1940.
49.   Dix, P.J. and Street, H.E. Sodium chloride-resistant cultured cell lines from *Nicotiana sylvestris* and
        Capsicum annuum. Plant Sci. Letters **5**, 231−237, 1975.
50.   Dore, C. Production de plantes homozygotes males et femelles a partir d'antheres d'asperge cultivee
        in vitro (*Asparagus officinalis* L.). *C.R. Acad. Sci. Paris e.* **278** (Serie D), 2135−2138, 1974.

51. Dublin, P. *Les haploides de Theobroma cacao L. diploidisation et obtention d'individus homozygotes.* Cafe Cacao The **18**, 83–96, 1974.

52. Dunwell, J.M. and Perry, M.E. The influence of *in vivo* growth conditions of *N. tabacum* plants on the *in vitro* embryogenic potential of their anthera. *John Innes Annual Report No. 64*, 1973.

53. Dunwell, J.M. and Sunderland, N. Anther culture of *Solanum tuberosum* L. *Euphytica* **22**, 317–323, 1973.

54. Durr, A. and Fleck, J. Production of haploid plants of *Nicotiana langsdorffii. Plant Sci. Lett.* **18**, 75–79, 1980.

55. Ecochard, R., Ramanna, M.S. and de Nettancourt, D. Detection and cytological analysis of tomato haploids. *Genetica* **40**, 181–189, 1969.

56. Engvild, K.C. Plantlet ploidy and flower-bud size in tobacco anther cultures. *Hereditas* **76**, 320–322, 1974.

57. Engvild, K.C., Linde-Laursen, I.B. and Lundqvist, A. Anther cultures of *Datura innoxia*: flower bud stage and embryoid level of ploidy. *Hereditas* **72**, 331–332, 1972.

58. Evans, D.A. and Gamborg, O.L. Effects of para-fluorophenylalanine on ploidy levels of cell suspension cultures of *Datura innoxia. Environ. Experiment. Bot.* **19**, 269–275, 1979.

59. Facciotti, D. and Pilet, P.E. Plants and embryoids from haploid *Nicotiana sylvestris* protoplasts. *Plant Sci. Lett.* **15**, 1–6, 1979.

60. Feaster, C.V. and Turcotte, E.L. Yield stability in doubled haploids of American Pima cotton. *Crop Sci.* **13**, 232–233, 1973.

61. Foroughi-Wehr, B., Wilson, H.M., Mix, G. and Gaul, H. Monoploid plants from anthers of a dihaploid genotype of *Solanum tuberosum* L. *Euphytica* **26**, 361–367, 1977.

62. Frandsen, N.O. Haploidproduktion aus einem kartofellzuchtmaterial mit intensiver wildarteinkreuzung. *Zuchter* **27**, 120–134, 1967.

63. Furner, I.J., King, J. and Gamborg, O.L. Plant regeneration from protoplasts isolated from predominantly haploid suspension cultures of *Datura innoxia* (Mill.). *Plant Sci. Lett.* **11**, 169–176, 1978.

64. Gabert, A.C. Factors influencing the frequency of haploids in the common potato (*Solanum tuberosum* L.) *Ph.D. Thesis*, Univ. of Wisconsin, Madison, 1963.

65. Gabert, A.C., Hougas, R.W. and Peloquin, S.J. Heritability of the pollinator effect on haploid frequency in the common potato, *Solanum tuberosum* L. *Agronomy Abstracts*, p. 80, 1963.

66. Gamburg, K.Z. and Vysotskaya, E.F. Effect of colchicine on growth of tobacco tissue in suspension culture. *Tsitologiya* **14**, 1188–1191, 1972.

67. Gelfant, S. Inhibition of cell division: a critical and experimental analysis. *Int. Rev. Cytol.* **14**, 1–39, 1963.

68. Greenleaf, W.H. Induction of polyploidy in *Nicotiana* by heteroauxin treatment. *J. Heredity* **29**, 451–464, 1938.

69. Gresshoff, P.M. and Doy, C.H. Development and differentiation of haploid *Lycopersicon esculentum* (tomato). *Planta* **107**, 61–170, 1972a.

70. Gresshoff, P.M. and Doy, C.H. Haploid *Arabidopsis thaliana* callus and plants from anther culture. *Australian J. Biol. Sci.* **25**, 259–264, 1972b.

71. Griffing, B. Efficiency change due to use of doubled-haploids in recurrent selection methods. *Theor. Appl. Genet.* **46**, 367–385, 1975.

72. Grimsley, N.H., Ashton, N.W. and Cove, D.J. The production of somatic hybrids by protoplast fusion in the moss, *Physcomitrella patens. Mol. Gen. Genet.* **254**, 97–100, 1977.

73. Guha, S. and Maheshwari, S.C. *In vitro* production of embryos from anther of *Datura. Nature* **204**, 497, 1964.

74. Guha, S. and Maheshwari, S.C. Cell division and differentiation of embryos in the pollen grains of *Datura in vitro. Nature* **212**, 97–98, 1966.

75. Guha, S. and Maheshwari, S.C. Development of embryos from pollen grains of *Datura in vitro. Phytomorphology* **17**, 454–461, 1967.

76. Guha, S., Iyer, R.D., Gupta, H. and Swaminathan, M.S. Totipotency of gametic cells and the production of haploids in rice. *Current Science* **39**, 174–176, 1970.

77. Guha-Mukherjee, S. Genotypic differences in the *in vitro* formation of embryoids from rice pollen. *J. Exp. Bot.* **24**, 139–144, 1973.

78. Guo, C. Effects of chemical and physical factors on the chromosome number in *Nicotiana* anther cultures. *In Vitro* **7**, 381–386, 1972.

79. Gupta, S.B. Duration of mitotic cycle and regulation of DNA replication in *Nicotiana plumbaginifolia* and a hybrid derivative of *N. tabacum* showing chromosome instability. *Can. J. Genet. Cytol.* **11**, 133–142, 1969.

80. Gupta, S.B. and Gupta, P. Selective elimination of *Nicotiana glutinosa* chromosomes in the F_1 hybrids of *N. suaveolens* and *N. glutinosa*. *Genetics* **73**, 605–612, 1973.

81. Heinz, D.J. and Mee, G.W.P. Colchicine-induced polyploids from cell suspension cultures of sugarcane. *Crop. Sci.* **10**, 696–699, 1970.

82. Hess, D. Versuche zur transformation an hoheren pflanzen: Induktion und konstante weitergabe der anthocyansynthese bei *Petunia hybrida*. *Z. Pflanzenphysiol.* **60**, 348–358, 1969.

83. Hess, D. Cell modification by DNA uptake. *Applied and Fundamental Aspects of Plant Cell, Tissue and Organ Culture*, 506–535. Reinert, J. and Bajaj, Y.P.S. (eds.). Springer-Verlag, Berlin, 1977.

84. Ho, K.M. and Jones, G.E. Mingo barley. *Can. J. Plant Sci.* **60**, 279–280, 1980.

85. Ho, K.M. and Kasha, K.J. Genetic control of chromosome elimination during haploid formation in barley. *Genetics* **81**, 263–275, 1975.

86. Holl, F.B. Molecular genetic modification of legumes. *Workshop on Uses of Molecular Genetic Modification on Eukaryotes*. Univ. Minn., 1975.

87. Hondelmann, W. and Wilberg, B. Breeding allmale-varieties of Asparagus by utilization of anther and tissue culture. *Z. Pflanzenzuchtg.* **69**, 19–24, 1973.

88. Hougas, R.W. and Peloquin, S.J. A haploid plant of the potato variety Katahdin. *Nature* **180**, 1209–1210, 1957.

89. Hougas, R.W., Peloquin, S.J. and Gabert, A.C. Effect of seed-parent and pollinator on frequency of haploids in *Solanum tuberosum*. *Crop Sci.* **4**, 593–595, 1964.

90. Hougas, R.W., Peloquin, S.J. and Ross, R.W. Haploids of the common potato. *J. Heredity* **47**, 103–107, 1958.

91. Howard, H.W. *Genetics of the potato, Solanum tuberosum* 126. Springer-Verlag, New York, 1970.

92. Hughes, K.W., Bell, S.L. and Caponetti, J.D. Anther-derived haploids of the African Violet. *Can. J. Bot.* **53**, 1442–1444, 1975.

93. Irikura, Y. Induction of haploid plants by anther culture in tuber bearing species and interspecific hybrids of *Solanum*. *Potato Res.* **18**, 133–140, 1975.

94. Irikura, Y. and Sakaguchi, S. Induction of 12-chromosome plants from anther culture in a tuberous *Solanum*. *Potato Res.* **15**, 170–173, 1972.

95. Iyer, R.D. and Raina, S.K. The early ontogeny of embryoids and callus from pollen and subsequent organogenesis in anther cultures of *Datura metel* and rice. *Planta* **104**, 146–156, 1972.

96. Jacobsen, E. Doubling dihaploid potato clones via leaf tissue culture. *Z. Pflanzenzuchtg.* **80**, 80–82, 1978a.

97. Jacobsen, E. Haploid, diploid, and triploid parthenogenesis in interspecific crosses between *Solanum tuberosum* interdihaploids and *S. phyreja*. *Potato Res.* **21**, 15–17, 1978b.

98. Jensen, C.J. Techniques in haploids. *Haploids in Higher Plants: Advance and Potential*, 153–190. Kasha, K.J. (ed.). Univ. of Guelph Press, Ontario, Canada, 1974.

99. Jorgensen, C.A. The experimental formation of heteroploid plants in the genus *Solanum*. *J. Genet.* **19**, 133–211, 1928.

100. Kamata, M. and Sakamoto, S. Production of haploid albino plants of Aegilops by anther culture. *Jap. J. Genet.* **47**, 61–63, 1972.

101. Kameya, T. and Hinata, K. Induction of haploid plants from pollen grains of *Brassica*. *Japan J. Breeding* **20**, 82–87, 1970.

102. Kasha, K.J. Haploids from somatic cells. *Haploids in Higher Plants: Advances and Potential*, 67–87. Kasha, K.J. (ed.), Univ. of Guelph Press, Ontario, Canada, 1974.

103. Kasha, K.J. and Kao, K.N. High frequency haploid production in barley (*Hordeum vulgare* L.). *Nature* **225**, 874–876, 1970.

104. Kasha, K.J. and Reinbergs, E. Achievements with haploids in barley research and breeding. *The Plant Genome*, 215–230. Davis, D.R. and Hopwood, D.A. (eds.). The John Innes Charity, Norwich, 1980.

105. Kasha, K.J., Reinbergs, E., Johns, W.A., Subrahmanyan, N.C. and Ho, K.M. Barley haploid studies. *Barley Genet. Newsl.* **2**, 36–41, 1972.

106. Kasperbauer, M.J., Buckner, R.C. and Springer, W.D. Haploid plants by anther-panicle culture of tall fescue. *Crop Sci.* **20**, 103—107, 1980.

107. Kasperbauer, M.J. and Collins, G.B. Reconstitution of diploid from leaf tissue of anther-derived haploids in tobacco. *Crop Sci.* **12**, 98—101, 1972.

108. Keller, W.A. and Armstrong, K.C. Stimulation of embryogenesis and haploid production in *Brassica campestris* anther cultures by elevated temperature treatments. *Theor. Appl. Genet.* **55**, 65—67, 1979.

109. Keller, W.A., Rajhathy, T. and Lacapra, J. *In vitro* production of plants from in *Brassica campestris*. *Can. J. Genet. Cytol.* **17**, 655—666, 1975.

110. Kermicle, J.L. Androgenesis conditioned by a mutation in maize. *Science* **166**, 1422—1424, 1969.

111. Kimber, G. and Riley, R. Haploid angiosperms. *Bot. Rev.* **29**, 480—531, 1963.

112. King, J., Horsch, R.B. and Savage, A.D. Partial characterization of two stable auxotrophic cell strains of *Datura innoxia* Mill. *Planta* **149**, 480—484, 1980.

113. Kleinhofs, A. and Behki, R. Prospects for plant genome modification by nonconventional methods. *Ann. Rev. Genet.* **11**, 79—101, 1977.

114. Kochhar, T., Sabharwal, P. and Engelberg, J. Production of homozygous diploid plants by tissue culture technique. *J. Heredity* **62** (1), 59—61, 1971.

115. Kohlenbach, H.W. and Geier, T. Embryonen aus *in vitro* kultivierten antheren von *Datura meteloides* Dun., *Datura wrightii* Regel und *Solanum tuberosum* L. *Z. Pflanzenphysiol.* **67**, 161—165, 1972.

116. Lacadena, J.R. Spontaneous and induced parthenogenesis and androgenesis. *Haploids in Higher Plants: Advances and Potential*, 12—32. Kasha, K.J. (ed.). Univ. of Guelph Press, Guelph, 1974.

117. Lange, W. Crosses between *Hordeum vulgare* L. and *H. bulbosum* L. II. Elimination of chromosomes in hybrid tissues. *Euphytica* **20**, 181—194, 1971.

118. Langton, F.A. Investigations with 2x and vegetatively doubled 4x potato clones. *Potato Res.* **16**, 323—324, 1973.

119. Lawyer, A.L., Berlyn, M.B. and Zelitch, I. Isolation and characterization of glycine hydroxamate-resistant cell lines of *Nicotiana tabacum*. *Plant Physiol.* **66**, 334—341, 1980.

120. Ledoux, L., Huart, R. and Jacobs, M. DNA-mediated genetic correction of thiamineless *Arabidopsis thaliana*. *Nature* **249**, 17—21, 1974.

121. Legg, P.D. and Collins, G.B. Genetic variances in a random-intercrossed population of burley tobacco. *Crop Sci.* **14**, 805—808, 1974.

122. Lindstrom, E.W. and Koos, K. Cytogenetic investigations of a haploid tomato and its diploid and tetraploid progeny. *Amer. J. Bot.* **18**, 398—410, 1931.

123. Lurquin, P.F. and Kado, C.I. *Escherichia coli* plasmid pBR313 insertion into plant protoplasts and into their nuclei. *Mol. Gen. Genet.* **154**, 113—121, 1977.

124. Magoon, M.L. and Khanna, K.R. Haploids. *Caryologia* **16**, 191—235, 1963.

125. Malhotra, K. and Maheshwari, S.C. Enhancement by cold treatment of pollen embryoid development in *Petunia hybrida*. *Z. Pflanzenphysiol.* **85**, 177—180, 1977.

126. Maliga, P., Brezonovits, A. and Marton, L. Streptomycin resistant plants from callus culture of haploid tobacco. *Nature New Biol.* **244**, 29—30, 1973.

127. Maliga, P., Xuan, L.T., Dix, P.J. and Cseplo, A. Antibiotic resistance in *Nicotiana*. *Plant Cell Cultures: Results and Perspectives*, 161—166. Elsevier/N. Holland, 1980.

128. Malmberg, R.L. Temperature sensitive variants of *Nicotiana tabacum* isolated from somatic cell culture. *Genetics* **92**, 215—221, 1979.

129. Malmberg, R.L. Biochemical, cellular, and developmental characterization of a temperature-sensitive mutant of *Nicotiana tabacum* and its second site revertant. *Cell* **22**, 603—609, 1980.

130. Mastrangelo, I.A. and Smith, H.H. Selection and differentiation of aminopterin resistant cells of *Datura innoxia*. *Plant Sci. Lett.* **10**, 171—179, 1977.

131. Matthews, B., Dray, S., Widholm, J. and Ostro, M. Liposome-mediated transfer of bacterial RNA into carrot protoplasts. *Planta* **145**, 37—44, 1979.

132. McComb, J.A. and McComb, A.J. The cytology of plantlets derived from cultured anthers of *Nicotiana sylvestris*. *New Phytol.* **79**, 679—688, 1977.

133. Melchers, G. Haploid higher plants for plant breeding. *Z. Pflanzenzuchtg.* **67**, 19—32, 1972.

134. Melchers, G. and Labib, G. Somatic hybridization of plants by fusion of protoplasts. I. Selection of light resistant hybrids of "haploid" light sensitive varieties of tobacco. *Molec. Gen. Genet.* **135**, 277—294, 1974.

135. Mendiburo, A.O. and Peloquin, S.J. Gene-centromere mapping by 4x — 2x matings in potatoes. *Theoretical and Applied Genetics* **54**, 177—180, 1979.

136. Meyer, J.R. and Justus, N. Properties of doubled haploids of cotton. *Crop Sci.* **1**, 462—464, 1961.

137. Moav, R. Genetic instability in *Nicotiana* hybrids. II. Studies of the Ws(pbg) locus of *N. plumbaginifolia* in *N. tabacum* nuclei. *Genetics* **46**, 1069—1088, 1961.

138. Mokhtarzadeh, A. and Constantin, M.J. Plant regeneration from hypocotyl- and anther-derived callus of berseem clover. *Crop Sci.* **18**, 567—572, 1978.

139. Montelongo-Escobedo, H. Factors influencing haploid frequency in tetraploid *Solanum tuberosum*. *Ph.D. Thesis*, Univ. of Wisconsin, Madison, 1968.

140. Montelongo-Escobedo, H. and Rowe, P.R. Haploid induction in potato: Cytological basis for the pollinator effect. *Euphytica* **18**, 116—123, 1969.

141. Morrison, G. The occurrence and use of haploid plants in the tomato with special reference to the variety Marglobe. *Proc. 6th Intern. Congr. Genet.* **2**, 137—139, 1932.

142. Muller, A.J. and Grafe, R. Isolation and characterization of cell lines of *Nicotiana tabacum* lacking nitrate reductase. *Molec. Gen. Genet.* **161**, 67—76, 1978.

143. Murton, L., Willems, G.J., Molendiyk, L. and Schilperoort, R.A. *In vitro* transformation of cultured cells from *Nicotiana tabacum* by *Agrobacterium tumefaciens*. *Nature* **277**, 129—131, 1979.

144. Nakamura, A., Yamada, T., Kadontani, N. and Itagaki, R. Improvement of flue-cured tobacco variety MC 1610 by means of haploid breeding method and investigations on some problems of this method. *Haploids in Higher Plants, Advances and Potential*, 297—298. Kasha, K.J. (ed.), Univ. of Guelph Press, Ontario, Canada, 1974a.

145. Nakamura, A., Yamada, T., Kadotani, N., Itagaki, R. and Oka, M. Studies on the haploid method of breeding in tobacco. *Sabrao J.* **6**, 107—131, 1974b.

146. Nakata, K. and Kurihara, T. Competition among pollen grains for haploid tobacco plant formation by anther culture. II. Analysis with resistance to tobacco virus (TMV) and wildfire diseases, leaf color, and leaf base shape characters. *Japan J. Breed.* **22**, 92—98, 1972.

147. Nei, M. The efficiency of the haploid method of plant breeding. *Heredity* **18**, 95—100, 1963.

148. Newcomer, E.H. A colchicine-induced homozygous tomato obtained through doubling clonal haploids. *Proc. Amer. Soc. Hort. Sci.* **38**, 610—612, 1941.

149. Niizeki, H. and Oono, K. Rice plants obtained by anther culture. *Les Cultures de Tissues de Plantes, Colloques Intern. du C.N.R.S., Paris* d. **193**, 251—257, 1971.

150. Nishiyama, I. Monosomic analysis of *Avena byzantina* C. Koch var. Kanota. *Z. Pflanzenzuchtg.* **63**, 41—55, 1970.

151. Nitsch, C. La culture de pollen sur millieu synthetique. *Compt. Rend. Acad. Sci., Paris* D, **278**, 1031—1034, 1974a.

152. Nitsch, C. Pollen culture: A new technique for mass production of haploid and homozygous plants. *Haploids in Higher Plants, Advances and Potential*, 123—135. Kasha, K.J. (ed.). Univ. Guelph, Press, Ontario, Canada, 1974b.

153. Nitsch, C. and Norreel, B. Factors favoring the formation of androgenetic embryos in anther culture. *Genes, Enzymes and Populations, Vol. 2*. 128—144. Srb, A. (ed.). Plenum Press, 1972.

154. Nitsch, C. and Norreel, B. Effect d'un choc thermique sur le pouvoir embryogene du pollen de *Datura* cultive dans l'anthere ou isole de l'anthere. *Compt. Rend. Acad. Sci.* Paris D, **276**, 303—306, 1973.

155. Nitsch, J.P. and Nitsch, C. Haploid plants from pollen grains. *Science* **163**, 85—87, 1969.

156. Nitsch, J.P. *La production in vitro d'embryons haploides: resultats et perspectives. Colloque internationaux C.N.R.S. no. 193, Les Cultures de Tissue de Plantes*, 282—294, 1971.

157. Nitsch, J.P. Haploid plants from pollen. *Z. Pflanzenzuchtg.* **67**, 3—18, 1972.

158. Nitzsche, W. and Wenzel, G. *Haploids in plant breeding*. Verlag Paul Parey, 1977.

159. Novak, F.J. and Vyskot, B. Cytology and pollen fertility of *Nicotiana tacacum* L. haploids derived from anther and tissue cultures. *Beitr. Biol. Pflanzen* **54**, 329—351, 1978.

160. Ohyama, K. and Nitsch, J.P. Flowering haploid plants obtained from protoplasts of tobacco leaves. *Plant Cell Physiol.* **13**, 229—236, 1972.

161. Ostergren, G. Polyploids and aneuploids of *Crepis capillaris* by treatment with nitrous oxide. *Genetica* 27, 54—64, 1954.

162. Ouyang, T.W., Hu, H., Chuang, C. and Tseng, C.C. Induction of pollen plants from anthers of *Triticum aestivum* L. cultured *in vitro*. *Scientia Sinica* 16, 79—90, 1973.

163. Owings, A.D., Sarvella, P. and Meyer, J.R. Twinning and haploidy in a strain of *Gossypium barbadense* L. *Crop Sci.* 4, 652—653, 1964.

164. Pagniez, M. and Demarly, Y. Obtention d'invididus androgenetiques par culture *in vitro* d'antheres de Ray-grass d'Itabe (*Lolium multiflorum* Lam.). *Ann. Amelior. Plantes* 29, 631—637, 1979.

165. Park, S.J., Walsh, E.J., Reinbergs, E., Song, L.S.P. and Kasha, K.J. Field performance of doubled haploid barley lines in comparison with lines developed by the pedigree and single seed descent methods. *Can. J. Plant Sci.* 56, 467—474, 1976.

166. Pelletier, G. Les conditions et les premiers stades de l'androgenese *in vitro* chez *Nicotiana tabacum*. *Mem. Soc. Botan. Fr. Colloq. Morphologie*, 261—268, 1973.

167. Peloquin, S.J., Hougas, R.W. and Gaberg, A.C. Haploidy as a new approach to the cytogenetics and breeding of *Solanum tuberosum*. *Chromosome Manipulations and Plant Genetics*, 21—28. Riley, R. and Lewis, K.R. (eds.), Oliver and Boyd, Edinburgh and London, 1966.

168. Pochard, E. and Dumas de Vaulx, R. La monoploidie chez le piment (*Capsicum annuum* L.). Realisation pratique d'un cycle de selection accelere par passage a l'etat monoploide en troisieme generation. *Z. Pflanzenzuchtg.* 65, 23—46, 1971.

169. Radojevic, L. *In vitro* induction of androgenic plants in *Aesculus hippocastanum*. *Protoplasma* 96, 369—374, 1978.

170. Raghavan, V. Induction of haploid plants from anther cultures of Henbane. *Z. Pflanzenphysiol.* 76, 89—92, 1975.

171. Raghavan, V. Role of the generative cell in androgenesis of Henbane. *Science* 191, 388—389, 1976.

172. Rajasekaran, K. and Mullins, M.G. Embryos and plantlets from cultured anthers of hybrid grapevines. *J. Exp. Bot.* 30, 399—407, 1979.

173. Raquin, C. and Pilet, V. Production de plantules a partir d'antheres de *Petunias* cultivees *in vitro*. *Compt. Rend. Acad. Sci.* Paris 247, 1019—1022, 1972.

174. Rashid, A. and Street, H.E. The development of haploid embryoids from anther cultures of *Atropa belladonna* L. *Planta* 113, 263—270, 1973.

175. Rashid, A. and Street, H.E. Growth, embryogenic potential and stability of a haploid cell line of *Atropa belladonna* L. *Plant Sci. Lett.* 2, 89—94, 1974.

176. Reinbergs, E., Park, S.J. and Kasha, K.J. The haploid technique in comparison with conventional methods as barley breeding. *Barley Genetics III,* 346—350. Gaud, H. (ed.), 1975.

177. Rieger, R., Michaelis, A. and Green, M.M. *A glossary of genetics and cytogenetics*, 507. Springer-Verlag, New York, 1968.

178. Rosati, P., Devreux, M. and Laneri, V. Anther culture of strawberry. *Hort. Science* 10, 119—120, 1975.

179. Rowe, P.R. Methods of producing haploids: Parthenogenesis following interspecific hybridization. *Haploids in Higher Plants: Advances and Potential*, 43—52. Kasha, K.J. (ed.). Univ. of Guelph Press, Ontario, Canada, 1974.

180. Sacristan, M.D. Karyotypic changes in callus cultures from haploid plants of *Crepis capillaris* (L.) Wallr. *Chromosoma* 33, 273—283, 1971.

181. Savage, A., King, J. and Gamborg, O.L. Recovery of a pantothenate auxotroph from a cell suspension culture of *Datura innoxia* Mill. *Plant Sci. Lett.* 16, 367—376, 1979.

182. Schieder, O. Regeneration of haploid and diploid *Datura innoxia* Mill. mesophyll protoplasts to plants. *Z. Pflanzenphysiol.* 76, 462—466, 1975.

183. Schieder, O. Attempts in regeneration of mesophyll protoplasts of haploid and diploid wild type lines, and those of chlorophyll-deficient strains from different *Solanaceae*. *Z. Pflanzenphysiol.* 84, 275—281, 1977.

184. Scholl, R.L. and Amos, J.A. Isolation of doubled-haploid plants through anther culture in *Arabidopsis thaliana*. *Z. Pflanzenphysiol.* 96, 407—414, 1980.

185. Scowcroft, W.R. Somatic cell genetics and plant improvement. *Advances in Agron.* 29, 39—81, 1977.

186. Sears, E.R. The aneuploids of common wheat. *Mo. Agric. Exp. Sta. Res. Bull.* 572, 58, 1954.

187. Sharp, W.R., Raskin, R.S. and Sommer, H.E. Haploids in *Lilium*. *Phytomorphology* **21**, 334–337, 1971.

188. Sharp, W.R., Raskin, R.S. and Sommer, H.E. The use of nurse culture in the development of haploid clones in tomato. *Planta* **104**, 357–361, 1972.

189. Shepard, J.F. Mutant selection and plant regeneration from potato mesophyll protoplasts. *Genetic Improvement of Crops*, 185–219. Rubenstein *et al.* (eds.). Univ. of Minn. Press, Minneapolis, 1980.

190. Shu-hua, M., Chung-Shen, K., Yao-lin, K., Au-tze, S., Shu-yung, K., Wen-liang, L., Yu-ying, W., Man-ling, C., Mao-kang, W. and Ling, H. Induction of pollen plants of maize and observations on their progeny. *Proceedings of Symposium on Plant Tissue Culture*, 23–34. Science Press, Peking, 1978.

191. Sinha, S., Roy, R.P. and Jha, K.K. Callus formation and shoot bud differentiation in anther culture of *Solanum surattense*. *Can. J. Bot.* **57**, 2524–2527, 1979.

192. Sink, K.C. Jr. and Padmanabhan, V. Anther and pollen culture to produce haploids: Progress and application for the plant breeder. *Hort. Science* **12**, 143–148, 1977.

193. Snape, J.W., Chapman, V., Moss, J., Blanchard, C.E. and Miller, T.E. The crossabilities of wheat varieties with *Hordium bulbosum*. *Heredity* **42**, 291–298, 1979.

194. Song, L.S.P., Park, S.J., Reinbergs, E., Choo, T.M. and Kasha, K.J. Doubled haploid vs. the bulk plot method for production of homozygous lines in barley. *Z. Pflanzenzucht.* **81**, 271–280, 1978.

195. Sopory, S.K. and Maheshwari, S.C. Production of haploid embryos by anther culture technique in *Datura innoxia*, a further study. *Phytomorphology* **22**, 87–90, 1973.

196. Subrahmanyam, N.C. Haploidy from *Hordeum* interspecific crosses. I. Polyhaploids of *H. parodii* and *H. procerum*. *Theor. Appl. Genet.* **49**, 209–217, 1977.

197. Subrahmanyam, N.C. Haploidy from *Hordeum* interspecific crosses. 2: Dihaploids of *H. brachy-antherum* and *H. depressum*. *Theor. Appl. Genet.* **55** 2, 139–144, 1979.

198. Subrahmanyam, N.C. Haploidy from *Hordeum* interspecific crosses. Trihaploids of *H. arizonicum* and *H. lechleri*. *Theor. Appl. Genet.* **56** 3, 257–263, 1980.

199. Subrahmayam, N.C. and Kasha, K.J. Selective chromosomal elimination during haploid formation in barley following interspecific hybridization. *Chromosoma* **42**, 111–125, 1973.

200. Subrahmanyam, N.C. and Kasha, K.J. Chromosome doubling of barley haploids by nitrous oxide and colchicine treatments. *Can. J. Genet. Cytol.* **17**, 573–583, 1975.

201. Sun, C.S., Wang, C.C. and Chu, Z.C. Cell division and differentiation of pollen grains in *Triticale* anthers cultured *in vitro*. *Scientia Sinica* **17** (1), 47–54, 1974.

202. Sunderland, N. Pollen plants and their significance. *New Scientist* **16**, 142–143, 1970.

203. Sunderland, N. Pollen and anther culture. *Plant Tissue and Cell Culture*, 205–238. Street, H.E. (ed.). Univ. California Press, 1973.

204. Sunderland, N. Anther culture as a means of haploid induction. *Haploids in Higher Plants – Advances and Potential*, 91–122. Kasha, K.J. (ed.). Univ. of Guelph, Ontario, Canada, 1974.

205. Sunderland, N. Comparative studies of anther and pollen culture. *Plant Cell and Tissue Culture – Principles and Applications*. Ohio State Univ. Press, Columbus, 1978.

206. Sunderland, N., Collins, G.B. and Dunwell, J.M. The role of nuclear fusion in pollen embryogenesis of *Datura innoxia* Mill. *Planta* **117**, 227–241, 1974.

207. Sunderland, N. and Wicks, F.M. Cultivation of haploid plants from tobacco pollen. *Nature* **224**, 1227–1229, 1969.

208. Sunderland, N. and Wicks, F.M. Embryoid formation in pollen grains of *Nicotiana tabacum*. *J. Expt. Bot.* **22**, 213–226, 1971.

209. Taylor, N.L., Anderson, M.K., Quesenberry, K.H. and Watson, L. Doubling the chromosome number of *Trifolium* species using nitrous oxide. *Crop Sci.* **16**, 516–518, 1976.

210. Thevenin, L. Haploids in *Asparagus* breeding. *Haploids in Higher Plants: Advances and Potentials*, 279. Kasha, K.J. (ed.), Univ. of Guelph Press, Ontario, Canada, 1974.

211. Thiebaut, J. and Kasha, K.J. Modification of the colchicine technique for chromosome doubling of barley haploids. *Can. J. Genet. Cytol.* **20**, 513–521, 1978.

212. Thomas, E. and Wenzel, G. Embryogenesis form microspores of *Brassica napus*. *Z. Pflanzenzuchtg.* **74**, 77–81, 1975.

213. Thomas, E., King, P.J. and Potrykus, I. Improvement of crop plants via single cells *in vitro*: an assessment. *Z. Pflanzenzuchtg.* **82**, 1–30, 1979.

214. Thompson, K.F. Homozygous diploid lines from naturally occurring haploids. 4. *Intern. Rapskongr.* Giessen, 119–124, 1974.

215. Tomes, D.T. and Collins, G.B. Factors affecting haploid plant production from *in vitro* anther cultures of *Nicotiana* species. *Crop Science* **16**, 837–840, 1976.

216. Tsay, H. and Tseng, M. Embryoid formation and plantlet regeneration from anther callus of sweet potato. *Bot. Bull. Academia Sinica* **20**, 117–122, 1979.

217. Turcotte, E.L. and Feaster, C.V. Haploids: High-frequency production from single-embryo seeds in a line of Pima cotton. *Science* **140**, 1407–1408, 1963.

218. Turcotte, E.L. and Feaster, C.V. Semigamy in Pima Cotton. *J. Heredity* **58**, 54–57, 1967.

219. Turcotte, E.L. and Feaster, C.V. Semigametic production of haploids in Pima cotton. *Crop Sci.* **9**, 653–655, 1969.

220. Turcotte, E.L. and Feaster, C.V. The origin of 2n and n sectors of ch chimeral Pima cotton plants. *Crop Sci.* **13**, 111–112, 1973.

221. Turcotte, E.L. and Feaster, C.V. Methods of producing haploids: Semigametic production of cotton haploids. *Haploids in Higher Plants: Advances and Potential*, 53–64. Kasha, K.J. (ed.), Univ. of Guelph Press, Ontario, Canada, 1974.

222. Turcotte, E.L. and Feaster, C.V. Comparison of doubled haploids and Pima S-5 in American Pima cotton. *The Plant Genome*, Davis, D.R. and Hopwood, D.A. (eds.). The John Innes Charity, Norwich 248, 1980.

223. Vyskot, B. and Novak, F.J. Experimental androgenesis *in vitro* in *Nicotiana clevelandii* Gray and *N. sanderae* Hort. *Theoret. Appl. Genet.* **44**, 138–140, 1973.

224. Wang, C.C., Chu, C.C., Sun, C.S. and Wu, S.H. The androgenesis in wheat (*Triticum aestivum*) anthers cultured *in vitro*. *Scientia sinica* **21**, 218–222, 1973a.

225. Wang, C.C., Chu, C.C., Sun, C.S., Wu, S.H., Yin, K.S. and Hsu, C. The androgenesis in wheat (*Triticum aestivum*) anthers cultivated *in vitro*. *Scientia Sinica* **16(2)**, 218–225, 1973b.

226. Wang, C.C., Sun, C.S. and Chu, Z.C. On the conditions for the induction of rice pollen plantlets and certain factors affecting the frequency of induction. *Acta Botanica Sinica* **16(1)**, 43–53, 1974.

227. Wang, Y.Y., Sun, C.S., Wang, C.C. and Chien, N.F. The induction of the pollen plantlets of *Triticale* and *Capsicum annuum* from anther culture. *Scientia Sinica* **16(1)**, 147–151, 1973c.

228. Wark, D.C. Doubled haploids in Australian tobacco breeding. *Proc. 3rd International Congress of SABRAO* **2**, 19–23, 1977.

229. Weatherhead, M.A. and Henshaw, G.G. The production of homozygous diploid plants of *Solanum verrucosum* by tissue culture techniques. *Euphytica* **28**, 765–768, 1979.

230. Wenzel, G., Hoffman, F. and Thomas, E. Anther culture as a breeding tool in rape. I. Ploidy level and phenotype of androgenetic plants. *Z. Pflanzenzuchtg.* **78**, 149–155, 1977.

231. Wenzel, G. and Thomas, E. Observations on the growth in culture of anthers of *Secale cereale*. *Z. Pflanzenzuchtg.* **72**, 89–92, 1974.

232. Wernicke, W. and Kohlenbach, H.W. Anther cultures in the genus Scopolia. *Z. Pflanzenphysiol.* **77**, 89–93, 1975.

233. Wernicke, W., Lorz, H. and Thomas, E. Plant regeneration from leaf protoplasts of haploid *Hyoscyamus muticus* L. produced via anther culture. *Plant Sci. Lett.* **15**, 239–250, 1979.

234. Yin, K., Hsu, Ch., Chu, Y., Pi, F.Y., Wang, S.T., Liu, T.Y., Chu, C.C., Wang, C.C. and Sun, C.S. A study of the new cultivar of rice raised by haploid breeding method. *Scientia Sinica* **19**, 227–242, 1976.

235. Zenkteler, M. *In vitro* production of haploid plants from pollen of *Atropa belladonna*. *Experientia* **27**, 1087, 1971.

236. Zenkteler, M. Development of haploid albinotic plants of *Bromus inermis* from anthers cultured *in vitro*. *Haploid Information Service* **15**, 6–7, 1975.

CHAPTER 13

BREEDING FOR ENHANCED PROTEIN

H.P. Müller
Institute of Genetics,
University of Bonn, FRG

INTRODUCTION

Cereal grains and legume seeds supply about 70% and 18%, respectively, to the plant protein requirement in human nutrition.[94] Consequently, publications being concerned with the improvement of the seed proteins are numerous and diverse, because of the fact that both groups are deficient in different essential amino acids. They include research-work on the genetic variability in the seed protein composition and content[1,3,4,15,42,48,52,63,66,88,114,127,130,178,179], as well as the manifold attempts to analyse the biochemical basis of the different compounds[18,19,20,23,75,82,83,126,127,128,189,190,191,192], the attempts to utilize mutant genes for analysing the inheritance of distinct biochemical traits[28,36,37,38,65,66,67,86,126], and the breeding experiments for elucidating the regulation of gene activities which are widely influenced by environmental factors[22,84,86,91,98,126,131]. The large variety of findings in this research field makes it necessary to pick out some crucial points in order to show some trends for further work. For explanation the examples selected are belonging to the family of *Leguminosae*, referring preferentially to economically important members of this family. This is due to the fact that legume seed proteins as characters of the respective species have been used extensively for estimating their ranges of variation. This work was done mainly within the frame-work of international research programs. The data obtained reveal numerous and highly interesting findings, contributing to the understanding of the basic mechanisms governing the synthesis and accumulation of seed proteins. The data also reveal that numerous other physiologically active nitrogenous constituents must be considered[1,3,9,13,15,20,52,91,96,98,110,114,119,156,168,179,188].

The seeds as differentiated organs constitute the final stage of an integrated developmental process. The genetic informations for storage proteins are repressed in all tissues except for certain developmental stages in the cotyledons when they are actively expressed[118,121,136] Analyses of the kinetics of protein accumulation in seeds reveal that numerous biological processes are involved in constituting protein quantity and quality among genotypes within species and greater taxonomic groups. Furthermore, these processes are influenced markedly by environmental factors, contributing to the site-to-site and year-to-year variations, which are well-known, while the effects of specific environmental factors have not been studied extensively.[64]

Total protein content

It is well established that the protein content of seeds is encoded in multigene families. In order to get an insight into the range of protein values found in seeds, they are firstly screened for total nitrogen content, as evaluated by the classical Kjeldahl method. Several genera of the *Leguminosae* which are important for human nutrition have been studied in detail, mainly with the aim of improving the seed protein quantity and quality. They include *Cajanus*, *Glycine*, *Phaseolus*, *Vigna*, *Vicia*, *Pisum*, *Lathyrus*, *Lens*, *Arachis*, *Cicer* and *Lupinus*. Values evaluated in different places are given in Table 13.1.

Considering the phenotypic expression of the trait seed protein content in the different leguminous genera on the one hand the great variability is evident, depending on the respective genotypes and environmental conditions. But on the other hand the correspondence of the seed protein percentages within different genera, grown at diverse locations is somewhat astonishing. Even if in some cases the environmental effects are reported to be greater than the genotypic effects[106,162] the data indicate a narrow genetic basis. The selection work on pulses, obviously, has led to a considerable conservation of the genetic informations for storage proteins. The term genotype was chosen because for most of the crop plants analysed the terms variety or cultivar are often without precise definition. They may involve clones or pure lines, selected hybrid lines from crosses between established varieties, land races and primitive forms.

The observed variation in seed protein contents by inducing mutations did not show large advantages with regard to an increase[15,66,69]. This character is, obviously, controlled by the interaction of many genes each of which exhibits small effects. The use of the terms "polygene" and "modifier gene" indicates the complexity of the matter. Therefore, the efficiency of selecting "positive mutants" will be generally lower than for traits controlled by single genes. Furthermore, selection within the variability of such quantitative characters is not easy because of the manifold genotype-environmental interactions, as evaluated from experimentally induced variation[9,15,36,65,69,70,86,130,141,144,170]. But the wide spectrum of induced variation of adaptive behaviour of genotypes offers an alternative with regard to breeding work with well-adapted genotypes[28,67,68,125,131]

Environment affects not only the seed protein content but also the yielding capacities of the genotypes under analysis. Both factors, therefore, and the relationships between them must be taken in account if improvement is considered. The manifold interactions between these

TABLE 13.1. Variation in total seed protein content in different leguminous genera

Genus	Seed protein content (%)		Number of genotypes analysed	References
Cajanus	(b)[(+)]	18–29	2279	3, 9, 98, 114, 144, 158, 186
Glycine	(a)	39–42	35	78, 84, 96, 98, 110, 144
	(b)	37–43		
	(c)	35–52		
Phaseolus	(a)	16–20	330	26, 36, 37, 38, 72, 75, 98, 101, 144, 200
	(b)	21–32		
	(c)	17–32		
	(d)	20–28		
	(e)	18–26		
Vigna	(a)	20–35	182	9, 13, 23, 88, 98, 114, 144, 170
	(b)	16–33		
	(d)	20–35		
Lathyrus	(b)	29–32	5	170
Lens	(b)	20–31	36	170, 186
Pisum	(b)	17–31	2467	4, 15, 65–70, 125, 130–132, 144, 179
	(c)	14–36		186
	(e)	16–33		
Vicia	(a)	25–34	831	86, 104, 110, 138, 141
	(c)	35–52		
	(e)	19–40		
Cicer	(a)	19–28	47	98, 114, 170, 186
	(b)	12–28		
	(d)	18–28		

[(+): Material grown in (a) Africa, (b) Asia, (c) N. America, (d) Latin America, (e) Europe].

factors, which shall not be discussed in detail, are the subject of numerous publications[14,15,17, 18,22,23,26,49,52,72,84,101,106,114,115,131,142,143,162,170,175,176,178,188]. The relationships between yield and seed protein percentages mainly show negative correlations. This is valid for *Phaseolus*: -0.45[109], -0.64[188]; *Vigna*: 0.14 (phenotypic correlation)[14], -0.38 (genotypic correlation within 11 varieties), -0.23 to -0.29 (phenotypic correlation)[17]; *Pisum*:

0.02 to 0.49[142,143]. The correlations are generally low, sometimes absent[70] or even positive[15,99]. With regard to the relationship between seed protein content and seed weight, another important component of yield, no consistent correlation was found: none in *Pisum*[69], a negative one in *Cicer, Phaseolus*[99], and a positive one in *Phaseolus* and *Glycine*[78,116]. Therefore, it should be possible to select and to develop lines which simultaneously show high yield and high protein content. Such lines have been selected in *Glucine*[78], indicating that an improvement of both characters seems to be realizable also in other legumes. The variation between varieties, mutants and recombinants derived from single varieties can be used effectively for selecting progenies with improved characters[37,38,66-69].

With regard to the genetic control of protein content and yield statistical methods have proven to be suitable tools to describe the effects of polygenes. The values obtained, expressed in terms of low and high additive variation and narrow-sense and broad-sense heritability, allow some prediction for selection work. In selected *Pisum*-lines the inheritance of the protein content shows an additive genetic variation[77,187]. Similar findings exist from *Phaseolus*[109,164]. The broad-sense heritability estimates range from 30.7–63% for *Phaseolus*[109], 29% for *Vigna*[14], and 78–79% for *Pisum*[187]. The heritability coefficients in the narrow-sense were 5–12% for *Phaseolus*, and 7.6% for *Pisum*.

Storage protein systems

The descriptive statistical methods give no answer to the question concerning the molecular basis of a polygene. Therefore, the experimental approach for the study of the genetic organization of quantitative inheritance consists in using more simple, clearly-marked analytical systems. The analysis of single proteins, for instance, includes the vicinity to primary gene products and excludes biological interactions of higher order. In this way molecular mechanisms can be analysed and the components of quantitative inheritance better understood.

During seed development storage proteins are synthesized and deposited in several differentiated tissue types within the seed. On the basis of their solubility two main types of proteins are found. The water-soluble proteins are referred to as albumins and the salt-soluble proteins as globulins[41,42]. The albumins are mainly metabolically active proteins, involved in different cellular activities, including the synthesis and degradation of the second group[5,7] which constitutes the bulk of storage proteins. They supply the germinating embryo with nitrogenous compounds[42,174]. The albumins also are involved in degradation processes during inhibition of seeds[5,134].

Albumin storage protein system

The albumins are incorporated during seed development in leguminous seeds like all other proteins[7,75,82,83,123,132,167]. Therefore, it is conceivable that they serve as storage proteins. Furthermore, the presence of variable amounts in seeds of different legumes indicates that this protein fraction must have more functions than presently attributed to it[66,71,82,83,132,134,154,155]. Considering the quantitative proportions of the albumin fraction as a percentage of total protein content, a large variability could be found. In seeds of different genotypes of *Phaseolus* the range was between 11 and 20%[75,115]. That range was even broader in *Pisum* genotypes: 13–24%[7,44,71,97,132,134,167]. Similar results were obtained for *Lupinus*[51]. This variation, notwithstanding the manifold genotype-environment interactions[69,97,157,191] should receive more attention with regard to the seed protein quality[4,44,134,167]. Furthermore, considering the fact that the albumin fraction makes up nearly a third of the total storage proteins this fraction needs more conceptual consideration, mainly with regard to its complex composition and the number of proteins involved[66,69,111,127,171,193,200]. This is valid especially for the quantitative relationships between the albumin and legumin fractions in seeds of different genotypes of *Pisum*. Both quantitative characters are negatively correlated. In wild forms of high legumin content corresponds to a low albumin content. This relation is reversed in field peas. They exhibit only a small variation in legumin content, but a wide range for albumin contents. Furthermore, round-seeded garden peas show the greatest variation in legumin content while wrinkled-seeded garden peas have the smallest variation for both components[167]. There exists a close relationship between the legumin content and the seed shape, which is controlled by the r_a locus[45]. These findings offer the possibility for studying the genetics of seed proteins more directly, also with regard to the microheterogenity existing between different proteins, as revealed by electrophoretic methods. Studies on the composition of the albumin fraction yielded, independently from external factors[57,58], highly reliable results. They are suitable for characterizing the genetic variability of different genotypes, on the level of species, varieties and mutants[48,73,89,93,107,126,133,147,195]. In this way also an identification of distinct genotypes will be possible. Moreover, electrophoretic analyses of the albumin fraction are useful for studying developmental steps in seed protein incorporation[75,102,185] and degradation[5]. The knowledge of these processes will be decisive with regard to the understanding of regulatory phenomena involved in seed protein biosynthesis.

Tests applied to determine the molecular sizes of the polypeptides constituting the albumin fraction by means of SDS gelelectrophoresis have demonstrated that it contains a few major and some additional minor components which obviously differ between species[44,71,93,97,102]. In this sector further research work is needed. These analyses tentatively establish assays for measuring gene expression in the protein fraction that contributes because of its amino acid composition essentially to the nutritive value of leguminous seeds. With an understanding of the amino acid sequences of the respective polypeptides of the albumin fraction it should be possible to know the "units" in which the storage protein system is encoded. This would be the first step envisaging the employment of genetic engineering techniques for improving the deficiencies in certain essential amino acids in legume seeds. But this will be research work in the years ahead. Meanwhile, we must limit our research to the biological reality. There is much that can be done, mainly with regard to further improvement of disease resistance, tolerance to different environmental extrenes, and yield potential, chiefly by employing conventional breeding methods.

In this connection *enzyme proteins*, mainly belonging to the albumin fraction, and especially their electrophoretically detectable *isozymes*[165], have proven a useful tool for selection work in screening programmes. Since the variant alleles are generally codominant[29,34,63,197] it is possible to identify and to distinguish heterozygotes from homozygotes. The isozyme patterns, which are not influenced by environmental factors[57,58,59,97,197], reflect directly a part of the genetic makeup of the respective genotypes, and their patters are representative. They can be used as sensitive biochemical markers at the molecular level.

So far, at least 14 enzyme systems were studied for identifying varietal differences in legumes. Some of them are monomorphic, the others polymorphic[57—59,95,124,129,198,199]. Apart from a functional meaning, the variation at enzyme loci has proven suitable for the characterization of species[129,174], varieties on a large scale[34,63,79—81,82,83,108], and mutants and recombinants of one species[124,127,128,132]. The intravarietal analysis of isozymes shows both qualitative and quantitative differences[34] although the quantitative aspect is still unclear in most cases. By applicating isozyme techniques it is possible to measure the genetic distances between cultivars of closely related species and to determine the genotype-specific isozyme patterns characterizing the level of inbreeding[58,198].

In connection with the possibilities of identifying genotypes from inbred lines and from progenies derived from crossings between them by means of isozyme analysis and to compare cultivars or cultivated species with one another, a further aspect for applying this technique in breeding work is of interest.

The comparison of the amylase isozyme patterns in cultivated and wild soybeans has shown that the development of cultivars has undoubtedly led to genetic uniformity[103]. Only in wild forms a relatively high genetic variation with regard to this trait was found.

Plant breeding has not only increased the productivity of crop plants, but at the same time narrowed the genetic variability as compared with land races or wild forms from which they were derived[15]. Concomitantly, in numerous plants, e.g. a disease-vulnerable gene pool was produced[29]. The various aspects concerning the maintenance of plant genetic resources are reviewed in detail[29]. According to numerous proposals the gene pools of primitive and wild forms must be explored in order to find genes for disease resistance, adaptive gene complexes etc. for reestablishing a broader genetic basis for further improvement of crop plants. Certainly, induced mutations have enlarged the variability considerably[15,28,64,67,69,86,130,132], but the exploration of the genetic variability available in wild and cultivated plant populations will become more and more important. In this connection the analysis of the levels of genic variation between genotypes can be performed by isozyme analysis[21,24,48,58,68,103,124,129, 165,195]. Isozymes are excellent genetic markers because they represent specific gene products and variants indicate single lesions[103]. Structural changes in comparable tissues, in which the same genes are expressed, may result in altered mobilities or in lacking isozymes[63] indicating changes in the amino acid sequence of the respective enzymes. Irrespectively from the until now poorly understood adaptive or selective significance of allozymes, strategies have been evolved for comparing gene pools of different plant genera[29]. With regard to legumes only few data are available[103,129,197]. More investigations in this research field are necessary for understanding the genetic systems responsible for the variability observed[29]. This would be useful for more systematic attempts in breeding programmes.

To the water-soluble proteins from leguminous seeds also belong the *protease inhibitors* which make up to 6% of the total seed protein[168]. Therefore, they cannot be neglected as a minor part of the storage proteins. The protease inhibitors are widespread within the legumes[25,62,98,111,120,133a] and constitute proteins of low molecular weights. They possibly exert a protective effect for plants because of their broad specificity towards animal and plant proteases. Some of them (Kunitz-, chymotrypsin-, and subtilisin inhibitors) have been well characterized[62,168]. The genetic control of the synthesis of some protease inhibitors and the existence of isoinhibitors have been proven[10,81,87,168,177]. Furthermore, there exists a genetically conditioned linkage of an acid phosphatase with the Kunitz trypsin inhibitor[81], indicating close relationships between enzymes and inhibitor proteins. The genotype-specific accumulation of these proteins in legume seeds must be taken into consideration for the evaluation of the nutritive value of seed proteins.

The same is valid for another group of water-soluble seed proteins of legumes, namely the *lectins (phytohemagglutinins)*. They constitute a group of heterogeneous proteins or glycoproteins showing remarkable chemical and biological properties[112,113,134,150,161,193] and a great variation[31,32,33,54,92]. They are able to interact with animal and plant cells[112] some of them are mitogenic[137,172]. In some cases they possess enzymatic activity[76]. They amount to 1—10% of the total seed protein content, their proportions varying according to the respective species[112,113,139,150]. Although water-soluble, the lectins cannot be attributed unanimously to a distinct seed protein fraction[54,76,92,139,150,153,161]. There are, obviously no relationships between the chemical properties of the lectins and their occurrence in distinct leguminous species[6,161]. Especially in the genus *Phaseolus* there is a great variation in biological effects of the lectins[31,32,33,54,92], obviously due to their subunit structure (isolectins)[76,92,150,153,161]. It can even be used for characterizing species and varieties[148]. Considering the structural similarities between the lectins in different legume genera, as revealed by ion exchange chromatography and electrophoresis[54,150], one should suppose that many of these proteins can be regarded as homologous proteins conserved as storage proteins[56]. The genetic control by single dominant genes is established[151]. The homozygous recessive genotypes produce no lectins.

The findings concerning the biological functions of the lectins are partly conflicting. This is valid for their function as important determinants of host range specificity in Rhizobium-legume-symbiosis[10,11,16,46,146]. But there are also effectively nodulated leguminous lines which lack lectin. Similar contradictory findings are available with regard to the protective function of lectins against insect attack. Furthermore, the nutritional toxicity of seed of some legume species for human beings is partly ascribed to high amounts of lectins[151,152,194].

A further group of albumin storage proteins must be mentioned. These are proteins which have proven to be allergenic. Only few data exist on this group of *allergenic proteins* which perhaps belong to the 2S fraction of seed proteins[43]. They are found in seed of different leguminous species[117]. Their wide distribution indicates that within the albumin fraction allergenic determinants exist which need further investigation in order to group these proteins according to their structures and precise functions.

The data compiled on the albumin storage protein system reveal a large diversity with regard to its components and their respective functions. Their interrelationships are not yet fully understood but they must be regarded in breeding for improved seed protein quality[18,100,101,169,173,178]. Considering the proportion of this fraction relative to the total protein content the significance becomes obvious. In this connection it should be mentioned that large quantities of protease inhibitors as well as lectins reduce the nutritional value of the seed[151,152], although these compounds contain in some cases high amounts of essential amino acids, also the deficient ones. Nevertheless, it seems to be promising to enlarge this storage protein fraction entirely because of its diversity and its composition.

Globulin storage protein system

The globulins as sets of organ- and tissue-specific gene products[75,118,121,136,189,192], coded for by the DNA of the embryo[122], constitute the major part of the legume seed storage proteins. They are quantitatively inherited[135] and their genetically controlled variation[68,81,107,108,190] as well as the influences by environment on their synthesis[12,22,157] are well known.

Their proportions vary according to the respective genera analysed, as shown in Table 13.2.

The globulins synthesized offer the possibility for improving the seed proteins in legumes, because they show a large qualitative and quantitative variation, as revealed mainly by electro-

TABLE 13.2. Globulin contents in different genera of legumes

Genus	Globulin content (% total protein content)	References
Glycine	90	105
Phaseolus	50–70	75, 89, 115, 159
Pisum	60–80	132, 167, 191, 192
Vicia	60–90	55, 118, 202
Lupinus	80	51, 169

phoretic methods[12,19,21,24,30,35,39,42,47,50,61,74,75,82,83,90,102,105,107,118,121,122, 133,135,136,163,167,189,191,192]. These methods allow genetic relationships in this group of proteins to be revealed.[30,74,81,90,106–108]

Structurally, the globulins are composed mainly of two types of oligomeric proteins, the 7–8S proteins, vicilin, and the 11–12S proteins, legumin[47]. In Glycine[82,83], Lupinus[169] and some other legumes exist a group of smaller globulin molecules with sedimentation coefficients near 2. Perhaps they correspond to the lectins, protease inhibitors and allergenic compounds. In Pisum a further storage protein was isolated, distinct from the 7S and 11S proteins, named convicilin[40].

According to their molecular weights and sedimentation coefficients vicilin-like and legumin-like proteins are found consistently in seeds of a wide range of leguminous species[47]. But their homology, the basis for convenient genetic comparisons, has been determined in only a few cases. Thus, the legumin of Pisum sativum and Vicia faba, the arachin of Arachis hypogaea and the glycinin of Glycine max have been shown to be homologous proteins[61]. There are indications that the same is valid for the legumin-like proteins of Phaseolus and Vigna. The vicilin-like proteins of Vicia, Pisum and Arachis also seem to be homologous. These findings indicate that the synthesis of the storage protein systems is controlled by similar multigene families, common to the legume family. This becomes clear if we consider the extensive microheterogeneity existing within the fractions. We need more knowledge on the amino acid sequences of storage protein subunits of known molecular weights in order to find and to compare repeat sequences. In this way we could learn more about the gene sequences in which the informations for these proteins are encoded. This would be, on the other hand, a prerequisite for improving leguminous proteins mainly with regard to deficient amino acids, applicating the methods of genetic engineering. But before being successful with this approach many questions must be answered, e.g. with regard to the genetic basis of the conservative protein structures in legume seeds, the possible effects of changes in amino acids on the structure of the proteins incorporated into the seeds and their stability, the possibilities of disturbing the metabolism by introduced genetic informations.

Of interest is the quantitative distribution of vicilin and legumin in leguminous seeds, because both fractions differ from one another with regard to their contents in the limiting amino acids methionine and cystine[18,35,47,75,85,105,122,133,149]. The legumin fraction is superior to the vicilin fraction in this respect. The legumin:vicilin ratios for some legume species are as follows: *Vicia* (2.30 − 4):$1^{18,35,133,202}$; *Pisum* (1 − 3):$1^{7,132,192}$; *Phaseolus* G1:G2 = 6:1^{75}; 1:9^{35}; in *Vigna* the 7S globulin constitutes the major protein fraction[35]. The ratios indicate genotype-specific differences which can vary between species and within species and varieties[47,133a,190,192,202]. There exists a sufficient variation with regard to this trait that it seems possible to achieve changes in the respective proportions of both storage protein fractions by breeding. Because it is known that, e.g. in peas the genetic information for the synthesis of legumin may be closely related to a gene responsible for a morphological character[45,136], selection of morphologically conspicuous phenotypes of legume species could be helpful for breeding. Considering the distribution of both fractions within leguminous seeds an improvement seems not to be easy. Their amounts differ considerably from one another. Even the enlargement of the embryo axis, showing a more balanced amino acid balance[140] than the cotyledons will hardly change the overall amino acid composition of the seeds because the embryo axis represents only 1% of the seed proteins.

The different legumin:vicilin ratios indicate that their incorporation is obviously controlled by different, genetically controlled regulatory mechanisms. This becomes evident by comparing the genetically controlled processes of seed maturation, as evaluated by comparing morphologically and biochemically the developmental stages in seed maturation[75,136,185,196]. These processes are influenced strongly by the time of flowering[60]. This underlines the significance of early flowering genotypes[67,131]. According to the findings of numerous authors the syntheses of legumin and vicilin follow genotype-specific patterns[35,75,102,122,123,133a,136,185,196,202]. The morphogenetic changes in seed development, the onset, the duration and the termination of protein biosynthesis show variation not only between species but also between varieties of one species[136]. Thus, genotypes showing differences in the rates of seed protein accumulation[35,123,191] should be envisaged for improving protein quality. Furthermore, the phenomenon of gene dosage effects as a consequence of the polyploid status of the cotyledonous cells should be considered.

Amino acid composition

The last section is devoted to the amino acid composition of leguminous seeds because its improvement is one of the urgent aims of breeding for protein quality.

The ranges of values found for the amino acid composition of the seeds of nutritionally used legume seeds show indeed large variations in the seeds of the following genera: *Cajanus*[9,158,160], *Glycine*[2,96,173], *Phaseolus*[1,2,53,96,99,100,101,115,152], *Vigna*[2,9,13,22], *Vicia*[138,145], *Pisum*[2,96], *Lens*[2], *Cicer*[2], *Lupinus*[51]. Generally one can state that the seed of *Pisum* and *Phaseolus* as compared with those of the other genera contain relatively high amounts of essential amino acids, mainly methionic. Compared with the amounts of essential amino acids of a standard protein[180] the limiting character of the sulphur containing amino acids becomes clear. But improving the seed proteins only with regard to these amino acids would cause a deficiency in the amino acid tryptophane. And this sequence could be extended to all other essential amino acids of the respective genotypes. Therefore, breeding programmes should not only envisage the first two amino acids, but the whole spectrum of essential amino acids. Reliable screening methods should regard this point. Furthermore, screening methods for amino acid composition should also include tests concerning the different constituents of the albumin storage protein system (protease inhibitors, allergenic compounds) from which it is known that they contain high amounts of sulphur amino acids. On the other hand, in the genus *Phaseolus* a great variability with regard to the methonine content was found[96] indicating that there is sufficient variation which can be used in plant breeding[49]. This variation should be exploited also in other legumes.

Considering the nutritive value of legume seed proteins the occurrence of non-protein amino acids, specific for legumes, should also be regarded[8]. Until now, data concerning the accumulation of these uncommon amino acids are scarce[8,181—184]. In the case of *Lens* the content in j-hydroxyarginin accounts for 76—91% of the free basic amino acids, and 0.76—1.7% of the dry matter, the corresponding values for hydroxyornithin are 3—7% and 0.04—0.08% respectively. The amounts vary according to the genotypes analysed. Therefore, they can be regarded as species-specific compounds. With regard to their physiological compatibility in animal and human nutrition no reliable data are available. But it is known that certain non-protein amino acids can be toxic. They are not destroyed by cooking[183].

Summarizing the diverse aspects of seed protein composition and quality as related to breeding for seed protein improvement, the findings available show that there are numerous encouraging data on quantitative and qualitative seed protein characters. But the knowledge of the regulatory processes involved in seed protein accumulation and the genetic components responsible for these traits is incomplete. This research field should attract more attention. The investigations should include numerous different genotypes, such as developmental mutants carrying genes for early flowering and photoperiod insensitivity, genotypes with altered flowering duration, alteration in seed size and structure. According to the rule of parallel variation such genotypes can be expected because striking differences in these characters are apparent between species and varieties.

REFERENCES

1. Adams, W.M. On the quest for quality in the field bean. Nutritional Improvement of Food Legumes by Breeding. Milner, M. (ed.), 143—149; PAG New York, 1973.
2. Altmann, P.L. and Dittmer, D.S. (eds.). Metabolism. Fed. Amer. Soc. Exp. Biology, Bethesda, Maryland, U.S.A., 1968.
2a. Argos, P., Pedersen, K., Marks, M.D. and Larkins, B.A. A structural model for Maize zein proteins. *J. Biol. Chem.* **257**, 9984—9990, 1982.
3. Aykroyd, W.R. and Doughty, J. Legumes in Human Nutrition. FAO Rome, 1964.
4. Bajaj, S. Biological value of legume proteins as influenced by genetic variation. Nutritional Improvement of Food Legumes by Breeding. Milner, M. (ed.), 223—232; PAG New York, 1973.
5. Basha, S.M.M. and Beevers, L. The development of proteolytic activity and protein degradation during the germination of *Pisum sativum* L. *Planta* **124**, 77—87, 1975.
6. Baumann, C.M. and Rüdiger, H. Interactions between the two lectins from *Vicia cracca*. FEBS Lett. **136**, 279—283, 1981.
7. Beevers, L. and Poulson, R. Protein synthesis in cotyledons of *Pisum sativum* L. I. Changes in cell-free amino acid incorporation capacity during seed development and maturation. *Plant Physiol.* **49**, 476—481, 1972.
8. Bell, E.A. Non-protein amino acids in the *Leguminosae*. In: Polhill, R.M. and Ravens, P.H. (eds.): Advances in Legume Systematics. II. 489—499; Royal Botanic Gardens, Kew; Ministry of Agriculture, Fisheries and Food, London, 1981.
9. Bhagwat, S.G., Bhatia, C.R., Gopalakrishna, T., Joshua, D.C., Mitra, R.K., Narahari, P., Pawar, S.E. and Thakare, R.G. Increasing protein production in cereals and grain legumes. Seed Protein Improvement in Cereals and Grain Legumes, II: 225—236; IAEA Vienna, 1979.
10. Bhuvaneswari, T.V. and Bauer, W.D. Role of lectins in plant-microorganism interactions. III. Influence of rhizosphere/rhizoplane culture conditions on the soybean lectin-binding properties of *Rhizobia*. *Plant Physiol.* **62**, 71—74, 1978.

11. Bhuvaneswari, T.V., Pueppke, S.G. and Bauer, W.D. Role of lectins in plant-microorganism inter-
     actions. I. Binding of soybean lectins to *Rhizobia*. *Plant Physiol.* **60**, 486–491, 1977.
12. Blagrove, R.J., Gillespie, J.M. and Randall, P.J. Effect of sulphur supply on the seed globulin com-
     position of *Lupinus angustifolius*. *Aust. J. Plant Physiol.* **3**, 173–184, 1976.
13. Bliss, F.A. Cowpeas in Nigeria. Nutritional Improvement of Food Legumes by Breeding. Milner, M.
     (ed.), 151–158; PAG New York, 1973.
14. Bliss, F.A., Barker, L.N., Franckowiak, J.D. and Hall, T.C. Genetic and environmental variation of
     seed yield, yield components, and seed protein quantity and quality of cowpeas. *Crop Sci.* **13**,
     656–660, 1973.
15. Blixt, S. Natural and induced variability for seed protein in temperate legumes. Seed Protein Im-
     provement in Cereals and Grain Legumes. II: 3–21; IAEA Vienna, 1979.
16. Bohlool, B.B. and Schmidt, E.L. Lectins: a possible basis for specifity in the *Rhizobium* – legume
     root nodule symbiosis. *Science* **185**, 269–271, 1974.
17. Bond, D.A. Yield and components of yield in diallel crosses between inbred lines of winter beans
     (*Vicia faba*). *J. Agric. Sci.* **67**, 325–336, 1966.
18. Boulter, D. Biosynthese, Struktur und Zusammensetzung der Samenproteine von *Vicia faba* im
     Hinblick auf die Auslese zur Verbesserung der Proteinqualität. Göttinger Pflanzenzüchter-Seminar
     2, 106–120, 1974.
19. Boulter, D. Structure and biosynthesis of legume storage proteins. Seed Protein Improvement in
     Cereals and Grain Legumes. I. 125–136; IAEA Vienna, 1979.
20. Boulter, D. Ontogeny and development of biochemical and nutritional attributes in legume seeds. In:
     Summerfield, R.J. and Bunting, A.H. (eds.): Advances in Legume Science, I: 127–134; Ministry
     of Agriculture, Fisheries and Food, London, 1980.
21. Boulter, D. and Derbyshire, E. Taxonomic aspects of the structure of legume proteins. In: Harborne,
     J.B., Boulter, D. and Turner, B.L. (eds.): Chemotaxonomy of the *Leguminosae*, 285–308; Aca-
     demic Press, London and New York, 1971.
22. Boulter, D., Evans, I.M. and Derbyshire, E. Proteins of some legumes with reference to environ-
     mental factors and nutritional value. *Qual. Plant.* **23**, 239–250, 1973.
23. Boulter, D., Evans, I.M., Thomson, A. and Yarwood, A. The amino-acid composition of *Vigna un-
     guiculata* (cow pea) meal in relation to nutrition. Nutritional Improvement of Food Legumes by
     Breeding. Milner, M. (ed.), 205–215; PAG New York, 1973.
24. Boulter, D., Thurman, D.A. and Derbyshire, E. A disc electrophoretic study of globulin proteins of
     legume seeds with reference to their systematics. *New Phytol.* **66**, 27–36, 1967.
25. Bowman, D.E. Isolation and properties of a proteinase inhibitor of navy bean. *Arch. Biochem. Bio-
     phys.* **144**, 541–548, 1971.
26. Bressani, R. Legumes in human diets and how they might be improved. Nutritional Improvement of
     Food Legumes by Breeding. Milner, M. (ed.), 15–42; PAG New York, 1973.
27. Bressani, R.L., Elias, G. and Valiente, A.T. Effect of cooking and of amino acid supplementation on
     the nutritive value of black beans (*Phaseolus vulgaris* L.). *Br. J. Nutr.* **17**, 69–78, 1963.
28. Brock, R.D. The role of induced mutations in plant improvement. *Rad. Bot.* **11**, 181–196, 1971.
29. Brown, A.H.D. Isozymes, plant population genetic structure and genetic conservation. *Theor. Appl.
     Genet.* **52**, 145–157, 1978.
30. Brown, J.W.S., Ma, Y., Bliss, F.A. and Hall, T.C. Genetic variation in the subunits of globulin-1
     storage protein of French bean. *Theor. Appl. Genet.* **59**, 83–88, 1981.
31. Brown, J.W.S., Osborn, T.C., Bliss, F.A. and Hall, T.C. Bean Lectins. Part 1: Relationships between
     agglutinating activity and electrophoretic variation in the lectin-containing G2/albumin seed
     proteins of French bean (*Phaseolus vulgaris* L.). *Theor. Appl. Genet.* **62**, 263–271, 1982a.
32. Brown, J.W.S., Osborn, T.C., Bliss, F.A. and Hall, T.C. Bean Lectins. Part 2: Relationships between
     qualitative lectin variation in *Phaseolus vulgaris* L. *Theor. Appl. Genet.* **62**, 361–367, 1982b.
33. Brücher, O., Wecksler, M., Levy, A., Palozzo, A. and Jaffé, W.G. Comparison of phytohaemagglutinins
     in wild beans (*Phaseolus aborigineus*) and in common beans (*Phaseolus vulgaris*) and their in-
     heritance. *Phytochem.* **8**, 1739–1743, 1969.
34. Buttery, B.R. and Buzzell, R.I. Properties and inheritance of urease isoenzymes in soybean seeds.
     Can. J. Bot. **49**, 1101–1105, 1971.

35. Carasco, J.F., Croy, R., Derbyshire, E. and Boulter, D. The isolation and characterization of the major polypeptides of the seed globulin of cowpea (*Vigna unquiculata* L. Walp) and their sequential synthesis in developing seeds. *J. Exp. Bot.* **29**, 309–323, 1978.

36. Crocomo, O.J., Gerald Lee, T.S., Derbyshire, E. and Boulter, D. Biochemical investigations on the seed proteins of a Brazilian variety and mutant of *Phaseolus vulgaris*. Seed Protein Improvement in Cereals and Grain Legumes, I, 217–229; IAEA Vienna, 1979.

37. Crocomo, O.J., Neto, A.T., Ando, A., Blixt, S. and Boulter, D. Breeding for improved protein content and quality in the bean (*Phaseolus vulgaris*): II. Further work in selections from spontaneous variation; new work on mutagenic treatments and the influence of added nitrogen levels. Seed Protein Improvement by Nuclear Techniques, 207–222; IAEA Vienna, 1978.

38. Crocomo, O.J., Neto, A.T., Blixt, S. and Mikaelsen, K. Breeding for protein in the bean (*Phaseolus vulgaris* L.): I. Inventory of some Brazilian varieties and a number of lines of differing origin. Evaluation of Seed Protein Alteration by Mutation Breeding, 197–212; IAEA Vienna, 1975.

39. Croy, R.R.D., Derbyshire, E., Krishna, T.G. and Boulter, D. Legumin of *Pisum sativum* and *Vicia faba*. *New Phytol.* **83**, 29–35, 1979.

40. Croy, R.R.D., Gatehouse, J.A., Tyler, M. and Boulter, D. The purification and characterization of a third storage protein (convicilin) from the seeds of pea (*Pisum sativum* L.) *Biochem. J.* **191**, 509–516, 1980.

41. Danielsson, C.E. Investigation of vicilin and legumin. *Acta chem. Scand.* **3**, 41–49, 1949.

42. Danielsson, C.E. Seed globulins of the *Gramineae* and *Leguminosae*. *Biochem. J.* **44**, 387–400, 1949.

43. Daussant, J., Ory, R.L. and Layton, L.L. Characterization of proteins and allergens in germinating castor seeds by immunochemical techniques. *J. Agric. Food Chem.* **24**, 103–107, 1976.

44. Davies, D.R. Variation in the storage proteins of peas in relation to sulphur amino-acid content. *Euphytica* **25**, 717–724, 1976.

45. Davies, D.R. The r_a locus and legumin synthesis in *Pisum sativum*. *Biochem. Genet.* **18**, 1207–1219, 1980.

46. Dazzo, F.B. and Brill, W.J. Regulation by fixed nitrogen of host-symbiont recognition in the *Rhizobium*-clover symbiosis. *Plant Physiol.* **62**, 18–21, 1978.

47. Derbyshire, E., Wright, D.J. and Boulter, D. Legumin and vicilin, storage proteins of legume seeds. *Phytochem.* **15**, 3–24, 1976.

48. Derbyshire, E., Müller, H.P., Carvalho, M.T.V. and Crocomo, O.J. Protein profiles of Brazilian beans (*Phaseolus vulgaris*) obtained by electrophotesis in slabs of polyacrylamide gel. *Energ. Nucl. Agric.* **3**, 100–109, 1981.

49. Dickson, M.H. and Hackler, L.R. Protein quantity and quality in high-yielding beans. Nutritional Improvement of Food Legumes by Breeding. Milner, M. (ed.), 185–192; PAG New York, 1973.

50. Dudman, W.F. and Millerd, A. Immunochemical behaviour of legumin and vicilin from *Vicia faba*: a survey of related proteins in the *Leguminosae* subfamily Faboideae. *Biochem. System. Ecol.* **3**, 25–33, 1975.

51. Duranti, M. and Cerletti, P. Amino acid composition of seed proteins of *Lupinus albus*. *J. Agric. Food Chem.* **27**, 977–978, 1979.

52. Evans, A.M. Genetic improvement of *Phaseolus vulgaris*. Nutritional Improvement of Food Legumes by Breeding. Milner, M. (ed.), 107–115; PAG New York, 1973.

53. Evans, I.M. and Boulter, D. Crude protein and sulphur amino acid contents of some commercial varieties of peas and beans. *J. Sci. Food Agric.* **31**, 238–242, 1980.

54. Felsted, R.L., Li, J., Pokrywka, G., Egorin, M.J., Spiegel, J. and Dale, R.M.K. Comparison of *Phaseolus vulgaris* cultivars on the basis of isolectin differences. *Int. J. Biochem.* **13**, 549–557, 1981.

55. Flink, J. and Christiansen, I. The production of a protein isolate from *Vicia faba*. *Lebensm.-Wiss. u. Technol.* **6**, 102–106, 1973.

56. Foriers, A., Wuilmart, C., Sharon, N. and Strosberg, A.D. Extensive sequence homologies among lectins from leguminous plants. *Biochem. Biophys. Res. Comm.* **75**, 980–986, 1977.

57. Gates, P. and Boulter, D. Nitrogen regime and isoenzyme changes in *Vicia faba*. *Phytochem.* **18**, 1789–1791, 1979a.

58. Gates, P. and Boulter, D. The use of seed isoenzymes as an aid to the breeding of field beans (*Vicia faba* L.) *New Phytol.* **83**, 783–791, 1979b.

59.   Gates, P. and Boulter, D.   The use of pollen isoenzymes as an aid to the breeding of field beans (*Vicia faba* L.). *New Phytol.* **84**, 501–504, 1980.

60.   Gbikpi, P.J. and Crookston, R.K.   Effect of flowering date on accumulation of dry matter and protein in soybean seeds. *Crop. Sci.* **21**, 652–655, 1981.

61.   Gilroy, J., Wright, D.J. and Boulter, D.   Homology of basis subunits of legumin from *Glycine max* and *Vicia faba*. *Phytochem.* **18**, 315–316, 1979.

62.   Gonzales, E., Callejar, A., Seidl, D. and Jaffé, W.G.   Subtilisin inhibitor activity in legume seeds. *J. Agric. Food Chem.* **27**, 912–913, 1979.

63.   Gorman, M.B. and Kiang, Y.T.   Models for the inheritance or several variant soybean electrophoretic zymograms. *The J. Hered.* **69**, 255–258, 1978.

64.   Gottschalk, W.   The flowering behaviour of Pisum genotypes under phytotron and field conditions. *Biol. Zbl.* **101**, 249–260, 1982.

65.   Gottschalk, W. and Müller, H.P.   Monogenic alteration of seed protein content and protein pattern in X-ray-induced *Pisum* mutants. Improving Plant Protein by Nuclear Techniques, 201–215; IAEA Vienna, 1970.

66.   Gottschalk, W. and Müller, H.P.   Quantitative and qualitative investigations on the seed proteins of mutants and recombinants of *Pisum sativum*. *Theor. Appl. Genet.* **45**, 7–20, 1974.

67.   Gottschalk, W. and Müller, H.P.   The reaction of an early-flowering *Pisum* recombinant to environment and genotypic background. Seed Protein Improvement in Cereals and Grain Legumes, I, 259–272; IAEA Vienna, 1979.

68.   Gottschalk, W. and Müller, H.P.   Seed proteins of *Pisum* mutants and recombinants. Qual. Plant. in press, 1982.

69.   Gottschalk, W., Müller, H.P. and Wolff, G.   Relations between protein production, protein quality and environmental factors in *Pisum* mutants. Breeding for Seed Protein Improvement Using Nuclear Techniques, 105–123; IAEA Vienna, 1975a.

70.   Gottshcalk, W., Müller, H.P. and Wolff, G.   The genetic control of seed protein production and composition. *Egypt. J. Genet. Cytol.* **4**, 453–468, 1975b.

71.   Grant, D.R., Sumner, A.K. and Johnson, J.   An investigation of pea seed albumins. *Can. Inst. Food Sci. Technol.* **9**, 84–91, 1976.

72.   Gridley, H.E. and Evans, A.M.   Prospects for combining high yield with increased protein production in *Phaseolus vulgaris* L. Seed Protein Improvement in Cereals and Legumes, II, 47–58; IAEA Vienna, 1979.

73.   Hall, T.C., McLeester, R.C. and Bliss, F.A.   Electrophoretic analysis of protein changes during the development of the French bean fruit. *Phytochem.* **11**, 647–649, 1972.

74.   Hall, T.C., McLeester, R.C. and Bliss, F.A.   Equal expression of the maternal and paternal alleles for the polypeptide subunits of the major storage protein of the bean *Phaseolus vulgaris* L. Plant *Physiol.* **59**, 1122–1124, 1977.

75.   Hall, T.C., Sun, S.M., Ma, Y., McLeester, R.C., Pyne, J.W., Bliss, F.A. and Buchbinder, B.U.   The major storage proteins of French bean seeds: characterization *in vivo* and translation *in vitro*. In: Rubenstein, I.R., Philips, R.L., Green, Ch.E. and Gengenbach, B.G. (eds.): The Plant Seed, Development, Preservation, and Germination, 3–26; Academic Press New York, London, Toronto, Sydney, San Francisco, 1979.

76.   Hankins, C.N. and Shannon, L.M.   The physical and enzymatic properties of a phytohemagglutining from mung beans. *J. Biol. Chem.* **253**, 7791–7797, 1978.

77.   Hanson, W.D., Probst, A.H. and Caldwell, B.E.   Evaluation of a population of soybean genotypes with implication for improving self-pollinated crops. *Crop Sci.* **7**, 99–103, 1967.

78.   Hartwig, E.E.   Breeding productive soybeans with a higher percentage of protein. Seed Protein Improvement in Cereals and Grain Legumes, II, 59–66; IAEA Vienna, 1979.

79.   Hildebrand, D.F. and Hymowitz, T.   The Sp1 locus in soybean codes for β-amylase. *Crop Sci.* **20**, 165–168, 1980a.

80.   Hildebrand, D.F. and Hymowitz, T.   Inheritance of β-amylase nulls in soybean seed. *Crop Sci.* **20**, 727–730, 1980b.

81.   Hildebrand, D.F., Orf, J.H. and Hymowitz, T.   Inheritance of an acid phosphatase and its linkage with the Kunitz trypsin inhibitor in seed protein of soybeans. *Crop Sci.* **20**, 83–85, 1980.

82. Hill, J.E. and Breidenbach, R.W. Proteins of soybean seeds. I. Isolation and characterization of the major components. *Plant Physiol.* **53**, 742–746, 1974a.

83. Hill, J.E. and Breidenbach, R.W. Proteins of soybean seeds. II. Accumulation of the major protein components during seed development and maturation. *Plant Physiol.* **53**, 747–751, 1974b.

84. Hiraiwa, S. and Tanaka, S. Effects of successive irradiation and mass screening for seed size, density and protein content of soybean. Seed Protein Improvement by Nuclear Techniques, 265–274; IAEA Vienna, 1978.

85. Hurich, J., Parzysz, H. and Przybylska, J. Comparative study of seed proteins in the genus *Pisum*. II. Amino acid composition of different protein fractions. *Genet. Pol.* **18**, 241–252, 1977.

86. Hussein, H.A.S. and Abdalla, M.M.F. Protein and yield traits of field bean mutants induced with gamma rays, EMS and their combination. Seed Protein Improvement by Nuclear Techniques, 253–264; IAEA Vienna, 1978.

87. Hymowitz, T. and Hadley, H.H. Inheritance of a trypsin inhibitor variant in seed protein of soybeans. *Crop Sci.* **12**, 197–198, 1972.

88. Imam, M.M. Variability in protein content of locally cultivated *Phaseolus* and *Vigna* ssp. Seed Protein Improvement in Cereals and Grain Legumes, II, 119–126; IAEA Vienna, 1979.

89. Ishino, K. and Ortega, M.L. Fractionation and characterization of major reserve proteins from seeds of *Phaseolus vulgaris*. *J. Agric. Food Chem.* **23**, 529–533, 1975.

90. Jackson, P., Boulter, D. and Thurman, D.A. A comparison of some properties of vicilin and legumin isolated from seeds of *Pisum sativum, Vicia faba* and *Cicer arietinum*. *New Phytol.* **68**, 25–33, 1969.

91. Jaffé, W.G. Factors affecting the nutritional value of beans. Nutritional Improvements of Food Legumes by Breeding. Milner, M. (ed.), 43–48; PAG New York, 1973.

92. Jaffé, W.G., Levy, A. and Gonzales, D.I. Isolation and partial characterization of bean phytohemagglutinins. *Phytochem.* **13**, 2685–2693, 1974.

93. Jakubek, M. and Przybylska, J. Comparative study of seed proteins in the genus *Pisum*. III. Electrophoretic patterns and amino acid composition of albumin fractions separated by gel filtration. *Genet. Pol.* **20**, 369–380, 1979.

94. Jalil, M.E. and Tahir, W.M. World supplies of plant proteins. Proteins in Human Nutrition. Porter, J.W.G. and Rools, B.A. (eds.), Academic Press, London, 35–46, 1973.

95. Johnson, G.B. Enzyme polymorphism and metabolism. *Science* **184**, 28–37, 1974.

96. Johnson, V.A. and Lay, C.L. Genetic improvement of plant protein. *J. Agric. Food Chem.* **22**, 558–566, 1974.

97. Johnston, A.W.B., Brewster, V. and Davies, D.R. Seed proteins of peas in relation to nitrogen fixation. *Ann. Bot.* **41**, 381–385, 1977.

98. Kaul, A.K. Mutation breeding and crop protein improvement. Nuclear Techniques for Seed Protein Improvement, 1–106; IAEA Vienna, 1973.

99. Kelly, J.D. and Bliss, F.A. Heritable estimates of percentage seed protein and available methionine and correlations with yield in dry beans. *Crop Sci.* **15**, 753–757, 1975a.

100. Kelly, J.D. and Bliss, F.A. Quality factors affecting the nutritive value of bean seed protein. *Crop Sci.* **15**, 757–760, 1975b.

101. Kelly, J.F. Increasing protein quantity and quality. Nutritional Improvement of Food Legumes by Breeding. Milner, M. (ed.), 179–184; PAG New York, 1973.

102. Khan, M.R.I., Gatehouse, J.A. and Boulter, D. The seed proteins of cowpeas (*Vigna unguiculata* L. Walp.). *J. Exp. Bot.* **31**, 1599–1611, 1980.

103. Kiang, Y.T. Inheritance and variation of amylase in cultivated and wild soybeans and their wild relatives. *The J. Hered.* **72**, 382–386, 1981.

104. King, K.W. Development of all-plant food mixture using crops indigenous to Haiti: amino acid composition and protein quality. *Econ. Bot.* **18**, 311–322, 1964.

105. Koshiyama, I. and Fukushima, D. Soybean globulins. In: Münts, K. (ed.): Seed Proteins of Dicotyledonous Plants, 21–43; Abh. Akad. Wiss. DDR, N4, Akad. Verlag Berlin, 1979.

106. Lantz, E.M., Gough, H.W. and Campbell, A.M. Effect of variety, location, and years on the protein and amino acid content of dried beans. *J. Agric. Food Chem.* **6**, 58–60, 1958.

107. Larsen, A.L.   Electrophoretic differences in seed proteins among varieties of soybean (*Glucine max.* (L.) Merill.). *Crop Sci.* **7**, 311–313, 1967.

108. Larsen, A.L. and Caldwell, B.E.   Inheritance of certain proteins in soybean. *Crop Sci.* **8**, 474–476, 1968.

109. Leleji, O.I., Dickson, M.H., Crowder, L.V. and Bourke, J.B.   Inheritance of crude protein percentage and its correlation with seed yield in beans, *Phaseolus vulgaris* L. *Crop Sci.* **12**, 168–171, 1972.

110. Leng, E.R.   University of Illinois international soybean program. Nutritional Improvement of Food Legumes by Breeding. Milner, M. (ed.), 101–106; PAG New York, 1973.

111. Liener, I.   Antitryptic and other antinutritional factors in legumes. Nutritional Improvement of Food Legumes by Breeding. Milner, M. (ed.), 239–258; PAG New York, 1973.

112. Liener, I.   Phytohemagglutinins (Phytolectins). *Ann. Rev. Plant Physiol.* **27**, 291–319, 1976.

113. Lis, H. and Sharon, N.   The biochemistry of plant lectins (phytohemagglutinins). *Ann. Rev. Biochem.* **42**, 541–574, 1073.

114. Luse, R.A. and Rachie, K.O.   Seed protein improvement in tropical food legumes. Seed Protein Improvement in Cereals and Grain Legumes, II, 87–104; IAEA Vienna, 1979.

115. Ma, Y. and Bliss, F.A.   Seed proteins of common bean. *Crop Sci.* **18**, 431–437, 1978.

116. Malhotra, V.V. and Singh, K.B.   Genetic variability and path coefficient analysis for protein in green gram *Phaseolus aureus* Roxb. *Egypt. J. Genet. Cytol.* **5**, 170–173, 1976.

117. Malley, A., Baecker, L., Mackler, B. and Perlman, F.   The isolation of allergens from the green pea. *J. Aggerg. Clin. Immunol.* **56**, 282–290, 1975.

118. McLeester, R.C., Hall, T.C., Sun, S.M. and Bliss, F.A.   Comparison of globulin proteins from *Phaseolus vulgaris* with those from *Vicia faba*. *Phytochem.* **2**, 85–93, 1973.

119. Meiners, J.P. and Litzenberger, S.C.   Breeding for nutritional improvement. Nutritional Improvement of Food Legumes by Breeding. Milner, M. (ed.), 131–141; PAG New York, 1973.

120. Mies, D.W. and Hymowitz, T.   Comparative electrophoretic studies of trypsin inhibitors in seed of the genus *Glucine*. *Bot. Gaz.* **134**, 121–125, 1973.

121. Millerd, A.   Biochemistry of legume seed proteins. *Ann. Rev. Plant Physiol.* **26**, 53–72, 1975.

122. Millerd, A., Simon, M. and Stern, H.   Legumin synthesis in developing cotyledons of *Vicia faba* L. *Plant Physiol.* **48**, 419–425, 1971.

123. Millerd, A., Thomson, J.A. and Schroeder, H.E.   Cotyledonary storage proteins in *Pisum sativum*. III. Patterns of accumulation during development. *Aust. J. Plant Physiol.* **5**, 519–534, 1978.

124. Müller, H.P.   Gene mapping on chromosomes and some aspects of gene regulation in eukaryotic cells. *Nucleus* **21**, 135–142, 1978a.

125. Müller, H.P.   The seed proteins in gene-ecological investigations. *Legume Res.* **2**, 29–40, 1978b.

126. Müller, H.P.   Die genetische Steuerung der Zusammensetzung von Samenproteinen. *Z. Pflanzenkrankh. Pflanzenschutz* **85**, 210–217, 1978c.

127. Müller, H.P.   The genetic control of seed protein polymorphism in *Pisum*. Proc. 1st Medit. Conf. Genet., 747–764, 1979.

128. Müller, H.P.   Biochemische Darstellung und gelelektrophoretische Charakterisierung der Samenproteine von Leguminosen. Göttinger Pflanzenzüchter-Seminar 4, 105–116, 1980.

129. Müller, H.P.   Genotype-specific zymograms after direct isoelectric focusing of cotyledonous tissue. 4th International Congress on Isozymes, Austin, Texas, 1982.

130. Müller, H.P. and Gottschalk, W.   Quantitative and qualitative situation of seed proteins in mutants and recombinants of *Pisum sativum*. Nuclear Techniques for Seed Protein Improvement, 235–253; IAEA Vienna, 1973.

131. Müller, H.P. and Gottschalk, W.   Gene-ecological investigations on the protein production of different *Pisum* genotypes. Seed Protein Improvement by Nuclear Techniques, 301–314; IAEA Vienna, 1978.

132. Müller, H.P. and Werner, S.   Seed protein characteristics of *Pisum* varieties, mutants and recombinants. In: Müntz, K. (ed.): Seed Proteins of Dicotyledonous Plants, 189–209; Abh. Akad. Wiss. DDR, N4, Akad. Verlag Berlin, 1979.

133. Müntz, K., Horstmann, C. und Scholz, G.   Proteine und Proteinbiosynthese in Samen von *Vicia faba* L. *Die Kulturpflanze* **20**, 277–326, 1972.

133a. Müntz, K., Bäumlein, H., Bassüner, R., Manteuffel, R., Püchel, M., Schmidt, P. und Wobus, W. Regulation von Biosynthese und Akkumulation der Reserveproteine während der Entwicklung pflanzlicher Samen. *Biochem. Physiol. Pflanzen* **176**, 401–422, 1981.

134. Murray, D.R. A storage role for albumins in pea cotyledons. *Plant, Cell and Environment* **2**, 221–226, 1979.

135. Mutschler, M.A. and Bliss, F.A. Inheritance of bean seed globulin content and its relationship to protein content and quality. *Crop Sci.* **21**, 289–294, 1981.

136. Mutschler, M.A., Bliss, F.A. and Hall, T.C. Variation in the accumulation of seed storage protein among genotypes of *Phaseolus vulgaris* (L.). *Plant Physiol.* **65**, 727–630, 1980.

137. Nag, S., Talukder, G. and Sharma, A. Plant lectins and their blastogenic and mitogenic activities. *Nucleus* **24**, 16–38, 1981.

138. Nagl, K. Breeding value of radio-induced mutants of *Vicia faba var. minor.* Seed Protein Improvement by Nuclear Techniques, 243–252; IAEA Vienna, 1978.

139. Orf, J.H., Hymowitz, T., Pull, S.P. and Pueppke, S.G. Inheritance of a soybean seed lectin. *Crop Sci.* **18**, 899–900, 1978.

140. Otoul, E. Répartition des principaux acides aminés dans les différentes parties de la graine d'un cultivar de *Phaseolus vulgaris* L. *Bull. Rech. agron. Gembloux* **4**, 287–301, 1969.

141. Pandey, M.P., Frauen, M. and Paul, C. Selection for methionine by GLC after CNBr treatment in a germplasm collection and mutagen-treated population of *Vicia faba* L. Seed Protein Improvement in Cereals and Grain Legumes, II, 37–46; IAEA Vienna, 1979.

142. Pandey, S. and Gritton, E.T. Inheritance of protein and other agronomic traits in a diallel cross of pea. *J. Amer. Soc. Hort. Sci.* **100**, 87–90, 1975.

143. Pandey, S. and Gritton, E.T. Observed and predicted response to selection for protein and yield in peas. *Crop Sci.* **16**, 289–292, 1976.

144. Panton, C.A., Coke, L.B. and Pierre, R.E. Seed protein improvement in certain legumes through induced mutation. Nuclear Techniques for Seed Protein Improvement, 169–171; IAEA Vienna, 1973.

145. Patel, K.M. and Johnson, J.A. Horsebean as protein supplement in breadmaking. I. Isolation of horsebean protein and its amino acid composition. *Cereal Chem.* **51**, 693–701, 1974.

146. Planqué, K. and Kijive, J.W. Binding of pea lectins to a glycan type polysaccharide in the cell walls of *Rhizobium leguminosarum.* *FEBS Lett.* **73**, 64–66, 1977.

147. Przybylska, J., Blixt, S., Hurich, J. and Zimniak-Przybylska, Z. Comparative study of seed proteins in the genus *Pisum.* I. Electrophoretic patterns of different protein fractions. *Genet. Pol.* **18**, 27–38, 1977.

148. Pueppke, S.G. Distribution of lectins in the Jumbo Virginia and Spanish varieties of the peanut, *Arachis hypogaea* L. *Plant Physiol.* **64**, 575–580, 1979.

149. Pusztai, A. and Watt, W.B. Glycoprotein II. The isolation and characterization of a major antigenic and non-hemagglutinating glucoprotein from *Phaseolus vulgaris.* *Biochim. Biophys. Acta* **207**, 413–431, 1970.

150. Pusztai, A. and Watt, W.B. Isolectins of *Phaseolus vulgaris.* A comprehensive study of fractionation. *Biochim. Biophys. Acta* **365**, 57–71, 1974.

151. Pusztai, A., Clarke, E.M.W. and King, T.P. The nutritional toxicity of *Phaseolus vulgaris* lectins. *Proc. Nutr. Soc.* **38**, 115–120, 1979a.

152. Pusztai, A., Clarke, E.M.W., King, T.P. and Stewart, J.C. Nutritional evaluation of kidney beans (*Phaseolus vulgaris*): Chemical composition, lectin content and nutritional value of selected cultivars. *J. Sci. Food Agric.* **30**, 843–848, 1979b.

153. Pusztai, A., Grant, G. and Stewart, J.C. A new type of *Phaseolus vulgaris* (cv. Pinto III) seed lectin: Isolation and characterization. *Biochim. Biophys. Acta* **671**, 146–154, 1981.

154. Raacke, I.D. Protein synthesis in ripening peas. 1. Analysis of whole seeds. *Biochem. J.* **66**, 101–110, 1957a.

155. Raacke, I.D. Protein synthesis in ripening peas. 2. Development of embryos and seed coats. *Biochem. J.* **66**, 110–116, 1957b.

156. Rachie, K.O. Improvement of food legumes in tropical Africa. Nutritional Improvement of Food Legumes by Breeding. Milner, M. (ed.), 83–92; PAG New York, 1973.

157. Randall, P.J., Thomson, J.A. and Schroeder, H.E. Cotyledonary storage proteins in *Pisum sativum*. IV. Effects of sulfur, phosphorus, potassium and magnesium deficiencies. *Aust. J. Plant Physiol.* 6, 11–24, 1979.

158. Reddy, L.J., Green J.M., Singh, U., Bisen, S.S. and Jambunathan, J. Seed protein studies on *Cajanus cajan*, *Atylosis* spp. and some hybrid derivates. Seed Protein Improvement in Cereals and Grain Legumes, II, 105–117; IAEA Vienna, 1979.

159. Romero, J., Sun, S.M., McLeester, R.C., Bliss, F.A. and Hall, T.C. Heritable variation in a polypeptide subunit of the major storage protein of the bean, *Phaseolus vulgaris* L. *Plant Physiol.* 56, 776–779, 1975.

160. Royes, W.V. Amino acid profiles of *Cajanus cajan* protein. Nutritional Improvement of Food Legumes by Breeding. Milner, M. (ed.), 193–196; PAG New York, 1973.

161. Rüdiger, H. Lektine, pflanzliche zuckerbindende Proteine. *Naturwiss.* 65, 139–244, 1978.

162. Rutger, J.N. Variation in protein content and its relation to the other characters in beans (*Phaseolus vulgaris* L.). Agron. Abstracts, Amer. Soc. Agron., 20, 1968.

163. Sahai, S. and Rana, R.S. Seed protein homology and elucidation of species relationships in *Phaseolus* and *Vigna* species. *New Phytol.* 79, 527–534, 1977.

164. Sandhu, S.S., Keim, W.F., Hodges, H.F. and Nyquist, W.E. Inheritance of protein and sulfur content in seeds of chickpeas. *Crop Sci.* 14, 649–654, 1974.

165. Scandalios, J.G. Isozymes in development and differentiation. *Ann. Rev. Plant Physiol.* 25, 225–258, 1974.

166. Scharpé, A. and van Parijs, R. The formation of polyploid cells in ripening cotyledons of *Pisum sativum* L. in relation to ribosome and protein synthesis. *J. Exp. Bot.* 24, 216–222, 1973.

167. Schroeder, H.E. Quantitative studies on the cotyledonary proteins in the genus *Pisum*. *J. Sci. Food Agric.* in press. 1982.

168. Seidl, D.S., Abreu, H. and Jaffé, W.G. Purification of a subtilisin inhibitor from black bean seeds. *FEBS Lett.* 92, 245–250, 1978.

169. Sgarbiere, V.C. and Galeazzi, M.A.M. Some physico-chemical and nutritional properties of a sweet lupin (*Lupinus albus var. multolupa*) protein. *J. Agric. Food Chem.* 26, 1438–1442, 1978.

170. Shaik, M.A.Q., Kaul, A.K., Mia, M.M., Choudhury, M.H. and Bhuiya, A.D. Screening of natural variants and induced mutants of some legumes for protein content and yielding potential. Seed Protein Improvement by Nuclear Techniques, 223–233; IAEA Vienna, 1978.

171. Sharma, R.P., Nandpuri, K.S. and Kumar, J.C. Mode of inheritance of ascorbic acid and protein content in pea (*Pisum sativum* L.). *Veg. Sci.* 1, 18–21, 1974.

172. Sharon, N. and Lis, H. Lectins: cell-agglutinating and sugar-specific proteins. *Science* 177, 949–964, 1972.

173. Sikka, K.C., Gupta, A.K., Singh, R. and Gupta, D.P. Comparative nutritive value, amino acid content, chemical composition, and digestability *in vitro* of vegetable and grain-type soybeans. *J. Agric. Food Chem.* 26, 312–316, 1978.

174. Simon, E.W. and Raja Harun, R.M. Leakage during seed inhibition. *J. Exp. Bot.* 23, 1076–1085, 1972.

175. Singh, O.P., Singh, R.B. and Singh, F. Combining ability of yield and some quality traits in pea (*Pisum sativum* L.). *Z. Pflanzenzüchtg.* 84, 133–138, 1980.

176. Singh, O.P., Singh, R.B. and Singh, F. Genetics of yield, seed weight and quality traits in peak (*Pisum sativum* L.). *Genet. Agr.* 35, 115–120, 1981.

177. Singh, L., Wilson, C.M. and Hadley, H.H. Genetic differences in soybean trypsin inhibitors separated by disc electrophoresis. *Crop Sci.* 9, 489–491, 1969.

178. Sjödin, J. Induzierte Variabilität von Qualitätseigenschaften der Ackerbohne. Göttinger Pflanzenzüchter-Seminar 2, 101–105, 1974.

179. Slinkard, A.E. Breeding and protein studies. Crop Development Centre, University of Saskatchewan S7N OWO, Canada, 1972.

180. Smith, M.H. The amino acid composition of proteins. *J. Theor. Biol.* 13, 261–282, 1966.

181. Sulser, H. und Stute, R. γ-Hydroxyargining und γ-Hydroxyornithin, zwei ungewöhnliche Aminosäuren in den Samen der Linse (*Lens culinaris* Med.). I. Nachweis und quantitative Bestimmung von γ-Hydroxyarginin und γ-Hydroxyornithin in Linsen. *Lebensmitt.-Wiss. u. Technol.* 7, 322–326, 1974a.

182. Sulser, H. und Sager, F. γ-Hydroxyargining und γ-Hydroxyornithin, zwei ungewöhnliche Aminosäuren in den Samen der Linse (*Lens culinaris* Med.). II. Zur Synthese und Struktur von γ-Hydroxyornithin und γ-Hydroxyarginin. *Lebensmitt.-Wiss. u. Technol.* **7**, 327–329, 1974b.

183. Sulser, H. and Sager, F. Identification of uncommon amino acids in the lentil seed (*Lens culinaris* Med.). *Experientia* **32**, 422–423, 1976.

184. Sulser, H., Beyeler, M. und Sager, F. γ-Hydroxyargining und γ-Hydroxyornithin, zwei ungewöhnliche Aminosäuren in den Samen der Linse (*Lens culinaris* Med.). III. Isolierung von γ-Hydroxyarginin mittles Ionenaustauschchromatographie. *Lebensmitt.-Wiss. u. Technol.* **8**, 161–162, 1975.

185. Sun, S.M., Mutschler, M.A., Bliss, F.A. and Hall, T.C. Protein synthesis and accumulation in bean cotyledons during growth. *Plant Physiol.* **61**, 918–923, 1978.

186. Swaminathan, M.S. and Jain, H.K. Food legumes in Indian agriculture. Nutritional Improvement of Food Legumes by Breeding. Milner, M. (ed.), 69–82; PAG New York, 1973.

187. Swiecicki, W.K., Kaczmarek, Z. and Surma, M. Inheritance and heritability of protein content in seeds of selected crosses of pea (*Pisum sativum* L.). *Genet. Pol.* **22**, 189–195, 1981.

188. Tandon, O.B., Bressani, R., Scrimshaw, N.S. and Le Beau, F. Nutrients in Central American beans. *J. Agric. Food Chem.* **5**, 137–142, 1957.

189. Thomson, J.A. and Doll, H. Genetics and evolution of seed storage proteins. Seed Protein Improvement in Cereals and Grain Legumes, I, 109–124; IAEA Vienna, 1979.

190. Thomson, J.A. and Schroeder, H.E. Cotyledons storage proteins in *Pisum sativum*. II. Hereditary variation in compounds of the legumin and vicilin fraction. *Aust. J. Plant Physiol.* **5**, 281–294, 1978.

191. Thomson, J.A., Millerd, A. and Schroeder, H.E. Genotype-dependent patterns of accumulation of seed storage proteins in *Pisum*. Seed Protein Improvement in Cereals and Grain Legumes, I, 231–240; IAEA Vienna, 1979.

192. Thomson, J.A., Schroeder, H.E. and Dudman, W.F. Cotyledonary storage proteins in *Pisum sativum*. I. Molecular herogeneity. *Aust. J. Plant Phys.* **5**, 263–279, 1978.

193. Toms, G.C. and Western, A. Phytohaemagglutinins. In: Harborne, J.B., Boulter, D. and Turner, B.L. (eds.): Chemotaxonomy of the Leguminosae, 367–462; Academic Press, London and New York, 1971.

194. Turner, R.H. and Liener, I.E. The effect of the selective removal of hemagglutinins on the nutritive value of soybeans. *J. Agric. Food Chem.* **23**, 484–487, 1975.

195. Waines, J.G. The biosystematics and domestication of peas (*Pisum sativum* L.). *Bull. Torrey Bot. Club* **102**, 385–395, 1975.

196. Walbot, V., Clutter, M. and Sussex, I.M. Reproductive development and embryogeny in *Phaseolus*. *Phytomorphology* **22**, 59–68, 1972.

197. Wall, J.R. and Wall, S.W. Isozyme polymorphisms in the study of evolution in *Phaseolus vulgaris* – *Phaseolus coccineus* complex of Mexiko. In: Markert, C.L. (ed.): Isozymes, IV, Genetics and Evolution, 287–305; Academic Press New York, San Francisco, London, 1975.

198. Weeden, N.F. and Gottlieb, L.D. Distinguishing allozymes and isozymes of phosphoglucoisomerases by electrophoretic comparisons of pollen and somatic tissues. *Biochem. Genet.* **17**, 287–296, 1979.

199. Weimberg, R. An electrophoretic analysis of the isozymes of malate dehydrogenase in several different plants. *Plant Physiol.* **43**, 622–628, 1968.

200. Wood, D.R., Nowick, E.A., Fabian, H.J. and McClean, P.E. Genetic variability and heritability of available methionine in the Colorado dry bean breeding programme. Seed Protein Improvement in Cereals and Grain Legumes, II, 69–85; IAEA Vienna, 1979.

201. Woolfe, J.A. and Hamblin, J. Within and between genotypes variation in crude protein content of *Phaseolus vulgaris* L. *Euphytica* **23**, 121–128, 1974.

202. Wright, D.J. and Boulter, D. The characterization of vicilin during seed development in *Vicia faba*. *Planta* **105**, 60–65, 1972.

CHAPTER 14

TISSUE CULTURE FOR DISEASE ELIMINATION AND GERMPLASM STORAGE

G.G. Henshaw
School of Biological Sciences, University of Bath,
Claverton Down, Bath, U.K.

INTRODUCTION

In vitro techniques involving the isolation and multiplication of non-adventitious meristems have a number of important features: their potential for rapid propagation, in comparison with conventional methods; their genetic stability, relative to techniques based on adventitious buds, and their efficiency at eliminating systemic pathogens. A potential for rapid propagation, alone, would be of limited value without some guarantee of trueness-to-type and freedom from plant health problems and it is the combined advantages of these techniques which are particularly significant, to the extent that commercial exploitation is now feasible with many species in which vegetative propagation is important (see Chapter 11). The economics of commercial exploitation are, however, complex and due weight must be given to all aspects of these techniques and it might therefore be important that these basic techniques can also be used for the long-term conservation of genotypes to be used either as germplasm in breeding programmes or as nuclear stocks for multiplication programmes. For these purposes, genetic stability and good plant health are, of course, critical but there is less need for rapid multiplication and, indeed, low multiplication rates can be advantageous.

DISEASE ELIMINATION

In vitro techniques rely on aseptic procedures for the elimination and exclusion of contaminating organisms which would otherwise damage the plant tissues growing in culture. These procedures are based, essentially, on a careful choice of initial explant which is usually subjected to a surface-sterilisation procedure involving immersion in a suitable decontaminating fluid, such as sodium hypochlorite solution (see Chapter 11). This approach can be quite adequate for many purposes, but it can only be regarded as a first line of defence where plant health is a major consideration, since certain contaminants, especially systemic bacterial, mycoplasmas and viruses, can remain undetected unless special indexing procedures are adopted.

According to needs, cultures are initiated from different parts of the plant, but where the aim is to eliminate systemic pathogens it is standard practice to take the shoot apical meristem together with two or three leaf primordia as the explant for the production of a meristem culture from which a plant might eventually be regenerated. This procedure is based on the work of Morel and Martin[24] showing that cultures derived from such explants are frequently free of viruses, even when viruses are known to have been present in the original plant. The approach was based on earlier observations of the non-uniform distribution of viruses in plants and, especially, their low concentration in or possible absence from meristems.[12,46] It is now known, however, that whilst this might be true for certain viruses it is certainly not universally true[11,25] and it has been suggested[22] that the effectiveness of the meristem culture procedure for the elimination of viruses arises from the cell injury at the time of excision which leads to the inactivation of the virus.

The choice of a shoot apical meristem, together with two or three leaf primordia, as the standard explant for elimination of systemic pathogens, is really a compromise between conflicting aims. Although there can be an inverse relationship between the size of explant and the success rate for virus elimination[25], there is conversely a direct relationship with the success rate for initiation of cultures. Where plant health is not a factor, larger explants present fewer culture problems and it is technically far less demanding to isolate large shoot tips with more leaf primordia or even complete buds or nodes. The chances of effective pathogen elimination are likely to be improved by the use of even smaller explants but then the actual operation involves considerably more technical skill. There is a greater chance of damage to the plant tissues and the culture requirements can become more critical. In fact, cultures have very rarely been initiated from isolated shoot apical meristems which have been completely separated from leaf primordia.

It is now apparent that the present *in vitro* techniques do not, alone, guarantee the elimination of systemic pathogens. The improvements that might be achieved by the use of very small explants do not compensate for the extra technical difficulties that are encountered and it is more effective to concentrate efforts on the reduction of pathogen levels in the mother plants from which explants are taken. It is an obvious precaution to isolate these plants from the pathogen vectors but, in addition, it has been shown that the chances of obtaining pathogen-free cultures are increased when the plants have been subjected to thermotherapeutic treatments. Thermotherapy, involving the growth of plants for several weeks at elevated temperatures in the range $36°-40°C$, is a recognized method for the elimination of plant viruses but, as with the *in vitro* methods, its effectiveness cannot be fully guaranteed.[41]

Chemotherapy, involving compounds such as Virazole[35], is also under investigation as a means of eliminating viruses from plant tissues, but again success cannot be fully guaranteed and the techniques might be most useful when used in combination with *in vitro* techniques. Although both thermotherapy and chemotherapy can be applied to plants before explants are taken for culture, it might be more effective to treat the cultures directly, since this would avoid the problem of initiating cultures from debilitated plants and, further, the treatments can be controlled more precisely and applied more uniformly.[41]

Since it is only acceptable to use therapy treatments or *in vitro* techniques in combination with appropriate indexing procedures, which can be both expensive and time-consuming, it is essential to eliminate pathogens from a relatively high proportion of plants if the overall process is to be an economic proposition. Increasingly, the evidence suggests that this situation is most readily achieved by the use of combined treatments involving both therapy and *in vitro* methods. The optimal balance between effort spend on indexing and on elimination will depend on the facilities that might be available.

Where *in vitro* methods are an integral part of the whole process it would be a considerable advantage if the indexing procedure could be applied directly to the cultures but, at present, this is not really possible, except for preliminary screening purposes, since normal symptoms might not be exhibited and, further, the pathogen might not be present at very low levels or in a latent state. Until more sensitive and reliable indexing procedures are available it is essential to regenerate plants for the application of the full range of tests. If the intention is to exploit further the *in vitro* methods for multiplication purposes or for the long-term storage of nuclear stocks or germplasm, it is necessary to re-isolate cultures for those purposes from plants which index negatively. This lengthy procedure is now used routinely with the potato germplasm collection at the International Potato Center, Peru[28] and a simpler alternative procedure, involving the retention of those cultures from which regenerated plants were satisfactory in the indexing tests, is not considered to be satisfactory because of the possible uneven distribution of contaminants within the cultures.

Although the *in vitro* approach to plant disease elimination is obviously complex and therefore expensive, there is increasing evidence of its effectiveness and Walkey[41] listed 42 economically important crop species which had been freed of virus by these methods. The process would not necessarily be economic with all of these species because the value of the virus-free product must be taken into account in the overall equation, but with potato, for example, the advantages seem to have been widely accepted and in 1977 Mellor and Stace-Smith[22] were able to list 134 virus-free cultivars which had been produced by *in vitro* methods.

Because of their special plant health problems, the vegetatively-propagated species present particular difficulties with regard to international distribution. For most species, seeds are the preferred propagules for movement from one country to another because of their relative freedom from pests and pathogens, and also because of their relatively small bulk, which itself reduces the risk of transmission of contaminants. As vegetative propagules, in contrast, are generally quite bulky and possibly less free of contaminants, there would be an advantage in using tissue cultures as an alternative system for international distribution on the grounds of size alone since the risks of disease transmission are proportionally reduced. There is also the further advantage that the same culture system can also be used for producing 'pathogen-tested' material for distribution and for rapid multiplication once quarantine requirements have been satisfied. These advantages have been discussed by Kahn[14] who has described the use of shoot-tip cultures for the exchange of asparagus and sugar-cane clones between East Africa and the USA. Roca *et al.*[28] have used similar methods to exchange potato germplasm with quarantine authorities from different countries and it was found that the passage through quarantine was considerably expedited.

GERMPLASM STORAGE

Objectives and options

It is essential with any plant breeding programme to maintain a constant flow of genes into the breeding population from other sources, such as locally-adapted primitive varieties and their wild relatives, in order to prevent a decline in the efficiency of that programme.[37] These genetic resources have, until fairly recently, been preserved in primitive agricultural systems and in natural habitats, but in recent years it has been recognized that this supply is threatened by the success of the breeders and the international exploitation of their new varieties, which leads to the disappearance of the old material varieties and landraces. Also, losses on natural habitats have led to the disappearance of locally-adapted populations of the wild ancestral species. Clearly, these changes are to some extent unavoidable but it is now accepted that breeding programmes will be impeded unless deliberate steps are taken to conserve representative samples of this invaluable germplasm.[5,6]

Genetic conservation strategy is complex but essentially it includes both *in situ* schemes — i.e. in the natural habitats and primitive agricultural systems — and *ex situ* gene banks. It is expected that the main burden of the task will be undertaken by the *ex situ* gene banks where germplasm will be stored mainly in the form of seed. There are, however, two important groups of crop species for which seed storage is not a feasible technique: the vegetatively-propagated species where the maintenance of clonal identity is critical and those species — mainly forest and fruit trees — which produce 'recalcitrant' seeds, unsuitable for long-term storage. This material can only be stored, therefore, in clonal form and, although conventional vegetative methods are generally practicable, they can be very expensive and they are vulnerable to losses through disease or other natural disasters. A valid alternative is now seen in the form of *in vitro* techniques, similar to those already described for multiplication (see Chapter 11) and plant health purposes, but suitably modified to reduce maintenance costs.[9] These modifications generally involve the manipulation of conditions to reduce growth rates.

In addition to this more conventional germplasm, which has been derived directly from crop species and their wild relatives, there will in future be an increasing amount of material produced from biotechnology programmes which also requires long-term storage. This will include mutants which are required in breeding work involving gene manipulation and cell culture techniques and it might be essential to conserve some of this material in cell culture form. Acceptable levels of genetic stability are, of course, an essential feature of any germplasm storage system but it is difficult to define standards and absolute stability is, for practical purposes, almost unattainable. It is quite clear, however, that certain plant tissue culture systems, without modification, would be quite unsuitable for germplasm storage purposes because of their known instability.[4] Generally speaking, it is the more disorganized tissue and cell cultures which tend to be genetically unstable, although there are apparently exceptions to this rule, e.g. *Lilium*[36]. Whether the mutation rate in these cultures is necessarily higher than that among the somatic cells of the intact plant is a debatable point, but the consequences are likely to be more serious since adventitious meristems could develop more readily from the mutant cells. In the more organized shoot-tip cultures, on the other hand, diplontic selection apparently tends to exclude mutant cells from the non-adventitious shoot meristems upon which proliferation is based, just as *in vivo* where such meristems form the essential germ line of the plant. A further advantage of shoot tip cultures is that their plant regenerative capacity is a very stable property whereas that of the more disorganized cultures can decline rapidly. The reasons for the instability of disorganized cultures are unknown but it probably arises from a number of factors, which could include the presence of mutants in the original explant, the production of further mutants *in vitro* and the unusual selection pressures operating *in vitro*. There is little evidence, at present, that these processes can be controlled easily.

For many plant species, therefore, shoot tip cultures are likely to provide the most satisfactory basis for an *in vitro* germplasm storage system and disorganized cultures should only be considered when it is known that adequate levels of genetic stability can be assured. For germplasm storage purposes, it is important that cultures can be established readily from a wide range of genotypes from a particular species. Although commercial successes indicate that this might be possible with shoot-tip cultures from many herbaceous species (see Chapter 11), such cultures have not really been exploited with woody species[3] many of which produce 'recalcitrant' seeds. Even though it might eventually be possible to produce shoot-tip cultures from a wider range of woody species there is no doubt that the cultures will be most readily produced from juvenile parts of the plant whereas valuable genotypes might only be identified at the adult stage when it might be too late to produce cultures. Many woody species during their development, however, lay down tissues such as sphaeroblasts[43] which retain their juvenile properties and these can be induced to produce adventitious shoots under certain conditions including hormonal treatments. Whether these structures are also genetically stable is not yet certain but, if so, they could probably be exploited to produce cultures which are suitable for germplasm storage.

Fully representative germplasm collections usually contain relatively large numbers of genotypes from any particular species or group of species. Where vegetative storage is essential, maintenance costs can be very high and *in vitro* methods will only be accepted as a valid alternative to the more conventional methods, involving field or glasshouse collections, if they are shown to have economic advantages. This means that the standard methods of culture maintenance, typically used for rapid multiplication purposes, are likely to be too costly because of the regular subculturing that is required. On purely economic grounds, therefore, it is important that methods for extending the subculturing intervals should be available. This could be achieved by reductions in growth rates which might also lead to an improvement in genetic stability since mutations generally occur during periods of DNA replication or mitosis. Preferably growth should be inhibited completely, but this particular aspect of plant tissue culture technology has received rather less attention than methods for the improvement of growth rates. It is, however, clear that there are difficulties in maintaining cultures in a completely non-growing state without the processes of senescence being initiated, unless methods producing a state of suspended animation can be employed. At present cryopreservation methods, involving storage at the temperature of liquid nitrogen ($-196°$), would seem to be the only practical means of achieving such a state and they have yet to be developed to the point at which they can be employed routinely with plant tissue culture systems which are suitable for germplasm storage.

Because of these difficulties, methods for maintaining minimal growth rates on a long-term basis also require development and it can be envisaged that eventually they will be used alongside cryopreservation methods in view of their complementary advantages. In many circumstances, minimal-growth storage involving shoot-tip cultures could be quite adequate in view of their morphogenetic and genetic stability and their use would be advantageous where it is important to be able to multiply material from store very quickly. Further, the utilised material would in time be replaced by the growth of the cultures and it is relatively easy to keep a visual check on the viability of cultures. In contrast cryopreserved stocks require virtually no attention, apart from the need to replenish liquid nitrogen in the storage vessels. Material taken from store, however, is not automatically replaced and viability is not easily checked, although theoretically this should not be necessary once a satisfactory procedure is established, since there should be no deterioration at such low temperatures, except for what are believed to be the relatively minor effects of background radiation.[52]

In summary, minimal growth methods for *in vitro* germplasm storage can be quite adequate and sometimes desirable where shoot-tip cultures are being employed. Such methods would not, however, be suitable for cell and tissue cultures where instability is suspected and, for them, cryopreservation would appear to be the only possible solution. Germplasm storage procedures involving shoot-tip cultures would seem to be most appropriate where the material is to be used in conventional breeding programmes because the risks of an unstable cell or tissue culture phase before and after cryopreservation are avoided. On the other hand, cryopreservation will be more suitable, or even essential, where cell or tissue cultures are used directly in biotechnology programmes or where shoot-tip culture is not a practical proposition.

Minimal-growth storage

There is no evidence from plant *in vitro* studies suggesting that shoot-tip cultures cannot be maintained indefinitely, even when growing continuously or intermittently. Unfortunately, very little is known about the developmental control of division in somatic plant cells and the methods for achieving minimal growth have been established in an empirical manner. Basically, four approaches have been used, either singly or in combination, involving modification of the physical, nutrient or gaseous conditions or the use of inhibitors. For a particular species various combinations of these methods can be used, with the choice probably depending on the economic situation for a particular germplasm collection. One of the simplest approaches involves a reduction in temperature from 25° which is standard for many species. In order to achieve a worthwhile reduction in growth rate it may be necessary with temperate species to use a temperature of 10°C or less, as for various potato species and cultivars.[44] With more tropical species such as sweet potato (*Ipomoea batatas*) chilling damage can occur below 15°C, but significant reductions in growth rate and improvements in the longevity of cultures can be achieved at temperatures around 18°C.[2] This is not necessarily a clear distinction between temperate and tropical species, as shoot-tip cultures of the tropical cocoyam (*Xanthosoma sagittifolium*) have been stored successfully for 12 months at 10°C.[1]

In addition to its effect on the growth rates of cultures, low temperature also has the effect of reducing the rate of medium evaporation. It is usually necessary to maintain some degree of gaseous exchange between cultures and the external atmosphere, otherwise toxic or anaerobic conditions can develop as oxygen levels fall and carbon dioxide levels rise. Also such a restriction of gaseous exchange can cause morphogenetic changes in the cultures, such as the production of stolons in shoot-tip cultures of potato[13,44], which might be due to the accumulation of ethylene. Provision for gaseous exchange with the external atmosphere inevitably means some loss of water vapour which can be serious with subculture intervals extending beyond one year. Evaporative rates are considerably reduced at a temperature of 10°C but it can still be useful to seal cultures with closures manufactured from substances, such as polypropylene or silicone rubber, which favour the diffusion of oxygen and carbon dioxide as opposed to water vapour. Evaporation is also less from media solidified with agar than from liquid media.

Long-term storage of cultures at temperatures below the ambient can be expensive both in terms of energy and the equipment required and there is a dependence on a reliable electricity supply. There are, therefore, circumstances under which it is advantageous to have alternative storage methods which are effective at ambient temperatures. One such method, which might have general applicability, involves the addition of substances such as mannitol to the medium in order to increase the osmolarity. Such substances are only metabolised very slowly, if at all, and experience with potato[45], yam[18] and cocoyam[1] has shown that approximately 0.2 M concentrations of these compounds severely inhibit growth rates without apparently causing injury on a long-term basis.

The effects of altering the nutrient conditions are less predictable and, compared with the standard media which are widely used, substantial excesses of many of the components seem to have toxic effects and deficiencies rapidly bring about senescence. The sucrose concentration can sometimes be more than doubled in comparison with the standard concentration of 3% (w/v) without damaging the cultures and their longevity may thus be increased, especially if the growth rate is at the same time inhibited in some other way. Conversely, reduced concentrations of sucrose can sometimes be used to increase longevity of cultures, under conditions where it seems likely that the cultures, as a result, become autotrophic to the extent that a slow growth rate can be maintained. With shoot-tip cultures of cocoyam, for example, the standard concentrations of sucrose in the medium (3% w/v) favour the development of relatively large leaves and the medium is rapidly depleted of nutrients, whereas low concentrations (1% w/v) restrict leaf growth and longevity is accordingly increased.[1]

Various growth regulators can be used to inhibit the growth of cultures but the effects of most would probably be deleterious in the long-term, especially as far as the genetic constitution of the cells is concerned. Certain growth regulators, however, such as abscisic acid, are apparently involved in the natural dormancy mechanisms of vegetative organs and, consequently, they might be expected to have less damaging effects. Westcott[45] found that abscisic acid at concentrations in the range of $10-20$ mg 1^{-1} could be used to inhibit the growth of potato shoot-tip cultures so that their longevity was increased.

As a general recommendation, it would seem that the most practical means of achieving minimal growth rates involve the manipulation of physical conditions such as the temperature or osmolarity of the growth medium. Manipulations of nutrient concentrations or addition of inhibitory growth regulators are likely to be of more limited value.

Cryopreservation

The first report of the successful cryopreservation of plant cells which had been grown in culture was by Quatrano.[27] It was known that certain cells with low water content, such as pollen, seeds and dormant buds and tissues from hardy species could be subjected to the temperature of liquid nitrogen without suffering damage[29] but it was clear that plant cells with higher water contents presented considerable problems because of the risk of injurious ice crystal formation. Quatrano[27] showed, for the first time, that a procedure involving the use of cryoprotectants and controlled cooling rates, similar to that which had been developed with microorganisms and animal cells, was also applicable to certain plant cells. Subsequently, it has also been shown that these techniques can be adapted successfully for use with cultured cells from a wide range of plant species, although certain types of cells remain recalcitrant.

In principle, there are several procedures which can help to overcome the cryopreservation problems caused by the high water contents of plant cells:

(a)   Cells might be selected at a developmental stage when their water contents are low;
(b)   Some of the water may be removed osmotically from the cells;
(c)   The cells might be treated with suitable cryoprotective substances;
(d)   Suitable cooling and thawing rates might be chosen.

A suitable combination of these methods must, however, be established empirically for every cell type, because of variation in their physical properties. Consequently, it is easier to establish a satisfactory protocol for a uniform population of cells than for a mixed population. Also assemblages of cells present greater problems than individual cells because of their bulk and the resulting difficulty in applying cooling rates and cryoprotectants uniformly. Cryopreservation techniques are, therefore, most readily applied to plant cells growing *in vitro*, since the environment and the physiological and physical properties of the cells can be controlled with a reasonable degree of precision. Finely dispersed suspension cultures are likely to provide the most suitable material because of the relatively small size of the cell aggregates and, if taken during the exponential phase of growth, the average cell size and their water content will be low. Most of the successes reported for plant cells have been achieved with this type of system (see reviews by Withers[48,49,50]).

The choice of an appropriate cooling rate is a critical factor in the establishment of a satisfactory cryopreservation protocol but, unfortunately, the mechanism of freezing damage during the cooling process is not fully understood. It is obvious, however, that it is a complex process and it is unlikely that any single theory will provide an adequate explanation.[23] In practice, it is found that the curve relating cell survival to the cooling rate has two peaks: a sharp optimum at relatively slow cooling rates, in the region of $1°C$ min^{-1}, and another peak with very rapid cooling rates, in excess of $1000°C$ min^{-1}, which continues to rise until the practical limits of rapid cooling are reached. As might be expected with such a pronounced bimodal response, it is necessary to provide different explanations for the two optimal cooling rates, although in both situations the effect is to minimise the damaging effects of intracellular ice crystal formation.

There is some agreement about the sequence of events when plant cells are cooled slowly to freezing temperatures[20,21] and it is believed that, for a number of physical reasons, the first ice formation will occur extracellularly.[23] The vapour pressure of the extracellular water will thus be lowered, causing water to flow osmotically from the cells, which will continue to dehydrate and shrink, as long as intracellular ice formation is delayed by supercooling or by the increased concentration of the solutes. According to the cooling rate and other factors, the dehydration process will continue either until all the available water is removed from the cells or until intracellular ice formation occurs. During this process of 'equilibrium freezing' damage may be caused by a number of factors, including ice crystal growth, increased solute concentration, over-dehydration, pH changes, plasmolysis. It is thought that the rather sharply defined optimal cooling rate is a result of the interaction between the protecting effects of dehydration and the damaging effects of increased solute concentration, etc.

The damaging effects of freezing during a slow-cooling process are, therefore, minimized by the inevitable removal of water from the cells. In contrast, when ultra-rapid cooling rates are employed there is no opportunity for water to leave the cell but with the faster rates of cooling larger numbers of smaller and, therefore, less damaging ice crystals are formed, a process which is sometimes known as vitrification.[17]

It is not only the freezing process which damages cells and it can be just as critical to selecting a thawing rate which is appropriate for a particular cooling rate. If the cell contains water frozen in a more or less amorphous state at $-196°C$, there is a transition from this state to a crystalline state at ca. $-130°C$ and recrystallization and ice crystal growth may occur at temperatures higher than ca. $-70°C$.[23] A fast rate of thawing will, therefore, reduce the opportunity for large damaging ice crystals to be formed. This is only an advantage when the cell contains small ice crystals as a result of ultra-fast freezing for example, and with highly dehydrated cells fast thawing can actually cause damage as a result of rehydration and deplasmolysis. Fast thawing is, therefore, essential for rapidly frozen cells and slow thawing may be beneficial for slowly frozen cells, although the thawing rate may not actually be critical for the latter.

In addition to the quite distinct cryopreservation procedures, based on either slow or rapid cooling rates, there is evidence to suggest it is possible to combine the advantages of both in a two-stage procedure. The first stage involves either step-wise cooling in a series of successively lower holding temperatures[32] or a continuous slow cooling rate throughout the same temperature range.[15,16] The temperature range chosen for the first stage depends on the physiological state of the cells and the cryoprotectant that is used and the aim is to achieve some protective dehydration before intracellular freezing is initiated in the second stage when the specimen is plunged directly into liquid nitrogen. The two-stage method is thought, therefore, to combine the protective dehydration of the slow-cooling method with the small ice crystal size of the rapid-cooling method.

With the freezing and thawing procedures that have been described it is essential to employ a suitable cryoprotectant treatment. The only exceptions are cells and tissues with low water contents such as pollen[42], seeds[26], and dehydrated seedling[39], and certain cold hardened tissues including twigs[31] and callus[30]. Almost any solute that is not toxic has some cryoprotective properties and a range of rather diverse compounds has been used successfully with plant and animal cells. As the molecular size of these compounds varies considerably and some penetrate the cell while others are probably non-penetrating, it is unlikely that a single theory can give an adequate explanation of their protective qualities. Various mechanisms have been suggested including a colligative effect by which the cryoprotectant lowers the concentration of potentially damaging substances in solution during the freezing process, lowering the freezing point of the intracellular water, osmotic dehydration, increased membrane permeability and the protection of macromolecules, especially those in membranes, from denaturation (see discussions by Meryman[23], and Mazur[20]).

The most commonly employed cryoprotectants with plant cell and tissue cultures are dimethyl sulphoxide and glycerol, used at approximately molar concentrations, and it has also been suggested[51] that the amino acid proline can be very effective. Frequently, it seems, these compounds are more effective when used in mixtures, rather than singly.[48] They are generally added more or less immediately before the freezing process, but they may be incorporated into a medium in which the cells are grown for several days before freezing.[15]

The actual cooling procedure can be carried out with the specimen immersed in a mixture of growth medium and cryoprotectant. This is only really suitable for slow-freezing procedures, since the volume of the medium does not allow uniform rapid cooling rates to be achieved, but it is convenient and frequently used for cell suspensions. Alternatively, the specimens, after cryoprotectant treatment, can be frozen 'dry', either enclosed within aluminium foil[47] or unprotected on some form of support, such as the tip of a hypodermic needle which is plunged directly into liquid nitrogen.[7] The latter procedure is only suitable for rapid-cooling procedures and, under certain circumstances, it could be difficult to maintain sterility.

The thawing procedure is usually relatively simple: for 'slow' thawing the specimen is usually transferred to air at room temperature and for 'rapid' thawing it is immersed in water or medium at a temperature of $35°-40°C$.

Although a great deal of attention has been given to the freezing and thawing aspects of cryopreservation there is evidence that the treatment of the specimens both before and after these stages can also be quite critical. It has been shown[7,10,38], for example, that shoot meristems of the potato, *Solanum goniocalyx* respond best to a rapid freezing procedure when they have been first incubated for 2 days on a normal growth medium. It was also shown with other potato species that the recovery after thawing is affected by both the physical conditions (light and temperature) and the medium composition (especially the complement of growth regulators). There is obviously a great deal of scope for treatments which produce the optimal physiological state in cells prior to freezing and which protect the cells during their recovery from what has undoubtedly been a drastic experience. Both aspects of the problem are discussed at length by Withers.[48]

The equipment that has been used for cryopreservation work can be relatively simple and inexpensive or much more sophisticated and costly. Generally speaking, apart from the facilities required for the *in vitro* work and the handling of liquid nitrogen, it is the controlled-cooling apparatus that is likely to be the most elaborate. Often, however, it is only used for the preliminary empirical investigations by which a suitable protocol is established. Thereafter, it is often possible to design a more simple piece of equipment for the routine application of the protocol, especially if it involves rapid cooling. In practice, if large numbers of diverse specimens are to be handled, investment in more automated equipment is probably justified. Various types of freezing units are described in a further review by Withers.[49]

The present situation regarding the possibility of cryopreservation techniques being used routinely for the conservation of plant germplasm is that far greater progress has been made with cell and callus cultures than with the various types of organized cultures (see list in review by Withers[49]). There seems to be agreement that procedures for cell cultures should be based on either slow or two-stage cooling, although it is obvious that 'fine tuning' is essential for both cooling rates and the cryoprotectant requirements. It is not unrealistic to suggest that, with further refinements in these basic methods, cryopreservation protocols could be devised for cell cultures from most plant species. As discussed earlier, however, this is not necessarily the most suitable type of culture for germplasm storage, except for material that is actually produced *in vitro*.

Unfortunately, there is less agreement about the most appropriate procedure for the cryopreservation of meristems and other types of organized cultures, which would be more suitable for germplasm storage. Here the technical problems are more severe because of the diversity of cell types, the size of the specimen and the need to maintain multicellular organization in addition to cell viability. Fewer successes have been reported and initially they were achieved with rapid cooling methods (Carnation: Seibert[33], Seibert and Wetherbee[34]; Potato: Grout and Henshaw[7]), although more recently two-stage methods have been successfully employed (Strawberry: Sakai *et al.*[32]; Kartha *et al.*[16]; Pea: Kartha *et al.*[15]; Potato: Towill[40]). The rapid-cooling method has the advantage of an extremely simple technical procedure but also the disadvantage of being less flexible than the methods which involve some form of protective dehydration, since there is less scope for adjustments to the cooling rate. This implies that rapid-freezing methods are likely to be more dependent on the availability of meristems in a suitable physiological state than the methods involving protective dehydration which might, within limits, be adjusted to suit the state.

In practice, whether it is more economic to adjust the state of the meristems or to adjust the freezing method will depend on many factors. The work with the rapid freezing method and potato meristems taken directly from glasshouse-grown plants has shown, for example, that success is dependent on the age of the plant and the season and, further, the conditions which suit one cultivar are not necessarily those which suit another (O'Hara, personal communication). On the other hand, the investigations of Kartha et al.[16] with a two-stage freezing method and strawberry meristems taken from cultures shows just how precise the adjustments to cooling rates must be: the optimal procedure involved continuous slow cooling at a rate of $0.84°C$ min^{-1} to $-40°C$ before transfer to liquid nitrogen and alternative cooling rates of $0.56°C$ min^{-1} and $0.95°C$ min^{-1} were considerably less effective. It is certainly arguable that neither of these approaches is yet a practical proposition for routine germplasm storage, where large numbers of genotypes, possibly with differing requirements, are involved.

Withers[49] has suggested that a two-stage method, involving stepwise cooling in the first stage before transfer to liquid nitrogen, might be of more general use since it is easier to determine precise cooling requirements and the risk of over-dehydration is reduced. This might well be true, but at this stage it is probably unwise to rule out any of the approaches which have so far achieved the rather limited success with meristems.

It is important that the meristems should survive the cryopreservation process in a state which enables them to continue normal growth, but it has been shown in electron microscope studies[8,15] that meristems do suffer some damage and it is quite possible that only a proportion of the original cells is involved in the re-growth of the meristem. Whether this is within the regenerative capacity of the original meristem or whether there is an intervening callus phase followed by adventitious production of a new meristem is a debatable question, which could have genetic implications. It is important, therefore, that the value of a particular cryopreservation procedure should also be judged by its ability to allow the regrowth of meristems in as normal a manner as possible. The different procedures, with their different effects on the shrinking and swelling of cells during the stages of freezing and thawing, might have rather different effects on the maintenance of organizational integrity.

SUMMARY AND GENERAL CONCLUSIONS

In vitro methods can now be used for the elimination and exclusion of pests and diseases from many vegetatively-propagated plant species. The same basic methods can also be used to reduce quarantine restrictions on the international transfer of such material, and for the multiplication stage from pathogen-tested nuclear stocks. The methods are not infallible and they must be used with appropriate phytosanitary precautions; their advantages compared with conventional methods are quantitative rather than qualitative and their use can lead to greater efficiency.

It is now generally accepted that *in vitro* methods also have a role in the overall strategy for germplasm conservation, especially with the vegetatively-propagated species and those which produce 'recalcitrant' seeds. Where possible, the risks of genetic instability should be minimized by the use of methods based on the non-adventitious multiplication of shoot meristems. Such methods are widely applicable with the herbaceous species, but less so among those which are woody. Maintenance costs can be reduced and genetic stability possibly improved by the lowering of growth rates. Minimal growth rates can be achieved quite readily by manipulation of the physical and nutrient conditions, but it is essential to use cryopreservation techniques for the complete suspension of growth. The latter techniques have now been developed to the point at which they are beginning to be used routinely with cell cultures, but meristem cultures present greater technical problems which have yet to be completely surmounted. The minimal growth and cryopreservation methods for *in vitro* germplasm storage are essentially complementary in their advantages and disadvantages and, ideally, both methods should be available for use with any particular species. Assumptions about the long-term genetic stability of material stored under *in vitro* conditions are largely based on the supposed stability of non-adventitious shoot meristems and evidence from the use of similar methods for large scale commercial multiplication; further genetic evidence is required but complete confidence can only come from the actual long-term use of the methods.

REFERENCES

1.   Acheampong, E. *Ph.D. Thesis*, University of Birmingham, England, 1982.
2.   Alan, J.J. *Ph.D. Thesis*, University of Birmingham, England, 1979.
3.   Bonga, J.M. Vegetative propagation: tissue and organ culture as an alternative to rooting cuttings. *New Zealand J. Forestry Sci.* 4, 253—260, 1974.
4.   D'Amato, F. The problems of genetic stability in plant tissue and cell culture. *Crop Genetic Resources for Today and Tomorrow*, 333—348. Frankel, O.H. and Hawkes, J.G. (eds.). Cambridge Univ. Press, Cambridge, 1975.
5.   Frankel, O.H. and Bennett, E. (eds.). *Genetic Resources in Plants — Their Exploration and Conservation*. Blackwell, Oxford, 1970.
6.   Frankel, O.H. and Hawkes, J.G. (eds.). *Crop Genetic Resources for Today and Tomorrow*. Cambridge Univ. Press, Cambridge, 1975.
7.   Grout, B.W.W. and Henshaw, G.G. Freeze-preservation of potato shoot-tip cultures. *Ann. Bot.* 42, 1227—1229, 1978.
8.   Grout, B.W.W. and Henshaw, G.G. Structural observations on the growth of potato shoot-tip cultures after thawing from liquid nitrogen. *Ann. Bot.* 46, 243—248, 1980.
9.   Henshaw, G.G. Plant tissue culture: its potential for dissemination of pathogen-free germplasm and multiplication of planting material. *Plant Health*, 139—147. Ebbels, D.L. and King, J.E. (eds.). Blackwell, Oxford, 1979.
10.   Henshaw, G.G., O'Hara, J.F. and Westcott, R.J. Tissue culture methods for the storage and utilization of potato germplasm. *Tissue culture for plant pathologists*, 71—76. Ingram, D. and Helgeson, J. (eds.). Blackwell, Oxford, 1980.
11.   Hollings, M. and Stone, O.M. Investigation of carnation viruses. I. Carnation mottle. *Ann. Appl. Biol.* 53, 103—118, 1964.
12.   Homes, F.O. Elimination of spotted wilt from a stock of dahlia. *Phytopathol.,* 38, 314, 1948.
13.   Hussey, G. and Stacey, N.J. *In vitro* propagation of potato (*Solanum tuberosum* L.). *Ann. Bot.* 48, 787—796, 1981.
14.   Kahn, R.P. Plant quarantine: principles, methodology, and suggested approaches. *Plant Health and Quarantine in International Transfer of Genetic Resources*, 289—307. Hewitt, W.B. and Chiarappa, L. (eds.). C.R.C. Press, Cleveland, 1977.

15.   Kartha, K.K., Leung, N.L. and Gamborg, O.L. Freeze preservation of pea meristems in liquid nitrogen and subsequent plant regeneration. *Plant Sci.* **15**, 7–15, 1979.

16.   Kartha, K.K., Leung, N.L. and Pahl, K. Cryopreservation of strawberry meristems and mass propagation of plantlets. *J. Amer. Soc. Hort. Sci.* **105**, 481–484, 1980.

17.   Luyett, B.J. Anatomy of the freezing process in physical systems. *Cryobiology*, 115–138. Meryman, H.T. (ed.). Academic Press, London, 1966.

18.   Mathias, S.F. *M.Sc. Thesis*, University of Birmingham, England, 1981.

19.   Mazur, P. Physical and chemical basis of injury in single-celled microorganisms subjected to freezing and thawing. *Cryobiology*, 214–316. Meryman, H.T. (ed.). Academic Press, London, 1966.

20.   Mazur, P. Freezing injury in plants. *Ann. Rev. Plant Physiol.* **20**, 419–448, 1969.

21.   Mazur, P., Leibo, S.P. and Chu, E.H.Y. A two-factor hypothesis of freezing injury. Evidence from chinese hamster tissue culture cells. *Exp. Cell Res.* **71**, 354–355, 1972.

22.   Mellor, F.C. and Stace-Smith, R. Virus-free potatoes by tissue culture. *Plant Cell, Tissue and Organ Culture*, 616–635. Reinert, J. and Bajaj, Y.P.S. (eds.). Springer-Verlag, Berlin, 1977.

23.   Meryman, H.T. (ed.). Review of biological freezing. *Cryobiology*, 2–106. Academic Press, London, 1966.

24.   Morel, G.M. and Martin, C. Guérison de Dahlia atteints d'une maladie à virus. *C.R. Acad. Sci.* **235**, 1324–1325, 1952.

25.   Mori, K. and Hosokawa, D. Localization of viruses in apical meristem and production of virus-free plants by means of meristem and tissue culture. *Acta Hort.* **78**, 389–396, 1977.

26.   Mumford, P.M. and Grout, B.W.W. Germination and liquid nitrogen storage of cassava seed. *Ann. Bot.* **42**, 255–257, 1978.

27.   Quatrano, R.S. Freeze-preservation of cultured flax cells utilizing DMSO. *Plant Physiol.* **43**, 2057–2061, 1968.

28.   Roca, W.M., Bryan, J.E. and Roca, M.R. Tissue culture for the international transfer of potato genetic resources. *Amer. Potato J.* **56**, 1–10, 1979.

29.   Sakai, A. and Noshiro, M. Some factors contributing to the survival of crop seeds cooled to the temperature of liquid nitrogen. *Crop Genetic Resources for Today and Tomorrow*, 317–326. Frankel, O.H. and Hawkes, J.G. (eds.). Cambridge Univ. Press, Cambridge, 1975.

30.   Sakai, A. and Sugawara, Y. Survival of poplar callus at super-low temperatures after cold acclimation. *Plant and Cell Physiology* **13**, 1129–1133, 1973.

31.   Sakai, A., Otsuka, K. and Yoshida, S. Mechanism of survival in plant cells at super-low temperatures by rapid cooling and re-warming. *Cryobiology*, **4**, 165–168, 1968.

32.   Sakai, A., Yamakawa, M., Sakato, D., Harada, T. and Yakuwa, T. Development of a whole plant from an excised strawberry runner apex frozen to −196°C. *Low Temp. Sci. Series B*, **36**, 31–38, 1978.

33.   Seibert, M. Shoot initiation from carnation shoot apices frozen to −196°C. *Science* **191**, 1178–1179, 1976.

34.   Seibert, M. and Wetherbee, P.J. Increased survival and differentiation of frozen herbaceous plant organ cultures through cold treatment. *Plant Physiol.* **59**, 1043–1046, 1977.

35.   Shepard, J.F. Regeneration of plants from protoplasts of potato virus X-infected tobacco leaves. II. Influence of virazole on the frequency of infection. *Virology* **78**, 261–266, 1977.

36.   Sheridan, W.F. Tissue culture of the monocot. *Lilium*. *Planta* **82**, 189–192, 1968.

37.   Simmonds, N.W. *Principles of Crop Improvement*, 408. Longman, London, 1979.

38.   Stamp, J.A. *M.Sc Thesis*, University of Birmingham, England, 1978.

39.   Sun, C.N. The survival of excised pea seedlings after drying and freezing in liquid nitrogen. *Bot. Gaz.* **119**, 234–236, 1958.

40.   Towill, L.E. *Solanum tuberosum*: a model for studying the cryobiology of shoot-tips in the tuber-bearing *Solanum* species. *Plant Sci. Lett.* **20**, 315–324, 1981.

41.   Walkey, D.G.A. *In vitro* methods for virus elimination. *Frontiers of Plant Tissue Culture 1978*, 245–254. Thorpe, T.A. (ed.). International Ass. for Plant Tissue Culture, Calgary, 1978.

42.   Weatherhead, M.A., Grout, B.W.W. and Henshaw, G.G.   Advantages of storage of potato pollen in liquid nitrogen. *Potato Research* **21**, 97–100, 1979.

43.   Wellensiek, S.J.   Rejuvenation of woody plants by formation of sphaeroblasts. *Proc. kon. ned. Akad. Wetensch. Series C,* **55**, 567–573, 1952.

44.   Westcott, R.J.   Tissue culture storage of potato germplasm. 1. Minimal growth storage. *Potato Research* **24**, 331–342, 1981a.

45.   Westcott, R.J.   Tissue culture storage of potato germplasm. 2. Use of growth retardants. *Potato Research* **24**, 343–352, 1981b.

46.   White, P.R.   Multiplication of the viruses of Tobacco and Aucuba mosaics in growing excised tomato root tips. *Phytopathology* **24**, 1003–1011, 1934.

47.   Withers, L.A.   Freeze-preservation of somatic embryos and clonal plantlets of carrot (*Daucus carota* L.). *Plant Physiology* **63**, 460–467, 1979.

48.   Withers, L.A.   Low temperature storage of plant tissue cultures. *Advances in Biochemical Engineering Vol. 18*, 102–150. Fiechter, A. (ed.). Springer-Verlag, Berlin, 1980a.

49.   Withers, L.A.   *Tissue culture storage for genetic conservation*, 91. *International Board for Plant Genetic Resources Technical Report*, IBPGR, Rome, 1980b.

50.   Withers, L.A.   Preservation of germplasm. *Perspectives in plant cell and tissue culture. International Review of Cytology. Suppl. IIB*, 101–136. Vasil, I.K. and Murphy, D.G. (eds.). Academic Press, New York, 1980c.

51.   Withers, L.A. and King, P.J.   Proline – a novel cryoprotectant for the freeze-preservation of cultured cells of *Zea mays* L. *Plant Physiol.* **64**, 675–678, 1979.

52.   Whittingham, D.G., Lyon, M.E. and Glenister, P.H.   Long-term storage of mouse embryos at -196°C: The effect of background radiation. *Genet. Res.* **29**, 171–178, 1977.

CHAPTER 15

BREEDING TOWARDS AN IDEOTYPE – AIMING AT A MOVING TARGET?

Stig Blixt
Weibullsholm Plant Breeding Institute,
Landskrona, Sweden

and

P.B. Vose
International Atomic Energy Agency, Vienna, at Centro de Energia
Nuclear na Agricultura, Piracicaba, S.P., Brazil

IDEOTYPE AND GENOTYPE

The origin of the ideotype concept is probably lost in prehistory when primitive Man came out of caves to start living in huts. At the same time the first primitive cultivation began around the settlements. Probably the women were the first cultivators and it takes little imagination to think that, in saving seeds for the next crop, those seeds came from the largest and most vigorous looking plants. Then began the whole concept of phenotypic selection for yield, which has dominated plant breeding practice down to the present time. Thus parent plants were chosen for crossing primarily on their high yielding ability, and the segregating progeny again selected for yield.

In fact, although it was comparatively recently that Donald[11] proposed the use of the word 'ideotype' for biological models expected to perform in a predictable manner within a defined environment, plant breeders have for a long time been defining and re-defining what they have considered to be their desired plant, and as farming systems and technology have changed these too have steadily changed thinking on ideal plant type. Thus Ishizuka[20] has noted that the continuously increasing yield of rice in Hokkaido Province, Japan, since 1890 has been accompanied by greatly increasing tillering and much reduced stem height. This provides the current rice model described by Jennings[21] and Beachell and Jennings[2] with semi-dwarf straw, high tillering and with short, thick highly angled leaves. In Washington State, USA, the semi-dwarf wheat selections in effect developed their own model, as Vogel et al.[30] noted that the characteristics of their highest yielding selections 'comprise a general plant pattern, and any selection differing distinctly from this pattern in one or more of those agronomic characteristics was notably less efficient in production'. In W. Europe improved tillage and drilling methods have made wheat establishment more secure and thus partly removed the need for tillering capacity. Thus MacKey[23] noted the trend towards decreased tillers with larger heads, while viewing models primarily as a concept for assisting decisions in breeding programmes. With corn (Zea mays) in the USA increasing yields have come through greater plant populations with decreasing row width, reducing tendency to tiller and putting a premium on types with angled leaves for good light interception.[13] Models are therefore nothing new, but we are now demanding more from them.

The ideotype can be seen as a specified model of a plant which is to produce a wanted product in a required amount and quality, in a way that minimizes the effect of environmental variability during growth and subsequent losses. In practice such a model is usually described in terms of phenotype, but theoretically it could also be visualized as a genotype. If the genotype is constructed from knowledge of the single allele effects, and the interaction effects between all single alleles present in the phenotype under a specified range of environmental conditions were known, together with their interaction with environmental parameters, the predicted performance of such an ideotype would be identical to the performance of a phenotype with that genotype.

Until now the genotypes of higher plants have been deduced mainly from studies of the phenotype. Therefore the mechanism for converting phenotype to genotype knowledge is readily available in the form of different methods for gene and linkage analyses used in plant genetics. In plant breeding work, however, most thinking along ideotype lines has been based on phenotype rather than genotype. From this point of view phenotype has often been studied in a rather simplistic manner.

In some discussions on model building, or on defining the specific features of an ideotype, it appears, whether consciously or not, that the designer of the model has regarded morphological characters as something which exist, without regard to the mechanism of their origin nor indeed the physiological significance of their being. As we see it, the morphological phenotypes are merely the visible physical expression of a biochemical or physiological message that has caused a change in phenotypic morphology. Such morphological change may or may not be accompanied by what we may term a continuing physiological expression, such as may reflect differences in photosynthesis, translocation capacity, nitrogen utilization, assimilate partitioning, or assimilate remobilization. As long ago as 1934 Boonstra[3] thought that the analysis of yield components had proved disappointing and believed that causes of differences in yield were to be found in studies of metabolic processes such as assimilation, absorption, transpiration, etc. of different genotypes. Since that time a number of authors[28,31,33,35] have emphasized the genetic basis of nutritional and physiological factors and their potential importance for increased yield.

Within crop species there are general underlying patterns to carbon dioxide, nitrogen and water economy. Nevertheless, considerable variation exists both between and within genera, and within this range lies our capacity for selection and recombination of physiological characters, both for improved efficiency and for specific situations. Probably in many, maybe most, cases we have not yet got the optimum combinations, and in this discussion we put forward the view that ideotypes, while useful as a guide, must be continually updated to match new knowledge, new possibilities and new requirements. We might note here that ideotypes are rather idealized targets to which we may aspire to work towards, bearing in mind that in practical terms the ideal may not be obtainable or even desirable. Donald[11] has sometimes been criticized for developing a wheat ideotype based on the concept of a single culm, on the grounds that a uniculm plant lacks adaptability and is unable to develop tillers to fill up a thin crop. In fact Donald himself recognized this, and that he was designing an idealised wheat ideotype for ideal circumstances which might not be achieved.

The phenotype as a whole being very complex and therefore difficult to handle, it has been subdivided into elements we usually refer to as characters. Now the entire plant, the phenotype, bears a direct relation to the genotype: the phenotype is the total result of the effects of all allelic interactions within the genotype and the interaction of these and the environment. This, of course, is the reason why the phenotype has always been a useful tool. If the division in characters is to mean more refinement and greater efficiency, then the direct relation to genetics must be kept and refined, too. In other words, we have to define the genetic component of that particular part of the phenotype and put some kind of measure on the interaction of this genetic part with the part of the genotype, and also delimit the environmental part. Many important standard characters are used today, for instance yield, though it is well known that yield is a very complex result of a multitude of elements of which the more obvious ones have often been called yield components. What then is a character?

For a long time genetics has gained most fundamental theoretical insight into genetic mechanisms from studying, both intensively and extensively, the much simpler procaryotic system.[36] These studies have, of course, equally benefited higher plant genetics. One result relevant in this context is the demonstration that the phenotype of higher plants can, and should, also be subdivided into characters corresponding to the effect of one gene or locus, where character expression is the effect of the specific allele of that locus in interaction with the rest of the genotype and the environment. But it has also left higher plant genetics — and plant breeders — in a situation where remarkably little seems to be known of the detailed gene-control of characters in crops of the utmost importance for mankind. However, a vast amount of information exists, describing, defining and interpreting such characters in different ways. It is therefore probable that much more genetic information could be made available, though it may not now be expressed in sufficiently clear genetical terms. One example may be given from taxonomy: most keys for the identification of plants, including crop plants, probably contain concealed genetic information; a good clear key character with small variation indicates a strong and relatively simple inheritance.[37,38] This is particularly true at the subspecific or lower levels, bearing in mind that 'species' is a very broad term with a very varying content.

There is no doubt that when the plant breeder presently defines a new superior variety in terms of phenotype, whether defined by morphology or physiology, he is contriving a new ideotype. To find this new ideotype, complementary or related phenotypes are combined in the hope that the wanted ideotype appears. How much of this process will be left to chance and how much can be controlled is due to how much genetic knowledge is available for the choice of parents and selection criteria. It is the coded sequences of DNA which are transferred, not the phenotypic characters. And what comes out of the new recombinations cannot be predicted by adding phenotypic effects, only by knowing allele-environment interactions.

THE GENETIC BASE

As usual, while one may see the road far ahead smooth and paved, the part we have to tread today may look full of holes and all uphill. After all, the number of genes in a higher plant may amount to several tens of thousands and it seems perhaps an impossible task. But perhaps not. There are some few thousands of genes, because of phylogeny, almost certainly the same as found in microorganisms. These are the genes working on the cellular level. A large group must therefore be considered in all probability to be common to all higher plants, for the same phylogenetic reason.[39] Thus, for such genes research on any microorganism or plant will have some degree of validity for all plant species. Plant orders, families and genera, in all probability share large numbers of genes that do not need to be studied separately in each species. This suggests that as we now have access to comparatively cheap computer facilities capable of handling millions of information bits, the task of handling at the first stage some ten thousand genes and recording some ten key data for each gene does not seem too difficult. To produce and collect this information may seem a big job, but so much data is already coming out every day that could be used but isn't. For instance, much plant physiological work results in the determination of genotypic data with respect to some or other character in different plant strains, although it may not be presented in genetical terms. It often takes very little additional work to establish the genetic situation, but first of all the experimental material, those particular genotypes, varieties, lines, etc. showing the contrasting characters, have to be stored in a genebank together with the information. Had this been done during the last 30 years, which wasn't possible of course as the facilities were not available, vast numbers of genes might already have been known. Particularly so, if this had been done in all pertinent fields of plant science.

In peas, the number of known genes has increased in a logarithmic fashion: one gene was recognizably described about the year 1200; 4 about 1670; 11 genes were known in 1900; about 53 in 1925; 94 in 1950; 235 in 1975 and 373 in 1980; another two thousand have been found but await naming and analysis.[40] It seems inevitable that development will be towards a situation where the majority of genes are known, and that is not too far away in the time-scale of human civilization.

We may conclude therefore that efforts spent on initially defining a character as the smallest possible element that can be phenotypically discerned and handled, will result in characters controlled by fewer genes. The fewer genes controlling the character then the more predictable will be the outcome of recombinations; and this predictability is a highly desirable quality which paves and smoothes the road.

Such a conclusion must greatly affect our concepts of germplasm conservation. We have just noted that the breeder gets best control over the breeding process in order to materialize an ideotype the more the characters in the germplasm he uses correspond to genes to be transferred and recombined. The information collected and preserved in germplasm collections, or genebanks, should therefore characterize the material, i.e. provide information on character expression as just defined. This type of information is always useful. Evaluating an accession for a number of standard complex agronomic characters such as yield, brewing quality, lodging resistance, etc. to find out what phenotype results from a certain environment, also helps in finding out if the accession corresponds directly to the ideotype in mind. It does not necessarily add to the possibility of using the material for materializing another ideotype. Moreover, the concept of germplasm conservation rests heavily on the belief that new ideotypes have to be contrived and materialized continuously to meet the changing and increasing demands for agricultural production, i.e. the continuous success of an increasingly efficient plant breeding.[41,42]

Much for the sake of convenience, we talk about different types of characters, morphological, physiological, biochemical, etc. This, of course, is very useful in cataloguing knowledge of characters, as long as it is crystal clear that this way of classifying characters is entirely artificial and done for the sake of cataloguing only. "Characters", as defined, must have a hereditary element but all characteristics result from allele/environment interactions. Thus, all characters begin as DNA-sequences, turned into various regulatory substances, interacting at the cellular level to form, through a number of steps or cycles, that which is observed or measured as a character.

Therefore, every character, whether it is visible, i.e. morphological, or measured through metabolic or physiological criteria should be expected to affect at least the biochemical constitution and the physiological reactions of the plant. There are certainly great differences in degree but probably not in quality. If we should distinguish different types of characters, it might be between those acting mainly at the cellular level and thus also mainly or entirely having an affect at the organ site, i.e. they are involved in the differentiation mechanism. It seems plausible that the latter affect is necessary to make a character morphologically visible.

THE PHYSIOLOGICAL BASE

The case against too strict morphological ideotypes has been well put by Evans[15] in relation to *Phaseolus*. She wrote: "Although at its best this is an attractive logical approach, at its worst it is unfortunately dogmatic and unrealistic. It seeks to rationalize the plant breeders' programme, to create the ideal plant and to suggest an ideal environment for it . . . . A model, of necessity, narrows the spectrum of a breeding programme, and we know that several models can lead to high yield. Plant design may be both difficult and profitless because the application of the principle of design may lead to a range of theoretical models. The question arises whether any of the principles of design can have substantial constancy of expression over a wide range of environment. Until such a situation is achieved, I think the breeder would be well advised to give greater priority to the adaptability concept rather than the ideotype concept."

The concept of adaptability is a good one, but it can be carried too far if one is concerned with selecting genotypes for extremely adverse environments. In general one might agree that physiological characters such as efficient photosynthesis or water use, and good 'architectural' characters such as relating to canopy structure, leaf angle and size, tillering habit, etc. should be introduced as good base characteristics of the general breeding material. However, if one is wishing to obtain material suitable for extreme conditions such as highly saline soils, or for highly acid soils, then the need to obtain rare tolerant material fundamentally transcends any other requirement. This is because without such basic tolerance, all other good yield attributes will be unavailing. This is well seen in the CYMMYT wheat varieties, characterised by Sonora, which has shown great adaptability in many parts of the World, but is quite unsuited to the characteristically acid soils of Brazil, for instance. The Michigan family of *Phaseolus* varieties related to Sanilac are similarly unsuited to acid soils though seemingly otherwise quite well adapted to many regions.

To a lesser extent a similar situation applies to soybean, sorghum and maize varieties intended for growing on calcareous soils with high pH, liable to iron deficiency. Proper choice of basic material having tolerance to typical iron deficiency situations can prevent subsequent field susceptibility. Drought and cold tolerance also demand basic material which has been predominantly chosen for these characteristics. This is emphasized here, because much of the yet remaining undeveloped agricultural land in the World is situated in regions or on soils where there are major stress problems.[12,18,26] Breeding goals must, in many cases, take these extreme conditions into account, and the more extreme the requirement, such as salt tolerance, then the more this must be a basic property of the breeding material.

We know that there are genotypic differences in the efficiency of nitrogen utilization and response, and differences in the capacity to remobilize and translocate nitrogen assimilates to developing grain. As we better appreciate how to use this variation, it can ultimately be employed in improving the effectiveness of N-fertilizer. At what stage this type of physiological variation should be brought into a breeding programme is not yet clear, but probably like stress tolerance it must be introduced at a very early stage as a basic property.

In general, the results of research directed towards improved photosynthesis have been disappointing in practical terms of improved varieties. Improvements in overall carbon assimilation have come from improved light interception through selecting for leaf type and angle, e.g. in rice. Moreover, most 'improvements' in yield have come about not through increasing the total biomass of a crop, but through increasing the proportion of economic yield, such as grain, in relation to straw (i.e. high 'harvest index'[10]). Nevertheless, this in itself is an achievement, as in some cases a smaller leaf area is, in effect, doing greater work, e.g. in rice, or else less photo-assimilate is going into producing excess straw. Here we should note that economic yield means different things to different Economies, e.g. dwarf sorghum may be appropriate for machine cultivation and harvesting in California, but not desired for peasant farming in Africa where the tall stalks can be used for fuel or hut construction.

Although modern high yielding wheats have larger leaves than the more primitive types they have less photosynthetic capacity per unit leaf area. One can almost say that variations in photosynthetic rate per unit leaf area has so far had no apparent effect on yield. But are we therefore to conclude that increases in grain yield will come up against a ceiling imposed by the impossibility of increasing the harvest index beyond a certain point?

It seems unreasonable to consider that variability in the photosynthetic process, i.e. basic photosynthetic efficiency, cannot ultimately be applied to increasing yield, but the limitation in many cases seems to be the capacity for assimilate translocation, and remobilization, and sink capacity. This is an area where we are aware of differences, but have so far made little attempt to investigate them deeply, let alone to exploit them in practical breeding. It could be that translocation or assimilate remobilization is in many cases a principal limiting factor on the photosynthetic process, so that variations in photosynthesis are not directly reflected by yield. In any case it is surely philosophically unacceptable to those concerned with plant improvement through the application of improved physiology, to assume readily the notion that all the variability that is presented to us at the various stages of the carbon assimilation process from photosynthetic reactions onwards, cannot be optimally synchronized to raise the overall efficiency of the process, independently of any possible improvement in harvest index.

This is not likely to come about as a quantum leap in efficiency, but more through the steady elimination of limiting factors in the system, and incorporating such improvements in the base material. Of course this has in effect been the method adopted, on the empirical basis of yield, by plant breeders for many years, but we are hopefully now reaching the stage where we can consciously select and improve the physiological components necessary to achieving an efficient carbon assimilation system.

As was emphasized recently[32] much more attention will have to be paid to biological nitrogen fixation, especially in developing countries where cost of N-fertilizer has become a major handicap. This factor could well modify our ideas of desirable physiological ideotypes, as we are only just beginning to approach, in a logical fashion, the problem of selecting for improved N_2-fixation in any crop. Even with the well known legume/*Rhizobium* nodule system we are not yet sure as to what features of the plant contribute to the most effective N_2-fixation, although we know that varietal differences exist. In the case of the newer associative N_2-fixing systems which are possible in rice, tropical grasses, sorghum, sugarcane and certain wheats, much work will need to be done before we understand the optimum physiology and biochemistry for supporting these systems.

Since Whyte's[34] pioneering book there have been quite a few symposia or special publications[5,6,7,8,9,14,16,19,25,27,29] in recent years devoted to crop physiology and related topics, although there have been somewhat pessimistic views[22] as to the feasibility of plant breeders incorporating specifically physiological characters in practical breeding programmes, due to the many other requirements that have to be taken into account. The difficulties are real enough, but our knowledge is continually advancing and cultivars are becoming more sophisticated. Therefore at this stage it is becoming not only more feasible to take account of specific factors but also necessary, as increased yields, at least in some crops, become harder to achieve.

IDENTIFYING THE IDEOTYPE

In contriving an ideotype, morphological characters have hitherto been principally important, for instance the dwarf or semi-dwarf wheats. The shortening of the culm, by shortening or exclusion of the internodes, or both, is most often determined by one gene, though many genes are now known to give about the same effect. The morphologically visible change is caused by shortening or exclusion of cell elements of the stem structure, while transport paths are changed in the stem, and root structure also seems to undergo the same changes. These changes must be preceded by changes in biochemical composition of the active parts of cells and/or organs controlled by the 'shortening' allele. As the phenotype is the sum-effect of allele/environment interactions such changes must be supposed to have also a changed physiological reaction. Thus, no single-gene mutant should be assumed to affect 'only' one character. It should rather be the rule to assume that any new allele introduced into the genotype must, if the physiological end result should seem unchanged, be accompanied with some degree of compensating changes at other loci.

Consider the idea of an ideal ideotype for a crop plant. The purpose, firstly, of growing a certain crop varies considerably over geographic and climatic areas, and through cultural and traditional background. Wheat, for instance, can be desired as a winter or spring crop, for white bread, spaghetti, or the 'chapatti' of India, for extensive or intensive cultivation, for mechanical or manual harvesting, for high or low fertilizer applications, for acid or basic soils, for tropical or temperate climates. Similarly with *Phaseolus* grown for dry beans, there is no generally accepted view as to what should constitute an ideotype, though attempts concentrating on morphology have been made.[1] For developed agriculture and modern machine harvesting a 'determinate' or 'bush' type is desirable. In the case of small scale peasant farming typical of some Latin American areas, *Phaseolus* is often grown in combination with maize, and in this case an indeterminate climbing type is wanted, because by using the maize as a support the leaves are not shaded by the tall growing corn nor smothered by weeds. Additional to the opposed growth habit type demanded by different systems, we probably have to take into account the ability to grow on acid soils, disease resistance, seed size and colour according to consumer preference, and a good capacity to support dinitrogen fixation.

Climate and soil may be fairly constant factors over large areas, but applied fertilizer levels, for instance, may certainly change considerably in the future. While developing countries have to make an effort to increase the use of fertilizers to reach rather average levels, developed countries have to make a similar effort to reduce sometimes rather excessive fertilizer levels and yet still maintain yield levels. Here, while one could possibly agree on the phenotypic or genotypic make up of the ideotype where both objectives meet, the genotypes needed for materializing this ideotype in recombinations with the locally adapted base material could certainly be rather different. We should also expect that the perceived ideotype would slowly but constantly move, in the one case towards performing better with higher, and in the other case, lower fertilizer levels, as supplies and costs increase respectively.

Regarding the actual use of the product, traditions break down and cultural patterns change and breeders who may be providing their wheat ideotypes with certain baking quality characteristics today will find the bakers demanding others tomorrow. The demand for short culm wheats brought about by combine harvesters, high fertilizer levels, and low straw utilization, may again change as straw becomes an interesting energy resource. In short, the demands for certain properties of the crop must be expected to change continuously and therefore the ideotype must also be continuously changing and developing. This emphasizes the need for information making it possible to form and reform ideotypes as demand changes, and also material to materialize these new ideotypes as they are conceived.

CONSTRUCTING AN IDEOTYPE

How then could a 'standard' ideotype be put together today? Our present knowledge of character gene control in most crop plants obviously comes mainly from phenotype 'building stones'. However, defining these building stones as small as possible would be very useful. For one thing, such a list would come very close to a descriptor list for genebank or germplasm collection use, but would not only be descriptive phenotypically but actually half-way towards a genetic description, i.e. the breeder could use it not only to see what the desired plant looks like but also how it might function in recombination and selection. So let us take the pea as an example,[43] and begin almost from the beginning.

The fertilized egg cell, the zygote, contains all the DNA and other genetic information to make a pea and as the first cell division is initiated this information is brought into effect and a number of characters, genes of great interest for the agricultural performance of the plant are being manufactured. Many of these will be known from microorganisms and molecular genetics, such as those controlling general energy turnover processes, ATP cycle, amino-acid production, etc. Much of this information will be available but it needs to be confirmed to be valid for the species, and variations, i.e. alleles, need to be related to particular accessions, or we cannot utilize the information. After a number of cell divisions differentiation sets in and at this point a number of characters, manifesting themselves much later in plant morphology, begin their action. It seems logical that some of these hundreds of genes known in the pea controlling different morphological characters, such as leaf shape, stem length, pod shape, etc. will already exert an influence on many physiological processes in the new plant.

Somewhere along the way the embryo will differentiate into root and shoot and now appears a major defect in our knowledge of pea genetics. Very little is known about roots as compared to shoots, a situation in common with most crop plants, yet some of the more important processes for agricultural performance of the pea plant are determined by basic root processes. There are the mechanisms of water and nutrient uptake and transport, for which no characters have been defined, and no genes are known; root morphology is virtually unknown ground, although some internode length genes are assumed to have a pleiotropic effect on root-length, but this has not been really proved, and of course, in the pea as in other legumes, there is the whole system of *Rhizobium* symbiosis and nitrogen fixation. It is in the last system that some information is now available: the genes *Nod* and *Noda* giving gene control of nodule number and the *Sym*-genes giving gene control of *Rhizobium* infection and development. For the rest, the only piece of information on roots seems to be the fact that the gene *age* is involved in controlling the geotropic response of the roots and maybe also shoots with some isoenzyme systems being identified, but not yet genetically analysed. The field is open for plant physiologists, microbiologists, and biochemists, to provide a 'standard list' of physiological and other mechanisms probably gene controlled as a base for a character list/gene list to serve as the geneticists check list. The fact that such a list might be very long should not deter us, as computers are useful tools to handle such a problem.

We may now turn to the more familiar aspect of the plant, the parts above ground. At least in some crop plants, including pea, maize, barley, and tomatoes, character definition with respect to some morphological and certain biochemical characters has actually been brought to the level of one character/one gene. In other words, it has been done and therefore it can be done again, and certainly with other characters. In the pea the list is actually so long that one example will have to suffice, so let us consider the 'character', yield of dry seeds, which is a function of the number of seeds and weight of seed per unit area.

Number of seeds is determined by several 'morphological' characters, such as potential number of pods per plant, and potential number of seeds per pod; as well as by a series of 'physiological' characters involved in energy production and transfer. The potential number of pods per plant can then be described in terms of potential number of branches per plant, potential number of generative nodes per plant, i.e. potential numbers of inflorescences and potential number of flowers per inflorescence. Of these, among the genes controlling potential number of branches the genes *Fr, Fru, Ram, Rms* are known; likewise for potential number of flowers we know the genes *Fn* and *Fna*; for potential number of nodes *Mie, Mier, Mine, Miu*; for numbers of vegetative nodes *Lf, Sn*. For potential number of seeds per pod one gene has probably been found but has not been finally analysed. For potential seed weight the genes *Sg-1* and *Sg-2* have been reported.

When the entire potential of the pea is allowed full expression the yield is of the order of

20 tons of dry seeds per ha. This means growing high, branching plants, of late maturing type, at distances of about 0.6 × 0.2 m and harvesting by hand. Under these conditions the characters, that is the genes controlling water and nutrient uptake, photoassimilation and other processes, can express themselves fully or almost so. The number of flowers per inflorescence can, for instance, be 6 or more, but even under seemingly ideal conditions more than 3 pods are rarely formed. Why then are average dry pea yields 4 t/ha rather than 20 t/ha? Simply because the ideotype the breeder has to work towards must provide for mechanical harvesting with the type of equipment traditionally used for cereals. The low peas required for this type of machine harvesting utilize only about 0.5 m^3 per m^2 for pod production whereas high peas grown with support may utilize 1.5 m^3 or more.

The ideotype for low peas for mechanical harvesting will clearly have fixed characters such as number of branches and number of inflorescences characteristic of certain genotypes; in any case, plant height is fixed and environmental restrictions will limit the expression of these characters. The breeder will have to concentrate on breeding for fullest possible expression through number of pods per inflorescence, number of seeds per pod and seed weight. In regard to number of pods per inflorescence and number of seeds per pod, there is still unused potential in present varieties. However, the main assimilate contribution to an inflorescence seems to come from the leaf below the inflorescence and some from the pods themselves, but this is clearly not enough. The ideotype for types with increased yield in dwarf peas should therefore include characters for increased photosynthesis and assimilate redistribution and transport within the plant as primary selection criteria.

The election can be based on a phenotype-derived ideotype, for instance plants with 10 instead of 8 seeds per pod. However, when selecting parents to cross and obtain recombinants for better photosynthetic efficiency and more efficient transport, the ideotype should be based on genotype. We need to combine efficiency in different systems to get recombinants with higher number of seeds per pod. To give an example: high sugar content, which is determined by the gene r, had been known since the 16th century. Crossing between high sugar content peas never resulted in higher sugar content selection, since all variation was modificative as the gene was the same. In 1962 it was found that in one line high sugar content was determined by another gene in another chromosome, r_b, and crosses $r \times r_b$ gave the recombinant rr_b double-sweet. If we list the information by phenotypic character 'sweet', there is no way of knowing whether a variety is r or r_b. Therefore, the ideotype of the wanted selection may well be 'double-sweet' but the information on parents must be in terms of genotype, r or r_b.

A similar approach will be possible for most of the major crops. For crops not so universally grown it will obviously take more time to gather the necessary genetic background and a more simplistic approach will still be necessary in the meantime.

FUTURE FOR IDEOTYPES

We have briefly considered a selection of factors governing the construction of an ideotype. Some of these it is possible to put into practice today, but some of them must await much additional basic work. Over ten years ago Sprague[28] noted that very little is known about biochemical causes of differential yield potential in plants, and we are only a little further forward today. Lack of knowledge of control mechanisms and their genetic base is probably still the major handicap at the present time, though knowledge is gradually accumulating.

Apart from drawing attention to the genotypic and essentially functional base of ideotypes we have, in line with much of the thinking behind this book, tended to emphasise the essentially physiological background of all morphological expression. This has been deliberate. Various of our contributors have drawn attention to the wealth of physiological variation that the breeder has potentially at his disposal. Much of this variation will have to be used both to achieve new and higher yield plateaux under favourable high input, high technology conditions, and on the other hand to extend useful cultivation to unfavourable soils and climates, and to areas with low inputs and low technology.

Incorporating all the potentially available and desirable agronomic, physiological and disease resistance attributes into a marketable variety is not easy, and that of course is why most practical breeding has concentrated on selection and crossing on a yield basis, coupled with defect elimination to incorporate attributes such as disease resistance and quality. Brady[4] has likened the development of an adapted variety to a rope in which the finished variety, represented by the rope, is built up from separate strands representing agronomic characters, disease and insect resistance, protein content and quality factors, and tolerance to adverse soils, low temperatures, flooding, etc. This is a nice analogy, but the main problem is which of the 'strands' should form the core of the 'rope'? Which character should receive emphasis?

Some will say agronomic characters, deriving from existing material of known good yield potential, and probably in most cases this will be correct. If, however, a fundamental requirement of the cultivar (ideotype) is that it should be adapted to very extreme conditions such as salinity or low pH soils, then probably the basic material should be primarily chosen for this character, bearing in mind how relatively rare these qualities are in crop plants, and how they affect the whole physiology of the plant. Selection for agronomic characteristics would then follow and disease resistance introduced through crossing and back crossing.

The development of mutation induction techniques has somewhat altered the 'incorporation priority' that might be given to certain characters. Differences in seed, grain, and flower colour are relatively easily obtained in developed cultivars without fundamentally altering the performance or type. If such characters are important for marketing they can if necessary be obtained at a late stage of development. Reduced straw length is one of the most common mutants, and the production of semi-dwarfs in cereals can almost be guaranteed. This means that it is no longer necessary to accept irrelevant 'genetic garbage' from other existing varieties merely to obtain the desired genes for short straw. This is well seen in rice where Hargrove et al.[17] have traced the ancestry of improved cultivars of Asian rice. In effect, these improved cultivars are so interrelated as to pose a potential disease susceptibility threat, although a main purpose behind many of the crosses was to retain or acquire short straw. That semi-dwarf mutants can be obtained readily on existing long straw rice cultivars has been shown by a number of workers, e.g. Rutger and Peterson[25] in California.

The rapidly developing tissue culture techniques, especially those concerned with haploid and protoplast technology are undoubtedly going to have a profound effect on future plant breeding, but where do they fit into a view of ideotypes? It is not entirely clear at present, although obviously until such time as tissue-derived material is regenerated to normal plants it cannot correspond to any ideotype, in the sense that is normally understood. The regenerated material may or may not have the characteristics of the ideotype, that is the goal. There is more or less the same element of chance involved as in mutation induction. Where tissue culture may prove to have significance in relation to cultivar development, is in altering our ideas of where we might introduce such basic characters as salt tolerance, cold tolerance, and tolerance to high

Al, Mn and herbicides. If these can be selected at a cellular level, then such attributes might be found and stabilised at a much later level of varietal development than suggested above.

Looking at ideotypes from a long term point of view, surely the most important aspect is not to get too firmly tied to any one concept, because the characteristics thought to be desirable today, may have to be revised in the light of new physiological and genetic information, different cultural techniques, changes in consumer/market requirements, changes in national economy, changes in pricing policy (not many major crops are subject to free market conditions), etc. Ideotypes are in part a challenge, and something to aspire to. It is therefore important in designing an ideotype to ensure that it is visualised as 'open-ended', to allow for changing circumstances, so that as our knowledge advances so should the ideotype roll forward.

REFERENCES

1. Adams, M.W. Plant architecture and physiological efficiency in the field bean. *Potentials of field beans and other food legumes in Latin America*, 266–278, Seminar Series No. 2E, CIAT, Colombia, 1973.

2. Beachell, H.M. and Jennings, P.R. Need for modification of plant type. *The mineral nutrition of the rice plant.* John Hopkins Press, Baltimore, 29–35, 1965.

3. Boonstra, A.E.H.R. Meded. LandbHogesch., Wageningen, 38, 1–99, 1934.

4. Brady, N.C. Rice responds to science. *Crop Productivity — Research Imperatives.* A.W.A. Brown *et al.* (eds.), 62–96, Kettering Foundation, Yellow Springs, Ohio, 1975.

5. Brown, A.W.A., Byerly, T.C., Gills, M. and San Pietro, A. (eds.). *Crop Productivity — Research Imperatives,* 399. Kettering Foundation, Yellow Springs, Ohio, 1975.

6. Burris, R.H. and Black, C.C. (eds.). *CO_2 Metabolism and Plant Productivity.* Univ. Park Press, Baltimore, 1976.

7. Carlson, P.S. (ed.). *The Biology of Crop Production.* Academic Press, New York, 1980.

8. Carson, E.W. (ed.). *The Plant Root and its Environment,* 601, Univ. Virginia Press, Charlottesville, 1974.

9. Cooper, J.P. (ed.). *Photosynthesis and Productivity in Different Environments.* Cambridge Univ. Press, 1975.

10. Donald, C.M. In search of yield. *J. Aust. Inst. Agric. Sci.* 28, 171–178, 1962.

11. Donald, C.M. The breeding of crop ideotypes. *Euphytica* 17, 385–403, 1968.

12. Drosdoff, M. Soil. *Review of Science and Technology,* 377–378, McGraw-Hill, 1974.

13. Duncan, W.G. Cultural manipulation for higher yields. *Physiological Aspects of Crop Yield,* 327–339, J.D. Eastin *et al.* (eds.), Amer. Soc. Agron., Madison, 1975.

14. Eastin, J.D., Haskins, F.A., Sullivan, C.Y. and Van Bavel, C.H.M. (eds.). *Physiological Aspects of Crop Yield,* 396, Amer. Soc. Agron., Madison, 1969.

15. Evans, A.M. Commentary on plant architecture and physiological efficiency in the field bean, by M.W. Adams. *Potentials of field beans and other food legumes in Latin America,* 279–286. Seminar Series No. 2E, CIAT, Colombia, 1973.

16. Evans, L.T. (ed.). *Crop Physiology — Some Case Histories.* Cambridge Univ. Press, 1975.

17. Hargrove, T.R., Coffman, W.R. and Cabanilla, V.L. Ancestry of improved cultivars of Asian rice. *Crop Sci.* 20, 721–727, 1980.

18. Heichel, G.H. Agricultural Production and Energy Resources. *Amer. Sci.* 64, 64–72, 1976.

19. Hurd, R.G. *et al.* (eds.). *Opportunities for increasing crop yields.* Pitman, London, 1980.

20. Ishizuka, Y. Engineering for higher yields. *Physiological Aspects of Crop Yield.* Eastin, J.D. *et al.* (eds.). Amer. Soc. Agron., Madison, 15–26, 1969.

21. Jennings, P.R. Plant type as a rice breeding objective. *Crop Sci.* 4, 13–16, 1964.

22. Langer, R.H.M. Physiological approach to yield determination in wheat and barley. *Field Crop Abstr.* 20, 101–106, 1967.

23. MacKey, J. The wheat plant as a model in adaptation to yield determination in wheat and barley. *Proc. 5th Yugoslav. Symp. on Research in Wheat, Novi Sad*, 1966.

24. Marcelle, R., Clÿsters, H. and van Pouke, M. *Photosynthesis and Plant Development*. W. Junk bv. Publ., The Hague, 1979.

25. Rutger, J.N. and Peterson, M.L. Research tool uses of rice mutant for increasing crop productivity. *Proc. Int. Symp. Induced Mutations as a Tool for Crop Plant Improvement*. IAEA-SM-251/13. IAEA, Vienna, 1981.

26. Sanchez, P.A. and Buol, S.W. Soils of the tropics and the world food crisis. *Sci.* **168**, 598–603, 1975.

27. Spiertz, J.H.J. and Kramer, T.L. (eds.). Crop Physiology and Cereal Breeding. Centre for Agricultural Publishing and Documentation, Wageningen, 1979.

28. Sprague, G.F. Germplasm manipulation in the future. *Physiological Aspects of Crop Yield*, 375. Eastin, J.D. *et al.* (eds.). Amer. Soc. Agron., Madison, Wisc., 1969.

29. Stone, J.F. (ed.). *Modification for More Efficient Water Use*, Elsevier, New York, 1975.

30. Vogel, O.A., Allan, R.E. and Peterson, C.J. Plant and performance characteristics of semidwarf winter wheats producing most efficiently in Eastern Washington. *Agron. J.* **55**, 397–398, 1963.

31. Vose, P.B. Varietal differences in plant nutrition. *Herbage Abstracts* **33**, 1–13, 1963.

32. Vose, P.B. Crops for all conditions. *New Scientist* **89**, 688–690, 1981.

33. Wallace, D.H., Ozbun, J.L. and Munger, H.M. Physiological genetics of crop yield. *Advances in Agronomy* **24**, 97–146, 1972.

34. Whyte, R.O. *Crop production and environment*. 2nd Ed. Faber and Faber, London, 1960.

35. Yoshida, S. Physiological aspects of grain yield. *Ann. Rev. Pl. Physiol.* **23**, 437–464, 1972.

36. Palese, P. and Roizman, B. (ed.). *Genetic variation of viruses*; Ann. New York Ac. Sci., 1980.

37. Makasheva, R.Kh. Grain legumes. Part 1, Pea; *Flora of cultivated plants*, Leningrad, Kolos, 1979.

38. Lehmann, C.O. Das morphologische System der Saaterbsen (*Pisum sativum* L. *sensu lat.* Gov. ssp. *sativum*). *Der Züchter* **24**, 316–317, 1954.

39. King, R.C. *Handbook of Genetics, Vol. 1–2*. New York, Plenum Press, 1974.

40. Blixt, S. Some genes of importance for the evolution of the pea in cultivation. *Proc. Conf. Broadening Genetic Base Crops, Wageningen, 1978*, Wageningen, Pudoc, 1979.

41. Blixt, S. *A genebank information system. Appendix V; Report of the First Governing Board Meeting, FAO/UNDP European Cooperative Programme for Conservation and Exchange of Crop Genetic Resources, Geneva 15–18 Dec., 1980.*

42. Blixt, S. *Germplasm documentation: The Future; FAO/UNEP/IBPGR Technical Conference on Crop Genetic Resources, Rome 6–10 April, 1981.*

43. Blixt, S. Mutation Genetics in *Pisum*. *Agri Hortique Genetica* **30**, 1–293, 1972.

INDEX